Morphological and Physiological Research on Fish

Morphological and Physiological Research on Fish

Guest Editors

Elena De Felice
Paola Scocco

MDPI

Basel • Beijing • Wuhan • Barcelona • Belgrade • Novi Sad • Cluj • Manchester

Guest Editors

Elena De Felice
School of Biosciences and
Veterinary Medicine
University of Camerino
Camerino
Italy

Paola Scocco
School of Biosciences and
Veterinary Medicine
University of Camerino
Camerino
Italy

Editorial Office
MDPI AG
Grosspeteranlage 5
4052 Basel, Switzerland

This is a reprint of the Special Issue, published open access by the journal *Animals* (ISSN 2076-2615), freely accessible at: https://www.mdpi.com/journal/animals/special_issues/Morphological_and_Physiological_Research_on_Fish.

For citation purposes, cite each article independently as indicated on the article page online and as indicated below:

Lastname, A.A.; Lastname, B.B. Article Title. *Journal Name* **Year**, *Volume Number*, Page Range.

ISBN 978-3-7258-6009-8 (Hbk)
ISBN 978-3-7258-6010-4 (PDF)
https://doi.org/10.3390/books978-3-7258-6010-4

Cover image courtesy of Paola Scocco

Contents

About the Editors

Elena De Felice

Elena De Felice is an Associate Professor of the Veterinary Anatomy at the University of Camerino, Italy. She graduated in Veterinary Medicine in 2008 from the University of Naples Federico II. In 2012, she obtained her PhD at the Faculty of Veterinary Medicine of the University of Messina. In 2014, she specialized in "Technology and the pathology of poultry, rabbit, and game species" at the University of Naples. She worked from 2009 to 2014 at the Zoological Station "A. Dohrn" in Naples, from 2015 to 2016 at the Tigem "Telethon Institute of Genetics and Medicine" in Pozzuoli, and in 2017, at the A.O.R.N. Cardarelli of Naples. At the University of Camerino, since 2017, she has taught the course Systematic and Comparative Veterinary Anatomy; and since 2021, the course Topographic Veterinary Anatomy in the Degree in Veterinary Medicine. She has also been teaching since 2018 at the International School of Advanced Studies; in addition, since 2021, she has taught at the Specialization School in Hygiene and the Control of Fishery and Aquaculture Products. She is the author and co-author of several publications in international journals and contributions to national and international conferences. Since 2017, in collaboration with the Zenolab laboratory of the University of Naples Federico II, she has worked with laboratory fish (zebrafish and killifish) while since 2020 with a mouse model, in collaboration with CNR of Monterotondo. Her main research field are sensory decays during aging and the study of molecules involved in the regulation of food intake and energy metabolism in animals of veterinary interest. She is a member of the Topical Advisory Panel of *Animals* and Associate Editor for Animal Nutrition and Metabolism (*Frontiers in Veterinary Science*).

Paola Scocco

Paola Scocco, graduated in Biological Sciences in 1986 (summa cum laude), is a Full Professor of the Veterinary Anatomy at the University of Camerino, Italy. Her research deal with the glycoconjugates, neuropeptides and adipokines, in the digestive and reproductive apparatus of lab and farm animal species. She applies anatomical competences to the conservative management of pastural ecosystems via the study of grazing animals and the consideration of the wild species. She was the Project Leader of the inter-territorial cooperation program "Laboratory of the Environment and Landscape-APE: Apennines Park of Europe" and the Coordinator of the morpho-functional group in the "Zootechnics and fire prevention" project of the Marche Region. She carries out many environmental and sustainability education activities, aimed at citizens and children, training activities for both teachers and stakeholders in the agro-zootechnical and environmental fields, and is Responsible for the Citizen Interactive Sustainability Lab. for actions in Public Engagement. She is the Vice President of the Italian Association of Veterinary Morphologists, and Member of the Italian Society of Veterinary Sciences, of the Italian Zoological Union and of the Eastern Alpine and Dinaric Society of Vegetation Ecology. She is a member of the Editorial Board of *Animals* and of *Veterinary Sciences* (MDPI).

animals

MDPI

Editorial

Morphological and Physiological Research on Fish

Paola Scocco and Elena De Felice *

School of Biosciences and Veterinary Medicine, University of Camerino (MC), 62032 Camerino, Italy; paola.scocco@unicam.it
* Correspondence: elena.defelice@unicam.it

The Special Issue "Morphological and Physiological Research on Fish" is dedicated to the application of morphological and physiological studies carried out on fish, attempting to provide information regarding the relevance of morphological studies and their relationships to functions. Indeed, fish not only represent the largest and most varied group of vertebrates but there are also numerous applications for these animals in both research and industry. Moreover, aquaculture is one of the world's most efficient and sustainable methods of producing high-quality protein and, in recent decades, fish have emerged as an interesting model system in biomedical research, due to the close similarities they have with mammals in terms of various basic mechanisms. Morphological studies provide us with context for the comprehension of the spatial organization and relationship between physiological and biochemical data, and the molecular machinery that is promptly being explained through molecular techniques directed at the genome, transcriptome, and proteome. As a result, new morphological and physiological investigations of fish would enhance our understanding of these creatures, with exceptional and interesting relevance both in basic and applied research.

Many authors contributed to the Special Issue with ten original articles and four reviews, and the range of topics covered extended from descriptive morphology to morphometry, from stress markers to toxic agents, from wild to aquaculture or model species, involving different organs and apparatus. Morphometrical and morphological studies were the most numerous contributions to this Special Issue.

Hemibarbus maculatus, a small freshwater fish of the family Cyprinidae, is a common economic fish in the midstream and downstream of the Jialing River, China). In order to resolve the difficulties in aquacultural cultivation of this fish (H. maculatus usually die quickly and do not grow to the same size as the wild ones)—Wu and colleagues [Contribution 1] tested the intestinal and liver digestive function of wild and cultured fishes. Histological staining and biochemical methods were used to compare the differences in morphological structure, goblet cells, argyrophil cells, lymphocytes, and Na+/K+ATPase in the intestine through hematoxylin eosin (H/E), periodic acid–Schiff in combination with the Alcian blue (AB-PAS) and silver leaching methods. To determine the gut absorption function, the Na+/K+-ATP enzyme's vitality was detected. For the observation of liver tissue, H/E, PAS, and oil red O staining were used for morphology, glycogen, and lipid observations. The findings revealed that cultured *Hemibarbus maculatus* had a higher fat content attached to the gut, lower Na+/K+ATPase activity in both the foregut and hindgut ($p < 0.01$), and a reduced number of goblet cells in the hindgut ($p < 0.01$) compared to their wild counterparts. The number of argyrophilic cells did not differ significantly between the two groups, but the number of lymphocytes was notably lower in the gut segments of the cultured fish. This indicates that the absorptive function and intestinal immunity are weaker in the cultured fishes. Additionally, the liver of cultured fish contained more glycogen and

 https://doi.org/10.3390/ani15111513

lipid, suggesting a decreased digestive function. In summary, the wild *Hemibarbus maculatus* exhibit better intestinal digestion, absorption, and lymphocyte levels than the cultured ones, which also show more hepatic lipopexia and glycogen accumulation. Future aquaculture practices should take these changes into account when addressing practical issues.

Among the most-used fish species in aquaculture is the Nile tilapia (*Oreochromis niloticus*), due to its rapid growth rate and its adaptation to a wide range of farming conditions. The aim of the work of Palladino and colleagues [Contribution 2] was an integrated description of esophagus and stomach morphology, crucial to gaining a deeper understanding of the various facets of this species' physiological development and health. Combining scanning electron microscopy and morphometry, the thickness of the superficial and glandular epithelium, the submucosa, and the circular and longitudinal musculature of the stomach were measured. The authors highlight the presence of five different zones in the stomach (1: the esophagus–gastric lumen passage; 2: the descending glandular portion; 3: the fundic portion; 4: the ascending glandular portion; and 5: the gastric–pyloric transition portion). Histochemical investigation, consisting of AB/PAS and AB stainings at different pH levels, showed the secretion of carboxylates glycoconjugates along the esophagus and sulphated complex carbohydrates in the stomach. This dataset could be helpful for research aimed at enhancing animal welfare and production rates by better understanding the morphological alterations brought about by various feed regimens.

Beluga (*Huso huso*) and sevruga (*Acipenser stellatus*) are among the most important species of sturgeon fish (Acipenseridae) inhabiting the Caspian Sea, with demand for their products, such as caviar, meat, skin, and cartilage, being particularly high. The creation of specially designed compound diets to support growth and product quality is a crucial component of the sturgeon farming sector that influences both production efficiency and the sector's long-term viability. However, the presence of digestive enzymes in various areas of the gastrointestinal tract is primarily responsible for the fish's capacity to digest complicated meals. Trypsin, as an alkaline protease, is able to hydrolyze protein residues and peptides to release free amino acids and small peptides for intestinal absorption; therefore, the activity of trypsin has been widely used as a valuable indicator of digestive capacity in fish. In their work, Zamani and coworkers [Contribution 3] aimed to characterize trypsin from beluga and sevruga for the first time. The study's findings indicate that the physicochemical and biochemical characteristics of trypsin from beluga and sevruga are consistent with information found in bony fish, and they could be used as a starting point for the development of in vitro tests to evaluate the digestibility of proteins in these archaic species.

The marbled flounder (*Pseudopleuronectes yokohamae*), which is distributed throughout Korea, southern Hokkaido in Japan, and the East China Sea, is a fish species with significant commercial value in both Korea and Japan, and it is gaining popularity as a new aquaculture target. The study by Cho and colleagues [Contribution 4] looked at the morphology, histology, and histochemistry of the digestive tract of *P. yokohamae*. The relative length of the gut (RLG) was measured, and samples from nine different regions of the digestive tract (the esophagus, cardiac stomach portion, fundic stomach portion, pyloric stomach portion, pyloric caeca, anterior intestine portion, mid intestine portion, posterior intestine portion, and rectum) were stained with Hansen's hematoxylin and 0.5% eosin (H/E) for histological observation, in particular targeting the length of the mucosal folds and the thickness of the muscularis externa, which were stained with Alcian blue/periodic acid–Schiff (AB/PAS) for observations of mucus-secreting goblet cells. In this study, marbled flounder had an RLG value of 1.54 ± 0.10, like omnivore fish. However, the digestive tract's form and structure resembled carnivorous fish. It possessed a simple stomach with six to nine pyloric caeca, and the mucosal folds of the digestive system had a generally branching architecture. Goblet cells were most abundant in the esophagus, followed by the anus, and CCK-producing cells, analyzed through immunohistochemistry, were detected at varying rates in the pyloric caeca and intestine, but not in the esophagus or the stomach. CCK-producing cells and goblet cells in the marbled flounder are effectively adapted to facilitate the ideal regulation of digestion. This work provides fundamental insights for further physiological and nutritional research.

The European sea bass, *Dicentrarchus labrax*, has great economic importance because it is one of the most important marine aquaculture fish species in the Mediterranean. Despite its widespread cultivation, this fish exhibits higher cortisol responses to stress than other commonly grown species in the Mediterranean. The purpose of the review by Samaras [Contribution 5] was to examine all previously published data on baseline and post-acute stress cortisol levels in this species. For this systematic review and meta-analysis, the Web of Science and Scopus databases were searched for studies reporting plasma or serum cortisol levels with no language or date constraints. The data were extracted directly for the reported results and were analyzed separately for basal and post-acute stress levels, as well as their standardized mean differences (SMDs) using random-effects meta-analyses. The current meta-analysis examined 69 studies. This meta-analysis concluded that there is significant variation between studies in both basal and post-stress concentrations. The test type seems to be one of the most important factors influencing heterogeneity. Studies employing ELISA assays had a considerably larger pooled effect size than studies using RIA assays at both the baseline and post-stress levels. As a result, it was impossible to definitively determine which assay type—ELISA or RIA—was more reliable in terms of reporting cortisol levels, but the current study supports the idea that the cortisol assay type should be carefully considered when organizing a study and interpreting the findings. Anesthesia also influenced cortisol levels, which caused the results to vary.

The Atlantic salmon business is predicted to expand internationally by 2–3% by 2030. The primary protein source in salmon diets is fishmeal (FM) (anchovies, pilchards, mackerel, herring, and blue whiting), which is derived from wild-caught marine fish whose natural stocks are in danger of depletion. It is therefore of fundamental importance to pick alternative feed ingredients that present a sufficient protein profile, guarantee a good health status and growth performance of Atlantic salmon, and are also environmentally sustainable. However, it is crucial to note that replacing FM with terrestrial plant elements may produce alterations and have a negative impact on the gastrointestinal tract. The utilization of plant components is limited in carnivorous fish species due to the presence of starch and structural carbohydrates, and a wide range of anti-nutritional factors. Diet can also have an impact on microbiota status, and many different plant-based proteins have been studied for their deleterious effects on the intestine. These changes may impair the animal's digestive functions, development, and, eventually, well-being. Aidos and colleagues [Contribution 6] provide an updated review of alternative protein sources, including their impact on intestinal health in terms of morphology and microbial composition. They reviewed the literature on soybean meal, peas, faba beans, wheat, maize meal, sunflower meal, rapeseed meal, and wheat gluten. They also present animal protein sources that may serve as viable alternatives. A good example was the poultry by-product meal (from poultry processing); also black soldier fly and yellow meal worm. Such information may serve as a foundation for the selection and development of Atlantic salmon feeds that ensure fish health and growth performance while having a minimal impact on the surrounding environment, both in terms of the depletion of the fish's natural stocks and pressure on terrestrial agriculture.

The review by De Marco and coworkers [Contribution 7] focuses on how feed additives containing prebiotics and probiotics affect fish microbiota. The number of microorganisms in a fish's gastrointestinal tract (GIT) has been estimated to be between 107 and 108 per gram. However, the GIT microbiota is highly changeable, and the typical variation in quantity and variety of microorganisms is influenced by different factors, including age, nutrition, host genetics, and the environment (freshwater or oceanwater). Different sections of the fish GIT have been identified, each with a distinct microbial density. Indeed, a certain number of bacteria can be detected throughout the gastrointestinal tract, but some areas are more colonized than others. Aside from being identified as a very good alternative to the use of antibiotics, the use of these feed additives in aquaculture represents a useful strategy in terms of improving the overall performance of the aquaculture sector because it affects a variety of factors including rearing water quality, food absorption, and digestion, and has been shown to have a positive effect on the growth performance of fish. Furthermore, the use of histological assays as a valid technique for assessing fish welfare was highlighted, as were insights into nutritional absorption capacity and pathogen reactivity in fish using gut morphological endpoints. In particular, in histological studies of the fish intestine, endpoints such as the length of the mucosal folds (villi), muscle thickness, and crypt depth can provide useful information on intestinal efficiency in terms of nutrient absorption. Meanwhile, assessing the number of goblet cells and leukocytes could help to evaluate the state of the immune system response, which is very important for the balance of the gut microbiota. The evaluation of these endpoints can be performed using both light and electron microscopy and provides useful information to accurately evaluate the effects of prebiotics and probiotics at the level of the fish gastrointestinal tissue. Overall, gaining a solid knowledge base on the appropriate dosing of feed additives for individual fish species is a promising strategy to significantly improve the aquaculture sector in terms of both quality and sustainable production.

In aquaculture, transportation is a crucial process for transferring juveniles from the hatchery to the commercial farm, where they are reared before being sent to the food market. Furthermore, in farms located far from the slaughterhouse, transportation becomes essential to convey the fish to the facility where they are sacrificed and prepared for sale. Transportation can be stressful for fish due to handling, air exposure, confinement, and reduced oxygen levels. Therefore, it is crucial to gain insights into optimal practices for organizing transportation, ensuring the well-being of these animals and minimizing excessive stress. Bortoletti and colleagues [Contribution 8] investigated the effects of 24 h transport on the stress response in meager (*Argyrosomus regius*) juveniles by using radioimmunoassay to measure muscle cortisol levels and immunohistochemistry in order to detect the presence and intracellular localization of heat shock protein 70 (HSP70), 4-hydroxy-2-nonenal (HNE), 3-nitrotyrosine (NT), and 8-hydroxy-20-guanosine (8-OHdG) in various organs and tissues. The muscle cortisol levels in fish transferred to the transport tanks were significantly higher than those of the controls. Regarding immunohistochemistry, positive results for the various markers were detected in the epidermis of the skin, intestine, and gill lamellae, as well as the gastric glands and hepatopancreas. However, no substantial variations in signal intensity were observed when comparing control and transported animals, except in the case of HNE. The observed stress response was mostly related to loading stress and the transport process, highlighting the significance of developing proper operating procedures to protect fish well-being during transport. The unaltered distribution of oxidative stress indicators across the control and transferred groups showed that the stress experienced may be within acceptable limits.

Cytoplasmic linker-associated protein-2 (CLASP2) is a member of the CLIP-associating proteins (CLASPs) family and is involved in the structure and function of microtubules and Golgi apparatus; however, there are no studies that have clearly defined its role during fish germ cell formation. Ricci and coworkers [Contribution 9] aimed to analyze clasp2 gene transcription levels during fish germ line cell formation in zebrafish (*Danio rerio*) and guppy

(*Poecilia reticulata*). Using real-time q-PCR, the authors found that clasp2 is transcribed in several tissues, such as brain, eye, and kidney tissues, in a comparable fashion between zebrafish and guppy; however, higher transcription was observed in the testis. By in situ hybridization in zebrafish, the authors found that clasp2 is specifically transcribed in the spermatozoa of adult testes. Interestingly, in the guppy, clasp2 is highly transcribed in spermatozeugmata. This study represents the first evidence of clasp2 expression in developing male germ line cells, providing insight that enables further investigation of this protein as a target in teleost fish meiotic cell division.

Bisphenol A (BPA) is a significant component utilized in the manufacture of technical plastic items. BPA is of relevance because of its possible detrimental effects on humans, as well as the fact that it is a prevalent pollutant in aquatic environments. The aim of the study by Smorodinskaya and colleagues [Contribution 10] was to determine the effects of different lethal and sublethal concentrations of BPA on zebrafish (*Danio rerio*) survival in an acute experiment investigating hematological parameters and hematopoiesis processes, as well as histological changes in the head kidney. The obtained data made it possible to determine the acute toxic limits of BPA for adult zebrafish. Concentrations exceeding 4 mg/L reliably resulted in fish mortality within 96 h, with an LC50 of 6.22 mg/L. Significant increases in the frequency of erythrocyte nuclear abnormalities were observed at BPA concentrations of 6 and 8 mg/L, which indicates a genotoxic effect. BPA's impact on fish peripheral blood parameters manifested as an increase in the number of erythrocytes (RBC) and immature erythrocytes, as well as a decrease in the number of lymphocytes. The most notable pathological changes in the head kidney's hematopoietic tissue included circulatory disturbances and the formation of inflammation/degradation foci. The conducted study makes it possible to conclude that BPA, at concentrations of 6 and 8 mg/L, influences the differentiation of cell elements in the erythroid series. Despite major advances in the study of hematopoiesis in lower vertebrates, including fish, morphological and quantitative assessment approaches still require refining. The authors of this work proposed a method for counting hematopoietic progenitor cells, focusing on elements from the erythropoietic and myelopoietic lineages. This method has potential uses in determining the impact of hazardous chemicals on hematopoiesis.

The poor quality of biological information, such as the relationship between length and weight parameters, could be a source of variability with a significant impact on stock assessment results. This relationship between the length and weight of individuals could differ between males and females as well as two separated islands and/or may be linked to the reproduction period of a species. Therefore, to estimate the morphometric relationships between total length and weight (the length–weight relationship, LWR) of each fish species according to potential spatial, temporal, and sex differences, Mahè and coworkers sampled 109 fish species (24,996 individuals) around the islands of Guadeloupe and Martinique from October 2021 to September 2022 [Contribution 11]. This is the first time that the LWR was estimated in the Atlantic Ocean for 16 species. There was a significant relationship between length and weight for all of the tested species. The sex effect on the LWR showed significant sexual dimorphism for 24 species. Additionally, a link between the temporal effect and the reproduction period was tested for 68 species, of which 35 presented significant differences relative to the annual quarter of sampling. Finally, the geographical effect (i.e., the difference between samples from around Guadeloupe and those from Martinique) was significant for 60 species. This island effect was significant for 25 species. These data and results will be essential for the stock assessment of Caribbean fishes.

Cadmium is one of the most toxic heavy metal contaminants in the environment and it can enter aquatic ecosystems via industrial waste. The use of chitosan to mitigate the negative effects of heavy metal contamination on aquaculture organisms is a common treatment strategy for eliminating heavy metal pollution in water. The aim of study by Zhang and colleagues [Contribution 12] was to investigate the effects of dietary chitosan supplementation on the muscle composition, digestion, lipid metabolism, and stress resistance, and their related gene expression, of genetically improved farmed (GIFT) juvenile tilapia (*Oreochromis niloticus*) subjected to cadmium stress. Compared to the control group, dietary supplementation with chitosan significantly increased the contents of crude protein and crude fat in the muscles and the activities of lipase, trypsin, and amylase in the intestines of juvenile GIFT exposed to cadmium-induced stress. These findings suggested that dietary supplementation with chitosan exhibited the potential to mitigate the adverse effects of cadmium-induced stress in terms of intestinal enzyme activity and improving digestive performance. In addition, dietary supplementation with chitosan could significantly upregulate the relative expression levels of hormone-sensitive lipase (hsl), peroxisome proliferator-activated receptor alpha (ppar*α*), lipoprotein lipase (lpl), malate dehydrogenase (mdh), carnitine palmitoyltransferase 1 (cpt-1), leptin (lep), fatty acid synthase (fas), stearoyl-CoA desaturase (scd), peroxisome proliferator-activated receptor gamma (pparγ), and sterol regulatory element-binding protein 1 (srebp1) genes in the livers of GIFT juveniles exposed to cadmium-induced stress, suggesting that dietary supplementation with chitosan could effectively alleviate lipid metabolism impairment while concurrently augmenting their lipid metabolism capacity. These findings may provide scientific insights into the use of chitosan in aquaculture operations and the management of heavy metal contamination in aquatic ecosystems.

As the Mediterranean Sea continues to undergo transformations, the study of teleost reproduction serves as a critical lens through which to comprehend the broader implications of human activities on marine ecosystems and work towards their preservation. Lombò and coworkers [Contribution 13] in their review analyzed three stressors, namely climate change, environmental pollution, and overfishing, which influence various levels of the reproductive process in this changing scenario.

To better understand the population structure and biological background of fishery *Conger myriaster*, an economically important marine fish, five meristic counts and seventeen morphological measurements from seven different geographical populations in coastal China were analyzed. The traditional morphometry method was applied to analyze the standardized measurements together with the meristic counts so as to provide supplementary information for the fishery biology, population assessment, and fishery resource protection of this fish. The results of the work of Xiao and Yang [Contribution 14] showed that the greatest divergence was observed between the Dalian and Qingdao populations, whereas the smallest difference was found between the Lianyungang and Zhoushan populations. Statistical differences in tail length (*TAL*) were detected between all populations. The morphological traits with high variation coefficient values were mostly related to body weight (*BW*), confirming greater potential variations in these weight-related traits. Principal component analysis (PCA) extracted seven principal components (PCs) with eigenvalues greater than 1, and the cumulative contribution rate was 72.790%. The results of cluster analysis, together with the PCA and DFAs (discriminant function analyses), supported separating the populations into three groups linked with their geographic distribution and their specific environment localization. Considering the particularity of the natural environment of the Bohai Sea and the sophisticated oceanic circulations of the Shandong Peninsula, the relationships of *C. myriaster* populations in the northwest Pacific Ocean along the China coast were closely related to their geographical distributions and oceanic circulations. Understanding the composition of a population in terms of its demographic characteristics, spatial distribution, and social organization is of great importance for policymakers, researchers, and organizations, enabling them to make informed decisions and develop effective strategies.

In conclusion, we can state that there is a very broad interest in the fish world, ranging from basic to applied research. Studies on fish morphology and physiology are critical to help us understand their adaptation, ecological roles, and evolutionary relevance. They are critical to conservation efforts, sustainable aquaculture, and fisheries management, guaranteeing biodiversity and encouraging responsible resource use. Exploring these elements provides scientists and researchers with significant insights into the resilience and functionality of aquatic life, which benefits both ecosystems and human enterprises. We believe that this Special Issue may be of interest, and we hope that this topic will be increasingly explored in the future.

Conflicts of Interest: The authors declare no conflicts of interest.

List of Contributions:

1. Wu, B.; Lei, H.; Zhen, J.; Zhao, L.; Song, B.; Zeng, Y. Comparative Research on Intestinal Functions of Wild and Cultured *Hemibarbus maculatus* in Jialing River. *Animals* **2023**, *13*, 189. https://doi.org/10.3390/ani13020189.
2. Palladino, A.; De Felice, E.; Attanasio, C.; Barone, C.M.A.; Crasto, A.; D'Angelo, L.; Giaquinto, D.; Lambiase, C.; Scocco, P.; Serrapica, F.; Maruccio, L. A Morphological and Ultrastructural Study of the Anterior Digestive Tract of Adult Nile Tilapia *Oreochromis niloticus*. *Animals* **2023**, *13*, 420. https://doi.org/10.3390/ani13030420.
3. Zamani, A.; Khajavi, M.; Kenari, A.A.; Nazarpak, M.H.; Solouk, A.; Esmaeili, M.; Gisbert, E. Physicochemical and Biochemical Properties of Trypsin-like Enzyme from Two Sturgeon Species. *Animals* **2023**, *13*, 853. https://doi.org/10.3390/ani13050853.
4. Cho, J.-H.; Park, J.W.; Ryu, Y.-W., Kim, K.-W., Hur, S.-W. Morphology, Histology, and Histochemistry of the Digestive Tract of the Marbled Flounder *Pseudopleuronectes yokohamae*. *Animals* **2023**, *13*, 936. https://doi.org/10.3390/ani13050936.
5. Samaras, A. A Systematic Review and Meta-Analysis of Basal and Post-Stress Circulating Cortisol Concentration in an Important Marine Aquaculture Fish Species, European Sea Bass, *Dicentrarchus labrax*. *Animals* **2023**, *13*, 1340. https://doi.org/10.3390/ani13081340.
6. Aidos, L.; Mirra, G.; Pallaoro, M.; Herrera Millar, V.R.; Radaelli, G.; Bazzocchi, C.; Modina, S.C.; Di Giancamillo, A. How Do Alternative Protein Resources Affect the Intestine Morphology and Microbiota of Atlantic Salmon? *Animals* **2023**, *13*, 1922. https://doi.org/10.3390/ani13121922.
7. De Marco, G.; Cappello, T.; Maisano, M. Histomorphological Changes in Fish Gut in Response to Prebiotics and Probiotics Treatment to Improve Their Health Status: A Review. *Animals* **2023**, *13*, 2860. https://doi.org/10.3390/ani13182860.
8. Bortoletti, M.; Fonsatti, E.; Leva, F.; Maccatrozzo, L.; Ballarin, C.; Radaelli, G.; Caberlotto, S.; Bertotto, D. Influence of Transportation on Stress Response and Cellular Oxidative Stress Markers in Juvenile Meagre (*Argyrosomus regius*). *Animals* **2023**, *13*, 3288. https://doi.org/10.3390/ani13203288.
9. Ricci, S.; Lazzari, M.; Maurizii, M.G.; Franceschini, V.; Milani, L.; Cacialli, P. Analysis of clasp2 Transcription Pattern in Male Germ Cells during Spermatogenesis: A Comparative Study in Zebrafish (*Danio rerio*) and Guppy (*Poecilia reticulata*). *Animals* **2023**, *13*, 3617. https://doi.org/10.3390/ani13233617.
10. Smorodinskaya, S.; Kochetkov, N.; Gavrilin, K.; Nikiforov-Nikishin, D.; Reznikova, D.; Vatlin, A.; Klimuk, A.; Odorskaya, M.; Nikiforov-Nikishin, A.; Ponomarev, A.; Marsova, M.; Danilenko, V. The Effects of Acute Bisphenol A Toxicity on the Hematological Parameters, Hematopoiesis, and Kidney Histology of Zebrafish (*Danio rerio*). *Animals* **2023**, *13*, 3685.. https://doi.org/10.3390/ani13233685.
11. Mahé, K.; Baudrier, J.; Larivain, A.; Telliez, S.; Elleboode, R.; Bultel, E.; Pawlowski, L. Morphometric Relationships between Length and Weight of 109 Fish Species in the Caribbean Sea (FrenchWest Indies). *Animals* **2023**, *13*, 3852. https://doi.org/10.3390/ani13243852.

12. Zhang, Q.; Xie, Y.; Zhang, Y.; Huang, E.; Meng, L.; Liu, Y.; Tong, T. Effects of Dietary Supplementation with Chitosan on the Muscle Composition, Digestion, Lipid Metabolism, and Stress Resistance of Juvenile Tilapia (*Oreochromis niloticus*) Exposed to Cadmium-Induced Stress. *Animals* **2024**, *14*, 541. https://doi.org/10.3390/ani14040541.
13. Lombó, M.; Giommi, C.; Zarantoniello, M.; Chemello, G. A Pretty Kettle of Fish: A Review on the Current Challenges in Mediterranean Teleost Reproduction. *Animals* **2024**, *14*, 1597. https://doi.org/10.3390/ani14111597.
14. Xiao, P.; Yang, T. Population Morphometry of *Conger myriaster* (*Anguilliformes*: Congridae) along the Coast of China: Implications for Population Structure and Fishery Management. *Animals* **2024**, *14*, 2007. https://doi.org/10.3390/ani14132007.

animals

MDPI

Article

Population Morphometry of *Conger myriaster* (Anguilliformes: Congridae) along the Coast of China: Implications for Population Structure and Fishery Management

Peiyi Xiao and Tianyan Yang *

Fishery College, Zhejiang Ocean University, Zhoushan 316022, China; xiaopeiyi@zjou.edu.cn
* Correspondence: yangty@zjou.edu.cn

Simple Summary: Understanding the composition of a population in terms of its demographic characteristics, spatial distribution, and social organization within a population is of great importance for policymakers, researchers, and organizations to make informed decisions and develop effective strategies. To better understand the population structure and fishery biological background of *Conger myriaster*, an economically important marine fish, five meristic counts and seventeen morphological measurements from seven different geographical populations along coastal China were analyzed by traditional morphometry method. The univariate and multivariate statistical analysis results revealed a geographical population structure of *C. myriaster*. In the face of undergoing long-term fishing pressure and massive catching efforts, biogeography-based fishery strategies for *C. myriaster* should be employed in the coastal waters of China.

Abstract: In this study, the traditional morphometry method was applied to analyze the standardized measurements together with the meristic counts so as to provide supplementary information for fishery biology, population assessment, and fishery resources protection of *C. myriaster*. The results of one-way analysis of variance (ANOVA) showed that the greatest divergence was observed between the Dalian and Qingdao populations, whereas the smallest difference was found between the Lianyungang and Zhoushan populations. Statistical difference in tail length (*TAL*) was detected between all populations. The morphological traits with high *C.D* values were mostly related to body weight (*BW*), confirming greater potential variations of these weight-related traits. Principal component analysis (PCA) extracted 7 principal components (PCs) with eigenvalues greater than 1, and the cumulative contribution rate was 72.790%. The results of cluster analysis, together with the PCA and DFA, supported separating the populations into three groups linked with their geographic distribution and their specific environment localization. Considering the particularity of the natural environment of the Bohai Sea and the sophisticated oceanic circulations of the Shandong Peninsula, the relationships of *C. myriaster* populations in the northwest Pacific Ocean along the China coast were closely related to their geographical distributions and oceanic circulations.

Keywords: *Conger myriaster*; morphology; population structure; fishery management; China coast

1. Introduction

Phenotypic plasticity is an important biological mechanism for adapting to perpetual environmental fluctuations, which refers to the degree of change in bio-phenotype relative to environmental variations and has more producing capability than one alternative form of morphology, physiological state, and/or behavior [1,2]. The most common manifestation of phenotypic plasticity is population differentiation from distinct regions or habitats, which can be caused by the intervention of several factors, such as genetic variation, geographic isolation (habitat fragmentation), historical change, human impact, and ecological dynamism (inter and intra-species relationships) [3–7]. As the most basic content of biological research, fish morphometry has been widely used in stock assessment, germplasm conservation, and genetic breeding nowadays, which lays the foundation for further research on comparative biology, population genetics, phylogeny, and adaptive evolution [8–11]. Morphometric measurements have been widely used to identify differences between fish populations. Turki-Missaoui et al. [12] used the traditional morphological method to detect obvious differentiation between *Sander lucioperca* populations in relatively similar environments, although they were derived from a single gene pool. Myoung and Kim [13] conducted a multivariate morphometric analysis of *Konosirus punctatus* and clarified two morphotypes on the Korean coasts. Turan et al. [14] studied the genetic and morphological variation of *Pomatomus saltatrix* based on morphometric and meristic analyses. The results detected three morphologically differentiated groups and the highest reclassification rate for the eastern Black Sea and northeastern Mediterranean Sea samples.

Conger myriaster (Brevoort, 1856) [15], commonly known as Whitespotted conger, is an economically valuable marine fish species widely distributed in the northwest Pacific, from the southern East China Sea to the coasts of Japan and Korean Peninsula [16], and occupies a significant proportion in import and export trades of fishery resources in Asian countries. As a warm-water demersal fish, *C. myriaster* mainly inhabits muddy and gravelly bottoms in coastal waters with depths of 5~50 m and possesses a critical position in the nearshore food web and marine ecosystem [17]. Recent studies showed that fishery resources of *C. myriaster* in the Yellow Sea and the East China Sea had undergone over-fishing pressure to a certain extent [18,19]. Meanwhile, environmental changes, together with human activities, are also assumed to induce resource fluctuations substantially, which can cause profound impacts on the population structure and distribution of this species. Most literature on *C. myriaster* mainly focused on early life history, feeding ecology, reproductive biology, and population genetics [20–25]. As for morphologic references on *C. myriaster*, only the path analysis of morphological traits on body weight [26] and the comparative morphological study of only two populations in coastal China were available [27].

Traditional morphometry was regarded as the most basic method of population identification and taxonomic classification before the emergence of barcoding and molecular ecology because morphological traits can reflect external resemblance and partial inherent genetic characteristics [3]. Understanding the population structure and biological background of target catches should be an essential prerequisite for designing and establishing fishery management policies [28]. Therefore, it was particularly important to carry out more detailed morphological studies of distinct *C. myriaster* populations so as to provide basic data for sustainable exploitation and utilization of germplasm resources. In the present study, the morphological characteristics of *C. myriaster* from seven different geographic locations along the coastal waters of China were comprehensively compared and analyzed for the first time; besides, the intraspecific morphological variations and differentiation patterns were investigated. The results will provide some supplementary information for fishery biology research of *C. myriaster*, as well as the basic data for the population assessment and protection of the fishery resources.

2. Materials and Methods

2.1. Material

Samples Collection and Morphometric Measurement

A total of 218 *C. myriaster* individuals were collected by the commercial trawling from 7 different geographical sites along coastal China: Dalian (DL), Weihai (WH), Rushan (RS), Qingdao (QD), Rizhao (RZ), Lianyungang (LYG) and Zhoushan (ZS) from August 2022 to May 2023 (Table 1; Figure 1).

Table 1. Sampling information of *C. myriaster*.

Sampling Sites	Sea Regions	Geographical Coordinates	Sampling Time	Sample Size	Total Length (TL)	
					Range (cm)	Mean ± SD (cm)
Dalian (DL)	The Bohai Sea	38°55′ N; 121°37′ E	2023-05	40	33.7–51.7	45.7 ± 3.9
Weihai (WH)		37°31′ N; 122°7′ E	2022-09	21	28.9–36.6	32.7 ± 2.2
Rushan (RS)		36°56′ N; 121°32′ E	2022-10	29	27.6–45.7	35.0 ± 4.3
Qingdao (QD)	The Yellow Sea	36°4′ N; 120°23′ E	2022-09	41	30.6–39.2	35.0 ± 2.1
Rizhao (RZ)		35°25′ N; 119°32′ E	2022-08	25	33.9–48.4	38.2 ± 3.3
Lianyungang (LYG)		34°36′ N; 119°13′ E	2023-04	32	30.6–50.4	39.1 ± 3.7
Zhoushan (ZS)	The East China Sea	29°59′ N; 122°12′ E	2022-10	30	30.6–45.1	40.8 ± 3.9

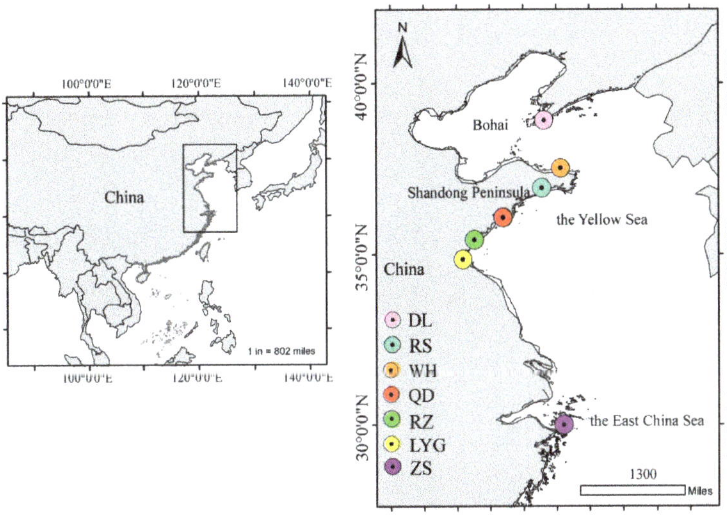

Figure 1. Sampling map of *C. myriaster* populations. (DL, WH, RS, QD, RZ, LYG, and ZS stand for the Dalian population, Weihai population, Rushan population, Qingdao population, Rizhao population, Lianyungang population, and Zhoushan population, respectively).

Iced fresh fishes were quickly transported to the Fishery Ecology and Biodiversity Laboratory (FEBL) of Zhejiang Ocean University of China. Then, experimenters immediately cleaned the samples in running water, drained them with absorbent paper, and placed them on a flat platform. Each individual was separately labeled with a specific code for identification, and the body weight (*BW*) was measured by using an electronic balance accurate to 0.01 g. The morphological analysis data comprised 5 vertebral counts and 17 morphological measurements (Table S1; Figure 2). Morphological measurements were always recorded from the left side of the fish body using a ruler and vernier caliper, with 0.1 cm and 0.01 mm precisions, respectively. Vertebral counts were carried out by removing the back muscles to expose the vertebrae. To ensure the reliability of this study, any sample that showed extensive damage or dysplasia was eliminated, meanwhile, all operations were performed by the same person.

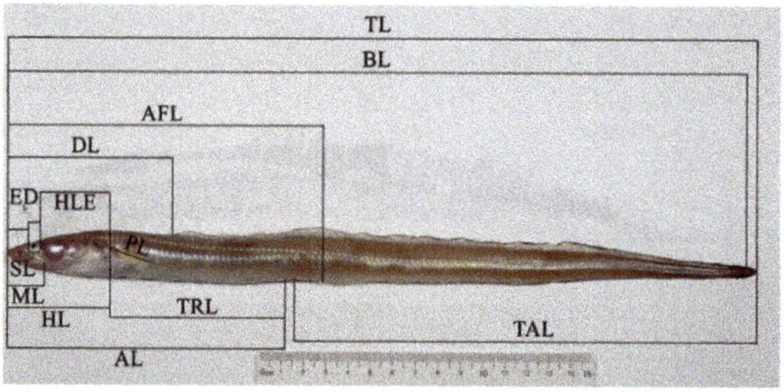

Figure 2. Analyzed morphological measurements of *C. myriaster* in this study. (The measurements *HB*, *EI*, *BD*, and *BB* are not marked in the figure due to the angle of view. Detailed information of analyzed meristic counts and morphological measurements are shown in Table S1).

2.2. Data Analysis

In order to eliminate variations resulting from allometric growth, morphometric measurement data were initially logarithmically converted. The standardization was conducted by an allometric method, as suggested by Elliott et al. (1995) [29]:

$$M_S = M_0\ (L_S/L_0)^{\ b} \tag{1}$$

$$b = \log_{10}{}^{M0}/\log_{10}{}^{L0} \tag{2}$$

In Formulas (1) and (2), M_S is the standardized measurement data, M_0 is the original measurement data, L_S is the overall mean of the original total length (*TL*) for all fish samples in each population, and L_0 is the original total length (*TL*).

The above-gathered data were made a preliminary arrangement in Microsoft Excel 2021, and then the variation coefficient (*C.D*) was calculated according to Mayr's 75% identification and division rules [30]:

$$C.D = |M_1 - M_2|/(S_1 + S_2) \tag{3}$$

In Formula (3), M_1 and M_2 represent the average values of measurement data in two populations S_1 and S_2 represent the standard deviations of measurement data in two populations.

Univariate and multivariate statistical analysis of the standardized data was performed by using one-way analysis of variance (ANOVA), hierarchical cluster analysis (HCA), discriminant function analyses (DFA), and principal component analysis (PCA), all of which were completed in SPSS 27.0 [31]. A univariate ANOVA was done to test whether there was any statistically significant difference for each morphometric character between pairwise populations. In ANOVA, Fisher's least significant difference (LSD) test was used to analyze measurement data with homogeneity of variances [32], while Tamhane's T2 test was used to analyze measurement data without homogeneity of variances [33]. Cluster analysis was applied to determine the closely related populations and draw the dendrogram based on the Euclidean distances. DFA and PCA were used to examine any phenotypic differences between populations. A stepwise discriminant procedure can reduce the number of variables in order to meet the requirement of reduced discriminant function analysis [34]. The discriminant equations were constructed, and the ability to differentiate populations was evaluated through original-validation tests [35–37]. PCA can also select morphometric data to extract several independent variables for population differentiation [38]. In order to more intuitively observe the differences among populations, OriginPro 2021 software was used to draw scatter plots based on the first and second principal components (PCs).

3. Results

3.1. The One-Way Analysis of Variance

One-way ANOVA results of both meristic counts and morphological measurements among seven *C. myriaster* populations are shown in Table S2. Twenty-two morphological indexes were statistically significant ($p < 0.05$) except for *DAV*/*TV*, *BL*, and *AL*. There were 17 morphological indexes with significant differences between the Dalian and Qingdao populations, but only 3 indexes between the Lianyungang and Zhoushan populations were significantly different. The *C.D* values of meristic counts and morphological measurements between two *C. myriaster* populations obtained in this study were all below 1.28, indicating that the diversity difference mainly originated from inter-populations. At the same time, the larger *C.D* values of *AFV*, *AFL*, *AL*, *TRL*, and *EI* implied greater variations in the longitudinal body axis of this species (Table S3). Considering the large *C.D* values of tail length (*TAL*) and no significant difference in the mean body length, the variations of fish bodies were affected by the seasonal predation and nutrient structure presumably.

3.2. The Principal Component Analysis

A total of 7 PCs with eigenvalues greater than 1 were extracted, and the cumulative contribution rate was 72.790%. The contribution rate of the PC1 accounted for 18.839% with the larger load values of *DV*, *AFV*, *ML*, *ED*, *EI*, and *BB*. The PC2 accounted for 16.120%, and *AV*, *AFV*, *AV*/*TV*, *AFV*/*TV*, *AFL*, and *TRL* attained the higher load values (Table 2). These indexes mainly reflected the head features and the vertebral count changes. Moreover, *ML*, *AFL*, *TRL*, and *BB* were also associated with the predation habit and barrel-shaped body structure of *C. myriaster*.

Table 2. Load values and contribution rates of seven principal component factors.

Index	Principal Component (PC)						
	1	2	3	4	5	6	7
TV	0.393	−0.149	−0.547	−0.104	0.640	−0.096	−0.091
DV	0.596	0.255	−0.568	0.428	−0.007	0.050	0.047
AV	0.482	0.641	−0.417	−0.131	0.159	0.029	−0.005
AFV	0.593	0.561	−0.371	−0.268	0.220	−0.080	−0.011
DAV	0.115	0.420	0.209	−0.779	0.240	−0.158	0.014
DV/*TV*	0.538	0.326	−0.456	0.497	−0.200	0.081	0.080
AV/*TV*	0.183	0.807	0.021	−0.049	−0.372	0.111	0.069
AFV/*TV*	0.357	0.786	0.037	−0.221	−0.293	−0.010	0.064
DAV/*TV*	−0.095	0.473	0.480	−0.673	−0.106	−0.096	0.060
BL	0.153	−0.331	−0.095	−0.125	0.075	0.176	0.716
TAL	0.293	−0.309	0.131	−0.254	0.274	0.339	0.006
AFL	−0.406	0.593	0.417	0.273	0.322	0.132	0.060
AL	−0.121	−0.058	0.060	0.205	−0.240	−0.606	0.324
DL	−0.089	0.456	0.363	0.572	0.174	0.202	−0.128
TRL	−0.492	0.594	0.127	0.181	0.037	0.299	0.049
HL	0.190	0.244	0.673	0.432	0.279	−0.167	−0.001
HB	0.436	−0.282	0.081	−0.089	−0.352	0.414	−0.141
SL	0.438	−0.129	0.178	−0.082	−0.220	−0.212	−0.520
ML	0.584	−0.025	0.184	0.242	−0.147	−0.384	0.081
ED	0.658	−0.134	0.296	−0.047	0.139	−0.165	0.057
EI	0.606	−0.192	0.217	0.050	0.043	0.126	−0.088
HLE	0.525	0.016	0.550	0.307	0.224	−0.098	0.068
BD	0.450	−0.162	0.433	−0.080	0.012	0.213	0.109
BB	0.578	−0.205	0.296	−0.095	−0.190	0.303	0.124
PL	0.483	−0.204	0.291	0.009	−0.035	−0.018	0.019
Eigenvalues	4.710	4.030	3.090	2.573	1.471	1.314	1.010
Contribution ratio (%)	18.839	16.120	12.361	10.291	5.884	5.256	4.038
Cumulative contribution ratio (%)	18.839	34.960	47.320	57.612	63.495	68.751	72.790

A scatter plot based on PCA showed that individuals in the Rushan and Rizhao populations were more dispersed than others, and both of them occupied broader areas. Individuals from the Lianyungang population almost encircled those of the Zhoushan population, and both of them possessed a considerable percentage of common area with the Rizhao population. All populations gathered with each other, and the overlapping area between the Weihai and Dalian populations was minimal (Figure 3).

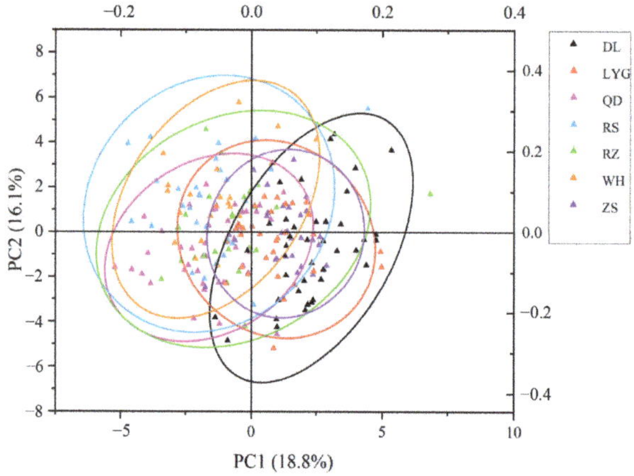

Figure 3. PCA scatter plot based on the first and second principal components. (DL, WH, RS, QD, RZ, LYG, and ZS stand for the Dalian population, Weihai population, Rushan population, Qingdao population, Rizhao population, Lianyungang population, and Zhoushan population, respectively).

3.3. The Cluster Analysis

Cluster analysis was carried out to display the similarity among populations. According to the Euclidean distances of 7 *C. myriaster* populations (Table S4), the relationship between Zhoushan and Lianyungang populations was the closest, with an Euclidean distance of only 3.261. While the Euclidean distance between Weihai and Zhoushan populations reached 17.710, showing quite a few differences. The dendrogram showed that the Dalian population had a distant relationship with other populations, and formed an independent branch. Individuals from the remaining six geographic locations belonging to the Yellow Sea and the East China Sea were generally co-clustered into two major branches, respectively, suggesting obvious differences between the two sea areas (Figure 4).

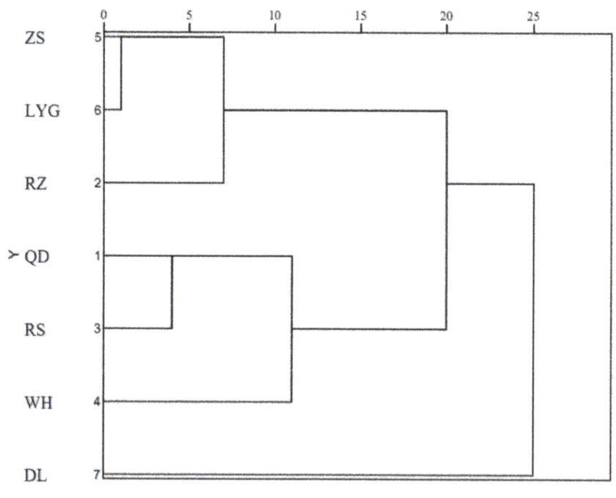

Figure 4. The dendrogram based on Euclidean distances of *C. myriaster* populations. (DL, WH, RS, QD, RZ, LYG, and ZS stand for the Dalian population, Weihai population, Rushan population, Qingdao population, Rizhao population, Lianyungang population, and Zhoushan population, respectively).

3.4. The Discriminant Function Analysis

The stepwise discriminant analysis was conducted according to the correlations between the independent variables and the dependent variables. In this study, a total of 9 characteristics (*AFV*, *AV/TV*, *BL*, *AFL*, *DL*, *TRL*, *EI*, *HLE*, and *PL*) were retained in the stepwise discriminant analysis, and 4 typical discriminant functions (DFs) with significant differences were obtained. The cumulative contribution rate of DF1 and DF2 reached 86.8%, which meant the functions included the most meaningful morphometric characteristics. Scatter plots based on DFA mainly divided the *C. myriaster* populations into three groups (Figure 5), which were almost consistent with their distribution of sea areas.

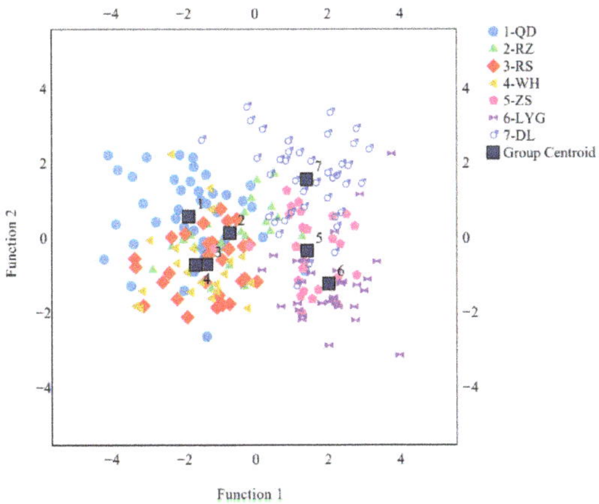

Figure 5. Discriminant analysis plot with 25 morphometric variables of *C. myriaster* populations. (DL, WH, RS, QD, RZ, LYG, and ZS stand for the Dalian population, Weihai population, Rushan population, Qingdao population, Rizhao population, Lianyungang population, and Zhoushan population, respectively).

The Dalian population was clearly distinguished from the other populations, forming an independent group with slight overlap, and the group centroid was the furthest from the other populations in either the vertical or horizontal axis. The populations in the Yellow Sea did not distribute in regularity. The Qingdao, Rushan, and Weihai populations clustered together and became another independent group. Despite being the most geographically distant, the Zhoushan (the northern East China Sea) and Lianyungang (the southern Yellow Sea) populations exhibited small differences and constituted the third group. Only the Rizhao population was a special one, which was slightly superposed with the latter two groups. These latter two groups were partially overlapped, and the distances of group centroids on the horizontal axis were greater than those on the vertical axis. Among all the populations, the group centroid of Qingdao and Lianyungang populations had the furthest distance on the horizontal axis, and the group centroid of Dalian and Lianyungang populations had the furthest distance on the vertical axis.

According to the results of discriminant analysis, the discriminant functions of each population were constructed as follows:

$$Y_{QD} = 10.122AFV + 2644.927AV/TV + 54.614BL + 0.522AFL + 5.098DL + 4.408TRL + 0.326EI + 2.318HLE + 1.623PL - 1858.731$$

$$Y_{RZ} = 11.757AFV + 2541.173AV/TV + 53.962BL + 0.405AFL + 5.264DL + 4.323TRL - 0.159EI + 2.430HLE + 1.765PL - 1865.746$$

$$Y_{RS} = 11.624AFV + 2580.945AV/TV + 53.880BL + 0.567AFL + 5.147DL + 4.352TRL - 0.270EI + 1.775HLE + 1.775PL - 1866.559$$

$$Y_{WH} = 11.550AFV + 2574.959AV/TV + 54.523BL + 0.646AFL + 5.253DL + 4.237TRL - 0.896EI + 2.360HLE + 1.542PL - 1894.879$$

$$Y_{ZS} = 13.101AFV + 2477.486AV/TV + 52.806BL + 0.625AFL + 4.790DL + 3.789TRL + 0.375EI + 2.411HLE + 2.480PL - 1845.895$$

$$Y_{LYG} = 13.587AFV + 2382.272AV/TV + 52.600BL + 0.704AFL + 4.847DL + 3.677TRL - 0.166EI + 2.037HLE + 3.294PL - 1840.668$$

$$Y_{DL} = 12.175AFV + 2606.883AV/TV + 52.974BL + 0.268AFL + 5.403DL + 3.769TRL + 0.452EI + 3.084HLE + 2.312PL - 1853.890$$

The discriminant accuracies are shown in Table S5, in which the discriminant accuracy of the Dalian population is the highest (72.50%). The DFA demonstrates that about 59.63% of inspected individuals are correctly classified into the original population on average.

4. Discussion

A sufficient degree of habitat isolation may lead to significant intraspecific phenotypic and genetic differentiation [39]. Reproductive isolation of fish populations can occur in different spawning areas due to hydrological characteristics that reduce or even prevent cross-regional migration [40,41]. In general, fishes are more susceptible to environmentally-induced morphological variation, allowing them to exhibit higher variations within and between populations than other vertebrates [42]. These morphological changes can be used as a basis for the distinction and management of different populations [37].

In this presently reported study, the multivariate analysis was employed to reflect the external morphology differences of *C. myriaster* in various areas of coastal China. The results of one-way ANOVA showed that the differences between the Dalian and Qingdao populations were the largest, and individuals in the Dalian population (the Bohai Sea) also held more significant differences from others. Previous studies revealed that the bottom fishery biological communities in the Yellow and Bohai Sea had been under serious disturbance in recent decades [43]. The environmental medley of factors might lead to great differences between these two populations located in different sea areas mentioned above. There were no significant differences in body length (*BL*) and length before anal (*AL*), but significant differences in tail length (*TAL*) were detected among all populations in ANOVA. Furthermore, the morphological traits with high variation coefficient values were mostly related to body weight (*BW*), which might be due to the special long-strip shape of eels. Tail length (*TAL*) reflecting the swimming ability of fishes can further affect their predatory ability and make these weight-related traits have greater potential variation [44].

PCA conclusions mainly manifested that the cumulative contribution rate of seven PCs' eigenvalue greater than 1 was less than 85%. Subsequent related studies could increase the traits analyzed or adopt other morphological analysis methods to conduct in-depth research according to actual situations [45]. Zou et al. (2020) [46] used mitochondrial control region sequences to assess the genetic diversity, population differentiation, and demographic history of *C. myriaster*. It was revealed that no clear genetic subdivision was found among populations, which is consistent with the conclusions of Ishikawa [21]. However, our results of multivariate analysis were generally consistent with the actual geographical distributions of each population. The cluster analysis together with the plots of PCA and DFA separated the populations into three groups: the north group (DL), the central group (WH, QD, and RS), and the south group (RZ, LYG, and ZS). A similar population structure of *Scomberomorus niphonius* from the eight spawning grounds along the Chinese coast was also detected, according to a morphometric truss network method [47]. It was speculated that the unique geographical location of the Bohai Sea, the only semi-enclosed inland sea, limited intraspecific communication and resulted in the specialization of morphological traits.

Consequently, the Dalian population occupied an isolated space and was relatively farther away across the straits. Coincidentally, strong morphometric and meristic differentiation resulted from the closed geographic structure, and environmental discreteness was also observed in horse mackerel (*Trachurus mediterraneus*) from the Marmara Sea by Turan [48]. The Shandong Peninsula is the largest peninsula in China, with a coastline length of more than 3000 km and accounting for about 1/6 of the total of China. The ambient conditions determined the composition and spatial distribution of macrobenthos in offshore areas of the Shandong Peninsula. The bottom temperature is a principal environmental element that is closely related to the macrozoobenthic community structure. The coastal currents, water masses, and continental shelf front interfere with the distribution of bottom temperature in the Yellow Sea [49].

As a bottom-dwelling eel with metamorphosis development, the habitat and migratory behaviors of *C. myriaster* are both affected by marine currents and the surroundings [50,51]. The literature evidenced that *C. myriaster* spawned along the Kyushu-Palau Ridge in the western North Pacific, even if it remained controversial [22,23]. Takai [52] suggested that the spawning grounds of this species were at the edge of the continental shelf in the East China Sea with the pattern of "a single spawning ground for a single population". Instead, Mochioka et al. [20] believed that Whitespotted congers living in East Asian waters had multiple composite spawning populations. Until recent years, researchers investigated the breeding areas and migration patterns of *C. myriaster* by utilizing the age-structured otolith chemistry profiles and confirmed again that *C. myriaster* in coastal China seas came from the same spawning area located in the western North Pacific [53]. The larvae and juveniles may be passively transported from their spawning area to the East China Sea together by the North Equatorial Current and the Kuroshio Current before being separated by the Yellow Sea Warm Current to different habitats during the subsequent life stages. Morphometric characters are susceptible to environmental influences throughout life. In addition, phenotypic variations between populations are easily influenced by local circumstances, often associated with the geographic regions that the species occupies throughout its ranges [54]. To name but a few examples, three independent stocks of *Harpadon nehereus* along the Indian coast were detected by the multiple morphological comparisons, and future policy regulations should consider collaboration among the maritime state governments since fishery on the west coast of India was replenished by a single stock [55]. The morphological variation of *Liparis tanakae* sampled from three localities surrounding the Korean peninsula was analyzed, and the canonical discriminant analysis clearly revealed three groups that separated according to locality [56]. Besides, the feeding relationships of marine organisms and their trophic positions in the food web are not always static but are affected by seasonal temperatures, showing a dynamic change process [57]. Therefore, the different sampling times are also likely to have a certain impact on fish morphological variation and differentiation. By comprehensively considering the multiple factors, we inferred that the population structure and spatial pattern of *C. myriaster* on the coast of China were possibly influenced directly or indirectly by the unique environmental elements, sloshing bottom fishery biological communities, coastal currents, water masses, and continental shelf front.

5. Conclusions

Summarizing the morphological multivariate data, we found that the relationships of *C. myriaster* populations in the northwest Pacific Ocean along China's coast were closely related to their geographical locations, where the population structure was most likely influenced by sophisticated oceanic circulations together with the various other environmental factors to a certain extent. As recent studies showed that marine resources in the Yellow and East China Seas were undergoing long-term fishing pressure and massive catching efforts, the demarcation of jurisdictional boundaries for *C. myriaster* must be fully considered, and biogeography-based fishery management units (FMUs) should be established to ensure sustainable utilization of *C. myriaster*.

Supplementary Materials: The following supporting information can be downloaded at: https://www.mdpi.com/article/10.3390/ani14132007/s1, Table S1: Analyzed meristic counts and morphological measurements of *C. myriaster* in this study; Table S2: One-way analysis of variance (ANOVA) results of seven *C. myriaster* populations; Table S3: The variation coefficient (*C.D*) values for each trait of pairwise populations of *C. myriaster;* Table S4: Euclidean distances of seven *C. myriaster* populations; Table S5: The accuracy of discriminant function analyses in each population.

Author Contributions: Conceptualization and writing—review and editing, T.Y.; methodology and writing—original draft preparation, P.X. All authors have read and agreed to the published version of the manuscript.

Funding: This research was supported by the Science and Technology Planning Project of Zhoushan (2022C41022) and the National Key Research and Development Program Project of China (2023YFD2401901).

Institutional Review Board Statement: The animal study protocol was approved by the Ethics Committee of Zhejiang Ocean University (approval code: ZJOU-ECAE20211878, approval date: 19 January 2022).

Informed Consent Statement: Not applicable.

Data Availability Statement: The data used in this study are available from the corresponding author upon reasonable request.

Acknowledgments: We would like to express our sincere thanks to Zhichuang Lu and Na Song for their helps in samples collecting.

Conflicts of Interest: The authors declare that they have no competing interests associated with the manuscript.

References

1. Kaneko, K. Relationship among phenotypic plasticity, phenotypic fluctuations, robustness, and evolvability; Waddington's legacy revisited under the spirit of Einstein. *J. Biosci.* **2009**, *34*, 529–542. [CrossRef] [PubMed]
2. West-Eberhard, M.J. Phenotypic plasticity and the origins of diversity. *Annu. Rev. Ecol. Syst.* **1989**, *20*, 249–278. [CrossRef]
3. Cadrin, S.X. Advances in morphometric identification of fishery stocks. *Rev. Fish Biol. Fish.* **2000**, *10*, 91–112. [CrossRef]
4. Imre, I.; McLaughlin, R.L.; Noakes, D.L.G. Phenotypic plasticity on brook charr: Changes in caudal fin induced by water flow. *J. Fish Biol.* **2002**, *61*, 1171–1181. [CrossRef]
5. Armstrong, D.; Kemp, P.S.; Kennedy, G.A.; Ladle, M.; Milner, N. Habitat requirements of Atlantic salmon and brown trout in rivers and streams. *Fish. Res.* **2003**, *62*, 143–170. [CrossRef]
6. Salgueiro, P.; Carvalho, G.; Collares-Pereira, M.J.; Coelho, M.M. Microsatellite analysis of genetic population structure of the endangered cyprinid *Anaecypris hispanica* in Portugal: Implications for conservation. *Biol. Conserv.* **2003**, *109*, 47–56. [CrossRef]
7. Silva, A. Morphometric variation among sardine (*Sardina pilchardus*) populations from the northeastern Atlantic and the western Mediterranean. *ICES J. Mar. Sci.* **2003**, *60*, 1352–1360. [CrossRef]
8. Dwivedi, A.K.; De, K. Role of morphometrics in fish diversity assessment: Status, challenges and future prospects. *Natl. Acad. Sci. Lett.* **2024**, *47*, 123–126. [CrossRef]
9. Lind, C.E.; Ponzoni, R.W.; Nguyen, N.H.; Khaw, H.L. Selective breeding in fish and conservation of genetic resources for aquaculture. *Reprod. Domest. Anim.* **2012**, *47*, 255–263. [CrossRef]
10. Frederich, B.; Liu, S.V.; Dai, C.F. Morphological and genetic divergences in a coral reef damselfish, *Pomacentrus coelestis*. *Evol. Biol.* **2012**, *39*, 359–370. [CrossRef]
11. Secor, D.H. Chapter Two—The Unit Stock Concept: Bounded Fish and Fisheries. In *Stock Identification Methods*, 2nd ed.; Elsevier: Amsterdam, The Netherlands, 2014; pp. 7–28. [CrossRef]
12. Turki-Missaoui, O.; M'Hetli, M.; Kraïem, M.M.; Chriki, A. Morphological differentiation of introduced pikeperch (*Sander lucioperca* L., 1758) populations in Tunisian freshwaters. *J. Appl. Ichthyol.* **2011**, *27*, 1181–1189. [CrossRef]
13. Myoung, S.H.; Kim, J.K. Population structure of the Korean gizzard shad, *Konosirus punctatus* (Clupeiformes, Clupeidae) using multivariate morphometric analysis. *Ocean Sci. J.* **2016**, *51*, 33–41. [CrossRef]
14. Turan, C.; Oral, M.; Öztürk, B.; Düzgüneş, E. Morphometric and meristic variation between stocks of Bluefish (*Pomatomus saltatrix*) in the Black, Marmara, Aegean and northeastern Mediterranean Seas. *Fish. Res.* **2006**, *79*, 139–147. [CrossRef]
15. Brevoort, J.C. *Notes on Some Figures of Japanese Fish Taken from Recent Specimens by the Artists of the U. S. Japan Expedition;* Creative Media Partners, LLC: Sacramento, CA, USA, 1856; pp. 253–288.
16. Chen, D.G.; Zhang, M.Z. *Marine Fishes of China;* China Ocean University Press: Qingdao, China, 2015; pp. 187–188. (In Chinese)
17. Kimura, Y.; Ishikawa, S.; Tokai, T.; Nishida, M.; Tsukamoto, K. Early life history characteristics and genetic homogeneity of *Conger myriaster* leptocephali along the east coast of central Japan. *Fish. Res.* **2004**, *70*, 61–69. [CrossRef]

18. Tokimura, M. Conger-eel fisheries and fisheries resources in the East China Sea. *Nippon Suisan Gakkaishi* **2001**, *67*, 125–126. [CrossRef]
19. Li, M.; Jiao, Y.; Bi, R.; Ren, Y. Population status and distribution of whitespotted conger (*Conger myriaster*) in Yellow Sea: An important migratory species along coastal China with limited data. *Fish Oceanogr.* **2020**, *29*, 32–45. [CrossRef]
20. Mochioka, N.; Tabeta, O.; Kubota, T. A pre-leptocephalus larva of *Conger myriaster* (family Congridae) collected from Suruga Bay, Central Japan. *Jpn. J. Ichthyol.* **1988**, *35*, 184–188. [CrossRef]
21. Ishikawa, S.; Kimura, Y.; Tokai, T.; Tsukamoto, K.; Nishida, M. Genetic variation in the mitochondrial and nuclear DNA of the Japanese conger *Conger myriaster*. *Fish. Sci.* **2001**, *67*, 1081–1087. [CrossRef]
22. Miller, M.J.; Yoshinaga, T.; Aoyama, J.; Otake, T.; Mochioka, N.; Kurogi, H.; Tsukamoto, K. Offshore spawning of *Conger myriaster* in the western North Pacific: Evidence for convergent migration strategies of anguilliform eels in the Atlantic and Pacific. *Naturwissenschaften* **2011**, *98*, 537–543. [CrossRef] [PubMed]
23. Kurogi, H.; Mochioka, N.; Okazaki, M.; Takahashi, M.; Miller, M.J.; Tsukamoto, K.; Ambe, D.; Katayama, S.; Chow, S. Discovery of a spawning area of the common Japanese conger *Conger myriaster* along the Kyushu-Palau Ridge in the western North Pacific. *Fish. Sci.* **2012**, *78*, 525–532. [CrossRef]
24. Kawazu, M.; Kameda, T.; Kurogi, H.; Yoda, M.; Ohshimo, S.; Sakai, T.; Tsukamoto, Y.; Mochioka, N. Biological characteristics of *Conger myriaster* during the initial stage of spawning migration in the East China Sea. *Fish. Sci.* **2015**, *81*, 663–671. [CrossRef]
25. Yang, H.; Shi, B.; Niu, H.X.; Zhang, D.Q.; Li, J. Advances and future prospects in *Conger myriaster* research. *Mar. Sci.* **2020**, *44*, 152–158. (In Chinese with English abstract)
26. Zhang, X.M.; Cheng, S.F. Path analysis and growth curve fitting of the morphological traits and body weight of *Conger myriaster*. *J. Fish. Sci. China* **2020**, *27*, 1167–1175. (In Chinese with English abstract)
27. Yin, J.; Mu, X.X.; Zhang, C.L.; Xu, B.D.; Xue, Y.; Ren, Y.P. Comparison of morphological and genetic characteristics of *Conger myriaster* populations in China. *J. Fish. China* **2020**, *44*, 358–367. (In Chinese with English abstract)
28. Ying, Y.P.; Chen, Y.; Lin, L.S.; Gao, T.X. Risks of ignoring fish population spatial structure in fisheries management. *Can. J. Fish. Aquat. Sci.* **2011**, *68*, 2101–2120. [CrossRef]
29. Elliott, N.G.; Haskard, K.; Koslow, J.A. Morphometric analysis of orange roughly (*Hoplostethus atianticus*) off the continental slope of Southern Australia. *Fish Biol.* **1995**, *46*, 202–220. [CrossRef]
30. Mayr, E.; Linsley, E.G.; Usinger, R.L. Methods and Principles of Systematic Zoology. *J. Mammal.* **1953**, *34*, 520–521.
31. Field, A. *Discovering Statistics Using SPSS*, 2nd ed.; Sage Publications: Thousand Oaks, CA, USA, 2005; pp. 127–169.
32. Fisher, R.A. *The Design of Experiments*; Oliver and Boyd: Edinburgh, UK, 1935; pp. 88 136.
33. Tamhane, A.C. A comparison of procedures for multiple comparisons of means with unequal variances. *J. Am. Stat. Assoc.* **1979**, *74*, 471–480. [CrossRef]
34. Jennrich, R.I. *Statistical Methods for Digital Computers III*; John Wiley & Sons, Inc.: Hoboken, NJ, USA, 1977; pp. 99–133.
35. Veasey, E.A.; Schammass, E.A.; Vencovsky, R.; Martins, P.S.; Bandel, G. Germplasm characterization of Sesbania accessions based on multivariate analyses. *Genet. Resour. Crop Evol.* **2001**, *48*, 79–91. [CrossRef]
36. Samaee, M.; Patzner, R.A.; Mansour, N. Morphological differentiation within the population of Siah mahi, *Capoeta capoeta gracilis*, (Cyprinidae, Teleostei) in a river of the south Caspian Sea basin: A pilot study. *J. Appl. Ichthyol.* **2006**, *25*, 583–590. [CrossRef]
37. Mir, J.I.; Sarkar, U.K.; Dwivedi, A.K.; Gusain, O.P.; Jena, J.K. Stock structure analysis of *Labeo rohita* (Hamilton, 1822) across the Ganga basin (India) using a truss network system. *J. Appl. Ichthyol.* **2013**, *29*, 1097–1103. [CrossRef]
38. Wold, S.; Esbensen, K.; Geladi, P. Principal component analysis. *Chemom. Intell. Lab. Syst.* **1987**, *2*, 37 52. [CrossRef]
39. Turan, C.; Erguden, D.; Gurlek, M.; Başusta, N.; Turan, F. Morphometric structuring of the anchovy (*Engraulis encrasicolus* L.) in the black, Aegean and Northeastern Mediterranean seas. *Turk. J. Vet. Anim. Sci.* **2004**, *28*, 865–871.
40. Hourston, A.S. Homing by Canada's west coast herring to management units and divisions as indicated by tag recoveries. *Can. J. Fish. Aquat. Sci.* **1982**, *39*, 1414–1422. [CrossRef]
41. Iles, T.D.; Sinclair, M. Atlantic herring: Stock discreteness and abundance. *Science* **1982**, *215*, 627–633. [CrossRef]
42. Wimberger, P.H. Plasticity of fish body shape. The effects of diet, development, family and age in two species of *Geophagus* (Pisces: Cichlidae). *Biol. J. Linn. Soc.* **1992**, *45*, 197–218. [CrossRef]
43. Zhao, Y.S.; Shan, X.J.; Yang, T.; Jin, X.S.; Wei, C. Biodiversity characteristics of bottom fishery communities in the sea Adjacent to the Miaodao Archipelago. *Prog. Fish. Sci.* **2022**, *43*, 132–147, (In Chinese with English abstract).
44. Triantafyllou, M.S.; Triantafyllou, G.S. An Efficient Swimming Machine. *Sci. Am.* **1995**, *272*, 64–70. [CrossRef]
45. Eisenhour, D. Systematics of *Macrhybopsis tetranema* (Cypriniformes: Cyprinidae). *Copeia* **1999**, *1999*, 969–980. [CrossRef]
46. Zou, C.C.; Wang, L.J.; Kong, L.M.; Wang, Y.J.; Wu, Z.H.; Xu, J.H.; Song, A.H.; Liu, H.; You, F. High levels of genetic diversity and connectivity of whitespotted conger *Conger myriaster* in the East China Coast. *Mar. Biodivers.* **2020**, *50*, 47. [CrossRef]
47. Jiang, Y.Q.; Zhang, C.; Ye, Z.J.; Xu, B.D.; Tian, Y.J.; Watanabe, Y. Stock structure analysis of the Japanese Spanish mackerel *Scomberomorus niphonius* (Cuvier, 1832) along the China coast based on truss network. *J. Ocean Univ. China* **2020**, *19*, 202–208. [CrossRef]
48. Turan, C. Stock identification of Mediterranean horse mackerel (*Trachurus medi-terraneus*) using morphometric and meristic characters. *ICES J. Mar. Sci.* **2004**, *61*, 774–781. [CrossRef]
49. Teng, A.K.; Zhang, M.S.; Yang, X.J.; Cui, W.Y.; Chen, C.; Zhu, L.; Shao, C.C.; Yu, Z.S. Study on assemblage structure of macrozoobenthos in the Yellow Sea. *Period. Ocean Univ. China* **2023**, *53*, 74–84, (In Chinese with English abstract).

50. Katayama, S.; Ishida, T.; Shimizu, Y.; Yamanobe, A. Seasonal change in distribution of conger eel *Conger myriaster* off the Pacific coast south of Tohoku, north-eastern Japan. *Fish. Sci.* **2004**, *70*, 1–6. [CrossRef]
51. Gorie, S.; Tanda, M.; Nagasawa, K. Habitat preference and feeding habits of juvenile white-spotted conger *Conger myriaster* in the eastern Seto Inland Sea, Japan. *Aquac. Sci.* **2010**, *58*, 167–179. [CrossRef]
52. Takai, T. Studies on the morphology, ecology and culture of the important apodal fishes, *Muraenesox cinereus* (Forsskal) and *Conger myriaster* (Brevoort). *J. Shimonoseki Coll. Fish.* **1959**, *8*, 209–555.
53. Mu, X.X.; Zhang, C.; Zhang, C.L.; Yang, J.; Ren, Y.P. Age-structured otolith chemistry profiles revealing the migration of *Conger myriaster* in China Seas. *Fish. Res.* **2021**, *239*, 105938. [CrossRef]
54. Swain, D.P.; Foote, C.J. Stocks and chameleons: The use of phenotypic variation in stock identification. *Fish. Res.* **1999**, *43*, 113–128. [CrossRef]
55. Pazhayamadom, D.G.; Chakraborty, S.K.; Jaiswar, A.K.; Sudheesan, D.; Sajina, A.M.; Jahageerdar, S. Stock structure analysis of 'Bombay duck' (*Harpadon nehereus* Hamilton, 1822) along the Indian coast using truss network morphometrics. *J. Appl. Ichthyol.* **2015**, *31*, 37–44. [CrossRef]
56. Myoung, S.H.; Ban, T.; Kim, J. Population structure of *Liparis tanakae* (Pisces, Liparidae) from Korea based on morphological and molecular traits. *Korean J. Fish. Aquat. Sci.* **2016**, *49*, 198–207, (In Korean with English abstract). [CrossRef]
57. Li, H.B.; Chen, X.; Denis, M.; Zhao, Y.; Huang, L.F.; Jiang, Z.G.; Zhang, W.C.; Xiao, T. Seasonal and spatial variation of pelagic microbial food web structure in a semi-enclosed temperate bay. *Front. Mar. Sci.* **2020**, *7*, 2296–7745. [CrossRef]

animals

MDPI

Article

Effects of Dietary Supplementation with Chitosan on the Muscle Composition, Digestion, Lipid Metabolism, and Stress Resistance of Juvenile Tilapia (*Oreochromis niloticus*) Exposed to Cadmium-Induced Stress

Qin Zhang [1,2,3,†], Yi Xie [1,2,3,†], Yuanhui Zhang [1,2,3], Enhao Huang [1,2,3], Liuqing Meng [1,2,3], Yongqiang Liu [1,2,3,*] and Tong Tong [1,2,3,*]

1 School of Marine Sciences and Biotechnology, Guangxi Minzu University, Nanning 530008, China; zhangqin@gxmzu.edu.cn (Q.Z.); xy463317779@163.com (Y.X.); zyh_sophie@163.com (Y.Z.); 15113161786@163.com (E.H.); mliuqing0713@126.com (L.M.)

2 Guangxi Key Laboratory for Polysaccharide Materials and Modifications, Nanning 530008, China

3 Guangxi Marine Microbial Resources Industrialization Engineering Technology Research Center, Nanning 530008, China

* Correspondence: liuyongqiang@gxmzu.edu.cn (Y.L.); tongtong@gxmzu.edu.cn (T.T.)

† These authors contributed equally to this work.

Simple Summary: Long-term exposure to cadmium can cause liver toxicity, kidney toxicity, bone toxicity, and cancer in aquatic animals, which seriously endangers the health of aquatic animals and the quality of aquatic products. Chitosan can be combined with cadmium in aquaculture water through adsorption, thus alleviating the toxic effects of cadmium on aquatic animals. In this study, juvenile tilapia was fed with a formulated feed containing five different levels (0%, 0.5%, 1.0%, 1.5%, and 2.0%) of chitosan for 60 days, while the water in all experimental groups contained a Cd^{2+} concentration of 0.2 mg/L. The results indicated that dietary chitosan supplementation could alleviate the effects of Cd^{2+} stress on the muscle composition, digestive enzymes, lipid metabolism, and stress resistance, and their related gene expression, of juvenile tilapia.

Abstract: The aim of this study was to investigate the effects of dietary chitosan supplementation on the muscle composition, digestion, lipid metabolism, and stress resistance, and their related gene expression, of juvenile tilapia (*Oreochromis niloticus*) subjected to cadmium (Cd^{2+}) stress. Juvenile tilapia with an initial body weight of 21.21 ± 0.24 g were fed with a formulated feed containing five different levels (0%, 0.5%, 1.0%, 1.5%, and 2.0%) of chitosan for 60 days, while the water in all experimental groups contained a Cd^{2+} concentration of 0.2 mg/L. The results showed that, compared with the control group (0% chitosan), the contents of crude fat and crude protein in the muscle, the activities of lipase, trypsin, and amylase in the intestine, as well as the relative expression levels of metallothionein (*mt*), cytochrome P450 1A (*cyp1a*), carnitine palmitoyltransferase-1 (*cpt-1*), peroxisome proliferator-activated receptor alpha (*pparα*), peroxisome proliferator-activated receptor gamma (*pparγ*), hormone-sensitive lipase (*hsl*), lipoprotein lipase (*lpl*), malate dehydrogenase (*mdh*), leptin (*lep*), fatty acid synthase (*fas*), sterol regulatory element-binding protein 1 (*srebp1*), and stearoyl-CoA desaturase (*scd*) genes in the liver of juveniles were significantly increased ($p < 0.05$). In conclusion, dietary chitosan supplementation could alleviate the effects of Cd^{2+} stress on the muscle composition, digestive enzymes, lipid metabolism, and stress resistance, and their related gene expression, of juvenile tilapia, and to some extent reduce the toxic effect of Cd^{2+} stress on tilapia.

Keywords: chitosan; cadmium; tilapia; stress resistance; lipid metabolism

1. Introduction

The element cadmium is one of the most toxic heavy-metal pollutants in the environment which can enter the aquatic ecosystem through industrial wastewater. Cadmium can pass through the food chain and accumulate within organisms at higher trophic levels, thereby inducing changes in biological physiology and affecting biological behavior [1]. The entry and accumulation of cadmium in organisms generally occur through respiration, skin absorption, and dietary intake, leading to toxic effects on the organisms [2]. Long-term exposure to cadmium can result in liver, kidney and skeletal toxicity and carcinogenicity in aquatic animals, posing a significant threat to their health and the quality of aquatic products [3]. The concentration of cadmium in several rivers in China generally ranged from 10 to 500 ng/L, with levels exceeding 1 mg/L reported in certain industrialized areas [4]. In accordance with the guidelines set by the US Environmental Protection Agency (USEPA), a permissible limit of 0.005 mg/L for cadmium (Cd(II)) in the water has been established [5]. At present, methods of removing cadmium in water pollution treatment are mainly physical and chemical methods and biological methods [6].

The utilization of chitosan for mitigating the deleterious impact of heavy-metal contamination on aquaculture organisms represents a prevalent treatment approach for the elimination of heavy-metal pollution in water [7]. An adsorption capacity of 10.0 mg/g for cadmium by chitosan beads has been reported (at pH 6–7) [8]. The environmentally friendly and sustainable nature of chitosan has garnered significant attention in the fields of water treatment and environmental protection. It is widely acknowledged that chitosan exhibits exceptional adsorption capabilities for heavy metals, thereby presenting a wide range of potential applications [9]. The results of various studies have demonstrated that low-molecular-weight chitosan has the capacity to adsorb cadmium in aquaculture water, thereby alleviating its toxic impact on aquatic organisms [10]. At the same time, the utilization of chitosan as a novel adsorbent demonstrates its exceptional affinity toward lead and copper ions in aqueous solutions, thereby showcasing remarkable adsorption capacity [11,12]. Additionally, chitosan exhibits the ability to adsorb cadmium ions in aqueous environments and acts as a natural marine prebiotic for enhancing the immune response and disease resistance of aquatic organisms, suppressing the proliferation of harmful microorganisms and accelerating wound healing. Its biodegradability and unique molecular structure are widely recognized within the aquaculture industry [10,11]. However, the long-term inclusion of chitosan in animal feed has been observed to induce weight reduction in fish. This phenomenon can be attributed to the intricate interplay between chitosan and the lipids, cholesterol, fat-soluble vitamins, and minerals present in fish feed, resulting in diminished nutrient absorption and subsequent loss of body mass [13]. Therefore, it is imperative to determine the optimal dosage of chitosan in fish feed.

The genetically improved farmed tilapia (GIFT, *Oreochromis niloticus*) is a genetically modified strain of tilapia that exhibits desirable traits, including exceptional salt tolerance, rapid growth rate, high meat yield, robust cold resistance, ease of domestication, and superior catchability, which have gained significant favor among both farmers and consumers [14,15]. Currently, research on tilapia primarily focuses on the physiological toxicity of heavy-metal pollution to fish. However, there is a limited number of studies investigating the potential mitigation of heavy-metal stress through chitosan supplementation in tilapia feed. The objective of this study was to investigate the effects of dietary chitosan supplementation on the muscle composition, digestive enzymes, lipid metabolism, and stress resistance, and their related gene expression, of juvenile tilapia under cadmium exposure. This study aims to provide scientific insights into the application of chitosan in aquaculture practices and the management of heavy-metal pollution in aquatic environments.

2. Materials and Methods

2.1. Experimental Materials

The cadmium chloride ($CdCl_2$) utilized in this study was of analytical purity and was procured from Sigma-Aldrich (Shanghai) Trading Co., Ltd., Shanghai, China. The chitosan used was food-grade with a deacetylation degree ≥ 95% and viscosity ranging from 100–200 mpa.s, procured from Shanghai Yichen Biotechnology Co., Ltd., Shanghai, China. The origin of chitosan was the marine crab shell. The chitosan had a particle size smaller than 74 μm (200 mesh sieve). All the feed materials were procured from Nanning Tongwei Feed Co., Ltd., Nanning, China, and they were animal food grade.

2.2. Experimental Diets

Based on prior research findings [10,14], five iso-nitrogen, iso-lipid, and iso-energy feeds were used in this study. The control diet was designated as the 0% group without chitosan supplementation and the remaining experimental diets were supplemented with chitosan at concentrations of 0.5%, 1.0%, 1.5%, and 2.0%, respectively designated as the 0.5% group, the 1.0% group, the 1.5% group and the 2.0% group. The composition of the experimental diets is shown in Table 1.

Table 1. Composition of the experimental diets for juvenile GIFT (g/100g of dried feed).

Ingredients	Chitosan Levels (%)				
	0	**0.5**	**1.0**	**1.5**	**2.0**
Chitosan	0.00	0.50	1.00	1.50	2.00
Soybean oil	2.00	2.00	2.00	2.00	2.00
Fish oil	2.00	2.00	2.00	2.00	2.00
Fish meal	8.00	8.00	8.00	8.00	8.00
Rapeseed meal	22.00	22.00	22.00	22.00	22.00
Soybean meal	33.00	33.00	33.00	33.00	33.00
Dextrin	24.39	23.89	23.39	22.89	22.39
Gelatin	5.00	5.00	5.00	5.00	5.00
Vitamins mixture [1]	1.00	1.00	1.00	1.00	1.00
Minerals mixture [2]	1.00	1.00	1.00	1.00	1.00
Choline chloride	0.50	0.50	0.50	0.50	0.50
Sodium chloride	0.50	0.50	0.50	0.50	0.50
Adhesive [3]	0.50	0.50	0.50	0.50	0.50
Attractant [4]	0.01	0.01	0.01	0.01	0.01
Preservative [5]	0.10	0.10	0.10	0.10	0.10
Proximate composition (%)					
Crude protein	33.88	33.88	33.88	33.88	33.88
Crude fat	7.35	7.35	7.35	7.35	7.35
Ash	6.96	6.96	6.96	6.96	6.96
Moisture	9.36	9.36	9.36	9.36	9.36
Crude fiber	5.56	5.56	5.56	5.56	5.56
Gross energy (Mcal/kg)	3.83	3.83	3.83	3.83	3.83

Note: [1] Vitamin mixture (IU or mg/kg of dried feed): vitamin A 2500 IU; vitamin D_3 1200 IU; vitamin K_3 60 IU; folic acid 5 mg; vitamin B_1 10 mg; vitamin B_2 10 mg; vitamin B_6 20 mg; vitamin B_{12} 0.15 mg; niacin 40 mg; calcium pantothenate 20 mg; inositol 150 mg; Biotin 0.2 mg; vitamin C 150 mg; vitamin E 60 mg. [2] Minerals mixture (mg/kg of dried feed): iron 15 mg; zinc 20 mg; manganese 2 mg; copper 1 mg; iodine 0.2 mg; selenium 0.05 mg; cobalt 0.25 mg; magnesium 0.06 mg; potassium 40 mg. [3] Adhesive: α-starch. [4] Attractant: nucleotides, betaine, amino acids, and taurine. [5] Preservative: sodium benzoate.

Different diets were made separately into pellets as follows. First, precise weighing of all raw materials based on the prescribed formula (Table 1). Second, except for fish oil and soybean oil, all solid ingredients underwent meticulous pulverization followed by passage through a 60-mesh sieve. They were then introduced into mixing equipment where they underwent comprehensive blending for a duration of 15 min. Third, the mixture was tempered and matured by a temperer, extruded by an expanding machine to make particles with a particle size of 1–1.5 mm, and then dried in a blast-drying oven until the water content was below 100 g/kg. Finally, the remaining fish oil and soybean oil were evenly distributed onto the feed particles through the employment of a vacuum-spraying technique. The finished feed was placed separately in a marked plastic bag and stored at −20 °C.

2.3. Experimental Fish, Acclimatization, and Culture

The juvenile GIFT used in this experiment was approved by the Ethics Committee of Guangxi Minzu University, Nanning, China (No. GXMZU-2022-001).

A total of 600 juvenile GIFT were sourced from the Guangxi Academy of Fishery Sciences' seed-breeding farm, Nanning, China. After a 15 min disinfection period using a potassium permanganate solution with a concentration of 20 mg/L, the juvenile GIFT were subsequently acclimated in the laboratory aquaculture system of Guangxi Minzu University for a duration of 14 days. The acclimatization conditions were water temperature 26 ± 1 °C, pH 7–8, dissolved oxygen > 9 mg/L, and a light–dark period of 12 h: 12 h. The juvenile GIFT in each group were fed three times a day at 9:00, 14:00, and 19:00 using a control diet (0% chitosan), and the daily feeding quantity was determined by feeding until the fish no longer exhibited feeding behavior at the feeding time.

After a 14 day acclimatization period, a total of 450 juvenile GIFT with an initial body weight of 21.21 ± 0.24 g were randomly selected and assigned to five groups with three replicates per group, making a total of 15 tanks with 30 fish in each tank for the formal experiment. The total volume of each tank is 1000 L, and the volume of the aquaculture water is 720 L. The juvenile GIFT were reared in the same breeding conditions and acclimatization conditions mentioned above.

According to the previous experimental findings of this study, when the concentration of Cd^{2+} in water exceeded 0.2 mg/L, there was a significant impact on the growth performance of juvenile GIFT. Therefore, a standardized concentration of Cd^{2+} at 0.2 mg/L was maintained in each aquaculture water sample (a total of 15 tanks).

During the whole experiment period, meticulous observations were conducted on the juvenile GIFT in each group to evaluate their condition and document activities such as feeding and body coloration. In the case of juvenile GIFT mortality, immediate removal and documentation procedures were implemented. A daily replacement of one-third of the aquaculture water was performed in each tank, accompanied by the addition of Cd^{2+} venom at an equivalent volume and concentration. Throughout the 60 day duration of the experiment, a relatively constant Cd^{2+} concentration at 0.2 mg/L was maintained in all tanks.

2.4. Sampling

Following a 60 day culture period, the juvenile GIFT were subjected to a 24 h fasting period. Subsequently, 18 fish were randomly selected from each experimental group (6 fish per tank). The sample fish were individually anesthetized using 200 mg/L ethyl 3-aminobenzoate mesylate (MS-222, Adamas, Shanghai Adamas Reagent Co., Ltd., Shanghai, China), and their muscles, intestines, and livers were taken and placed in labeled sample bags and stored in liquid nitrogen. After all the samples were sampled, they were preserved at −80 °C in an ultra-low-temperature refrigerator for subsequent analysis.

2.5. Determination of Digestive Enzyme Activity

The activities of α-amylase, lipase, and trypsin in the intestines of juvenile GIFT were determined by the microplate method using an ELISA analyzer (RT-6100, Rayto, Shenzhen, China) and assay kits. The assay kits were made by Nanjing Jiancheng Bioengineering Institute (Nanjing, China), and all the instruction manuals can be found and downloaded at http://www.njjcbio.com/ (accessed on 1 January 2024).

The determination of α-amylase was performed by starch–iodine colorimetry method. α-amylase unit definition is a s follows: Each milligram of protein in the tissue underwent substrate interaction at a temperature of 37 °C for a duration of 30 min, and the hydrolysis of 10 mg starch was designated as one unit of α-amylase activity (U/mgprot).

The determination of α-amylase was made by the colorimetric method. Lipase unit definition was conducted as follows. The reaction system operated at a temperature of 37 °C, where each gram of protein in the tissue underwent substrate reaction for a duration of 1 min. The utilization of 1 μmol of substrate corresponds to one unit of lipase activity (U/mgprot).

The determination of trypsin was made by the ultraviolet colorimetry method. Trypsin unit definition was performed as follows: the trypsin present in each milligram of protein in the tissue caused a change in absorbance of 0.003 per minute in a pH of 8.0 and 37 °C environment, exhibiting one unit of trypsin activity (U/mgprot).

2.6. Determination of Muscle Composition

Muscle crude protein was determined by the Kjeldahl nitrogen determination method (GB/T 6432-2018, Chinese standards) [15]. Muscle crude fat was determined by the Soxhlet extraction method (GB/T 6433-2006, Chinese standards) [16]. Muscle moisture was determined by the oven-drying constant weight method at 105 °C (GB/T 6435-2014, Chinese standards) [17]. Muscle ash was obtained by the burning method in a muffle furnace at 550 °C (GB/T 6438-2007, Chinese standards) [18].

2.7. Determination of Gene Expression

The expression levels of sonic hedgehog (*shh*), heat shock protein (*hsp*), metallothionein (*mt*), cytochrome P450 1A (*cyp1a*), carnitine palmitoyltransferase-1 (*cpt-1*), peroxisome proliferator-activated receptor alpha (*pparα*), hormone-sensitive lipase (*hsl*), lipoprotein lipase (*lpl*), malate dehydrogenase (*mdh*), leptin (*lep*), peroxisome proliferator-activated receptor gamma (*pparγ*), fatty acid synthase (*fas*), sterol regulatory element-binding protein 1 (*srebp1*), squalene epoxidase (*sqle*), and stearoyl-CoA desaturase (*scd*) genes in the liver of juvenile GIFT were determined using real-time quantitative polymerase chain reactions (RT-qPCRs). *β-actin* was employed as the internal reference gene. Forward and reverse primers were designed based on the mRNA sequences of tilapia available in the National Center for Biotechnology Information (NCBI) database, and synthesized by Shanghai Sangon Bioengineering Technology Service Co., Ltd., Shanghai, China. Detailed primer information is provided in Table 2.

The RT-qPCR method of Liu et al. was applied [19], and the brief steps were as follows. First, the total RNA was extracted from the liver of juvenile GIFT using the Steady Pure Universal RNA Extraction Kit produced by Accurate Biology Biotechnology Engineering Ltd., Changsha, China. The instructions provided in the kit offer additional information. The RNA quantity and purity were determined using spectrophotometry, specifically by assessing the absorbance ratio of 260:280 nm with an ND-2000 spectrophotometer (Thermo, Waltham, MA, USA). The integrity of the RNA was evaluated through gel electrophoresis on a 1% (*w*/*v*) agarose TAE gel stained with Gel RedTM nucleic acid stain (UVP, Upland, CA, USA). Second, 1 μL of total RNA was reverse transcribed into cDNA using the Evo M-MLV reverse transcription kit (Accurate Biology Biotechnology Engineering Ltd., Changsha, China). The reverse transcription process consisted of incubation at 30 °C for 10 min, followed by 42 °C for 15 min, and a final step at 95 °C for 5 min. After cooling to 5 °C for an additional 5 min, the reaction mixture containing 2 × Taqman PCR mix (25 μL), upstream primer (10 μM) (2.5 μL), downstream primer (10 μM) (2.5 μL), DNA templates (1 μL), and ddH_2O (19 μL) was prepared in a total volume of 50 μL per reaction cycle as specified by the manufacturer's instructions provided with the kit from Accurate Biology Biotechnology Engineering Ltd., Changsha, China. After the PCR reaction, the cDNA sample was obtained. Third, the RT-qPCR assays were conducted using a LightCycler 96 RT-qPCR Detection System (Roche, Basel, Switzerland) and the SYBR Green Pro Taq HS-qPCR kit produced by Accurate Biology Biotechnology Engineering Ltd., Changsha, China. The RT-qPCR reactions were carried out with 2 μL of gDNA Clean Reaction Mix Ver.2, 4 μL of 5 × Evo M-MLVRT Reaction Mix Ver.2, 1 μL of Forward primer, 1 μL of Reverse primer, 2 μL of cDNA, and 10 μL of RNase free water, in a total reaction volume of 20 μL. The thermal cycling conditions for the RT-qPCR reactions included an initial denaturation step at 95 °C for 10 s followed by amplification with 40 cycles consisting of denaturation at 95 °C for 60 s; annealing at 60 °C for 30 s; and extension at 72 °C for 90 s.

The specificity of the reactions was confirmed by analyzing the melting curve obtained during heating to 95 °C.

Table 2. Forward and reverse primers of detected genes for RT-qPCR.

Gene	Primer Sequence (5′→3′)	Amplicon Size (bp)	Gene Bank
β-actin [1]	F: TGACCCAGATCATGTTTGAGACC R: CTCGTAGATGGGTACTGTGTGGG	146	XM_031811226.1
shh [2]	F: GGGAGAGGCAGACTGTAGAGATAGC R: GACAAGCAGATGAGACCGACCAAC	125	XM_003439222.5
hsp [3]	F: CAAGGTGATTTCAGACGGAGGGAAG R: GCCTCTGCGATCTCCTTCATCTTC	123	XM_003442456.5
mt [4]	F: AACGCCAGCATCACTCGGAAC R: GCGGCAGGAACACTCACTCTTG	84	YP_003587621.1
cyp1a [5]	F: AGAGTCAGTAGGCACAGTGTCCATC R: GGGGCAAGTTGTTCCGATCAGAG	129	NM_001279489.1
cpt-1 [6]	F: ATTGGCAGGACAGCGACTACATTG R: GGAAGGAGGTGAAGGGTCATCTAGG	143	XM_019362661.2
pparα [7]	F: GTGGCTGCTATTATCTGCTGTGGAG R: CTGGGGAAAAGGAAGGTGTCATCTG	140	XM_019346353.2
hsl [8]	F: CAAGCGGCATCAGTCAGGAATAGG R: CTCAACTCGGGGTCAATGGCATAC	80	XM_005463937.4
lpl [9]	F: CTTCAGCCAGAACCAGCAGAGC R: GTCGGTGGTGATGAGGAAGGATTG	142	NM_001279753.1
mdh [10]	F: GGTGCTCGCTTCTTGTGGACAG R: GACGGCCTCATTCTCATCTTCTTCC	121	XM_005450070.4
lep [11]	F: GAAGTGGATCGCTGAGCATCTGG R: CCATCCAAGCAGACCGTGACTATG	129	XM_005449522.4
pparγ [12]	F: GTACACGGAGGCTACACGGAAAC R: CTGCTTCTGCTGAACGAGACTGAC	139	XM_019358463.2
fas [13]	F: AAGCCTTGTGTGCCTTCATCCAG R: TCCCTGTGAGCGGAGGTGATTAG	133	XM_003454056.5
srebp1 [14]	F: GAACAGCAGCCGACAGATCACTC R: TACAGCAGCCATTAACGAGCAAGTC	116	XM_005473610.4
sqle [15]	F: CTGACGGGAGGAGGGATGAGTG R: CATACAGGTCGGGAATGCTCTTGAG	82	XM_003453510.5
scd [16]	F: ACAAGCTCTCCGTGCTGGTCAT R: GCAGAGTTGGGACGAAGTAGGC	102	XM_005471382.2

Note: F: forward primer. R: reverse primer. [1] *β-actin*: internal reference gene; [2] *shh*: sonic hedgehog; [3] *hsp*: Heat shock protein; [4] *mt*: metallothionein; [5] *cyp1a*: cytochrome P450 1A; [6] *cpt-1*: carnitine palmitoyltransferase-1; [7] *pparα*: peroxisome proliferator-activated receptor alpha; [8] *hsl*: hormone-sensitive lipase; [9] *lpl*: lipoprotein lipase; [10] *mdh*: malate dehydrogenase; [11] *lep*: leptin; [12] *pparγ*: peroxisome proliferator-activated receptor gamma; [13] *fas*: fatty acid synthase; [14] *srebp1*: sterol regulatory element-binding protein 1; [15] *sqle*: squalene epoxidase; [16] *scd*: stearoyl-CoA desaturase.

The $2^{-\Delta\Delta Ct}$ method [20] was applied to calculate the relative expression levels of *shh*, *hsp*, *mt*, *cyp1a*, *cpt-1*, *pparα*, *hsl*, *lpl*, *mdh*, *lep*, *pparγ*, *fas*, *srebp1*, *sqle*, and *scd* genes.

2.8. Data Calculation and Statistics

The data underwent initial processing in Microsoft Excel 2023 (Version number: 16.78 [23100802], Microsoft Corporation, Washington, WA, USA). Subsequently, a one-way analysis of variance (ANOVA) was performed using the IBM SPSS 26 software package (International Business Machines Corporation, Armonk, NY, USA). Normality and homogeneity of variances among the groups were evaluated before conducting the statistical tests. For comparisons between groups, Duncan's multiple-range test was utilized. Statistical significance was set at $p < 0.05$ level and results are reported as mean ± standard error (mean ± SE).

3. Results

3.1. Effects of Dietary Supplementation with Chitosan on the Muscle Composition of Juvenile GIFT Exposed to Cadmium-Induced Stress

Compared to the control group (0% chitosan), dietary supplementation with chitosan significantly increased the contents of crude protein and crude fat in the muscle of juvenile GIFT exposed to cadmium-induced stress ($p < 0.05$). The 1.5% group exhibited the highest contents of crude protein and crude lipid, and there were no significant differences compared to the 0.5%, 1.0%, and 2.0% groups ($p > 0.05$), respectively. However, there were no significant differences observed in the moisture and ash content in their muscles ($p > 0.05$), as shown in Table 3.

Table 3. Effects of dietary supplementation with chitosan on the muscle composition of juvenile GIFT exposed to cadmium-induced stress (%/wet weight).

Index	Chitosan Levels (%)					*p*-Values
	0	0.5	1.0	1.5	2.0	
Moisture (%)	73.17 ± 1.23	72.67 ± 1.11	72.42 ± 1.37	72.48 ± 0.74	73.02 ± 0.63	0.401
Crude protein (%)	22.72 ± 0.25 [b]	23.49 ± 0.30 [a]	23.42 ± 0.37 [a]	23.71 ± 0.43 [a]	23.69 ± 0.23 [a]	0.000
Crude fat (%)	6.97 ± 0.16 [b]	7.38 ± 0.12 [a]	7.43 ± 0.10 [a]	7.58 ± 0.08 [a]	7.50 ± 0.18 [a]	0.000
Ash (%)	1.40 ± 0.02	1.40 ± 0.01	1.37 ± 0.01	1.37 ± 0.01	1.39 ± 0.01	0.297

Note: All above data are mean ± SE ($n = 3 \times 3 \times 3$), which means 3 parallel groups for each chitosan level, 3 fish samples for each parallel group, and 3 measured times for each fish sample. Different superscript letters in the same row indicate significant differences among the data ($p < 0.05$).

3.2. Effects of Dietary Supplementation with Chitosan on the Intestinal Digestive Enzyme Activities of Juvenile GIFT Exposed to Cadmium-Induced Stress

Compared to the control group (0% chitosan), dietary supplementation with chitosan significantly increased the activities of lipase, trypsin, and α-amylase in the intestine of juvenile GIFT exposed to cadmium-induced stress ($p < 0.05$). The 1.5% group exhibited the highest activities of lipase and trypsin, and there were no significant differences compared to the 0.5%, 1.0%, and 2.0% groups ($p > 0.05$). The 2.0% group exhibited the highest activity of α-amylase, and there were no significant differences compared to the 0.5%, 1.0%, and 1.5% groups ($p > 0.05$), as shown in Table 4.

Table 4. Effects of dietary supplementation with chitosan on the intestinal digestive enzyme activities of juvenile GIFT exposed to cadmium-induced stress.

Index	Chitosan Levels (%)					*p*-Values
	0	0.5	1.0	1.5	2.0	
Lipase (U/mgprot) [1]	0.12 ± 0.01 [c]	0.27 ± 0.01 [b]	0.28 ± 0.01 [b]	0.33 ± 0.02 [a]	0.32 ± 0.01 [a]	0.000
Trypsin (U/mgprot) [2]	8.21 ± 1.39 [b]	18.31 ± 1.55 [a]	18.58 ± 0.94 [a]	19.24 ± 0.58 [a]	18.24 ± 1.69 [a]	0.000
α-amylase (U/mgprot) [3]	1592.97 ± 100.05 [b]	2764.49 ± 322.01 [a]	2815.82 ± 185.95 [a]	2912.66 ± 367.67 [a]	2995.89 ± 249.55 [a]	0.000

Note: [1] α-amylase unit definition: each milligram of protein in the tissue undergoes substrate interaction at a temperature of 37 °C for a duration of 30 min, and the hydrolysis of 10 mg starch is designated as one unit of α-amylase activity (U/mgprot). [2] Lipase unit definition: the reaction system operates at a temperature of 37 °C, where each gram of protein in the tissue undergoes substrate reaction for a duration of 1 min. The utilization of 1 μmol of substrate corresponds to one unit of lipase activity (U/mgprot). [3] Trypsin unit definition: the trypsin present in each milligram of protein in the tissue causes a change in absorbance of 0.003 per minute by a pH 8.0 and 37 °C environment, exhibiting one unit of trypsin activity (U/mgprot). All above data are mean ± SE ($n = 3 \times 3 \times 3$), which means 3 parallel groups for each chitosan level, 3 fish samples for each parallel group, and 3 measured times for each fish sample. Different superscript letters in the same row indicate significant differences among the data ($p < 0.05$).

3.3. Effects of Dietary Supplementation with Chitosan on the Relative Expression Levels of Stress-Resistance Genes in the Liver of Juvenile GIFT Exposed to Cadmium-Induced Stress

Compared to the control group (0% chitosan), dietary supplementation with chitosan at concentrations of 1.0%, 1.5%, and 2.0% significantly up-regulated the relative expression levels of the heat shock protein (*hsp*) and sonic hedgehog (*shh*) genes in the liver of juvenile GIFT exposed to cadmium-induced stress ($p < 0.05$). The 1.5% group exhibited the highest relative expression levels of *hsp* and *shh* genes, which were significantly higher than those in the 1.0% group ($p < 0.05$), but no significant differences were observed compared to the 2.0% group ($p > 0.05$). However, the relative expression levels of the *hsp* and *shh* genes in the 0.5% group did not exhibit significant differences compared to those in the control group ($p > 0.05$), as shown in Figure 1.

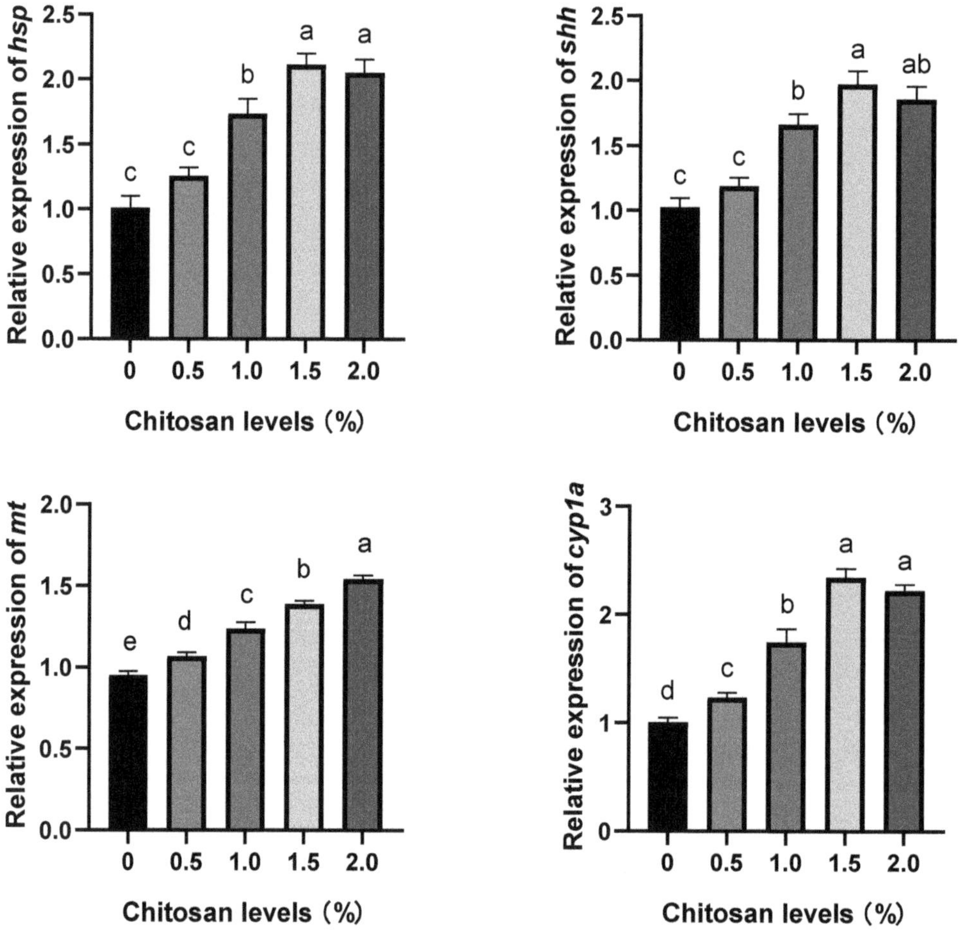

Figure 1. Effects of dietary supplementation with chitosan on the relative expression levels of stress-resistance genes in the liver of juvenile GIFT exposed to cadmium-induced stress, where *hsp* indicates heat shock protein, *shh* indicates sonic hedgehog, *mt* indicates metallothionein, and *cyp1a* indicates cytochrome P450 1A. All above data are mean ± SE ($n = 3 \times 3 \times 3$), which means 3 parallel groups for each chitosan level, 3 fish samples for each parallel group, and 3 measured times for each fish sample. Different superscript letters in the same figure indicate significant differences among the data ($p < 0.05$).

Compared to the control group (0% chitosan), dietary supplementation with chitosan at concentrations of 0.5%, 1.0%, 1.5%, and 2.0% significantly up-regulated the relative expression level of the metallothionein (*mt*) gene in the liver of juvenile GIFT exposed to cadmium-induced stress ($p < 0.05$). The 2.0% group exhibited the highest relative expression level of the *mt* gene, which was significantly higher than that in the 0.5%, 1.0%, and 1.5% groups ($p < 0.05$), as shown in Figure 1.

Compared to the control group (0% chitosan), dietary supplementation with chitosan at concentrations of 0.5%, 1.0%, 1.5%, and 2.0% significantly up-regulated the relative expression level of the cytochrome P450 1A (*cyp1a*) gene in the liver of juvenile GIFT exposed to cadmium-induced stress ($p < 0.05$). The 1.5% group exhibited the highest relative expression level of the *cyp1a* gene, which was significantly higher than those in the 0.5% and 1.0% groups ($p < 0.05$), but no significant difference was observed compared to the 2.0% group ($p > 0.05$), as shown in Figure 1.

3.4. Effects of Dietary Supplementation with Chitosan on the Relative Expression Levels of Lipid Metabolism Genes in the Liver of Juvenile GIFT Exposed to Cadmium-Induced Stress

Compared to the control group (0% chitosan), dietary supplementation with chitosan at concentrations of 0.5%, 1.0%, 1.5%, and 2.0% significantly up-regulated the relative expression levels of hormone-sensitive lipase (*hsl*), peroxisome proliferator-activated receptor alpha (*pparα*), lipoprotein lipase (*lpl*), malate dehydrogenase (*mdh*), carnitine palmitoyltransferase 1 (*cpt-1*), leptin (*lep*), fatty acid synthase (*fas*), stearoyl-CoA desaturase (*scd*), peroxisome proliferator-activated receptor gamma (*pparγ*), and sterol regulatory element-binding protein 1 (*srebp1*) genes in the liver of juvenile GIFT exposed to cadmium-induced stress ($p < 0.05$). The 2.0% group exhibited the highest relative expression level of the *hsl* gene, which was significantly higher than that of the 0.5%, 1.0%, and 1.5% groups ($p < 0.05$). The 1.5% group exhibited the highest relative expression levels of *pparα* and *pparγ* genes, which were significantly higher than those in the 0.5% and 1.0% groups ($p < 0.05$), but no significant differences were observed in the 2.0% group ($p > 0.05$). The 1.5% group exhibited the highest relative expression levels of *lpl* and *fas* genes, which were significantly higher than those in the 0.5% group ($p < 0.05$), but no significant differences were observed in the 1.0% and 2.0% groups ($p > 0.05$). The 2.0% group exhibited the highest relative expression levels of the *mdh* gene, which was significantly higher than those in the 0.5% and 1.0% groups ($p < 0.05$), but no significant difference was observed in the 1.5% group ($p > 0.05$). The 2.0% group exhibited the highest relative expression levels of the *cpt-1*, *lep*, and *srebp1* genes, which were significantly higher than those in the 0.5% group ($p < 0.05$), but no significant differences were observed in the 1.0% and 1.5% groups ($p > 0.05$). The 1.5% group exhibited the highest relative expression level of the *scd* gene, which was significantly higher than those of the 0.5%, 1.0%, and 2.0% groups ($p < 0.05$), as shown in Figure 2.

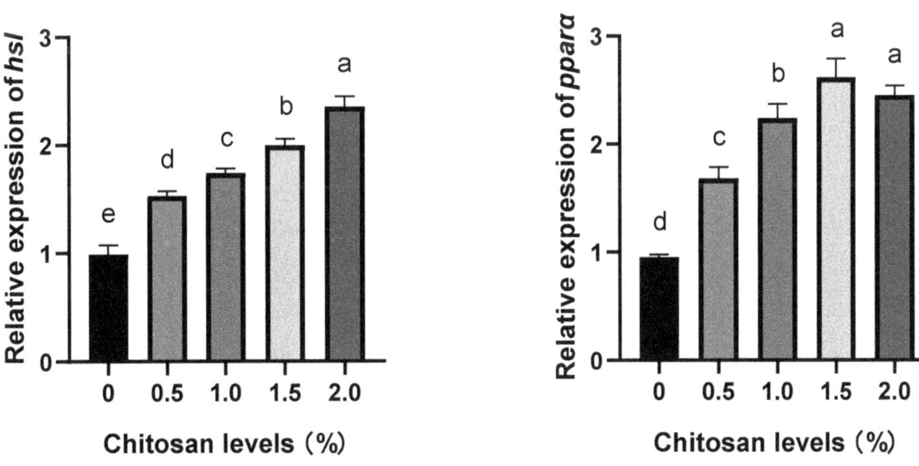

Figure 2. *Cont.*

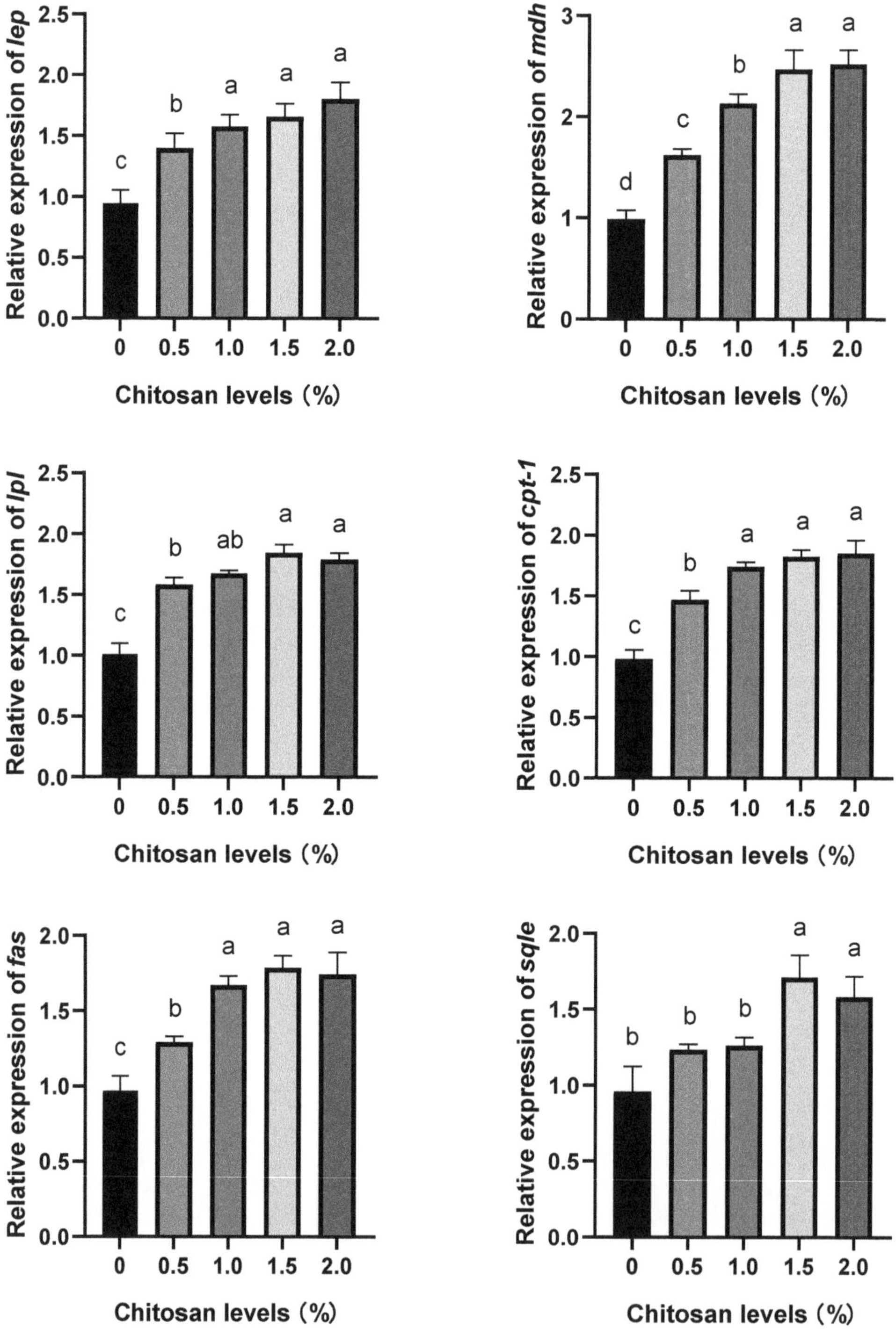

Figure 2. *Cont.*

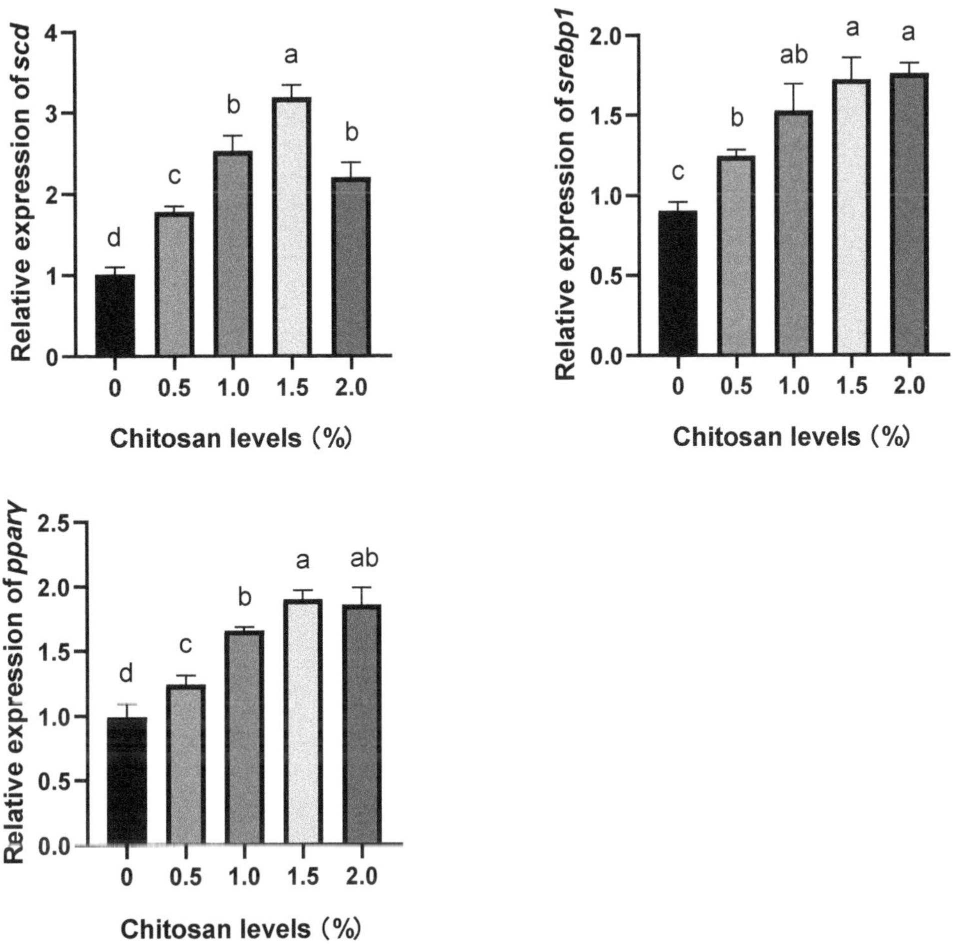

Figure 2. Effects of dietary supplementation with chitosan on the relative expression levels of lipid metabolism genes in the liver of juvenile GIFT exposed to cadmium-induced stress, where *hsl* indicates hormone-sensitive lipase, *pparα* indicates peroxisome proliferator-activated receptor alpha, *lpl* indicates lipoprotein lipase, *mdh* indicates malate dehydrogenase, *cpt-1* indicates carnitine palmitoyltransferase 1, *lep* indicates leptin, *fas* indicates fatty acid synthase, *scd* indicates stearoyl-CoA desaturase, *sqle* indicates squalene epoxidase, *pparγ* indicates peroxisome proliferator-activated receptor gamma, and *srebp1* indicates sterol regulatory element-binding protein 1. All above data are mean ± SE ($n = 3 \times 3 \times 3$), which means 3 parallel groups for each chitosan level, 3 fish samples for each parallel group, and 3 measured times for each fish sample. Different superscript letters in the same figure indicate significant differences among the data ($p < 0.05$).

Compared to the control group (0% chitosan), dietary supplementation with chitosan at concentrations of 1.5% and 2.0% significantly up-regulated the relative expression levels of squalene epoxidase (*sqle*) gene in the liver of juvenile GIFT exposed to cadmium-induced stress ($p < 0.05$). The 1.5% group exhibited the highest relative expression level of the *sqle* gene, which was significantly higher than that of the 0.5% and 1.0% groups ($p < 0.05$), but no significant difference was observed in the 2.0% group ($p > 0.05$). However, the relative expression levels of the *sqle* gene in the 0.5% and 1.0% groups did not exhibit significant differences compared to that in the control group ($p > 0.05$), as shown in Figure 2.

4. Discussion

The present study demonstrated that, compared to the control group, dietary supplementation with chitosan significantly increased the contents of crude protein and crude fat in the muscles of juvenile GIFT exposed to cadmium-induced stress. These findings suggest that dietary supplementation with chitosan could exert a regulatory influence on muscle composition. The possible reasons could be attributed to the following factors. First, the presence of cadmium in the aquatic environment induced oxidative stress in fish through the generation of reactive oxygen species (ROS), and the fish's physiological response involved an increased production of anions, hydroxyl radicals, and hydrogen peroxide. As a result, the fish expended significant energy on nutrient metabolism and ROS elimination within its system [21,22]. Second, due to the influence of cadmium-induced stress, the structural and functional integrity of biomacromolecules and various tissues and organs in fish were subjected to varying degrees of oxidative damage, leading to a reduction in nutrient absorption capacity and utilization rate as well as a decline in the fat and protein content of the fish [23,24]. Third, chitosan, a naturally occurring cationic polymer, demonstrated the capacity to chelate with cadmium in fish, thereby ameliorating its toxic effects. Moreover, chitosan exhibited an ability to sequester anions and scavenge hydroxyl free radicals, hydrogen peroxide, and other ROS, forming flocculates that alleviate oxidative stress-induced harm [25]. Fourth, the fish dietary supplementation with chitosan had the potential to enhance hepatic RNA repair, bolster protein synthesis, and regulate nutrient absorption and utilization in fish [26,27]. Fifth, chitosan exhibited the potential to augment the enzymatic activity involved in lipid and protein metabolism, thereby facilitating the proficient synthesis of lipids and proteins [28,29]. The dietary supplementation with chitosan resulted in a significant enhancement of crude protein content in yellow catfish (*Pelteobagrus fulvidraco*), as demonstrated by Li et al.'s study [30]. Likewise, Gheytasi et al. examined the impact of chitosan supplementation on rainbow trout (*Oncorhynchus mykiss*) and found notable enhancements in body composition, specifically an elevation in crude protein content relative to the control group [31].

The activities of lipase, trypsin, and amylase in the intestine can serve as indicators of the digestive efficiency of fish to a certain extent [32]. In this study, compared to the control group, the dietary supplementation with chitosan significantly increased the activities of lipase, trypsin, and amylase in the intestines of juvenile GIFT exposed to cadmium-induced stress. These findings suggested that dietary supplementation with chitosan exhibited potential for mitigating the adverse effects of cadmium stress on intestinal enzyme activity and improving digestive performance. The possible reasons could be attributed to the following factors. First, the accumulation of cadmium in fish leads to an increase in the abundance of harmful microbiota, such as spirochaetes and other pathogenic bacteria which colonize the intestinal tract of fish and induce diseases. Consequently, this resulted in a reduction in the proportion of beneficial bacteria within the fish gut microbiota, leading to the adhesion, damage, and cellular swelling of intestinal tissues. Therefore, this disrupted the activity and composition of digestive enzymes within the intestines [33,34]. Second, dietary supplementation with chitosan has been shown to augment the proliferation of beneficial intestinal bacteria, such as bifidobacteria and lactic acid bacteria, thereby modulating the composition of the gut microbiota. Consequently, this leads to enhanced intestinal metabolism, improved decomposition of toxic products under cadmium stress, heightened activity of digestive enzymes in the intestine, and increased digestive capacity for food [35,36]. Third, the chitosan molecule possesses a rich array of hydroxyl, acetyl, amino, and other functional groups, facilitating its adsorption onto the intestinal surface to establish a high-molecular-weight barrier film that effectively impeded the absorption of toxic substances by the intestine [35,37]. Fourth, chitosan possessed the capacity to induce the synthesis of anti-inflammatory compounds within the intestinal tract, thereby facilitating prompt restoration of intestinal damage and promoting recovery of food digestion and absorption capabilities. Additionally, the chemical constituents of chitosan contributed to fortifying the gut barrier function in fish while stimulating the secretion of beneficial substances and enhancing the production of digestive enzymes by intestinal cells [38–40]. Liu et al. administered chitosan-coated microfeed to large yellow croaker (*Larimichthys crocea*), resulting in a significant improvement in the intestinal digestive performance of the fish after 30 days [41]. Similarly, Asmaa et al. observed enhanced digestive performance in Nile tilapia (*Oreochromis niloticus*) through dietary supplementation with chitosan [36].

Heat shock protein (*hsp*) is a class of stress-inducible protein that is synthesized by organisms in response to adverse environmental stimuli [42]. Sonic hedgehog (*shh*) signaling molecules play a crucial role in regulating cellular development, and this pathway can activate tissue self-repair mechanisms [43]. Cytochrome P450 1A (*cyp1a*) serves as the primary activating enzyme involved in cell growth and development [44]. Metallothionein (*mt*) is an inducible non-enzymatic protein that exists within organisms [45]. The present study demonstrated that, compared to the control group, the dietary supplementation with chitosan could significantly up-regulate the relative expression levels of *hsp*, *shh*, *mt*, and *cyp1a* genes in the liver of juvenile GIFT exposed to cadmium-induced stress. These findings suggest that dietary supplementation with chitosan could effectively improve the stress resistance of GIFT. The possible reasons could be attributed to the following factors. First, cadmium exposure compromised the antioxidant and immune systems of fish, leading to inflammation [46]. Concurrently, heavy-metal stress disrupted the activity of antioxidant enzymes in fish by increasing ROS, resulting in oxidative stress, tissue, and organ damage, as well as diminished stress resistance [47]. Second, the chitosan molecular chain possessed numerous active groups with lone pair electrons, which could effectively coordinate with cadmium ions through coordination bonds. Consequently, a stable chelating polymer was formed, thereby ameliorating the toxicity of heavy-metal cadmium in fish [48–50]. Third, the expression of the *shh* signaling pathway was intricately linked to inflammation in vivo. Chitosan exhibited the capability to activate the *shh* signaling pathway in fish, thereby eliciting immune system stimulation for self-repair [51]. Fourth, the activation of the transcription factor NF-κB by chitosan occurred within the cytoplasmic milieu of damaged tissues, where it selectively bound to specific nuclear binding sites. This intricate interaction orchestrated the precise regulation of stress-resistance genes, thereby facilitating the proficient repair and restoration of compromised tissues and organs under cadmium-induced stress [52,53]. Fifth, exposure to cadmium stress resulted in an elevation of heavy-metal ion concentration within the body, leading to the induction of *mt* production. Chitosan facilitates an expedited formation of *mt*, which is complexed with cadmium ions which enables their non-toxic excretion from the system. This reduction in cadmium content mitigated adverse effects caused by cadmium stress on fish [54,55]. In the study by Brol et al., it was observed that the expression levels of *cyp1a* and glutathione peroxidase genes were significantly up-regulated in juvenile Pacific white shrimp (*Litopenaeus vannamei*) following chitosan administration to alleviate salt stress. This enhancement of defense mechanism and stress resistance under the dietary supplementation with chitosan conditions suggested potential benefits for juvenile Pacific white shrimp [56]. Sea cucumber (*Apostichopus japonicas*) was fed a dietary supplementation with chitosan by Wang et al., and after 8 weeks, it was observed that the expression of *hsp* significantly increased, resulting in enhanced stress resistance in the sea cucumber [57].

The regulation of lipid metabolism genes on the synthesis of enzymes involved in fat metabolism facilitates both lipogenesis and lipolysis, thereby playing a pivotal role in maintaining the homeostasis of lipid metabolism [58]. The present study demonstrated that compared to the control group, dietary supplementation with chitosan could significantly up-regulate the relative expression levels of hormone-sensitive lipase (*hsl*), peroxisome proliferator-activated receptor alpha (*pparα*), lipoprotein lipase (*lpl*), malate dehydrogenase (*mdh*), carnitine palmitoyltransferase 1 (*cpt-1*), leptin (*lep*), fatty acid synthase (*fas*), stearoyl-CoA desaturase (*scd*), peroxisome proliferator-activated receptor gamma (*pparγ*), and sterol regulatory element-binding protein 1 (*srebp1*) genes in the liver of juvenile GIFT exposed to cadmium-induced stress. These findings suggested that dietary supplementation with chitosan could effectively alleviate lipid metabolism impairment in juvenile GIFT exposed to cadmium stress, while concurrently augmenting their lipid metabolism capacity under such conditions. The possible reasons could be attributed to the following factors. First, the exposure to cadmium stress resulted in tissue and organ damage, disturbances in lipid metabolism, as well as suppression of the expression of genes associated with lipid metabolism in juvenile GIFT [59]. Second, the presence of chitosan enhanced the activity of digestive enzymes in the intestine, thereby facilitating nutrient absorption and promoting lipid metabolism [60]. Undoubtedly, the antibacterial and anti-inflammatory properties inherent in chitosan molecules significantly contributed to the preservation of intestinal and hepatic tissue integrity, thereby facilitating lipid metabolism within the organism [61]. Third, the chitosan–fat complex formed by chitosan in the digestive tract synergistically interacted with fatty acids and other substances, thereby augmenting the intestinal absorption of bile acids, facilitating the repair of fish intestinal damage induced by cadmium stress, and promoting the digestion and absorption of accumulated lipids in the fish intestine [62,63]. Fourth, the activation of chitosan on *ppar* could effectively suppress the NF-κB signaling pathway, thereby attenuating systemic inflammation, and facilitating in vivo fatty acid synthesis [64,65]. Fifth, the expression of *pparα*, *pparγ*, *lpl*, *scd*, and *fas* genes exhibited a decline at a chitosan concentration of 2.0%, potentially attributed to the fish's inherent susceptibility, inadequate absorption, and utilization of high chitosan levels by the digestive system, and deposition of chitosan in the body leading to digestive burden and adverse effects, thereby impeding further promotion of lipid metabolism [66]. Similar investigations have been conducted, such as Liu et al.'s study on large yellow croakers raised with chitosan-coated microfeed, which up-regulated the expression of lipid metabolism genes in the fish [41].

5. Conclusions

In conclusion, dietary chitosan supplementation could alleviate the effects of Cd^{2+} stress on the muscle composition, digestive enzymes, lipid metabolism, and stress resistance, and their related gene expression, of juvenile tilapia, and to some extent reduce the toxic response of tilapia to Cd^{2+} stress. These results can provide scientific insights for the application of chitosan in aquaculture practices and the management of heavy-metal pollution in aquatic environments.

Author Contributions: Conceptualization, Q.Z., Y.L. and T.T.; Data curation, Q.Z., Y.X., Y.L. and T.T.; Formal analysis, Q.Z., Y.X. and Y.L.; Investigation, Q.Z., Y.X., Y.Z., E.H., L.M., Y.L. and T.T.; Methodology, Q.Z., Y.X., Y.Z., E.H., L.M., Y.L. and T.T.; Project administration, Q.Z., Y.L. and T.T; Supervision, Q.Z.; Funding acquisition, Q.Z., Y.L. and T.T.; Validation and visualization, Q.Z., Y.X., Y.Z., E.H., L.M., Y.L. and T.T.; Writing—review and editing, Q.Z., Y.X., Y.L. and T.T.; Writing original draft preparation, Y.X. All authors have read and agreed to the published version of the manuscript.

Funding: This research was funded by the Research Capacity Enhancement Project for Young and Middle-aged Teachers in Guangxi Universities (2023KY0174); grants from the Scientific Research Foundation for the Introduced Talents of Guangxi Minzu University (2021KJQD17); the Joint Funds of the National Natural Science Foundation of China (U20A2064); the Shandong Provincial Key Research and Development Programs (2019JZZY020710); the Innovation-driven Development Special Fund Project of Guangxi (AA17204044); the Grants from the Scientific Research Foundation for the Introduced Talents of Guangxi Minzu University (2018KJQD14).

Institutional Review Board Statement: All animal experiments were approved by the Ethics Committee of Guangxi Minzu University, Nanning, China (No. GXMZU-2022-001), and this research does not contain any studies with human participants.

Informed Consent Statement: Not applicable.

Data Availability Statement: The original contributions presented in the study are included in the article, and further inquiries can be directed to the corresponding author(s).

Conflicts of Interest: The authors declare that they have no known competing financial interests or personal relationships that could have appeared to influence the work reported in this paper.

References

1. Genchi, G.; Sinicropi, M.; Lauria, G.; Catalano, A. The Effects of Cadmium Toxicity. *Int. J. Environ. Res. Public Health* **2020**, *17*, 3782. [CrossRef]
2. Paíga, P.; Sousa, S.; Vera, J.; Bitencourt, L.; Vieira, J.; Jorge, S.; Silva, J.; Correia, M.; Domingues, V.; Delerue-Matos, C. Multi-residue analysis of fifty pesticides in river waters and in wastewaters. *Environ. Sci. Pollut. Res.* **2021**, *28*, 66787–66803. [CrossRef]
3. Abou-Kassem, D.; Abd El-Hack, A.; Taha, A.; Ajarem, J.; Maodaa, S.; Allam, A. Detoxification Impacts of Ascorbic Acid and Clay on Laying Japanese Quail Fed Diets Polluted by Various Levels of Cadmium. *Animals* **2020**, *10*, 372. [CrossRef] [PubMed]
4. Cheng, C.; Ma, H.; Deng, Y.; Feng, J.; Jie, Y.; Guo, Z. Oxidative stress, cell cycle arrest, DNA damage and apoptosis in the mud crab (*Scylla paramamosain*) induced by cadmium exposure. *Chemosphere* **2021**, *263*, 128277. [CrossRef] [PubMed]
5. Elanchezhiyan, S.; Karthikeyan, P.; Rathinam, K.; Farzana, M.; Park, C. Magnetic kaolinite immobilized chitosan beads for the removal of Pb(II) and Cd(II) ions from an aqueous environment. *Carbohydr. Polym.* **2021**, *261*, 117892. [CrossRef]
6. Godiya, C.; Park, B. Removal of bisphenol A from wastewater by physical, chemical and biological remediation techniques. A review. *Environ. Chem. Lett.* **2022**, *20*, 1801–1837. [CrossRef]
7. Othman, Z.; Mackey, H.; Mahmoud, K. MXene/chitosan/lignosulfonate (MCL) nanocomposite for simultaneous removal of Co(II), Cr(VI), Cu(II), Ni(II) and Pb(II) heavy metals from wastewater. *2D Mater.* **2023**, *10*, 020224. [CrossRef]
8. Lan, Z.; Lin, Y.; Yang, C. Lanthanum-iron incorporated chitosan beads for adsorption of phosphate and cadmium from aqueous solutions. *Chem. Eng. J.* **2022**, *448*, 137519. [CrossRef]
9. Begum, S.; Yuhana, N.; Saleh, N.; Kamarudin, N.; Sulong, A. Review of chitosan composite as a heavy metal adsorbent: Material preparation and properties. *Carbohydr. Polym.* **2022**, *259*, 117613. [CrossRef]
10. Mohan, K.; Rajan, K.; Ganesan, R.; Ganesan, A.; Divya, D.; Johansen, J.; Zhang, S. Chitin, chitosan and chitooligosaccharides as potential growth promoters and immunostimulants in aquaculture: A comprehensive review. *Int. J. Biol. Macromol.* **2023**, *251*, 126285. [CrossRef]
11. Bhusare, S.; Sardar, P.; Sahu, P.; Shamna, N.; Kumar, P.; Paul, M.; Jana, P.; Raghuvaran, N.; Bhavatharaniya, U. Bile acid improves growth, lipid utilization and antioxidative status of genetically improved farmed tilapia (*Oreochromis niloticus*) fed with varying protein-lipid diets reared in inland saline water. *Anim. Feed Sci. Technol.* **2023**, *303*, 115677. [CrossRef]
12. Adene, C. Effects of Pretilachor Pyribenzoxim Pollution on the Water Quality, Serum Biochemical Indices, and Behavioural Response of *Oreochromis niloticus* Juveniles. *Asian J. Res. Agric. For.* **2023**, *9*, 119–127. [CrossRef]
13. Aranaz, I.; Alcántara, A.; Civera, M.; Arias, C.; Elorza, B.; Caballero, A.; Acosta, N. Chitosan: An Overview of Its Properties and Applications. *Polymers* **2021**, *13*, 3256. [CrossRef]
14. Ibrahim, R.; Elshopakey, G.; Abdelwarith, A.; Younis, E.; Ismail, S.; Ahmed, A.; El-Saberf, M.; Abdelhamid, A.; El-Murr, A.; Rahman, A. Chitosan neem nanocapsule enhances immunity and disease resistance in nile tilapia (*Oreochromis niloticus*). *Heliyon* **2023**, *9*, e19354. [CrossRef] [PubMed]
15. *GB/T 6432-2018*; Determination of Crude Protein in Feeds-Kjeldahl Method. Standardization Administration of the P.R.C: Beijing, China, 2018.
16. *GB/T 6433-2006*; Determination of Crude Fat in Feeds. Standardization Administration of the P.R.C: Beijing, China, 2006.
17. *GB/T 6435-2014*; Determination of Moisture in Feeds. Standardization Administration of the P.R.C: Beijing, China, 2014.
18. *GB/T 6438-2007*; Determination of Ash in Feeds. Standardization Administration of the P.R.C: Beijing, China, 2007.
19. Liu, Y.; Huang, E.; Xie, Y.; Meng, L.; Liu, D.; Zhang, Z.; Zhou, J.; Zhang, Q.; Tong, T. The Effect of Dietary Lipid Supplementation on the Serum Biochemistry, Antioxidant Responses, Initial Immunity, and mTOR Pathway of Juvenile Tilapia (*Oreochromis niloticus*). *Fishes* **2023**, *8*, 535. [CrossRef]
20. Schmittgen, T.; Livak, J. Analyzing real-time PCR data by the comparative CT method. *Nat. Protoc.* **2008**, *3*, 1101–1108. [CrossRef] [PubMed]
21. Falahatgar, D.; Javadian, S.; Bahram, S.; Bahrekazemi, M. EDTA detoxifies heavy metals on exposed beluga (*Huso huso*) with pollution stress: Growth performance, immunohaematology, blood biochemistry and antioxidant activity. *Aquac. Res.* **2021**, *52*, 4336–4349. [CrossRef]
22. Lee, J.; Jo, A.; Lee, D.; Choi, C.; Kang, J.; Kim, J. Review of cadmium toxicity effects on fish: Oxidative stress and immune responses. *Environ. Res.* **2023**, *236*, 116600. [CrossRef]
23. Bacou, E.; Walk, C.; Rider, S.; Litta, G.; Perez-Calvo, E. Dietary Oxidative Distress: A Review of Nutritional Challenges as Models for Poultry, Swine and Fish. *Antioxidants* **2021**, *10*, 525. [CrossRef]
24. Lall, S.; Kaushik, S. Nutrition and Metabolism of Minerals in Fish. *Animals* **2021**, *11*, 3510. [CrossRef]
25. El-Naggar, M.; Salaah, S.; El-Shabaka, H.; El-Rahman, F.; Khalil, M.; Suloma, A. Efficacy of dietary chitosan and chitosan nanoparticles supplementation on health status of Nile tilapia, *Oreochromis niloticus*. *Aquac. Rep.* **2021**, *19*, 100628. [CrossRef]
26. Mushawwir, A.; Permana, R.; Darwis, D.; Puspitasari, T.; Pengerteni, D.; Nuryanthi, N.; Suwarno, N. Enhancement of the liver histologic of broiler induced by irradiated chitosan (IC). *Proc. Int. Conf. Phys. Nucl. Sci. Technol.* **2021**, *2381*, 020046.
27. Hilițanu, L.; MititeluTarțău, L.; Bogdan, M.; Buca, B.; Păuna, A.; Pavel, L.; Pelin, A.; Meca, A.; Popa, G. The Use of Chitosan-Coated Nanovesicles in Repairing Alcohol-Induced Damage of Liver Cells in Mice. *Medicina* **2022**, *58*, 762. [CrossRef] [PubMed]

28. Sharma, D.; Arora, S.; Singh, J. Smart thermosensitive copolymer incorporating chitosan–zinc–insulin electrostatic complexes for controlled delivery of insulin: Effect of chitosan chain length. *Int. J. Polym. Mater. Polym. Biomater.* **2020**, *69*, 1054–1068. [CrossRef] [PubMed]
29. Ahmed, K.; Khan, M.; Siddiqui, H.; Jahan, A. Chitosan and its oligosaccharides, a promising option for sustainable crop production—A review. *Carbohydr. Polym.* **2020**, *227*, 115331. [CrossRef] [PubMed]
30. Li, S.; Li, C.; Wu, S. Dietary chitosan modulates the growth performance, body composition and nonspecific immunity of juvenile yellow catfish (*Pelteobagrus fulvidraco*). *Int. J. Biol. Macromol.* **2022**, *217*, 188–192. [CrossRef]
31. Gheytasi, A.; Shekarabi, S.; Islami, H.; Mehrgan, M. Feeding rainbow trout, Oncorhynchus mykiss, with lemon essential oil loaded in chitosan nanoparticles: Effect on growth performance, serum hemato-immunological parameters, and body composition. *Aquac. Int.* **2021**, *29*, 2207–2221. [CrossRef]
32. Assan, D.; Kuebutornye, F.; Hlordzi, V.; Chen, H.; Mraz, J.; Mustapha, U.; Abarike, E. Effects of probiotics on digestive enzymes of fish (finfish and shellfish); status and prospects: A mini review. *Comp. Biochem. Physiol. Part B Biochem. Mol. Biol.* **2022**, *257*, 110653. [CrossRef]
33. Kakade, A.; Sharma, M.; Salama, E.; Zhang, P.; Zhang, L.; Xing, X.; Yue, J.; Song, Z.; Nan, L.; Su, Y.; et al. Heavy metals (HMs) pollution in the aquatic environment: Role of probiotics and gut microbiota in HMs remediation. *Environ. Res.* **2023**, *223*, 115186. [CrossRef]
34. Cheaib, B.; Seghouani, H.; Llewellyn, M.; Vandal-Lenghan, K.; Mercier, P.; Derome, N. The yellow perch (*Perca flavescens*) microbiome revealed resistance to colonisation mostly associated with neutralism driven by rare taxa under cadmium disturbance. *Anim. Microbiome* **2021**, *3*, 3. [CrossRef]
35. Rahman, A.; Shin, J.; Whang, C.; Jung, W.; Yoo, D.; Seo, C.; Cho, B.; Jon, S. Bilirubin Nanomedicine Rescues Intestinal Barrier Destruction and Restores Mucosal Immunity in Colitis. *ACS Nano* **2023**, *17*, 10996–11013. [CrossRef]
36. Asmaa, S.; Adham, A.; Samar, S.; Naiel, M. Dietary combination of chitosan nanoparticle and thymol affects feed utilization, digestive enzymes, antioxidant status, and intestinal morphology of *Oreochromis niloticus*. *Aquaculture* **2020**, *515*, 734577.
37. Zhou, J.; Li, M.; Chen, Q.; Li, X.; Chen, L.; Dong, Z.; Zhu, W.; Yang, Y.; Liu, Z.; Chen, Q. Programmable probiotics modulate inflammation and gut microbiota for inflammatory bowel disease treatment after effective oral delivery. *Nat. Commun.* **2022**, *13*, 3432. [CrossRef]
38. Hendawy, O.; AlSanea, M.; Mohammed, E.; Rahman, H.; Hassan, Y.; Elshaarawy, R.; Khedr, A. Alginate-chitosan-microencapsulated tyrosols/oleuropein-rich olive mill waste extract for lipopolysaccharide-induced skin fibroblast inflammation treatment. *Int. J. Pharm.* **2023**, *643*, 123260. [CrossRef]
39. Xiong, M.; Li, Y.; He, H.; Hao, S.; Fang, P.; Xu, M.; Chen, Y.; Chen, Y.; Yu, S.; Hu, H. Cyclosporine A-loaded colon-targeted oral nanomicelles self-assembly by galactosylated carboxymethyl chitosan for efficient ulcerative colitis therapy. *Eur. J. Pharm. Biopharm.* **2023**, *189*, 152–164. [CrossRef] [PubMed]
40. Lu, X.; Chang, X.; Zhang, H.; Wang, J.; Qiu, K.; Wu, S. Effects of Dietary Rare Earth Chitosan Chelate on Performance, Egg Quality, Immune and Antioxidant Capacity, and Intestinal Digestive Enzyme Activity of Laying Hens. *Polymers* **2023**, *15*, 1600. [CrossRef] [PubMed]
41. Liu, J.; Xu, W.; Liu, Y.; Wang, Y.; Zhang, J.; Wang, Z.; Mai, K.; Ai, Q. Effects of Chitosan-Coated Microdiet on Dietary Physical Properties, Growth Performance, Digestive Enzyme Activities, Antioxidant Capacity, and Inflammation Response of Large Yellow Croaker (*Larimichthys crocea*) Larvae. *Aquac. Nutr.* **2022**, *2022*, 4355132. [CrossRef] [PubMed]
42. Roy, A.; Tamuli, R. Heat shock proteins and the calcineurin-crz1 signaling regulate stress responses in fungi. *Arch. Microbiol.* **2022**, *204*, 240. [CrossRef] [PubMed]
43. Velikic, G.; Maric, D.; Maric, D.; Supic, G.; Puletic, M.; Dulic, O.; Vojvodic, D. Harnessing the Stem Cell Niche in Regenerative Medicine: Innovative Avenue to Combat Neurodegenerative Diseases. *Int. J. Mol. Sci.* **2024**, *25*, 993. [CrossRef]
44. Esteves, F.; Rueff, J.; Kranendonk, M. The Central Role of Cytochrome P450 in Xenobiotic Metabolism—A Brief Review on a Fascinating Enzyme Family. *J. Xenobiot.* **2021**, *11*, 94–114. [CrossRef] [PubMed]
45. Jorge, D.; Lilia, L.; Miguel, A.; Méndez-Estrada, R.; Felix-Portillo, M.; Yepiz-Plascencia, G. Cloning, expression, purification and biochemical characterization of recombinant metallothionein from the white shrimp *Litopenaeus vannamei*. *Protein Expr. Purif.* **2020**, *166*, 105511.
46. Choudhury, C.; Mazumder, R.; Biswas, R.; Sengupta, M. Cadmium exposure induces inflammation through the canonical NF-κB pathway in monocytes/macrophages of *Channa punctatus* Bloch. *Fish Shellfish Immunol.* **2021**, *110*, 116–126. [CrossRef] [PubMed]
47. Yadav, R.; Kumar, R.; Gupta, R.; Kaur, T.; Kour, A.; Kaur, S.; Rajput, A. Heavy metal toxicity in earthworms and its environmental implications: A Review. *Environ. Adv.* **2023**, *12*, 100374. [CrossRef]
48. Bailei, L.; Jeevithan, E.; Wenhui, W. Recent Advancement of Molecular Structure and Biomaterial Function of Chitosan from Marine Organisms for Pharmaceutical and Nutraceutical Application. *Appl. Sci.* **2020**, *10*, 4719.
49. Dornjak, L.; Ostojić, K.; Klaser, T.; Urlić, I.; Rogina, A. Boric Acid Modified Chitosan Scaffolds Chemically Crosslinked by Genipin. *Kem. U Ind. Časopis Kemičara I Kem. Inženjera Hrvat.* **2022**, *71*, 691–698.
50. Alamrani, N.; Almutairi, F.; Alotaibi, F.; Alenazi, D.; Monier, M.; Abdel-Latif, D.; Elsayed, N. Developing thiosemicarbazide-modified/ion-imprinted chitosan for selective cadmium ion biosorption. *Mater. Today Chem.* **2023**, *30*, 101547. [CrossRef]
51. Mukarram, M.; Ali, J.; Dadkhah, A.; Kurjak, D.; Kačík, F.; Ďurkovič, J. Chitosan-induced biotic stress tolerance and crosstalk with phytohormones, antioxidants, and other signalling molecules. *Front. Recent Dev. Plant Sci.* **2023**, *14*, 1217822. [CrossRef]

52. Varghese, L.; Thomas, G. Chitosan triggers tolerance to Pythium myriotylum infection in ginger (*Zingiber officinale*) by modulating multiple defense signaling pathways. *Physiol. Mol. Plant Pathol.* **2023**, *125*, 101985. [CrossRef]
53. Araújo, L.; Matos, H.; Facchi, D.; Almeida, D.; Gonçalves, B.; Monteiro, J.; Martins, A.; Bonafé, E. Natural carbohydrate-based thermosensitive chitosan/pectin adsorbent for removal of Pb(II) from aqueous solutions. *Int. J. Biol. Macromol.* **2021**, *193*, 1813–1822. [CrossRef]
54. Liang, S.; Zhang, T.; Fu, X.; Zhu, C.; Mou, H. Partially degraded chitosan-based flocculation to achieve effective deodorization of oyster (*Crassostrea gigas*) hydrolysates. *Carbohydr. Polym.* **2020**, *234*, 115948. [CrossRef]
55. Le, L.; Giang, N.; Chien, P.; Trinh, X.; Long, N.; Anh, L.; Nga, P.; Zhang, X.; Nam, S.; Heo, C. Enhancement of Wound Healing Efficacy by Chitosan-based Hydrocolloid on Sprague Dawley Rats. *Int. Inst. Anticancer. Res.* **2023**, *37*, 1052–1064. [CrossRef]
56. Brol, J.; Müller, L.; Prates, E.; Farias, B.; Pedrosa, V.; Pinto, L.; Cadaval, T.; Tesser, M.; Wasielesky, W.; Ventura-Lima, J. Dietary chitosan supplementation in Litopenaeus vannamei reared in a biofloc system: Effect on antioxidant status facing saline stress. *Aquaculture* **2021**, *544*, 737034. [CrossRef]
57. Wang, R.; Li, X.; Tang, J.; Liu, J.; Yin, H.; Li, R.; Ye, S. Dietary effect chitosan nanoparticles on growth performance, immunity and resistance against Vibrio splendidus in the sea cucumber *Apostichopus japonicas*. *Aquac. Rep.* **2023**, *30*, 101625. [CrossRef]
58. Grabner, G.; Xie, H.; Schweiger, M.; Zechner, R. Lipolysis: Cellular mechanisms for lipid mobilization from fat stores. *Nat. Metab.* **2021**, *3*, 1445–1465. [CrossRef]
59. Mu, H.; Wan, W.; Song, J.; Kuang, R.; Deng, T. Mitochondrial Lipid Peroxidation and Microsomal Drug-metabolizing Enzyme Activity of Rat Hepatotoxicity under Heavy Metals from Slag Waste Exposure. *Cell Biochem. Biophys.* **2023**, *81*, 285–298. [CrossRef]
60. Jie, J.; Debora, T.; Àngels, R.; Roher, N. Nanodelivery Systems as New Tools for Immunostimulant or Vaccine Administration: Targeting the Fish Immune System. *Biology* **2015**, *4*, 664–696. [CrossRef]
61. Rosidah, Y. Mini-Review: The Role of Chitosan in Aquaculture Fish Health Management. *Asian J. Fish. Aquat. Res.* **2022**, *9*, 18197. [CrossRef]
62. Stanek, M.; Mazurkiewicz, J.; Rawski, M.; Bogucka, J.; Ziółkowska, E.; Dankowiakowska, A.; Kierończyk, B. Effect of chitosan on common carp (*Cyprinus carpio*) fry growth performance, feed utilization and nutriphysiological status. *Aquac. Rep.* **2023**, *30*, 101622. [CrossRef]
63. Mei, Q.; Hu, J.; Huang, Z.; Fan, J.; Huang, C.; Lu, Y.; Wang, X.; Zeng, Y. Pretreatment with chitosan oligosaccharides attenuate experimental severe acute pancreatitis via inhibiting oxidative stress and modulating intestinal homeostasis. *Acta Pharmacol. Sin.* **2021**, *42*, 942–953. [CrossRef]
64. Tzeng, H.; Liu, S.; Chiang, M. Antidiabetic Properties of Chitosan and Its Derivatives. *Mar. Drugs* **2022**, *20*, 784. [CrossRef]
65. Zhang, X.; Liang, S.; Gao, X.; Huang, H.; Lao, F.; Dai, X. Protective Effect of Chitosan Oligosaccharide against Hydrogen Peroxide-Mediated Oxidative Damage and Cell Apoptosis via Activating Nrf2/ARE Signaling Pathway. *Neurotox. Res.* **2021**, *39*, 1708–1720. [CrossRef]
66. Koochacksaraei, R.; Dastar, B.; Samadi, F.; Ebrahimi, P. Investigating of Antioxidant Protective Effects of Shrimp Shells Extracted Chitosan in Broiler Chickens. *World's Poult. Sci. J.* **2020**, *8*, 77–81.

animals

MDPI

Article

Morphometric Relationships between Length and Weight of 109 Fish Species in the Caribbean Sea (French West Indies)

Kélig Mahé [1,*], Jérome Baudrier [2], Angela Larivain [3], Solène Telliez [1], Romain Elleboode [1], Elise Bultel [4] and Lionel Pawlowski [3]

1 IFREMER, Unité HMMN, 150 quai Gambetta, BP 699, 62321 Boulogne-sur-Mer, France; solene.telliez@ifremer.fr (S.T.); romain.elleboode@ifremer.fr (R.E.)
2 IFREMER, Délégation pour les Antilles, Station de Martinique, Unité Biodivenv, 97231 Le Robert, France; jerome.baudrier@ifremer.fr
3 IFREMER, UMR DECOD, LTBH, 56100 Lorient, France; angela.larivain@ifremer.fr (A.L.); lionel.pawlowski@ifremer.fr (L.P.)
4 IFREMER, Unité HISSEO, 29280 Brest, France; elise.bultel@ifremer.fr
* Correspondence: kelig.mahe@ifremer.fr; Tel.: +33-321-995-602

Simple Summary: Poor quality of biological information, such as the relationship between length and weight parameters, could be a source of variability with a significant impact on stock assessment results. This relationship between the length and weight of individuals could differ between males and females as well as two separated islands and/or linked to the reproduction period according the studied species.

Abstract: In total, 109 fish species (24,996 individuals) were sampled around Guadeloupe and Martinique Islands from October 2021 to September 2022 to estimate the morphometric relationships between total length and weight (Length Weight Relationship: LWR) of each fish species according to potential spatial, temporal and sex differences. Of these species, this is the first time that the LWR was estimated in the Atlantic Ocean for 16 species. There is a significant relationship between length and weight for all tested species. For 83 tested species, the sex effect on the LWR showed significant sexual dimorphism for 24 species. Additionally, a link between the temporal effect and the reproduction period was tested for 68 species, of which 35 presented significant differences relative to the annual quarter of sampling. Finally, the geographical effect (i.e., the difference between samples from around Guadeloupe Island and those from Martinique Island) was significant for 60 species. This island effect was significant for 25 species.

Keywords: GLM; L–W relationship; sexual dimorphism; fork length; temporal effect; Guadeloupe; Martinique

1. Introduction

Biological information, such as body length and weight, is a pre-requisite in fishery sciences to build stock assessments based on length-at-age and weight-at-age data [1]. However, both length and weight data are not necessarily acquired during sampling. Consequently, the characterization of the length–weight relationship (LWR) allows readily usable estimates of each of these values. Researchers using morphometric data may not have enough information should an accurate conversion between different metrics be unavailable [2,3]. Moreover, this relationship is a sustainable proxy for "fatness" and "general well-being", like the condition factor [4–6]. The exponential function is the main method to characterize the LWR of fish, which depends on environmental conditions and the physiological state of the individual [1,4,7,8]. The fatness and consequently the outline shape of the fish species can be analyzed from the growth coefficient, which presents values ranging from 2.5 to 4 [1,9–12]. To weigh a fish, total weight (W_T) is generally the most commonly used metric. Occasionally, the gutted weight can also be used; however, there is often a conversion factor to obtain the proportion of total weight that is represented by this proxy.

Conversely, when measuring the length of a fish, there are many different potential lengths according to the anatomy of sampled species, especially the shape of the tail (i.e., with or without tail filaments). However, the two main metrics of measuring fish for research, resource management and regulation purposes are total length (TL: measured from the tip of the jaw or the snout to the farthest tip of the tail with the tail squeezed) and fork length (FL: measured from the tip of the jaw or the snout to the center of the fork in the tail).

The West Indies extend northward from South America in a long volcanic arc of over a dozen major islands and hundreds of small islets (from the Dominican Republic in the north to the Grenadines in the south of the Lesser Antilles). These islands are among the 34 identified global biodiversity hotspots [13,14] and are home to a high number of endemic genera and species [15]. The islands are of various sizes, and the largest is Guadeloupe Island at 1706 km^2. The length–weight relationships of fish species belonging to this biodiversity hotspot are relatively unknown. The literature provides sparse LWR for some local species from studies in adjacent regions, but generally from different marine environments several hundreds if not of thousands of kilometers distant (e.g., Florida or Brazil). These great distances likely make the use of published biological parameters unreliable to tackle the growth dynamics of local species. In the West Indies, a few specific studies have focused on a single species, such as the lionfish [16–22], which is an invasive species. Some papers have reported the values for several species, but these are often several years or even decades old (i.e., from the 1970s and 1980s). Consequently, this study focused on the length–weight relationships for 109 fish species caught in the waters of the French West Indies. There are four French islands (Guadeloupe, Martinique, Saint Barth and Saint Martin), and the two main islands are in the middle of the long arc with Guadeloupe Island in the north and Martinique Island in the south, separated by around 190 km, with Dominica Island in between (Figure 1). The potential influence of factors, such as sampling period (i.e., sampling quarter), sampling location/island (i.e., Guadeloupe and Martinique Islands) and physiological effect (i.e., the potential sexual dimorphism between males and females), on length–weight relationships was evaluated.

Figure 1. Map of the Lesser Antilles with the French West Indies (Guadeloupe and Martinique Islands) where all fish species were caught from September 2021 to August 2022.

2. Materials and Methods

Among all species, only those with a minimum sample number of 10 individuals were analyzed. This amount often provides a sufficiently precise relationship, but this can be completed with further sampling in the future. The results are presented in two groups. One group includes the species showing higher precision in the analysis (i.e., specimen number > 30). For the other group, only preliminary results are presented in Supplementary Tables due to the small number of specimens (i.e., number < 30). Consequently, 109 fish species were analyzed (*Acanthurus bahianus, A. chirurgus, A. coeruleus, A. tractus, Albula vulpes, Balistes capriscus, B. vetula, Canthidermis sufflamen, Melichthys niger, Xanthichthys ringens, Tylosurus crocodilus, Caranx bartholomaei, C. crysos, C. latus, C. lugubris, C. ruber, Chloroscombrus chrysurus, Decapterus punctatus, Selar crumenophthalmus, Selene brownii, S. vomer, Seriola dumerili, S. rivoliana, Trachinotus goodei, Chaetodipterus faber, Chaetodon capistratus, C. sedentarius, C. striatus, Dactylopterus volitans, Diodon holocanthus, Elops saurus, Diapterus auratus, Eucinostomus argenteus, Gerres cinereus, Anisotremus surinamensis, Brachygenys chrysargyreum, Haemulon aurolineatum, H. bonariense, H. carbonarium, H. flavolineatum, H. parra, H. plumierii, H. sciurus, H. striatum, Holocentrus adscensionis, H. rufus, Myripristis jacobus, Neoniphon marianus, Kyphosus sectatrix, Bodianus rufus, Clepticus parrae, Halichoeres radiatus, Lachnolaimus maximus, Etelis oculatus, Lutjanus analis, L. apodus, L. buccanella, L. cyanopterus, L. griseus, L. jocu, L. mahogoni, L. synagris, L. vivanus, Ocyurus chrysurus, Pristipomoides macrophthalmus, Rhomboplites aurorubens, Malacanthus plumieri, Aluterus monoceros, A. scriptus, Cantherhines macrocerus, C. pullus, Mulloidichthys martinicus, Pseudupeneus maculatus, Gymnothorax moringa, Acanthostracion polygonius, Lactophrys triqueter, Polydactylus oligodon, Polydactylus virginicus, Holacanthus tricolor, Abudefduf saxatilis, Heteropriacanthus cruentatus, Priacanthus arenatus, Scarus iseri, S. taeniopterus, S. vetula, Sparisoma aurofrenatum, S. chrysopterum, S. frondosum, S. rubripinne, S. viride, Larimus breviceps, Umbrina coroides, Scomberomorus regalis, Pterois volitans, Scorpaena plumieri, Alphestes afer, Cephalopholis cruentata, Cephalopholis fulva, Epinephelus guttatus, E. striatus, Paranthias furcifer, Rypticus saponaceus, Archosargus rhomboidalis, Calamus bajonado, C. calamus, C. penna, C. pennatula, Sphyraena barracuda* and *Synodus intermedius*) (Table S1). The total sample number per species varied from 10 individuals for *S. regalis* to 1501 individuals for *O. chrysurus*. A total of 24,996 marine fishes caught around Guadeloupe and Martinique Islands were sampled from October 2021 to September 2022 (except the first quarter of 2022). These 2 French islands are located in the archipelago of West Indies Islands, separating the Atlantic Ocean from the Eastern part of the Caribbean Sea. Guadeloupe and Martinique Islands are located 187 km apart with the Dominica Island between these two French islands (Figure 1). Consequently, the sampling from two islands could be interpreted as a potential geographical factor that could influence the morphological characteristics of the fish. All specimens were obtained from landings of fishing boats with commercial and non-commercial species, covering all lengths, including those below the legal size with authorization from the French government to perform the first large sampling for stock assessment. In the non-commercial fishes, there were discarded species due to low price and contaminated by ciguatera poisoning. All specimens were weighed (one individual weight: total, $W_T \pm 0.1$ g) and measured (two individual lengths: total, TL $\pm$ 0.5 cm and fork, FL $\pm$ 0.5 cm). Sex was determined by macroscopic gonad observation for 83 species, but these biological data were not observed for 26 species. All specimens were analyzed fresh. For 26 species, the gonad observation was too difficult to obtain results with a high accuracy. Among the 24,996 individuals, 32 families belonging to Actinopterygii were represented (Acanthuridae, Albulidae, Balistidae, Belonidae, Carangidae, Chaetodontidae, Dactylopteridae, Diodontidae, Elopidae, Gerreidae, Haemulidae, Holocentridae, Kyphosidae, Labridae, Lutjanidae, Malacanthidae, Monacanthidae, Mullidae, Muraenidae, Ostraciidae, Polynemidae, Pomacanthidae, Pomacentridae, Priacanthidae, Scaridae, Sciaenidae, Scombridae, Scorpaenidae, Serranidae, Sparidae, Sphyraenidae and Synodontidae). Among these Actinopterygii families, several families had a large number of species, such as Carangidae (13 species, n = 2086), Lutjanidae

(13 species, n = 4927), Haemulidae (10 species, n = 3344), Scaridae (8 species, n = 3885) and Serranidae (7 species, n = 1117) (Table S1).

All data were plotted to identify and delete the potential outliers. After this preliminary control step, the data were used to characterize the LWR for each species (Equation (1)):

$$W_T = a \cdot TL^b \quad (1)$$

The body shape coefficient is expressed as a, and the growth coefficient is represented by b [1,4,23]. The difference between this observed growth coefficient for each fish species and the theoretical value (b = 3.0) was tested with a one-sample *t*-test at a significance level of 0.05 to verify whether it was significantly different from the theoretical value of 3.0. The relationship between the two length measurements (Length–Length Relationship [LLR]) is also analyzed (Equation (2)):

$$TL = c \cdot FL + d \quad (2)$$

A complete generalized linear model (GLM) was used to tested the potential effects of the explanatory variables sex (S; females and males; 83 tested species), sampling period (Q; Quarter 4 2021; Quarter 2 2022 and Quarter 3 2022; 68 tested species) and geographical location/sampling island (G; 13506 individuals caught around Guadeloupe Island and 11490 individuals caught around Martinique Island) (Supplementary Table S2) on the LWR of each species (Equation (3)):

$$\log W_T \sim (\log TL) + (\log TL) \cdot S + (\log TL) \cdot Q + (\log TL) \cdot G + \varepsilon \quad (3)$$

For hermaphrodite species (i.e., change sex from female to male; Table S1), the difference between males and females for these species was directly influenced by the difference of length and age ranges between sexes. Consequently, for these hermaphrodite species, represented by 4 families Labridae, Pomacanthidae, Scaridae and Serranidae (*Bodianus rufus*, *Clepticus parrae*, *Lachnolaimus maximus*, *Holacanthus tricolor*, *Scarus iseri*, *Scarus taeniopterus*, *Scarus vetula*, *Sparisoma aurofrenatum*, *S. rubripinne*, *S. viride*, *Alphestes afer*, *Cephalopholis cruentata*, *C. fulva*, *Epinephelus guttatus* and *E. striatus*), the potential effect of sex was not tested.

The Fulton's condition factor (Kc) and the relative condition factor (Kn) [4] were applied to compare the observed total weight at the given length to the predicted total weight from the LWR (Equations (4) and (5)):

$$Kc = 100 \cdot W_T / TL^3 \quad (4)$$

$$Kn = W_T / (a \cdot TL^b) \quad (5)$$

All statistical analyses (with significant effect at $p < 0.05$) were carried out in the R statistical environment (version 4.0.2), specifically using the CAR and GGPLOT2 packages [24,25].

3. Results

The number of measured specimens and all information (mean ± standard deviation; the minimum and the maximum) for the three individual morphometric parameters (total length, fork length and total weight) are shown in Table S1 for each fish species. Measured total length ranged from 6.5 cm (*B. capriscus*) to 119 cm (*S. barracuda*). The individual total weight range ranged from 5 g (*H. aurolineatum*) to 8750 g (*C. pennatula*) (Supplementary Table S1). However, all LWR values are presented because there were either no previous results or very old data for several species. Among the 109 species, this study is the first time that the LWR was estimated for 16 species (*A. bahianus*, *H. striatum*, *N. marianus*, *K. sectatrix*, *B. rufus*, *C. parrae*, *H. radiates*, *C. pullus*, *P. oligodon*, *S. taeniopterus*, *S. vetula*, *S. aurofrenatum*, *S. frondosum*, *P. furcifer*, *R. saponaceus* and *S. intermedius*) and that the LLR was calculated for

7 species (*A. bahianus, H. striatum, G. moringa, A. polygonius, S. taeniopterus, S. aurofrenatum, S. frondosum*) in the Atlantic Ocean for more than 6 individual specimens (Table 1 and Table S3).

Table 1. Relationship between weight and body length for species with a specimen number greater than 30. Each species is identified (First Data with the star indicated that this information was presented for the first time), and this study is the first time that the relationship between two measurements was calculated with an individual number greater than 6. Here, *p*-values are reported for each LWR and LLR and for potential explanatory variables (sex, sampling quarter and area). The coefficients a and b of the LWR and the coefficients c and d of the LLR are presented, separately (value ± standard deviation). If the sample number was <5, no data are reported. Blue cells indicate significant *p*-values.

Family	Latin Name	N	W/TL							TL/FL			
			First Data	*p*-Value	a	b	Explanatory Factors			First Data	*p*-Value	c	d
							Sex	Quarter	Area				
Acanthuridae	*Acanthurus bahianus*	1187	*	<0.001	0.0222	2.9270	0.0645	<0.001	<0.001	*	<0.001	1.1532	−0.4007
	Acanthurus chirurgus	314		<0.001	0.0228	2.9702	0.2458	<0.001	0.0510		<0.001	1.0269	0.8089
	Acanthurus coeruleus	986		<0.001	0.0397	2.8139	0.1470	0.0498	<0.001		<0.001	1.1739	−0.9288
	Acanthurus tractus	99		<0.001	0.0341	2.7909	0.1870	no data	no data		<0.001	1.1564	−0.3920
Albulidae	*Albula vulpes*	74		<0.001	0.0057	3.1297	0.9329	0.0522	0.2420		<0.001	1.1570	0.6678
Balistidae	*Balistes vetula*	107		<0.001	0.1758	2.2849	0.1510	0.7050	0.0870		<0.001	1.4830	−3.3020
	Melichthys niger	62		<0.001	0.0570	2.7445	0.0163	<0.001	0.0130		<0.001	1.1289	−1.5303
Carangidae	*Caranx bartholomaei*	127		<0.001	0.0131	2.9573	0.9860	<0.001	<0.001		<0.001	1.2026	−0.3080
	Caranx crysos	273		<0.001	0.0303	2.7060	0.0182	<0.001	0.2920		<0.001	1.2414	−1.3354
	Caranx latus	203		<0.001	0.0188	2.8886	0.9980	0.6420	0.6440		<0.001	1.1715	0.6950
	Caranx ruber	907		<0.001	0.0056	3.1653	0.2457	<0.001	0.4613		<0.001	1.1614	0.7201
	Selar crumenoph-thalmus	47		<0.001	0.0035	3.4038	0.0050	<0.001	no data		<0.001	1.1146	0.1312
	Seriola rivoliana	404		<0.001	0.0082	3.0890	<0.001	<0.001	0.4230		<0.001	1.1138	1.0095
	Chaetodon capistratus	96		<0.001	0.0747	2.6228	0.7630	0.0570	0.9050		<0.001	1.1406	−0.9840
	Chaetodon striatus	87		<0.001	0.0328	3.0113	0.0300	0.1557	0.4154		<0.001	1.1131	−0.8654
Dactylopteridae	*Dactylopterus volitans*	82		<0.001	0.0045	3.2507	0.0239	0.0048	<0.001		<0.001	1.0623	−0.2675
Gerreidae	*Diapterus auratus*	370		<0.001	0.0494	2.6299	0.1454	0.0860	no data		<0.001	1.3193	−1.7101
	Gerres cinereus	439		<0.001	0.0233	2.8247	0.0040	0.2156	0.6586		<0.001	1.3045	−1.0524

Table 1. *Cont.*

Family	Latin Name	N	W/TL							TL/FL			
			First Data	*p*-Value	a	b	Explanatory Factors			First Data	*p*-Value	c	d
							Sex	Quarter	Area				
Haemulidae	*Anisotremus surinamensis*	34		<0.001	0.0350	2.8270	0.7670	0.7550	no data		<0.001	1.1485	−0.4158
	Brachygenys chrysargyreum	261		<0.001	0.0015	3.5320	<0.001	<0.001	0.9056		<0.001	1.1581	−0.4070
	Haemulon aurolineatum	965		<0.001	0.0120	3.0448	0.0499	0.1806	0.3396		<0.001	1.0785	0.6476
	Haemulon bonariense	75		<0.001	0.0179	2.9584	0.0523	0.2718	no data		<0.001	1.0453	1.3000
	Haemulon carbonarium	133		<0.001	0.0269	2.8301	0.9411	0.9103	0.6760		<0.001	1.0924	0.5993
	Haemulon flavolineatum	444		<0.001	0.0181	2.9435	0.3211	<0.001	0.1764		<0.001	1.0424	1.2720
	Haemulon parra	290		<0.001	0.0251	2.8600	0.8203	<0.001	<0.001		<0.001	1.1174	−0.1802
	Haemulon plumierii	781		<0.001	0.0148	3.0160	0.3113	<0.001	<0.001		<0.001	1.1386	−0.3127
	Haemulon sciurus	131		<0.001	0.0263	2.8472	0.3704	0.6789	0.8320		<0.001	1.1144	−0.0610
	Haemulon striatum	230	*	<0.001	0.0152	2.9246	0.4230	0.7560	no data	*	<0.001	1.0880	0.7281
Holocentridae	*Holocentrus adscensionis*	1044		<0.001	0.0227	2.7794	<0.001	<0.001	<0.001		<0.001	1.0937	2.7445
	Holocentrus rufus	1189		<0.001	0.1205	2.1885	<0.001	0.1290	0.0160		<0.001	1.1037	2.3760
	Myripristis jacobus	93		<0.001	0.0269	2.8789	0.2940	0.0236	<0.001		<0.001	1.0512	1.5363
Kyphosidae	*Kyphosus sectatrix*	81	*	<0.001	0.0153	3.0448	0.2423	0.9156	0.6055		<0.001	1.1355	−0.2492
Labridae	*Bodianus rufus*	35	*	<0.001	0.0795	2.5023	no data	no data	0.9328		<0.001	1.2720	−4.7375
Lutjanidae	*Etelis oculatus*	87		<0.001	0.0153	2.8384	0.2012	0.4230	0.0940		<0.001	1.2478	−0.1504
	Lutjanus analis	223		<0.001	0.0106	3.0677	0.7902	0.6895	0.6997		<0.001	1.1310	−0.9373
	Lutjanus apodus	201		<0.001	0.0193	2.9615	0.2956	0.4244	0.0205		<0.001	1.0878	−0.5060
	Lutjanus buccanella	934		<0.001	0.0100	3.1056	0.0090	<0.001	0.0420		<0.001	1.0976	−0.3253
	Lutjanus griseus	223		<0.001	0.0158	2.9645	0.7357	0.8046	0.0459		<0.001	1.0737	−0.7727
	Lutjanus mahogoni	152		<0.001	0.0081	3.1582	0.7490	<0.001	<0.001		<0.001	1.0382	0.6515
	Lutjanus synagris	278		<0.001	0.0165	2.9424	0.1243	0.1829	0.1247		<0.001	1.0958	−0.5297
	Lutjanus vivanus	623		<0.001	0.0128	2.9912	<0.001	<0.001	0.0047		<0.001	1.1091	−0.4535
	Ocyurus chrysurus	1501		<0.001	0.0153	2.8503	0.1864	0.0217	<0.001		<0.001	1.2636	−1.0014
	Rhomboplites aurorubens	653		<0.001	0.0145	2.9466	0.7792	<0.001	0.0019		<0.001	1.1084	−0.1796
	Aluterus scriptus	49		<0.001	0.0019	3.2708	0.6717	0.2308	0.8046		no data	no data	no data
	Cantherhines macrocerus	272		<0.001	0.0056	3.3534	0.0014	0.0980	<0.001		no data	no data	no data
	Cantherhines pullus	34	*	<0.001	0.0362	2.8053	no data	no data	no data		no data	no data	no data

Table 1. *Cont.*

Family	Latin Name	N	W/TL							TL/FL			
			First Data	*p*-Value	a	b	Explanatory Factors			First Data	*p*-Value	c	d
							Sex	Quarter	Area				
Mullidae	*Mulloidichthys martinicus*	543		<0.001	0.0085	3.0918	0.1891	0.0138	0.0002		<0.001	1.1587	0.4023
	Pseudupeneus maculatus	406		<0.001	0.0111	3.0419	0.0018	<0.001	0.3271		<0.001	1.0989	2.7970
Muraenidae	*Gymnothorax moringa*	44		<0.001	0.0061	2.7333	0.1225	0.6521	0.2816	*	<0.001	1.1000	0.0056
Ostraciidae	*Acanthostracion polygonius*	84		<0.001	0.0522	2.7156	0.0320	0.2361	0.7632	*	<0.001	1.0013	0.0143
	Lactophrys triqueter	72		<0.001	0.0622	2.7725	0.5302	0.4754	no data		<0.001	1.0045	0.0113
Pomacanthidae	*Holacanthus tricolor*	72		<0.001	0.0756	2.6444	0.0341	<0.001	0.6526		<0.001	1.1075	−1.2626
	Abudefduf saxatilis	93		<0.001	0.0417	2.7690	0.1196	0.7918	no data		<0.001	1.0355	1.6459
Priacanthidae	*Heteropriacanthus cruentatus*	70		<0.001	0.0322	2.7937	0.0581	0.6366	0.4549		<0.001	1.0264	−0.5200
	Priacanthus arenatus	85		<0.001	0.0772	2.4984	0.1452	0.0012	0.0091		<0.001	1.1112	−1.9655
Scaridae	*Scarus iseri*	39		<0.001	0.0248	2.9126	0.4427	no data	no data		<0.001	1.0218	−0.3942
	Scarus taeniopterus	329	*	<0.001	0.0101	3.2045	<0.001	0.0061	0.0022	*	<0.001	1.0198	−0.3683
	Sparisoma aurofrenatum	758	*	<0.001	0.0374	2.7505	0.0020	<0.001	0.6702	*	<0.001	1.0941	−1.3135
	Sparisoma chrysopterum	850		<0.001	0.0147	3.0299	<0.001	<0.001	0.9707		<0.001	1.1556	−1.9630
	Sparisoma frondosum	44	*	<0.001	0.0171	2.9970	no data	no data	no data	*	<0.001	1.1461	−2.0712
	Sparisoma rubripinne	1486		<0.001	0.0457	2.7454	<0.001	<0.001	<0.001		<0.001	1.0608	−1.0393
	Sparisoma viride	357		<0.001	0.0694	2.5944	<0.001	<0.001	<0.001		<0.001	1.2526	−4.3414
	Umbrina coroides	33		<0.001	0.0092	3.1329	no data	no data	no data		no data	no data	no data
Scorpaenidae	*Pterois volitans*	449		<0.001	0.0131	3.0009	0.8033	0.0001	<0.001		<0.001	1.0099	0.3873
	Scorpaena plumieri	49		<0.001	0.0306	2.9222	0.7582	0.0434	0.9689		<0.001	1.0078	0.5323
Serranidae	*Alphestes afer*	101		<0.001	0.0074	3.2861	0.2136	0.0055	0.2495		<0.001	1.0016	−0.0286
	Cephalopholis cruentata	60		<0.001	0.0322	2.7612	0.5269	0.9870	0.2416		<0.001	1.0100	0.1642
	Cephalopholis fulva	679		<0.001	0.0208	2.9266	0.9365	<0.001	0.1535		<0.001	1.0099	0.1584
	Epinephelus guttatus	220		<0.001	0.0092	3.1367	0.0235	<0.001	0.5750		<0.001	1.0989	0.3421
Sparidae	*Archosargus rhomboidalis*	362		<0.001	0.0230	2.9560	0.6310	0.1649	no data		<0.001	1.1433	−0.8086
	Calamus calamus	44		<0.001	0.0088	3.2120	0.8551	0.6276	no data		<0.001	1.1338	0.3011
	Calamus pennatula	131		<0.001	0.0236	2.9191	0.1427	<0.001	0.3989		<0.001	1.1229	0.3017

The LWR for each species consistently showed a significant correlation between two morphological parameters (Table 1 and Table S3). The relationship between the coefficients of LWR also showed a significant correlation between body shape coefficient and the growth coefficient (Figure 2). Only nine sampled species exhibited a b coefficient value equal to 3.0. Conversely, 65 species showed that the body became more elongated throughout life (i.e., b < 3.0), while individuals became thicker (i.e., b > 3.0) over the lifespan in 36 species.

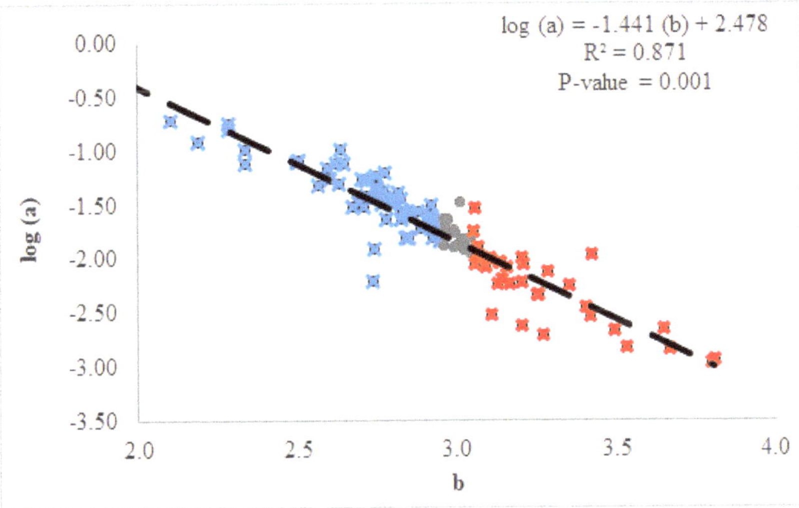

Figure 2. Plot of length–weight relationships (log a vs. b) in 109 fish species sampled from September 2021 to August 2022 around the French West Indies (Guadeloupe and Martinique Islands) (negative allometric growth with b < 3: blue cross; isometric growth with b = 3.0: grey circle; positive allometric growth with b > 3: red cross).

First, the influence of sex was estimated for the 83 species. The results (i.e., the slopes of the LWR) identified significant differences between females and males for 24 species (i.e., 29% of all species: *M. niger*, *C. crysos*, *S. crumenophthalmus*, *S. rivoliana*, *C.striatus*, *D. volitans*, *G. cinereus*, *H. aurolineatum*, *B. chrysargyreum*, *H. adscensionis*, *H. rufus*, *K. sectatrix*, *L. buccanella*, *L. vivanus*, *C. macrocerus*, *P. maculatus*, *A. polygonius*, *H. tricolor*, *S. taeniopterus*, *S. aurofrenatum*, *S. chrysopterum*, *S. rubripinne*, *S. viride* and *E. guttatus*; Table 1 and Table S3). For each species showing a significant difference between the sexes, the LWR was calculated for each sex (Supplementary Table S4). Second, a temporal effect (i.e., the sampling quarter on the slopes of the LWR) was detected for 35 species (i.e., 51% of 68 tested species). Finally, geographical differences in the LWR between Guadeloupe and Martinique Islands were tested for 60 species, and this effect was significant for 25 species (42%). For 103 species where the fork length was measured, the total length–fork length relationship was always significant ($p < 0.05$; Table 1). To understand the difference between species and/or potential explanatory variables (sex, sampling quarter and island), condition factors (Kc and Kn) were calculated (Figures 3 and 4). First, comparing Kn and Kc conditions factors on the same dataset, the trend for each factor was not comparable to another factor (Figure 3). Second, to analyze the condition factor according to geographical location/sampling island, sex, and sampling period, only Kn was used (Figure 4) The condition factor Kn was greater for females than males for 61 species, while 22 species presented the opposite trend (Figure 3). There is no clear trend among sampling quarters. Finally, the condition factor of fish from Martinique Island was often higher than that for those from Guadeloupe Island (i.e., 49 species vs. 22 species; Figure 4).

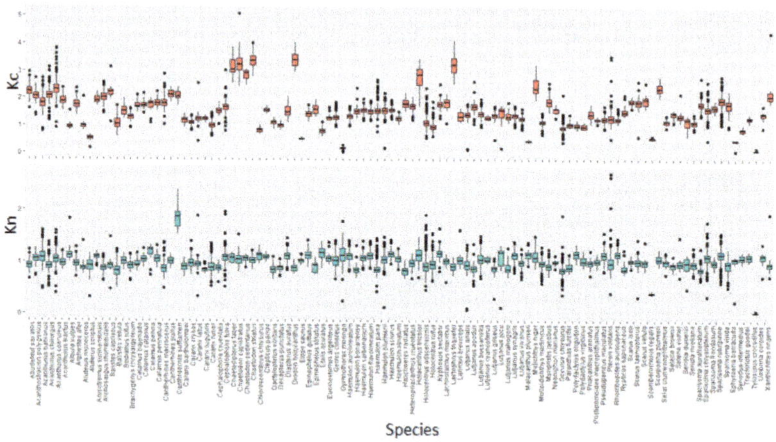

Figure 3. Boxplot of condition factors (Fulton's factor: Kc; Relative condition factor: Kn) for each species sampled from September 2021 to August 2022 around the French West Indies.

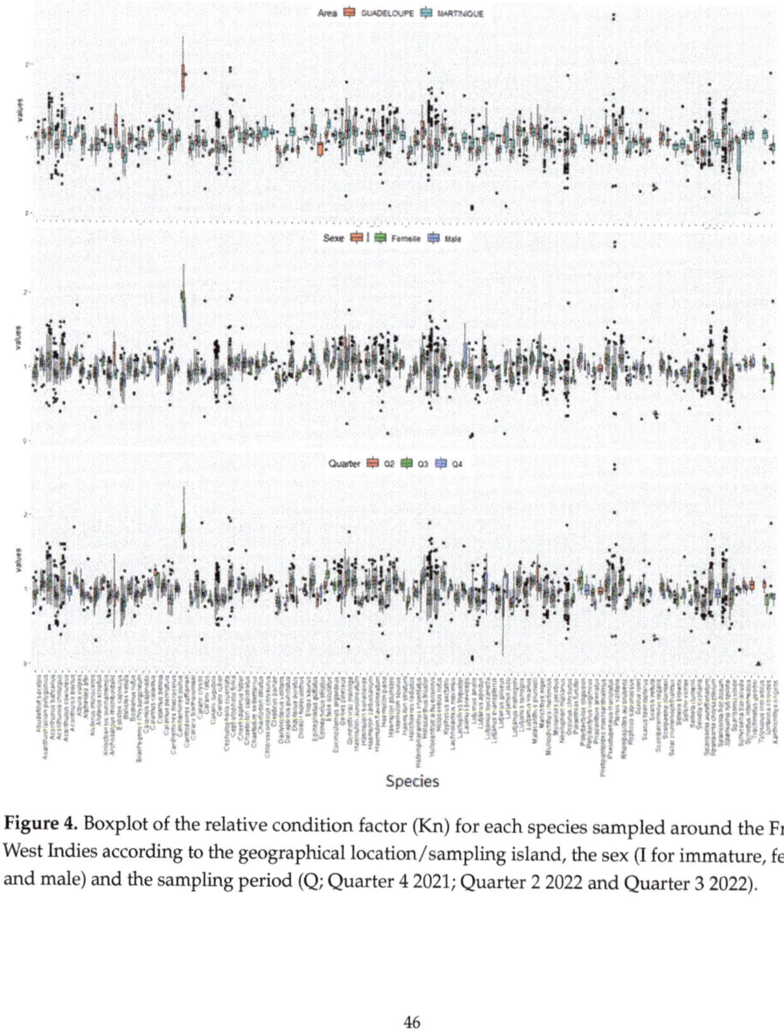

Figure 4. Boxplot of the relative condition factor (Kn) for each species sampled around the French West Indies according to the geographical location/sampling island, the sex (I for immature, female and male) and the sampling period (Q; Quarter 4 2021; Quarter 2 2022 and Quarter 3 2022).

4. Discussion

Morphometric relationships between length and weight were established for 109 fish species (n = 24,996 individuals). This variety of species is substantially higher than those observed in the temperate waters (i.e., Northeastern Atlantic Ocean [26], 45 species for 31,167 individuals) but closer to those observed in other warm water areas (i.e., the Gulf of Mexico [17], 174 fish species for 7503 individuals; Réunion Island (Indian Ocean) [27], 123 fish species for 10,218 individuals). This number is within the ranges of the count of fish species observed in the fish assemblage of the Caribbean arc, where between 63 and 167 fish species were reported per island [28,29]. These studies identified all commercial and non-commercial species without a minimum threshold number, whereas this value was fixed at 10 for this LWR study. However, this study on length–weight relationships showed that fish species diversity is higher than observed in previous studies on biological information in the Caribbean Sea (n = 37 species, Gulf of Salamanca, Columbia [30]; n = 20 species, Gulf of Salamanca [18]; n = 36 species, deep-sea fishes in Colombia [20]; n = 66 species, delta of the Atrato River, Colombia [21]; n = 22 species, freshwater fish in Venezuela [22]). Our results showed that the body length–weight relationship for all tested species was significant, which is consistent with previous studies carried out in the Northeastern Atlantic Ocean [26,31,32], Greek waters [33], the Persian Gulf [34], the Aegean Sea [35] and the Indian Ocean [27]. These results confirmed that, even with small numbers of sampled individuals, there is a significant relationship between length and weight, which allows us to switch from one morphological metric to the other and vice versa. For many species, this study either presented the first length–weight relationship for the Caribbean Sea (i.e., LWR only available for other geographical areas), or updated the LWR in the Caribbean Sea, which may be several decades old. Finally, for 16 species (*A. bahianus*, *H. striatum*, *N. marianus*, *K. sectatrix*, *B. rufus*, *C. parrae*, *H. radiates*, *C. pullus*, *P. oligodon*, *S. taeniopterus*, *S. vetula*, *S. aurofrenatum*, *S. frondosum*, *P. furcifer*, *R. saponaceus* and *S. intermedius*), this was the first time that a length–weight relationship has been identified in the scientific literature, providing basic data for monitoring fish populations. Among these species, 8 species had a higher precision sampling number (n > 30; Supplementary Figure S1). Similarly, the LLR showed significant correlations between fork and total lengths, which corroborated previous studies [36,37]. For all species with different tail shape, this result allows us to switch from one length to the other and vice versa. As with the LWR analysis, this is the first time that the LLR is calculated for 7 species (*A. bahianus*, *H. striatum*, *G. moringa*, *A. polygonius*, *S. taeniopterus*, *S. aurofrenatum* and *S. frondosum*) for which total length can be transformed into fork length or vice versa without introducing bias into the measurements. Among 109 analyzed species, 72 species had more than 30 individuals measured and a good coverage of the known size range for each species. For the remaining 37 species, results are preliminary and can be updated in the future. In this study, a value of 3.0 (i.e., isometric growth) for the allometric coefficient b was observed for only 9 species (8.2% of the total number of species, Figure 2). The b values ranged between 2.0 and 4.0, confirming the previous results for the large number of species [38]. For this reason and the correlation between two coefficients of LWR, two condition factors were used to identify the sources of difference among LWR. Traditionally, the Fulton's condition factor (Kc) based on isometric growth (b = 3.0) is applied, but the relative condition factor (Kn), defined by allometric growth (b < 3 or b > 3), is more representative for the wild fish species. Comparing the trend for both condition factors, there were several differences among these 109 species. As Kn directly uses the values of LWR coefficients extracted from the sampling data, it will be preferable to use the allometric condition factor Kn instead of the Fulton' isometric condition factor (Kc) [4,39]. The value of Kn is compared to a threshold value of 1.00, above which fish are in good health and grow in favorable conditions, limiting the stress of individuals (i.e., good range of environmental conditions, available prey and low density of predators) [4,40–43]. Within the sampled species, there are several potential factors that main explain differences in the LWR, such as the physiological aspect (ontogenic effect, sexual dimorphism, reproduction period vs. other months) and/or the environment (difference of habitats between islands) [26].

The LWR was first tested for a physiological effect on the difference between males and females. Among the 109 species analyzed, the influence of sex was estimated for 83 species, and 24 of those species presented significant sexual dimorphism (29%). According to the value of relative condition factor (Kn), the sexual dimorphism showed that females are often heavier than males of the same length as previously shown for Northeastern Atlantic Ocean species [26,44]. In the present study, the effect of sampling quarter was investigated because it could be directly related to reproduction period (gonad development and spawning period) [6,45,46]. The sampling period showed no clear trend (significant effect for 35 species among 68 tested species). For the species with a significant effect, the sampling quarter could be linked to the reproduction period or sexual rest period (i.e., for each species, Table S4). The females were heaviest just before and during the spawning period, particularly due to the degree of gonad development [1,4,7,8]. Finally, this study focused on geographical effects by sampling around Guadeloupe Island in the north and around Martinique Island in the south. Among 60 species, this effect was significant for 25 species (42%). The condition factor of fish from Martinique Island was often higher than those from Guadeloupe Island. Consequently, at the north and south of Dominica islands, the environmental conditions could be different, with better conditions and growth of the fish in the south around Martinique Island than that noted for those around Guadeloupe Island. This geographical difference could be directly linked to habitat characteristics explained by environmental factors, such as temperature and salinity, but also to feeding activities (e.g., food availability and feeding rate) [4,6–8,26,45,46].

5. Conclusions

The length–weight relationships reported here for 109 commercial fish species from 24,996 individuals caught around Guadeloupe and Martinique Islands by sampling from October 2021 to September 2022, with potential spatial, temporal and sex differences, have yielded updated information for some species and provided new information on these relationships between biological parameters (total weight, total and fork lengths) for 16 species for the Atlantic Ocean, especially the Caribbean Sea. This study showed a moderate difference in size linked to sex, but a higher discrepancy appeared due to geographical and temporal effects. The sampling quarter (linked to the reproduction period), the sampling island, and sexual dimorphism showed significant effects on the LWR for 35, 25, and 24 species, respectively. This dataset was assembled with special French authorizations to analyze fish under the commercial size and individuals of species contaminated by ciguatera poisoning. These data and results will be essential for stock assessment of Caribbean fishes.

Supplementary Materials: The following supporting information can be downloaded at: https://www.mdpi.com/article/10.3390/ani13243852/s1, Figure S1: Total Length/Total Weight relationship of species for the first time in the Caribbean Sea; Table S1: Fish metrics (total length: TL, fork length: FL and total weight: WT) of the 109 fish species caught around the French West Indies (Guadeloupe and Martinique Islands). For each species, the total number of individuals (N) has been specified. Species with an asterisk (*) are hermaphrodite. Table S2: Number of fish analysed per species among the 109 fish species caught around the French West Indies (Guadeloupe and Martinique Islands) by each modality (sex, sampling quarter and French Island) (Ind. = indeterminate). Table S3: Preliminary relationship between weight and body length for species with the number specimen smaller than 30. For each species, this study is identified if it was a first time that the relationship between two measures was calculated in the Atlantic Ocean with the individual number higher than 6 (The star in the First data column). P-value for the relationship between weight and total body length (W/TL) and for the influence of sex, sampling quarter and sampling area with the coefficients of the LWR a and b (value, Standard Deviation). *p*-value and the coefficients c and d of the total length and fork length relationship (no data when the factor was not tested because there was only one modality with n > 5 or when the fork length was not available; Grey cells indicate that a factor appears to have a significant effect (p-value < 0.05)). Table S4: Parameters a and b for the relationship between total weight and body length by sex for each species showing significant sexual dimorphism. Table S5: All individual biological data (Latin name; Island; Sex; Sampling Date; W (g); FL (cm); TL (cm)) of 109 fish species caught around the French West Indies (Guadeloupe and Martinique Islands).

Author Contributions: Conceptualization, K.M. and J.B.; methodology, A.L., S.T. and K.M.; validation, J.B., K.M., R.E., E.B. and L.P.; formal analysis, K.M.; investigation, J.B., L.P. and K.M.; writing—original draft preparation, K.M. and L.P.; supervision, K.M.; funding acquisition, E.B. All authors have read and agreed to the published version of the manuscript.

Funding: This research was funded by the Agence Française de Développement (AFD; by the Accobiom project n° AFD CZD1097).

Institutional Review Board Statement: This study was conducted by European Commission recommendation 2007/526/EC, on revised guidelines for the accommodation and care of animals used for experimental and other scientific purposes. All specimens were from landings commercial fishing boats with commercial and non-commercial species, covering all lengths, even including below the legal size with the authorization from the French government to realize the first large sampling for stock assessment. Ethical review and approval were waived for this study because the fish were obtained from fisheries and were already dead when the calcified structures were extracted.

Informed Consent Statement: Not applicable.

Data Availability Statement: The data presented in this study are available in Table S5.

Acknowledgments: We would like to express our gratitude to all the people, the scientific staff, students and fishers involved in the collection, provision and laboratory processing of the samples analyzed in this study. We would especially thank Kirsteen MacKenzie, a native English speaker, for her valuable help in correcting this manuscript.

Conflicts of Interest: The authors declare no conflict of interest.

References

1. Froese, R. Cube law, condition factor and weight-length relationships: History, meta-analysis and recommendations. *J. Appl. Ichthyol.* **2006**, *22*, 241–253. [CrossRef]
2. Pol, M.V.; Szymanski, M.J.; Chosid, D.M.; Salerno, D. Fork length-total length conversions for haddock and pollock. *N. Am. J. Fish. Manag.* **2011**, *31*, 427–430. [CrossRef]
3. Hansen, M.J.; Chavarie, L.; Muir, A.M.; Howland, K.L.; Krueger, C.C. Variation in fork-to-total length relationships of North American lake trout populations. *J. Fish Wildl. Manag.* **2020**, *11*, 263–272. [CrossRef]
4. Le Cren, E.D. The length-weight relationship and seasonal cycle in gonad weight and condition in the perch (*Perca fluviatilis*). *J. Anim. Ecol.* **1951**, *20*, 201–219. [CrossRef]
5. Tesch, F.W. Age and growth. In *Methods for Assessment of Fish Production in Fresh Waters*; Ricker, W.E., Ed.; Blackwell Scientific Publications: Oxford, UK, 1968; pp. 93–123.
6. Weatherley, A.H.; Gill, H.S. *The Biology of Fish Growth*; Academic Press: London, UK, 1987.
7. Pauly, D. *Gasping Fish and Panting Squids: Oxygen, Temperature and the Growth of Water-Breathing Animals*; Excellence in Ecology; International Ecology Institute: Oldendorf, Germany, 2010; Volume 22.
8. Mozsar, A.; Boros, G.; Sály, P.; Antal, L.; Nagy, S.A. Relationship between Fulton's condition factor and proximate body composition in three freshwater fish species. *J. Appl. Ichthyol.* **2015**, *31*, 315–320. [CrossRef]
9. Hile, R. Age and growth of Cisco, Leuhthys artedi (Le Suer) in the lake of north-eastern high lands. *Bull. U.S. Bur. Fish.* **1936**, *48*, 211–317.
10. Martin, W.R. *The Mechanics of Environmental Control of Body Form in Fishes*; University of Toronto Biological Studies; University of Toronto Press: Toronto, ON, Canada, 1949; Volume 56.
11. Pauly, D.; Gayanilo, F.C., Jr. *A Bee: An Alternative Approach to Estimating the Parameters of a Length-Weight Relationship from Length Frequency Samples and Their Bulk Weights*; NAGA ICLARM: Manila, Philippines, 1997.
12. Froese, R. Length-weight relationships for 18 less-studied fish species. *J. Appl. Ichthyol.* **1998**, *14*, 117–118. [CrossRef]
13. Myers, N. Biodiversity hotspots revisited. *BioScience* **2003**, *53*, 916–917. [CrossRef]
14. Conservation International. Biodiversity Hotspots. 2018. Available online: http://www.biodiversityhotspots.org (accessed on 18 September 2023).
15. Mittermeier, R.A.; Robles Gil, P.; Hoffmann, M.; Pilgrim, J.; Brooks, T.; Mittermeier, C.G.; Lamoreaux, J.; Da Fonseca, G.A.B. *Hotspots Revisited*; CEMEX, S.A. de C. V.: Mexico City, Mexico, 2004; 391p.
16. Munro, J.L. *Caribbean Coral Reef Fishery Resources*; Monographs; The WorldFish Center: Penang, Malaysia, 1983; p. 12310.

17. Bohnsack, J.A.; Harper, D.E. *Length-Weight Relationships of Selected Marine Reef Fishes from the Southeastern United States and the Caribbean*; NOAA Technical Memorandum NMFS-SEFC; U. S. Department of Commerce: Miami, FL, USA, 1988; Volume 215, pp. 1–31.
18. García, C.B.; Duarte, L.O. Length-based estimates of growth parameters and mortality rates of fish populations of the Caribbean Sea. *J. Appl. Ichthyol.* **2006**, *22*, 193–200. [CrossRef]
19. Jahson, B.; Alemu, I. The status and management of the lionfish, *Pterois* sp. in Trinidad and Tobago. *Mar. Pollut. Bull.* **2016**, *109*, 402–408. [CrossRef]
20. De La Hoz-M, J.; Motta, J.; Paramo, J. Length-weight relationships for 36 deep-sea fish in the Colombian Caribbean Sea. *J. Appl. Ichthyol.* **2016**, *32*, 1356–1359. [CrossRef]
21. Correa-Herrera, T.; Jiménez-Segura, L.F.; Barletta, M. Fish species from a micro-tidal delta in the Caribbean Sea. *J. Fish Biol.* **2016**, *89*, 863–875. [CrossRef] [PubMed]
22. Rodríguez-Olarte, D.; Taphorn, D.; Agudelo-Zamora, H. Length-weight relationships of fishes from western Caribbean freshwater drainages of Venezuela. *J. Appl. Ichthyol.* **2019**, *35*, 600–604. [CrossRef]
23. Ricker, W.E. Computation and interpretation of the biological statistics of fish populations. *Bull. Fish. Res. Board Can.* **1975**, *191*, –382.
24. Fox, J.; Weisberg, S. *An {R} Companion to Applied Regression*, 2nd ed.; Sage: Thousand Oaks, CA, USA, 2011.
25. R Core Team. *R: A Language and Environment for Statistical Computing*; R Foundation for Statistical Computing: Vienna, Austria, 2022.
26. Mahé, K.; Bellamy, E.; Delpech, J.P.; Lazard, C.; Salaun, M.; Verin, Y.; Coppin, F.; Travers-Trolet, M. Evidence of a relationship between weight and total length of marine fish in the Northeastern Atlantic Ocean: Physiological, spatial and temporal variations. *J. Mar. Biol. Assoc. U. K.* **2018**, *98*, 617–625. [CrossRef]
27. Roos, D.; Taconet, J.; Gentil, C.; Brisset, B.; Evano, H.; Aumond, Y.; Huet, J.; Lepetit, C.; Boymond-Morales, R.; Rungassamy, T.; et al. Variation of the relationships between lengths and weights applied to 123 fish species observed at Réunion Island (Indian Ocean). *Afr. J. Mar. Sci.* **2022**, *44*, 171–180. [CrossRef]
28. Bouchon-Navaro, Y.; Bouchon, C.; Louis, M.; Legendre, P. Biogeographic patterns of coastal fish assemblages in the West Indies. *J. Exp. Mar. Biol. Ecol.* **2005**, *315*, 31–47. [CrossRef]
29. Ruttenberg, B.; Caselle, J.E.; Estep, A.J.; Johnson, A.E.; Marhaver, K.L.; Richter, L.J.; Sandin, S.T.; Vermeij, M.J.A.; Smith, J.E.; Grenda, D.; et al. Ecological assessment of the marine ecosystems of Barbuda, West Indies: Using rapid scientific assessment to inform ocean zoning and fisheries management. *PLoS ONE* **2018**, *13*, e0189355. [CrossRef]
30. Duarte, L.O.; Garcia, C.B.; Sandoval, N.; von Schiller, D.; Melo, G.; Navajas, P. Length-weight relationships of fishes from the Gulf of Salamanca, Colombia. *Naga ICLARM Q.* **1999**, *22*, 34–36.
31. Silva, J.F.; Ellis, J.R.; Ayers, R.A. *Length-Weight Relationships of Marine Fish Collected from around the British Isles*; Science Series Technical report; Cefas Publications: Lowestoft, UK, 2013; Volume 150.
32. Wilhelms, I. *Atlas of Length-Weight Relationships of 93 Fish and Crustacean Species from the North Sea and the North-East Atlantic*; Thünen Working Paper; Johann Heinrich von Thünen Institute, Federal Research Institute for Rural Areas, Forestry and Fisheries: Braunschweig, Germany, 2013; Volume 12.
33. Petrakis, G.; Stergiou, K.I. Weight-length relationship for 33 fish species in Greek waters. *Fish. Res.* **1995**, *21*, 465–469. [CrossRef]
34. Naderi, M.; Zare, P.; Azvar, E. Length-weight relationships for five stingray species from the Persian Gulf. *J. Appl. Ichthyol.* **2013**, *29*, 1177–1178. [CrossRef]
35. Moutopoulos, D.K.; Stergiou, K.I. Length-weight and length-length relationships of fish species from the Aegean Sea (Greece). *J. Appl. Ichthyol.* **2002**, *18*, 200–203. [CrossRef]
36. Echeverria, T.; Lenarz, W.H. Conversions between total, fork, and standard lengths in 35 species of Sebastes from California. *Fish. Bull.* **1984**, *82*, 249–251.
37. Gaygusuz, O.; Gürsoy, C.; Özulu, M.; Serhan Tarkan, A.; Acpnar, H.; Bilge, G.; Filiz, H. Conversions of total, fork and standard length measurements based on 42 marine and freshwater fish species (from Turkish Waters). *Turk. J. Fish. Aquat. Sci.* **2006**, *6*, 79–84.
38. Froese, R.; Tsikliras, A.C.; Stergiou, K.I. Editorial note on weight-length relations of fishes. *Acta Ichthyol. Piscat.* **2011**, *41*, 261–263. [CrossRef]
39. Ragheb, E. Length-weight relationship and well-being factors of 33 fish species caught by gillnets from the Egyptian Mediterranean waters off Alexandria. *Egypt. J. Aquat. Res.* **2023**, *49*, 361–367. [CrossRef]
40. Anderson, R.O.; Newmann, R.M. Length weight and associated structural indices. In *Fisheries Techniques*; Murphy, B.R., Willis, D.W., Eds.; American Fisheries Society: Bethesda, MD, USA, 1996.
41. Muchlisin, Z.A.; Fransiska, V.; Muhammadar, A.A.; Fauzi, M.; Batubara, A.S. Length-weight relationships and condition factors of the three dominant species of marine fishes caught by traditional beach trawl in Ulelhee Bay, Banda Aceh City, Indonesia. *Croat. J. Fish.* **2017**, *75*, 104–112. [CrossRef]
42. Muchlisin, Z.A.; Musman, M.; Siti-Azizah, M.N. Length-weight relationships and condition factors of two threatened fishes, *Rasbora tawarensis* and *Poropuntius tawarensis*, endemic to Lake Laut Tawar, Aceh Province, Indonesia. *J. Appl. Ichthyol.* **2010**, *26*, 949–953. [CrossRef]

43. Jisr, N.; Younes, G.; Sukhn, C.; Mohammad, H.; El-Dakdouki, M.H. Length-weight relationships and relative condition factor of fish inhabiting the marine area of the Eastern Mediterranean city, Tripoli-Lebanon. *Egypt. J. Aquat. Res.* **2018**, *44*, 299–305. [CrossRef]
44. Morato, T.; Afonso, P.; Lourinho, P.; Barreiros, J.P.; Santos, R.S.; Nash, R.D.M. Length-weight relationships for 21 coastal fish species of the Azores, northeastern Atlantic. *Fish. Res.* **2001**, *50*, 297–302. [CrossRef]
45. Bagenal, T.B.; Tesch, F.W. Age and growth. In *Methods for Assessment of Fish Production in Fresh Waters, Volume 3*; Begenal, T., Ed.; IBP Handbook; Blackwell Science Publications: Oxford, UK, 1978; pp. 101–136.
46. Wootton, R.J. *Ecology of Teleost Fish*; Chapman and Hall: London, UK, 1990.

animals

MDPI

Article

The Effects of Acute Bisphenol A Toxicity on the Hematological Parameters, Hematopoiesis, and Kidney Histology of Zebrafish (*Danio rerio*)

Svetlana Smorodinskaya [1,2], Nikita Kochetkov [1,2,*], Kirill Gavrilin [2], Dmitry Nikiforov-Nikishin [1,2], Diana Reznikova [1,3], Aleksey Vatlin [1,4], Anastasia Klimuk [1,2], Maya Odorskaya [1], Alexei Nikiforov-Nikishin [2], Andrey Ponomarev [2], Maria Marsova [1,*] and Valery Danilenko [1]

1 Laboratory of Bacterial Genetics, Vavilov Institute of General Genetics Russian Academy of Sciences, 119333 Moscow, Russia; kler.smo@gmail.com (S.S.); niknikdl@rambler.ru (D.N.-N.); reznikova.da@phystech.edu (D.R.); vatlin_alexey123@mail.ru (A.V.); klimukanastasia27@gmail.com (A.K.); maya_epifanova@mail.ru (M.O.); valerid@vigg.ru (V.D.)

2 Faculty of Biotechnology and Fisheries, Moscow State University of Technologies and Management (FCU), 73, Zemlyanoy Val Str., 109004 Moscow, Russia; k.gavrilin@yandex.ru (K.G.); 9150699@mail.ru (A.N.-N.); ponomarev777@inbox.ru (A.P.)

3 Phystech School of Biological and Medical Physics, Moscow Institute of Physics and Technology, Institutsky Lane 9, 141700 Dolgoprudny, Russia

4 Institute of Ecology, Peoples' Friendship University of Russia (RUDN University), 117198 Moscow, Russia

* Correspondence: samatrixs@gmail.com (N.K.); marsova@vigg.ru (M.M.)

Simple Summary: Bisphenol A (BPA) is an organic compound used in the manufacture of plastic products and is considered a common pollutant in the aquatic environment. Its impact on aquatic organisms is manifested in several reactions, particularly in changes in peripheral blood and disorders of hematopoiesis. In the present study, the effects of high concentrations of BPA on several indices of zebrafish blood were investigated. Differences in the effects of high and low concentrations on blood composition and disorders in hematopoietic organs were identified. The genotoxic effect of BPA at concentrations of 6 and 8 mg/L was confirmed. The test parameters most affected by the toxicant, such as the number of young erythrocytes in peripheral blood, nuclear anomalies of erythrocytes, and disturbances in hematopoiesis processes in the head kidney, expressed as changes in the ratio of immature/undifferentiated cell elements, were determined.

Abstract: In this study, the results of evaluating the acute toxicity of Bisphenol A on *Danio rerio* are presented, encompassing peripheral blood parameters, the composition of hematopoietic cells of erythroid and myeloid lines in the head kidney, and data from histological studies. The LC50 values of Bisphenol A for adult zebrafish individuals for 12, 24, and 48–96 h were determined, which were 18.04, 7.55, and 6.22 mg/L, respectively. The study includes data on the morphology and quantitative frequency of specific cells in the hematopoietic tissue of the head kidney, along with the consideration of adaptive mechanisms in hematopoiesis under BPA exposure. The application of polynomial regression analysis to reveal the concentration–effect relationship for some hematological and histological parameters was demonstrated. Significant increases in the frequency of erythrocyte nuclear abnormalities were observed at BPA concentrations of 6 and 8 mg/L, which indicates a genotoxic effect. BPA's impact on fish peripheral blood parameters manifested as an increase in the number of erythrocytes (RBC) and immature erythrocytes, as well as a decrease in the number of lymphocytes. The most notable pathological changes in the head kidney's hematopoietic tissue included circulatory disturbances and the formation of inflammation/degradation foci, as confirmed by histopathologic indices. At BPA concentrations of 2 and 4 mg/L, the observed changes were compensated for by hematopoietic adaptation mechanisms; however, at concentrations of 6 and 8 mg/L, acute systemic toxicity was evident.

Keywords: LC50; hematopoiesis; head kidney; histology; genotoxicity; Bisphenol A; *Danio rerio*

1. Introduction

Bisphenol A (BPA) is an important component used in the production of technological plastic products, such as pipes, epoxies, and coatings, and in various food packaging [1,2]. The global production level of BPA has been steadily increasing, rising between 1% and 5% per year, indicating a steady demand [3]. The particular interest in BPA is due to its potential harmful effects on humans [4] as well as on aquatic ecosystems. It can enter water bodies through wastewater treatment plant effluents, industrial effluents from petrochemical plants, and landfill leachates [5]. Some species of aquatic organisms, living near plant effluents or river and marine sediments, may be exposed to high concentrations of BPA on a continuous basis [6], thus making it necessary to conduct in-depth studies on its effects.

Numerous studies have that BPA has a negative impact on the environment and aquatic organisms. The toxic properties of BPA are well enough studied, both on invertebrates and fish [6]. Many researchers have noted the genotoxic properties of this compound [7,8], as well as its negative effects on endocrine functions [9], the nervous system, and the digestive and excretory organs in fish [10]. The most significant effect that BPA has may be on the reproductive performance of organisms due to its estrogenic activity [11]. BPA is able to remain active in the aquatic environment for extended periods of time, accumulating in water bodies and aquatic organisms [12]. As a consequence, its acute and chronic toxicity to aquatic species can vary considerably, as acute toxicity is most often manifested through short-term exposure to the substance and is expressed in serious disorders that can lead to death of the organism. During the determination of the acute toxic effect of a pollutant on organ and tissue systems, it is advisable to conduct the study using high concentrations (i.e., lethal and sublethal).

Hematological and biochemical indicators are often used as prognostic and diagnostic methods to assess fish health because of their versatility and high sensitivity [13]. It is worth noting that alterations in the morphological pattern of blood can manifest quite rapidly after the ingestion of a toxicant by the fish organism. Many toxicants induce toxic anemia in fish, the causes of which can be both the disruption of the gill apparatus and changes in the processes of erythropoiesis and deposition of cellular elements [14]. If the toxic effects do not lead to mortality, the blood composition of the fish undergoes significant changes as a result of the activation of the organism's compensatory reactions. Adaptive responses in the qualitative and quantitative composition of peripheral blood have been studied for many fish species [15,16]. However, there are some discrepancies in the methodology used to account for these changes [17,18].

Hematopoiesis is a complex process in which hematopoietic stem cells, which are the most immature elements of the hematopoietic lineage, proliferate and differentiate into different groups of hematopoietic progenitor cells [19]. The action of the toxicant on the differentiation processes in fish blood cells will be expressed in the disruption of cell kinetics (in the ratio of cells in the hematopoietic tissue, presence in the bloodstream, and deposition locations). For this reason, quantitative indicators of hematopoiesis can be used as a marker of toxic effects.

Laboratory fish species can be used to determine the mechanisms of action of toxicants, as well as target organs. *Danio rerio* is a common test subject for studying the toxicity of various compounds [20] and is also actively used by researchers in studying the processes of hemocytopoiesis [21]. The most active hematopoietic processes in zebrafish occur in the head section of the kidney, although immature cells of the hematopoietic series can be found in the spleen and some other organs (i.e., mucosa-associated lymphoid tissues) [22]. During the phases of cell division and differentiation, cells can be exposed to cytotoxic effects; the visible manifestations of which are the formation of various nuclear abnormalities found in mature cells [23]. The micronucleus test, performed on the peripheral blood erythrocytes of fish, assesses the frequency of occurrence of such nuclear abnormalities in red blood cells and allows the estimation of the genotoxic effect of the examined substance [24]. The micronucleus test is widely used to assess clastogenic and aneugenic effects both in vitro and in ecotoxicological studies [25].

Thus, the aim of this study was to determine the effects of different lethal and sublethal concentrations of Bisphenol A on fish survival in an acute experiment, hematological parameters, and hematopoiesis processes, as well as histological changes in the head kidney of *Danio rerio*.

2. Materials and Methods

2.1. Animal Maintenance

Danio rerio (wild type), 4–5 months old, were kept in 300 L aquariums with 200 fish in each aquarium, equipped with mechanical and biological filtration systems, with 10% water replacement per day. Fish maintenance and care were carried out according to the recommendations of Westerfield [26]. Prior to the experiment, fish were fed with Tetra Min Flakes XL (Spectrum Brands, Melle, Germany).

For the study, individuals of both sexes without visible injuries were selected, with an average size of 2.12 $\pm$ 0.3 cm, and a weight of 0.29 $\pm$ 0.03 g. During the acute experiment, fish feeding was not performed.

2.2. Acute Toxicity Bioassay

To determine lethal concentrations of Bisphenol A (BPA), LC50 (Lethal concentration 50) values for 12–96 h were used. The studies were performed according to OECD Test no. 203 with some modifications. Bisphenol A (2,2-bis(4-hydroxyphenyl)propane) with 99% purity was purchased from Sigma-Aldrich (St. Louis, MO, USA). The concentration range for the acute experiment was selected based on data from the literature on other fish species and included: 0.5, 1.0, 2.0, 4.0, 6.0, 8.0, 12.0, 16.0, and 20.0 mg/L [27,28]. Experiments were carried out in triplicate in 50 L aquaria with 20 zebrafish individuals placed each (n = 20 $\times$ 3). Temperature (24 °C), light regime (12:12 h), and hydrochemical parameters (pH 7.3 $\pm$ 0.3; O_2 7.9 $\pm$ 0.2; NH_4 < 0.05; NO_2 0.15 $\pm$ 0.02; NO_3 3.1 $\pm$ 0.9) in the experimental tanks at the beginning of the acute experiment were the same as the maintenance aquariums. Changes in behavior, clinical signs, and mortality were monitored throughout the experiment.

Since BPA is poorly soluble in water, dimethyl sulfoxide (DMSO, Tula Pharmaceutical Factory LLC, Tula, Russia) was used as the standard solvent (OECD Test no. 203, [29]) at a concentration of 1% (v/v) to prepare a stock solution, following the approach of other authors [30]. In the control group (CTR), fish were exposed to DMSO alone.

2.3. Preparation of Peripheral Blood and Kidney Smears

After the end of the acute experiments, four zebrafish individuals (n = 4) were sampled from the control and experimental groups with Bisphenol A concentrations (2 mg/L or 0.32 LC50; 4 mg/L or 0.64 LC50; 6 mg/L or 0.96 LC50; and 8 mg/L or 1.28 LC50) to prepare blood smears and the hematopoietic tissue of the head kidney. Before sampling biological material, fish were anesthetized in an MS-222 solution (20 mg/L, LGC Mikromol, Luckenwalde, Germany) for 3 min. Blood was collected for the preparation of peripheral blood smears from the caudal vein according to the technique previously described in [31]. Immediately after blood collection, the number of erythrocytes and leukocytes was counted using a hemocytometer [32].

After preparing the blood smears, the dissection of experimental fish was performed following the methodology outlined by Ivanovski et al. [33]. Therefore, upon opening the abdominal cavity of the zebrafish, the organs of the gastrointestinal tract were removed to provide access to the head section of the mesonephros. Kidney tissue samples were extracted using a flat-tipped curette, after which the tissue was spread on a slide and allowed to dry for 12 h.

The prepared smears were stained according to the standard technique [34]. Briefly, after drying for 4 min, smears were fixed in Nikiforov's mixture (*v*/*v*, methyl alcohol/diethyl ether) and exposed for 2 min, followed by eosin staining with methylene blue for 4 min (Mein-Grunwald Giemsa Staining). Then, the preparations were washed and dried, followed by additional staining with Romanovsky–Giemsa azure–eosin for 10 min. The obtained smears were washed again in containers with distilled water and dried for 30 min.

Smears of blood and the head kidney were viewed under an Olympus BX53 light microscope (Olympus Corporation, Tokyo, Japan) with Carl Zeiss ERc 5s (Zeiss, Oberkochen, Germany) and ToupCam 16.0 MP (ToupTek Photonics, Hangzhou, China) ocular attachments using ZEN lite (Zeiss, Germany) and ToupCam view 16.0 (ToupTek Photonics, China) software.

2.4. Assessment of Hemopoiesis Activity and Composition of Peripheral Blood

The following cellular elements were counted in peripheral blood smears: mature and young (juvenile) forms of erythrocytes, lymphocytes, granulocytes, monocytes, and thrombocytes. At least 40 fields of view were examined for each blood sample. The total number of cells counted for one preparation was at least 2500. Peripheral blood cellular elements were counted and determined according to generally accepted methods [13].

Erythropoietic (basophilic erythroblast, polychromatic erythroblast, orthochromatic erythroblast, and erythrocyte) and myelopoietic (promyelocyte, myelocyte, metamyelocyte, bacillary granulocyte, segmented granulocyte) cellular elements, as well as blast cell forms (i.e., erythroblasts, myeloblasts/granuloblasts), were counted on the obtained head kidney smears according to the characteristics described by Ivanova [35], Fijan [36], and Kondera [37]. Cells with features of several undefinable elements were referred to as unclassified cells. Basophilic, eosinophilic, and neutrophilic forms were counted together when counting differentiated forms of granulocytes; the distinction was made based on the morphology of the nucleus (band and segmented form) in accordance with the work of Wlasow and Dabrowska [38]. At least 20 fields of view were examined per smear, and at least 500 cells were counted. Additionally, morphometric parameters (cell, nuclei length/width; Table S1) were assessed using the ImageJ program (Wayne Rasband, NIH; [39]) to determine the different morphological stages of hematopoietic stem cells and hematopoietic progenitor cells.

2.5. Micronucleus Test

Erythrocytes with pronounced nuclear abnormalities were counted in prepared fish peripheral blood smears according to the methods [24,40]. In addition to micronuclei (MN), the following nuclear abnormalities were considered: notched nuclei (NN), lobed nuclei (LN), and blebbed nuclei (BN). The detection of several types of nuclear abnormalities allows an increase in the reliability of the genotoxicity assessment [31]. To accurately identify micronuclei and other nuclear anomalies, a number of distinguishing characteristics were used: identical morphology of the nucleus and micronuclei, a small distance from the nucleus, size from 1/16 to 1/3 of the nucleus, and minimal difference in color from the nucleus [34].

2.6. Histological Preparations, Histomorphometric Analysis and Semi-Quantitative Assessment

To obtain histologic sections, 4 individuals were selected from each experimental group. Three sections were made from each selected fish ($n = 4 \times 3$). The fish were anesthetized in MS-222 solution (10 mg/L) and then fixed in Davidson's solution for 24 h. Tissue samples were subsequently dehydrated in a series of graded alcohols and embedded in paraffin. Next, total serial sections (4 μm) were prepared in the frontal plane and stained with hematoxylin and eosin (H&E) and periodic acid-Schiff (PAS). The histological sections were prepared and stained following the method described by Suvarna [41].

Histologic sections were examined using the equipment described in Section 2.3. The ImageJ program was used to measure morphometric parameters on histological sections of the kidney. The measurements were performed according to the method described in the previous paper [42].

The histopathologic index (HI) was calculated according to Bernet [43]. Briefly, for each of the 4 types of reaction patterns (circulatory changes, regressive and progressive changes, inflammatory response), a number of histopathologic abnormalities were defined, each corresponding to its own importance factor (ranging from 1 to 3) reflecting the effect of histopathologic changes on organ functions. The value of each histopathologic abnormality was set between 0 and 5, where 0 indicates the absence of changes in the entire slide area and 5 represents the presence of changes in 80–100% of the slide [44]. The index of each reaction pattern was then calculated by multiplying the histopathologic abnormality score by its importance factor and summing them for each reaction pattern.

2.7. Statistical Analysis

Comparison data of the analyzed variables are presented as mean ± SD. Statistical significance was determined using nonparametric tests (Kruskal–Wallis test, Mann–Whitney U-test), based on the distribution of the data and the homogeneity of variances (assessed with the Shapiro–Wilk and Levene tests). A p-value < 0.05 was considered statistically significant. The correlation between the studied parameters and toxicant concentration was determined using Pearson correlation. In the case of obtaining a heterogeneous structure of parameter dependence on toxicant concentration and a low regression coefficient, polynomial regression was used.

The median lethal concentration was calculated graphically using probit analysis [45]. The 95% confidence intervals (CI) for the obtained LC50 values were also determined.

Statistical analyses were conducted using GraphPad Prism version 9.0 software (GraphPad, San Diego, CA, USA) and R software (v3.5.2)/RStudio [46,47].

3. Results

3.1. Determination of LC50

We conducted the acute experiment in a wide range of concentrations of Bisphenol A, allowing us to establish the calculated LC50 values (Table 1). At the maximum studied concentrations (16 and 20 mg/L), mortality began to occur at 12 h of the experiment. The acute toxic properties of the experimental toxicant were manifested on the first day at concentrations above 8 mg/L. At concentrations of 6 and 8 mg/L, fish death was observed at 48 h, while no further deaths were found during the remaining period of the experiment.

Table 1. Mean mortality of *Danio rerio* (in %) in the acute experiment during exposure in BPA solutions and LC50 values.

Concentration, mg/L	Survival Rate, %				LC50 Value (CI), mg/L
	12 h	24 h	48 h	96 h	
Control	100	100	100	100	LC_{50}^{12} = 18.04 (17.81–18.28); LC_{50}^{24} = 7.55 (7.38–7.72); $LC_{50}^{48/96}$ = 6.22 (6.11–6.33)
0.5	100	100	100	100	
1	100	100	100	100	
2	100	100	100	100	
4	100	100	96.6	96.6	
6	100	83.3	56.6	56.6	
8	100	40	13.3	13.3	
12	100	0	0	0	
16	76.6	0	0	0	
20	26.6	0	0	0	

BPA concentrations below 6 mg/L did not result in significant fish deaths for the entire duration of the experiment. The obtained LC50 values at 24 and 48/96 h are shown in Figure 1, and the numerical data and conversion to mg/L are summarized in Table 1. The LC50 value at 48 and 96 h was 6.22 mg/L (CI 6.11–6.33).

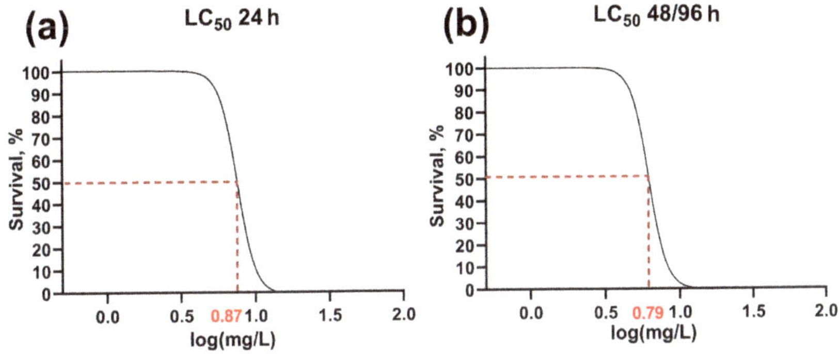

Figure 1. Determination of LC50 values of Bisphenol A for *Danio rerio* adults for 24 (**a**) and 48/96 h (**b**).

3.2. Hematopoiesis in the Head Kidney

The effect of Bisphenol A exposure on the processes of hematopoiesis was studied after a 96 h exposure at concentrations of 2, 4, 6, and 8 mg/L. The general appearance of *D. rerio* head kidney smears and the cells present at different stages of cytodifferentiation are shown in Figure 2b. The occurrence of erythroid elements in the control group was 50.2%, relative to the other cells studied. Basophilic and orthochromatic erythroblasts (14.70 and 14.58%) had the highest occurrence among the blast forms of erythrocytes. The occurrence of erythrocytes in the head kidney smears of the control fish was 7.34%. The total number of myeloid elements was 44.47%. Metamyelocytes were most frequently detected in the smears, the occurrence of which was 14.4%. The list of abbreviations used for hematopoietic cells, hematological, and histological parameters is given in Table S6.

Blast cells (erythroblasts, granulo/myeloblasts) had a large nucleus occupying most of the cell, which could be shifted to the edge, and the cytoplasm was brightly basophilic staining (Figure 2c). Some cells had an oval/kidney-shaped nucleus and small vacuoles in the cytoplasm.

Morphological changes in the erythroid cell during development included: a decrease in the degree of basophilia of the cytoplasm; a decrease in the size of the cell and nucleus (as a result of chromatin condensation) and the nucleus/cytoplasm area ratio; a change in the cell shape toward ellipsoidal. The basophilic erythroblast had a smaller size than the blast cells (Table S1), a larger cytoplasm volume, and also showed bright basophilic staining (Figure 2d). The cell nucleus was round shaped, with granular chromatin. The polychromatic erythroblast had a lighter cytoplasm (light blue/gray), with a rounded or oval nucleus and individual clumps of chromatin (Figure 2e). This erythrocyte precursor also showed a decrease in cell area compared to the previous stages. The last erythrocyte precursors distinguishable on kidney smears, the orthochromatic erythroblast, had an oval nucleus with compacted chromatin, and the cytoplasm was pink or gray in color (Figure 2f). Compared to the mature erythrocyte, the nucleus was less condensed.

The promyelocytes had slightly basophilic granular cytoplasm, and an oval-shaped heterochromatin nucleus shifted to the periphery, occupying about half of the cell (Figure 2h). The myelocytes had a larger cytoplasm size and, consequently, a larger nucleus/cytoplasm area ratio (Figure 2i). The oval or deformed nucleus also had predominantly heterochromatin content with small clumps of euchromatin. Metamyelocytes were characterized by a more elongated nucleus that could occupy up to half of the cell (Figure 2j). The cytoplasm had either basophilic staining with granular inclusions or weakly acidophilic staining with less granulation. Mature granulocytes/myelocytes had a deformed nucleus of two types:

band form (Figure 2k) and segmented form (Figure 2l). The ratio of cytoplasm and nucleus area, as well as cytoplasm staining, could vary depending on the cell type.

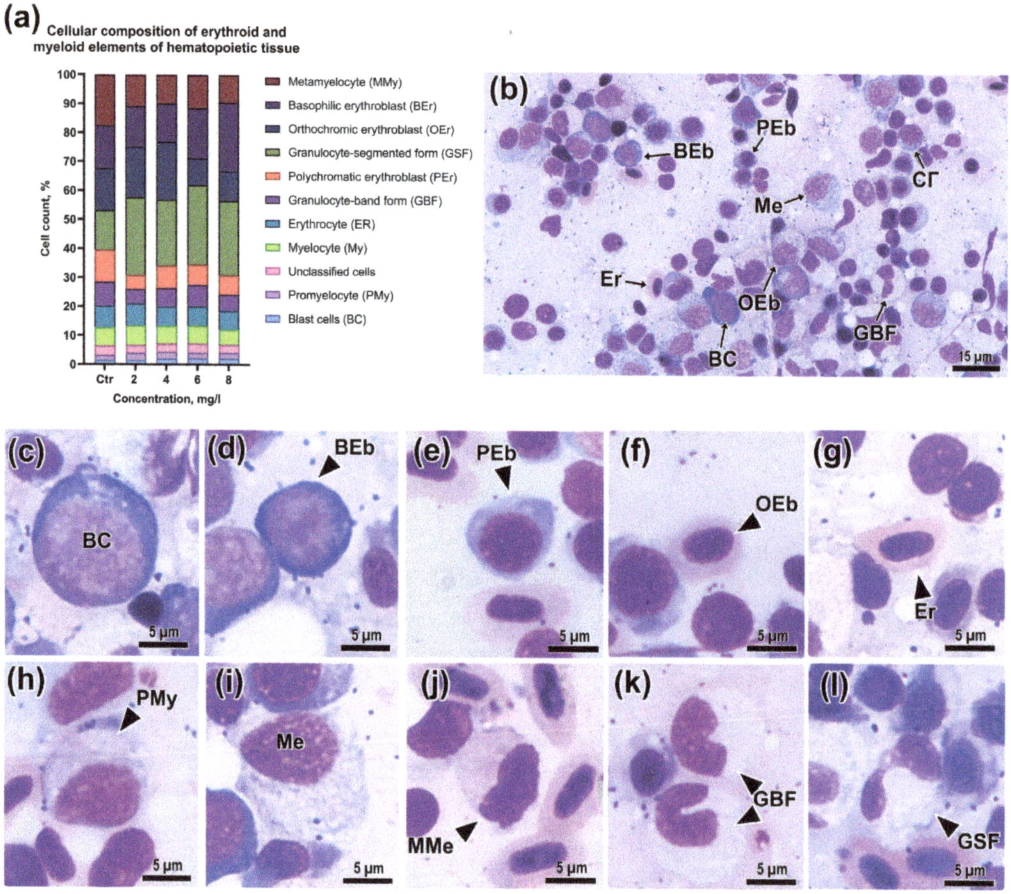

Figure 2. (**a**) Ratio of cellular elements of erythroid and myeloid lineages of hematopoietic tissue of *Danio rerio* head kidney: (**b**) microphotographs of individual stages of hematopoietic cells (**b–l**). BC—blast cell, BEb—basophilic erythroblast; PEb—polychromatic erythroblast; OEb—orthochromatic erythroblast; Er—erythrocyte; PMe—promyelocyte; Me—myelocyte; MMe—metamyelocyte; GBF—band form granulocyte; GSF—segmented form granulocyte. May–Grunwald–Giemsa staining. Scale bar: 15 µm (**b**) and 5 µm (**c–l**).

At high concentrations of the toxicant (6 and 8 mg/L), significant changes in the ratio of erythropoietic and myelopoietic cells were observed (Table 2; Figure S1). The maximum concentration of the toxicant led to a substantial increase in the number of basophilic erythroblasts ($p < 0.05$). At the same time, there was a significant decrease in the occurrence of orthochromatic erythroblasts relative to the concentrations of 2 and 4 mg/L ($p < 0.05$). All studied BPA concentrations above 4 mg/L resulted in a considerable increase in the occurrence of polychromatic erythroblasts ($p < 0.05$). The number of mature erythrocytes in the hematopoietic tissue of the head kidney of experimental *D. rerio* individuals did not change significantly in any of the experimental groups.

Table 2. Relative abundance of cellular elements in smears of hematopoietic tissue of *Danio rerio* head kidney during acute exposure to Bisphenol A. Significance ($p < 0.05$) from the Kruskal–Wallis test. Superscript letters (a–c) indicate statistical significance between different experimental groups.

Hematopoetic Cell, %	BPA Concentration, mg/L				
	Ctr	2	4	6	8
Blast cells (BC)	1.55 ± 0.15^{b}	1.57 ± 0.26^{ab}	2.06 ± 0.86^{ab}	2.2 ± 0.39^{a}	2.13 ± 0.32^{a}
Basophilic erythroblast (BEb)	14.7 ± 1.66^{b}	14.02 ± 1.7^{b}	13.19 ± 1.8^{b}	17.29 ± 1.7^{ab}	23.7 ± 1.79^{a}
Polychromatic erythroblast (PEb)	13.58 ± 1.7^{b}	26.9 ± 2.88^{a}	22.7 ± 1.07^{ab}	27.35 ± 0.92^{a}	25.65 ± 3.8^{a}
Orthochromatic erythroblast (OEb)	14.58 ± 3.1^{ab}	17.49 ± 2.5^{a}	20.04 ± 1.09^{a}	9.22 ± 1.01^{b}	10.19 ± 2.2^{b}
Erythrocyte (ER)	7.34 ± 1.45	7.56 ± 0.55	6.66 ± 0.81	6.61 ± 0.91	6.45 ± 1.13
Promyelocyte (PMy)	1.74 ± 0.39	2.46 ± 0.96	2.43 ± 0.71	1.99 ± 0.08	2.02 ± 0.45
Myelocyte (My)	6.37 ± 1.11	6.91 ± 1.63	6.31 ± 0.69	6.38 ± 1.04	5.22 ± 0.68
Metamyelocyte (MMy)	17.4 ± 1.78^{a}	10.76 ± 2.2^{b}	9.83 ± 0.58^{b}	11.5 ± 1.28^{ab}	9.49 ± 2.4^{b}
Granulocyte-band form (GBF)	8.36 ± 1.48^{a}	4.97 ± 0.9^{c}	6.2 ± 0.42^{abc}	7.38 ± 1.26^{ab}	5.66 ± 0.63^{c}
Granulocyte-segmented form (GSF)	11 ± 1.38^{a}	4.78 ± 0.97^{c}	7.83 ± 0.65^{ab}	7.11 ± 0.71^{bc}	6.71 ± 0.94^{bc}
Unclassified cells	3.14 ± 0.18	2.5 ± 0.5	2.54 ± 1.17	2.87 ± 0.79	2.68 ± 0.21

BPA had a meaningful effect on the occurrence of myeloid cells, leading to a decrease in their number compared to the control. A significant effect was detected on the number of cells such as metamyelocytes and segmented and band form granulocytes (Table 2). Thus, a substantial decrease in the relative occurrence of two forms of granulocytes was observed at a concentration of 2 mg/L ($p < 0.05$).

3.3. Peripheral Blood Composition

An assessment of the effects of acute toxic exposure to BPA on peripheral blood showed significant changes in the composition and number of erythrocytes and leukocytes (Table S2). The number of red blood cells (RBC) at concentrations of 6 and 8 mg/L considerably exceeded the control values ($p < 0.05$; Figure 3a). Meanwhile, the occurrence of young red blood cells in peripheral blood was also significantly higher than the control values at the respective pollutant concentrations ($p < 0.05$; Figure 3c). Young erythrocytes in peripheral blood, compared to mature ones, had a rounder nucleus, with large chromatin lobules, and weak basophilic staining. Mature erythrocytes had a more pronounced oval-disc shape with an oval nucleus and eosinophilic cytoplasm. It is worth noting that the number of white blood cells (WBC) recorded at the concentration of 8 mg/L was not significantly different from the control; however, it was substantially lower than at concentrations of 4 and 6 mg/L ($p < 0.05$; Figure 3b). The relative number of lymphocytes decreased with increasing BPA concentration, differing meaningfully from the control only at the maximum concentration of the toxicant ($p < 0.05$; Figure 3e). In turn, the occurrence of granulocytes and monocytes had an opposite trend: their number increased significantly with increasing BPA concentration (Figure 3d).

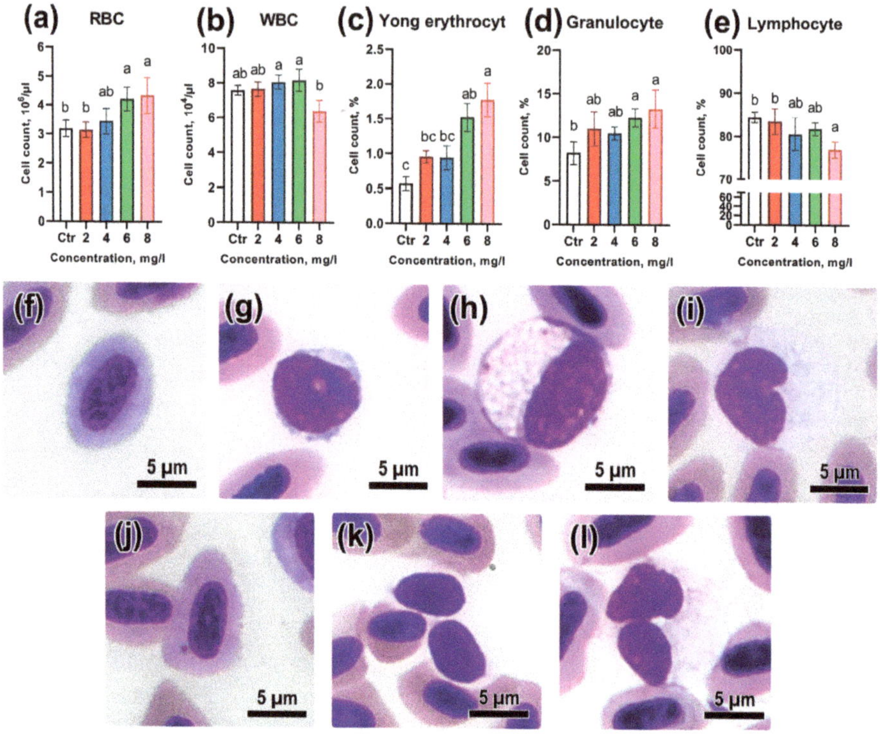

Figure 3. (**a–e**) Cell composition of peripheral blood of *Danio rerio* under acute exposure to Bisphenol A: (**f,j**)—young erythrocytes at different stages of differentiation; (**g**)—lymphocyte; (**h,l**)—granulocytes with different morphology of the nucleus; (**i**)—monocyte; (**k**)—thrombocyte. Significance ($p < 0.05$) from the Kruskal–Wallis test. Superscript letters (a, b) indicate statistical significance between different experimental groups. May–Grunwald–Giemsa staining. Scale bar of 5 µm.

3.4. Assessment of Genotoxicity

Evaluation of the occurrence of nuclear abnormalities in erythrocytes showed that high concentrations of BPA cause a significant increase in the number of erythrocytes with abnormal nucleus morphology (Table S3). The total number of nuclear abnormalities in the control group was at the level of 10.3‰, which is consistent with earlier data on the occurrence of spontaneous nuclear abnormalities in zebrafish [31]. The relative number of micronuclei at concentrations of 6 and 8 mg/L was considerably higher than the control values and values at concentrations of 2 and 4 mg/L ($p < 0.05$; Figure 4e). The occurrence of other nuclear abnormalities in the groups during BPA exposure at concentrations of 6 and 8 mg/L was also significantly higher than in the controls ($p < 0.05$; Figure 4b,c). Total nuclear abnormality (TNA) counts showed that the genotoxic effect of BPA was reliably evident starting at a concentration of 6 mg/L ($p < 0.05$; Figure 4d).

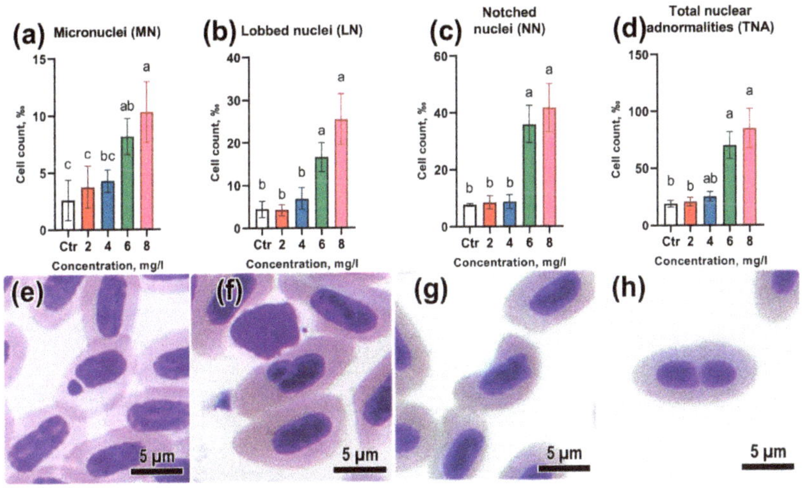

Figure 4. (**a**–**d**) Relative abundance of *Danio rerio* erythrocytes with nuclear abnormalities under acute exposure to Bisphenol A: (**e**) micronuclei; (**f**) blebbed nuclei; (**g**) lobed nuclei; (**h**) notched nuclei. Significance ($p < 0.05$) from the Kruskal–Wallis test. Superscript letters (a–c) indicate statistical significance between different experimental groups. May–Grunwald–Giemsa staining. Scale bar of 5 μm.

3.5. Histologic Parameters of the Kidneys

The study of histological sections of *D. rerio* head kidney after exposure to a BPA solution allowed us to establish a number of pathological changes, the significance of which increased with the concentration of the toxicant. Sections of the head kidney of control fish showed hematopoietic tissue consisting of various hematopoietic progenitor cells with sinusoidal spaces between them (Figure 5a,b). In the caudal direction, sections of the nephron, represented by renal tubules and glomeruli, could be seen.

In the group exposed to a BPA concentration of 2 mg/L, there was an increase in the sinusoid width (Figure 5c), as well as an increase in the occurrence of PAS-positive cells (Figure 5d). Increasing the concentration to 4 mg/L revealed an elevated occurrence of circulatory abnormalities, expressed as the enlargement of sinusoidal spaces and blood congestion (Figure 5e). Additionally, foci of inflammation, associated with the disruption of hematopoietic tissue structure and the occurrence of necrotic cells, were observed in some areas (Figure 5f). Necrotic cells from the hematopoietic tissue had chromophoric cytoplasm and disformed karyorrhexis or pyknotic nuclei. A concentration of 6 mg/L led to a further increase in the number of foci of inflammation and necrosis of the hematopoietic tissue. Areas with blood congestion in sinusoids had a diffuse distribution (Figure 5g,h). The occurrence of PAS-positive cells was substantially lower than at concentrations of 2 and 4 mg/L. Significant architectural and structural tissue alterations were expressed in necrosis of hematopoietic tissue and impaired circulation (blood congestion, sinusoidal dilation) were observed at the maximum studied concentration of BPA.

Studying the kidney tissue of *D. rerio* allowed us to reveal disturbances in the structure of nephrons. On sections from control individuals, it was possible to distinguish nephron sections, namely: proximal, distal, and glomeruli, which were separated by islets of hematopoietic tissue (Figure 6a). In PAS staining, the proximal tubules had a distinct PAS-positive brush border.

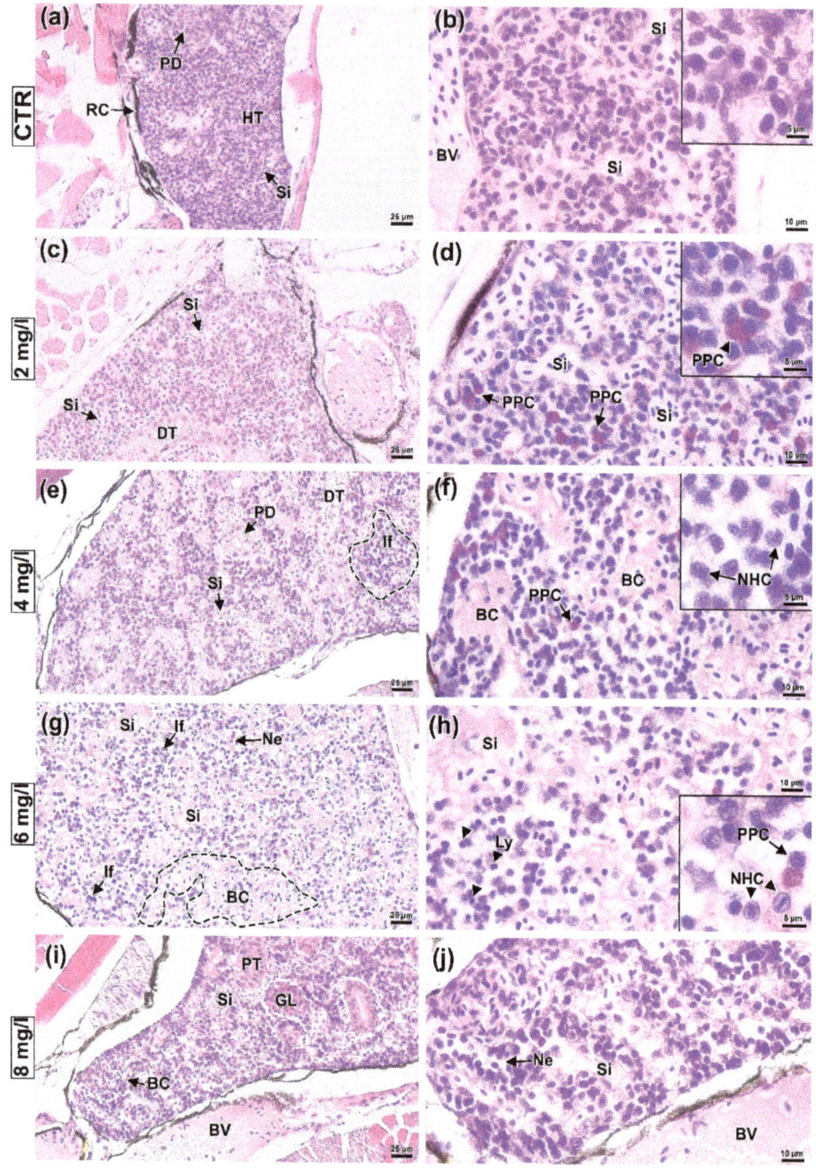

Figure 5. Histologic abnormalities in the hematopoietic tissue of *Danio rerio* head kidney exposed to BPA. CRT (**a**,**b**): normal structure of hematopoietic tissue (HT) surrounded by the renal capsule (RC) and including sinusoidal spaces (Si), blood vessels (BV), and areas of distal (DT), proximal tubules (PT), and glomeruli (GL). In the 2 mg/L group (**c**,**d**): sinusoidal spaces dilated, as well as an increase in the number of PAS-positive cells (PPC) in hematopoietic tissue, were observed. In the 4 mg/L group (**e**,**f**): enlarged sinusoids with evidence of blood congestion (BC), as well as small foci of inflammation (If, dashed line) in hematopoietic tissue, including necrotic cells (NHC). In the 6 mg/L group (**g**,**h**): significant sinusoidal enlargement with blood congestion (dashed line) throughout the head kidney; foci of inflammation and necrosis (Ne) associated with lymphocyte-like cells (Ly). In

the 8 mg/L group (**i**,**j**): architectural and structural tissue alterations expressed as impaired circulation (sinusoidal dilation, blood congestion) and necrosis of hematopoietic tissue. H&E (**a**–**c**,**e**,**i**,**j**) and PAS (**d**,**f**–**h**) staining. Scale bar of 25 μm (**a**) and 10 μm (**b**–**f**).

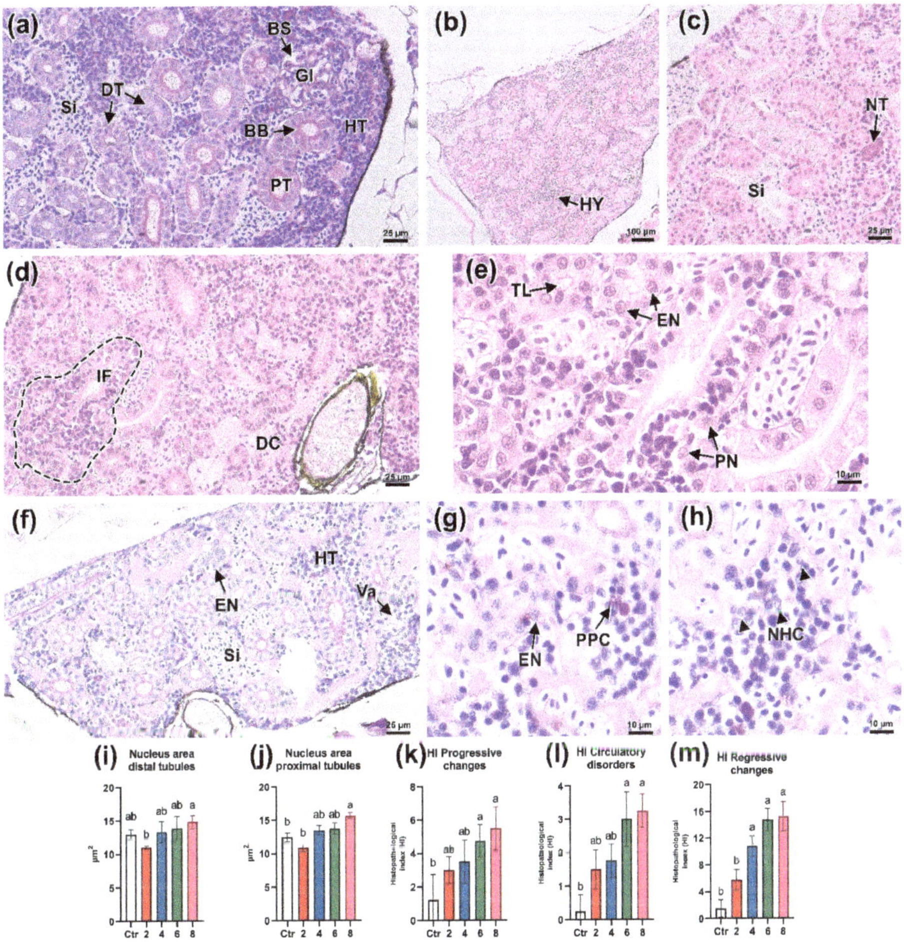

Figure 6. Effect of BPA on nephron structure, hematopoietic tissue, morphometric parameters (**i**,**j**) and histological index (**k**–**m**) of *Danio rerio* trunk kidney. CRT (**a**): normal kidney morphology, distal (DT) and proximal tubules (PT) with a PAS-positive brush border (BB), glomerulus (Gl), and the hematopoietic tissue (HT) and sinusoidal spaces (Si) separating them can be distinguished on the slides. In the 2 mg/L group (**b**,**c**): a significant increase in sinusoid size indicating organ hyperemia (HY); some sections showed an increased occurrence of neonephrogenic tubules (NT). In the 4 mg/L group (**d**,**e**): inflammation foci (If, dashed line) in the area of distal tubule destruction with pycnotic nuclei (PN); marked blood congestion (BC) was seen throughout the slide; individual proximal tubules showed enlarged nuclei (EN) and narrowing of the lumen (TL). In the 6 mg/L group (**f**–**h**): increased nucleus area and vacuolization of renal tubule epithelium (VA); structural abnormalities of hematopoietic tissue as a result of circulation disruption (sinusoidal dilation) as well as necrotic hematopoietic cells (NHC, arrowhead); individual PAS-positive cells (PPC) were found in areas with degradation of hematopoietic tissue. H&E (**b**–**e**) and PAS (**a**,**f**–**h**) staining. Scale bars are 100 μm (**b**), 25 μm (**a**,**c**,**d**,**f**), and 10 μm (**e**,**g**,**h**). Significance ($p < 0.05$) from the Kruskal–Wallis test. Superscript letters (a, b) indicate statistical significance between different experimental groups.

Exposure to a concentration of 2 mg/L manifested in a significant enlargement of the sinusoidal spaces filled with erythrocytes, observed throughout the entire slide area, which may indicate hyperemia (Figure 6b). Additionally, an increase in the occurrence of neonephrogenic tubules was detected, the epitheliums of which had long oval nuclei showing few chromatins and a bright basophilic staining of the cytoplasm (Figure 6c). At this concentration, a slight decrease in the area of renal tubule epithelial nuclei was also observed (Figure 6i,j). At a concentration of 4 mg/L, in addition to the above changes, the presence of foci of inflammation located near the degradation sites of renal tubules was found (Figure 6d). Cubic epithelium with pycnotic nuclei were detected in these areas. The presence of narrowing of the lumen of some tubules, as well as an increase in the area of epithelial nuclei should be noted (Figure 6e).

Histopathological changes in the kidneys of *D. rerio* at 6 mg/L were comparable to those at lower concentrations. Changes in the structure of tubules were expressed in vacuolization of cubic epithelium, as well as an increase in the nuclei area (Figure 6f,g). In the hematopoietic tissue, foci of degradation were common and were expressed in the presence of pyknotic cells and also a small number of PAS-positive cells (Figure 6h).

The histopathologic disturbances described for lower concentrations were most pronounced when exposed to 8 mg/L BPA. Circulation disorders, namely sinusoidal dilation and blood stasis were detected throughout the whole area of the organ (Figure 7a). Renal tubules showed signs of hypertrophy (Figure 7b), expressed by a significant increase in the area of the epithelial cell nucleus ($p < 0.05$; Figure 6i,j; Table S4). Thus, a large number of neonephrogenic tubules were found in the organ stroma (Figure 7c). Alterations of glomerulus morphology were manifested in a significant decrease in their area relative to the control at a concentration of 2 mg/L ($p < 0.05$; Figure 7f). At the same time, no differences in the area of Bowman's space were revealed (Figure 7g).

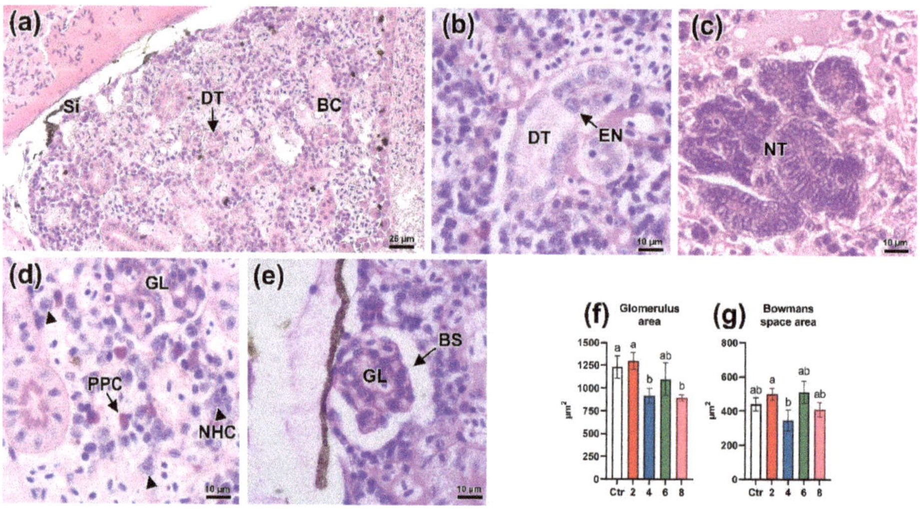

Figure 7. Effect of 8 mg/L BPA on nephron structure, hematopoietic tissue and morphometric parameters (**f**,**g**) of *Danio rerio* trunk kidney. (**a**–**e**): Significant circulatory disturbances manifested by sinusoidal dilation (Si) and blood congestion (BC); narrowing of the lumen and enlargement of the nuclei (EN) of the epithelium of some distal tubules (DT); increased occurrence of neonephrogenic tubules (NT); necrosis of areas of hematopoietic tissue associated with necrotic nuclei (NHC, arrowhead) and PAS-positive cells (PPC); decreased area of some glomeruli (GL) with an increase in Bowman's space (BC). H&E (**a**,**c**) and PAS (**b**,**d**,**e**) staining. Scale bars of 25 µm (**a**) and 10 µm (**b**–**e**). Significance ($p < 0.05$) from the Kruskal–Wallis test. Superscript letters (a, b) indicate statistical significance between different experimental groups.

Calculation of the histopathologic index showed that concentrations of 6 and 8 mg/L led to a significant increase in the number and severity of progressive (HIpc) and regressive (HIrc) changes ($p < 0.05$; Figure 6f; Table S5). These changes were predominantly expressed in alterations of hematopoietic tissue (presence of cells with pyknotic nuclei, increased number of PAS-positive cells, and general disturbances of organ histoarchitectonics) and structure of nephron sections (necrosis/inflammation of tubules, vacuolization of epithelium). Similar dynamics were also observed in the index of circulation disturbance (HIcd), which, at high concentrations of the toxicant, manifested itself in the expansion of sinusoidal spaces and blood congestion. It should be noted that the regressive changes at a concentration of 2 mg/L were comparable to the control and differed considerably from the value at higher concentrations.

3.6. Identifying the Concentration–Effect Relationship

To identify the concentration–effect relationship for the hematological and hematopoietic parameters of *D. rerio*, linear and polynomial regression analysis of the data with the most significant differences from the control was performed (Figure 5a,f). The most pronounced effect of BPA was on erythroid cell forms: the linear correlation coefficient for basophilic erythroblasts was 0.4945 (Figure 8b), and its value was considerably higher when a polynomial was used ($R^2 = 0.9602$). The occurrence of orthochromatic erythroblasts was less dependent on BPA concentration in both linear and polynomial regression analyses ($R^2 = 0.3385$ and 0.4247, respectively; Figure 8c). The number of metamyelocytes decreased as dependent on toxicant concentration, and this trend was more clearly demonstrated by the polynomial analysis ($R^2 = 0.5660$; Figure 8d).

The occurrence rate of erythrocyte nuclear abnormalities (MN, LN, and NN) had a strict positive linear relationship with BPA concentration (Figure 8f; Figure S2). At the same time, the correlation between the total number of nuclear abnormalities (TNA) and toxicant concentration was more significant using polynomial analysis ($R^2 = 0.8328$; Figure 8e). The parameters of the fish's peripheral blood, such as the relative abundance of lymphocytes ($R^2 = 0.8787$) and young erythrocytes ($R^2 = 0.9663$), had a significant positive linear relationship with BPA concentration (Figure 8g,i). At the same time, the relative number of lymphocytes decreased with increasing toxicant concentration ($R^2 = 0.6357$; Figure 8h).

The linear regression model revealed a significant correlation between a number of *D. rerio* kidney and BPA concentration morphometric parameters (Figure 9a,d). Thus, the regression coefficient for the distal tubule wall thickness was 0.8080 (Figure 9c). In turn, the glomerular area had a smaller relationship with toxicant concentration ($R^2 = 0.5787$; Figure 9b). A similar dependence was found for the other parameters. However, the application of the polynomial model did not demonstrate a more relevant regression coefficient.

The histopathological index of regressive changes, which had the largest contribution to the organ index, had comparable regression coefficients in both the linear ($R^2 = 0.9502$) and polynomial regression analyses ($R^2 = 0.9741$; Figure 9e). This is attributed to the similar frequency and prevalence of histopathologic abnormalities at BPA concentrations of 6 and 8 mg/L.

For other studied parameters, the application of linear and polynomial regression analysis did not reveal a significant relationship with the studied BPA concentrations.

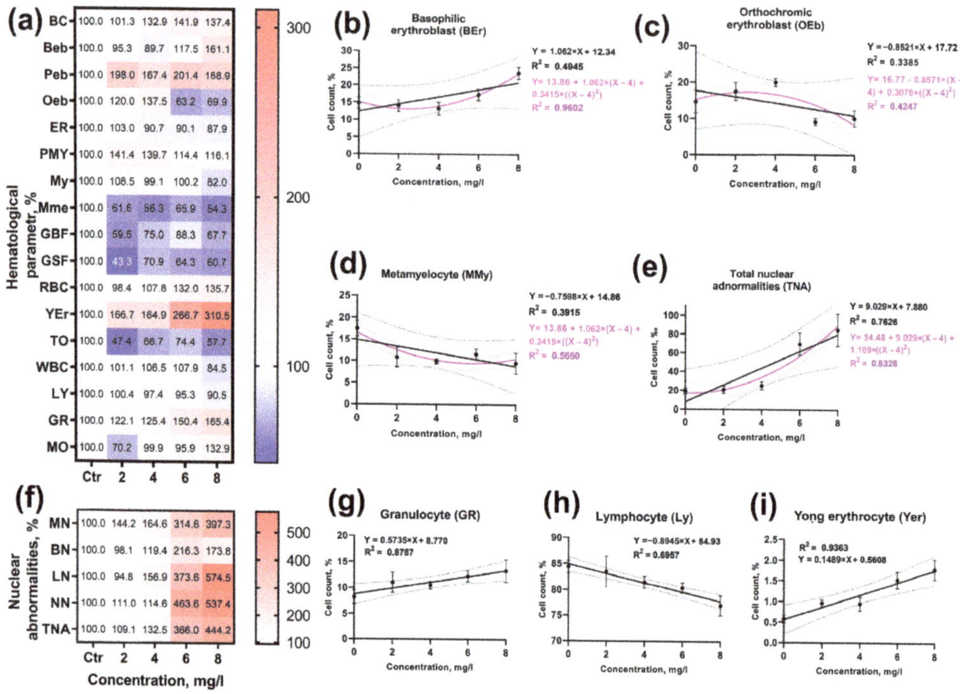

Figure 8. Concentration–effect relationship of BPA on hematological and hematopoietic parameters of *Danio rerio*. (**a**,**f**): matrix of changes in different parameters compared to the control in %; (**b**–**e**): comparison of linear and polynomial regression of hematopoietic parameters; (**g**–**i**): linear regression of hematological parameters.

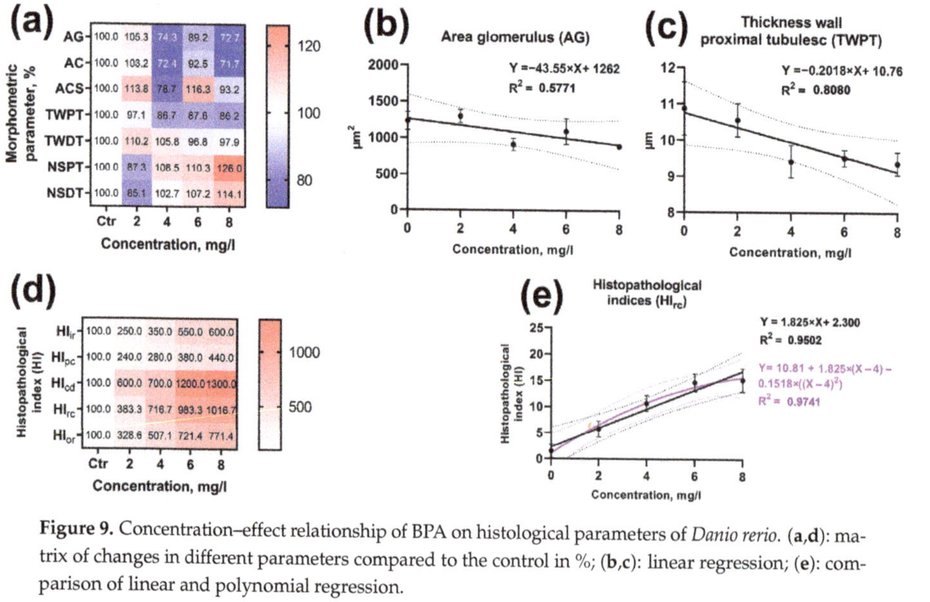

Figure 9. Concentration–effect relationship of BPA on histological parameters of *Danio rerio*. (**a**,**d**): matrix of changes in different parameters compared to the control in %; (**b**,**c**): linear regression; (**e**): comparison of linear and polynomial regression.

4. Discussion

4.1. Acute BPA Toxicity to Adult Zebrafish

Obtained data made it possible to determine the acute toxic effect limits of BPA for adult zebrafish. Concentrations exceeding 4 mg/L reliably resulted in fish mortality within 96 h, with an LC50 of 6.22 mg/L. Acute toxicity of Bisphenol A was observed with fish mortality during the first 48 h of the experiment, with no further mortality occurring between 48 and 96 h. The lethal concentration in this study aligns with data from other researchers. For instance, medaka (*Oryzias latipes*) exhibited an LC50 of 9.4 mg/L for 96 h [48], while tilapia (*Oreochromis mossambicus*) showed an LC50 of 6.48 mg/L [49]. Carp (*Cyprinus carpio*) had an LC50 of 4.6 mg/L [50], and the spotted snakehead (*Channa punctatus*) had an LC50 of 7.62 mg/L [51].

In the case of the early developmental stages of zebrafish (embryo and larva), half of the individuals died within 72 h at concentrations of 8.22 and 10.94 mg/L, respectively [28]. These values are slightly higher than the ones obtained in this study for adult fish. The variations in lethal concentration values may be attributed to differences in fish species characteristics and the use of different solvents, such as acetone.

Based on the results of the histological study, the cause of fish mortality was attributed to the systematic toxic effects of BPA on hematopoiesis and excretory functions. In addition to kidney disorders, visible signs of gill lamellae atrophy were observed (see Figure S3). Similar mechanisms of action have been reported for many toxicants, including heavy metals, pesticides, and nanoparticles, which exhibit acute toxicity at high concentrations [52]. In surviving individuals, as a result of the adaptation processes, changes in the composition of the hematopoietic tissue cells and peripheral blood were observed.

4.2. Hematopoietic Parameters of the Head Kidney

The morphological stages of erythroid cell development found in the head kidney of zebrafish align with the descriptions provided by earlier authors [37,53,54]. However, it is worth noting that some morphological differences in myeloid cells, specifically in promyelocytes and myelocytes, were comparable to the descriptions by Kondera et al. [37]. In this study, myelocytes exhibited the largest size among all granulocyte precursors, which deviates from Kondera's description, but consistent with the one reported by Ivanova [35].

The relative abundance of early stages of erythrocyte differentiation observed in smears of zebrafish head kidney is consistent with findings from previous studies. For instance, the numbers of polychromatic and orthochromatic erythroblasts obtained in this study (13.58% and 14.58%) are in agreement with data from several researchers [37,38,54,55]. These studies reported similar ranges, with percentages varying from 12.78% to 19.36% for polychromatic erythroblasts and 7.12% to 12.84% for orthochromatic erythroblasts. Notably, a significant variation was observed in the occurrence of mature erythrocytes (ranging from 0.36% to 12.44%), which is likely attributed to differences in the methodology used for preparing head kidney smears. The occurrence of myeloid hematopoietic progenitor cells also aligns with previous data [56–58], demonstrating an incremental increase as cells progress through the differentiation stages.

It is also important to consider that the variations in the frequency of hematopoietic progenitor cells occurrence may be attributed to species-specific differences [13,15,37] as the mentioned studies encompass a wide range of species (such as *Cyprinus carpio*, *Ictalurus punctatus*, *Acipenser dabryanus*, *Scorpaena porcus*, etc.).

Despite significant advancements in the study of hematopoiesis in lower vertebrates, including fish, there is an ongoing need for further refinement of morphological and quantitative assessment methods [17]. In our study, we introduced an approach for counting hematopoietic progenitor cells, with particular emphasis on elements from the erythropoietic and myelopoietic series. However, it is important to note that our myeloid cell counting did not involve categorization based on staining affinity (neutrophils, basophils, eosinophils). This described approach has the potential for application in evaluating the effects of toxic substances on hematopoiesis.

The conducted study allows us to conclude that BPA, at concentrations of 6 and 8 mg/L, influences the differentiation of cell elements in the erythroid series. Moreover, the impact of low concentrations of the toxicant was observed on various cell types, including orthochromatic erythroblasts, metamyelocytes, and differentiated forms of granulocytes. The increase in erythropoiesis activity as a response to stress factors, including toxic exposure, is a common phenomenon observed in all vertebrate organisms. An extensively studied factor affecting the number of red blood cells in peripheral blood is anemia, which triggers the release of stored red blood cells and an increased proliferation of new ones, consequently enhancing blood oxygen saturation [14,59]. Given that the primary route of BPA entry into a fish's body is through the gills [6], it can be hypothesized that the disruption of their function due to toxic effects might lead to anemia and, in turn, stimulate erythropoiesis.

Oxidative stress could represent another mechanism of BPA's toxic effects on erythropoiesis. Previous studies have demonstrated that BPA-induced inflammation directly correlates with an increase in the levels of reactive nitrogen and oxygen species [12,60]. The excess of these free radicals has the potential to cause cytotoxic damage to various body tissues, including blood cells and the hematopoietic tissue of zebrafish. This phenomenon may also account for the reduced occurrence of certain myeloid cells across all experimental BPA concentrations, as they may migrate to sites of nonspecific immune response. Given that granulocytes in fish have a short life cycle and a high renewal rate [58], prolonged exposure to BPA may result in significant alterations in their occurrence within the hematopoietic tissue of the head kidney.

The endocrine-disruptive effects of BPA, particularly its estrogenic activity [9], may also extend to the impairment of the immune-endocrine components through its impact on the hypothalamus–pituitary–adrenal axis, which plays a role in the regulation of myeloid and lymphoid cell activity [61,62]. In fish, the head kidney serves multiple functions, including hematopoietic, excretory, and potentially endocrine functions (e.g., the adrenal gland [63]). The disruption of these functions, along with other elements of the head kidney, could amplify the toxic effects of BPA.

Similar alterations in erythropoiesis, resembling signs of anemia resulting from cytotoxic damage to red blood cells, have been observed with other toxicants such as cadmium [64], copper nanoparticles [65], dioxin [66], and nitrite [14].

4.3. *Genotoxicity*

Numerous studies have demonstrated the genotoxic effects of BPA on both aquatic and higher vertebrates [7,51,67,68]. In this study, using a micronucleus test on fish erythrocytes, acute exposure to BPA at concentrations of 6 and 8 mg/L resulted in a significant increase in the frequency of nuclear abnormalities. Examination of nuclear abnormalities revealed that lobed nuclei and notched nuclei were significantly more common than other types of alterations. The absence of significant differences from the control at concentrations of 2 and 4 mg/L suggests the lack of a genotoxic effect in the conditions of short-term experiments. However, in long-term experiments, these concentrations could also lead to disruptions in the nuclear apparatus of erythrocytes.

For instance, Akram et al. [8] demonstrated that the number of nuclear abnormalities in a BPA concentration of 1.5 mg/L significantly increased only after 45 to 60 days of exposure. Similarly, in the study by Afzal et al. [7], a meaningful increase in the number of erythrocytes with nuclear abnormalities was observed at concentrations greater than 3 mg/L after 30 days of exposure. In an acute experiment lasting 96 h, BPA resulted in a significant increase in micronuclei frequency in *Channa punctatus* at concentrations of 1.91 and 3.81 mg/L [51]. The researchers mentioned above attribute the genotoxicity of BPA to increased free radical generation and oxidative stress. The variations in the genotoxic effect of BPA may be related to species-specific differences in erythropoiesis and the sensitivity of fish.

The formation of nuclear abnormalities occurs as a result of cell division from acentric chromosome fragments and/or chromatid fragments in the anaphase of mitosis [25], and the frequency of these abnormalities may be regulated by the rate of erythropoiesis [23]. Thus, the increase in blast forms of erythrocytes in the head kidney may be directly related to the increased frequency of abnormal erythrocytes in the peripheral blood of zebrafish. According to some authors [69,70], the formation of nuclear abnormalities after exposure to a toxicant typically occurs within 1–2 days. This suggests that visible abnormalities of the nuclear apparatus occur at the latter stages of erythrocyte differentiation. In our study, basophilic and polychromatic erythroblasts did not exhibit morphologic disturbances in the nucleus, even when exposed to concentrations of 6 and 8 mg/L. This might be due to the fact that the MN can be reincorporated into the main nucleus during subsequent mitosis or destroyed by the autophagy process [71,72].

The increased frequency of nuclear anomalies, including lobed nuclei and notched nuclei, detected in the study may indicate the effect of BPA on the frequency of structural chromosomal aberrations (i.e., clastogenesis). One of the main mechanisms of action of clastogenic agents is the breaking of DNA strands, which, in the case of incorrect repair, are able to form acentric chromatids [73]. Toxicants that induce oxidative stress with short-term action have been shown to specifically lead to clastogenic abnormalities [74]. The ability to induce nuclear disorders has also been observed for other compounds in the phenolic group [75].

The results obtained in this study demonstrate that considering several morphological types of nuclear abnormalities in the blood of zebrafish allows for the simultaneous screening of pollutants in conjunction with other hematological indicators.

4.4. Peripheral Blood Composition

Under stress or toxic effects, changes in the composition of peripheral blood occur most rapidly. The compensatory response of the fish organism to the negative effects of stress factors is manifested in the release of cellular elements, hormones, and enzymes into the peripheral blood and changes in its chemical composition [76]. In our study, exposure to high concentrations of BPA resulted in significant changes in both the number of red blood cells and the leukocyte ratio. Thus, the observed fluctuations in the number of RBCs and the relative occurrence of immature erythrocytes directly correlate with data on changes in the relative occurrence of erythroblasts in the head kidney of zebrafish.

The release of immature forms of erythrocytes only partially compensated for the toxic effects of high BPA concentrations, leading to systemic disorders and the death of some fish by the end of the acute experiment. In contrast, at concentrations of 2 and 1 mg/L, no such effect was observed, suggesting that the fish organism can withstand such pollutant levels during short-term exposure. Existing studies present conflicting data regarding the effect of BPA on the number of red and white blood cells. For example, in *Oncorhynchus mykiss* there is an increase in RBC counts, while in *Vimba vimba* [77] and *Channa punctatus* [51] there is a decrease. It is possible that the toxicity of BPA increases with rising temperatures in the aquatic environment.

Changes in the composition of white blood elements, while accompanied by an increase in the number of granulocytes and monocytes at high concentrations, did not lead to a significant increase in the WBC count in the peripheral blood of fish. On the contrary, their number significantly decreased at a concentration of 8 mg/L. This decrease may be related to the disruption of differentiation processes in hematopoietic organs and/or their mobilization to participate in nonspecific immune responses in other tissues. Similar results were obtained by Holeyappa et al. [78] in *Cyprinus carpio* under chronic exposure to BPA.

The number of thrombocytes significantly decreased at concentrations of 2, 4, and 8 mg/L. The decrease in their quantity at the lowest concentrations may have been caused by disruptions in the blood coagulation system. However, it is important to note that, as pointed out by Witeska [79], blood coagulation under stress is not always directly related to an increase in the number of thrombocytes in the bloodstream of fish.

4.5. Histological Parameters of Head and Trunk Kidney

Examination of the histological structure of the head kidney of zebrafish revealed the presence of pathological abnormalities at all studied concentrations. Significant disturbances in the structure of hematopoietic tissue, which were observed starting from the concentration of 4 mg/L, are directly associated with changes in the ratio of hematopoietic cells. The most notable pathological changes included impaired circulation and the occurrence of inflammation/necrosis foci. Nephrotoxic effects of BPA have also been reported by many authors across a wide range of concentrations. Among the histopathologic abnormalities of fish hematopoietic tissue, necrosis and a reduction in the amount of hematopoietic tissue have been identified [7,78,80] which is in agreement with our results. As noted earlier, the action of BPA on hematopoietic tissue is primarily associated with the induction of oxidative stress. The excretion of BPA as part of bile acids through the intestine [81] and the low toxicity of its metabolites [82] suggest that renal damage is associated with the persistence of its activity in blood and tissue fluids. Additionally, renal impairment may be related to the continuous ingestion of the toxicant through the gills from the surrounding aquatic environment [6].

The kidney plays a crucial role in excretion and osmoregulation and has been reported by several researchers to be primarily affected by toxicants from the phenolic group [83]. Toxicants enter the kidney through the bloodstream [84]. In addition to its impact on hematopoietic tissue, BPA also affected the structure of nephrons. Even at a concentration of 2 mg/L, significant circulatory alterations were observed, which could be attributed to hyperemia. As the concentration of the toxicant increased, circulatory alterations progressed, potentially indicating disturbances in the integrity of vascular and capillary walls. Changes in renal tubules, such as the enlargement of cubic epithelial nuclei, vacuolization, and the narrowing of the tubule lumen, represent the initial response of the tissue to the presence of a toxicant in the filtrate [85]. Similar histopathologic changes have been observed during exposure to various pollutants.

The increased occurrence of neonephrogenic tubules under the influence of BPA suggests the activation of renal tissue regeneration mechanisms. According to some authors [86], these areas are where cell proliferation occurs to form new renal tubules. However, the supposed induction of neonephrogenesis under the influence of BPA warrants further study.

The head portion of the trunk kidney, which combines hematopoietic and excretory functions, is particularly susceptible to the acute effects of BPA. Therefore, it can be concluded that high concentrations of BPA have an acute nephrotoxic effect on zebrafish.

4.6. Concentration–Effect Relationship

Polynomial regression analysis can, in certain cases, unveil more intricate relationships between the concentration of a toxicant and changes in a test parameter than linear regression [87]. Polynomial regression is actively employed in determining the extent of heavy metal accumulation in the tissues of aquatic organisms [88], as well as in establishing the most effective concentrations of feed additives [89].

Utilizing this method, it was observed that, for some immature forms of erythrocytes, a correlation existed between BPA concentration and the relative cell number. This correlation was marked by the absence of BPA influence at low/sublethal concentrations (2 and 4 mg/L) and a significant increase or decrease in the number of cells at higher/lethal concentrations (6 and 8 mg/L). This pattern was particularly pronounced for basophilic erythroblasts and the total number of nuclear abnormalities, both of which showed a significant increased at higher concentrations of the toxicant.

In the case of parameters such as the number of nuclear abnormalities and the relative abundance of lymphocytes, granulocytes, and young erythrocytes, linear regression analysis revealed a significant concentration–effect relationship. Additionally, a linear dependence on the concentration of BPA was found for certain morphometric kidney parameters. It is worth noting that the presence of a linear relationship between the occurrence of nuclear abnormalities and the concentration of the toxicant suggests the presence of clastogenic genotoxicity of BPA [90].

The obtained results indicate that the acute toxic effect of BPA is most pronounced at high concentrations (6 and 8 mg/L). However, for several parameters, such as the number of young erythrocytes and the relative number of lymphocytes, a linear dependence on the concentration of the toxicant was observed. Similar linear concentration-dependent effects were also noted when studying the impact of BPA on hematological parameters in *Oncorhynchus mykiss* and *Vimba vimba* [77].

According to Canesi et al. [6], a nonlinear dose–response relationship is often observed for compounds with endocrine-disruptive effects. Additionally, toxicogenomics data indicate that BPA is characterized by an inverted U-shaped pattern in the dose–response relationship [6], which is partially supported by the results obtained in this work. To establish more accurate relationships between BPA concentration and the indicators of aquatic organisms, further research is needed, encompassing both low concentrations (sublethal/occurring in the environment) and high concentrations.

Differences in the effects of BPA concentrations on the composition of hematopoietic cells may be attributed to variations in the timing and stages of differentiation of blast forms in the erythroid and myeloid lineages. This variability can also be explained by the inhibitory effects of low concentrations of BPA and the cytotoxic effects of higher concentrations. However, the differentiation of individual hematopoietic tissues occurs at different rates, and as a result, short-term exposure to a stress factor could lead to a significant shift in their ratio [91].

4.7. Hematological Parameters in the Evaluation of Endocrine-Disruptive Substances

The results obtained in this work and the data from other researchers show that BPA can have a significant effect on the processes of hematopoiesis. Exposure to high concentrations of BPA disturbs the processes of development and function of fish blood cells, which is expressed not only in changes in the kinetics of blood development and formation of nuclear anomalies in erythrocytes but also in the appearance of histopathological disorders in hematopoietic organs (head kidney). Thus, these abnormalities may reduce the ability of fish to respond to infections and other physiological stressors [14,54]. In addition, due to the fact that BPA is able to accumulate in tissues, it poses a danger not only to aquatic animals but also to humans (through diet) [4,92].

For these reasons, when searching for new ways to alleviate the toxic effect of endocrine-disruptive substances (e.g., the use of antioxidants, sorbents, probiotics, food/feed additives) [93–95], the indicators of the cellular composition of hematopoietic organs and peripheral blood can act as significant test parameters.

5. Conclusions

Bisphenol A at concentrations of 6 and 8 mg/L resulted in the mortality of a portion of *Danio rerio* individuals during the acute experiment (96 h). This mortality can be attributed to systemic toxic effects, which include disruptions in hematopoiesis and osmoregulation. These concentrations also led to a substantial increase in the occurrence of erythrocyte nuclear abnormalities, indicating a potential genotoxic effect. Notably, significant changes in the composition of hematopoietic cells were observed in head kidney smears across all studied BPA concentrations. These changes are in line with alterations in the number of cells in peripheral blood and the histopathologic changes observed in the hematopoietic tissue of zebrafish head kidney. This alignment is further confirmed through the calculation of histopathologic indices.

The identified toxic effects of BPA, as described using both linear and polynomial models, revealed relatively weak responses to low concentrations of the toxicant, with a sharp change occurring at concentrations of 6 and 8 mg/L. This suggests that the limits of adaptation responses in zebrafish are within the range of 4–6 mg/L, slightly below the obtained LC50 (96 h) values. In summary, the results of this study demonstrate that the systemic effects of high BPA concentrations primarily manifest in hematopoietic organs and the cell composition of the blood.

Supplementary Materials: The following supporting information can be downloaded at: https://www.mdpi.com/article/10.3390/ani13233685/s1, Table S1: Cell and nucleus size of various hematopoietic cells in control *Danio rerio*; Figure S1: Relative abundance of cellular elements of hematopoietic tissue of *Danio rerio* head kidney during acute exposure to Bisphenol A; Table S2: Number and relative occurrence of cellular elements of peripheral blood of *Danio rerio* during acute exposure to Bisphenol A; Table S3: Relative abundance of *Danio rerio* erythrocytes with nuclear abnormalities under acute exposure to Bisphenol A; Table S4: Morphometric parameters of kidney of experimental *Danio rerio* under acute exposure to bisphenol A; Table S5: Histopathological indices of kidney of experimental *Danio rerio* under acute exposure to bisphenol A; Table S6: Abbreviation of histopathological indexes and morphometric parameters used in the work; Figure S2: Linear concentration–effect relationship of BPA on nuclear abnormalities of *Danio rerio*; Figure S3: Histological sections of *Danio rerio* gills in control and experimental groups. (a) control group gills, including primary (PL) and secondary lamellae (SL), as well as basal cells (BC) and respiratory epithelium; (b) 6 mg/L: at this concentration, a decrease in the thickness of secondary lamellae and respiratory epithelium (RE) as well as basal cells was observed; (c, d) 8 mg/L: in addition to the above-mentioned abnormalities, areas with curvature of secondary lamellae (LC) were observed. H&E staining Scale bars 25 μm.

Author Contributions: Conceptualization, D.N.-N. and K.G.; data curation, S.S.; formal analysis, M.O. and A.P.; investigation, S.S., N.K. and D.N.-N.; methodology, S.S., K.G. and N.K.; project administration, M.M.; resources, S.S. and N.K.; supervision, A.N.-N. and V.D.; validation, D.R., M.M. and A.V.; visualization, N.K. and A.K.; writing—original draft, S.S., N.K., K.G. and D.N.-N.; writing—review and editing, N.K., D.N.-N. and M.M. All authors have read and agreed to the published version of the manuscript.

Funding: This work was supported by the Russian Science Foundation (project No. 23-16-00123).

Institutional Review Board Statement: This study was conducted in accordance with the guidelines of the Local Ethics Commission of the Scientific and Technical Council of the Moscow State University of Technology and Management (approval number 1, 15 January 2023).

Informed Consent Statement: Not applicable.

Data Availability Statement: Data are contained within the article and Supplementary Materials.

Conflicts of Interest: The authors declare no conflict of interest.

References

1. Staples, C.A.; Dome, P.B.; Klecka, G.M.; Oblock, S.T.; Harris, L.R. A Review of the Environmental Fate, Effects, and Exposures of Bisphenol A. *Chemosphere* **1998**, *36*, 2149–2173. [CrossRef] [PubMed]
2. Ackerman, L.K.; Noonan, G.O.; Heiserman, W.M.; Roach, J.A.; Limm, W.; Begley, T.H. Determination of bisphenol A in US infant formulas: Updated methods and concentrations. *J. Agric. Food Chem.* **2010**, *58*, 2307–2313. [CrossRef] [PubMed]
3. Corrales, J.; Kristofco, L.A.; Steele, W.B.; Yates, B.S.; Breed, C.S.; Williams, E.S.; Brooks, B.W. Global Assessment of Bisphenol a in the Environment. *Dose-Response* **2015**, *13*, 155932581559830. [CrossRef]
4. Rezg, R.; El-Fazaa, S.; Gharbi, N.; Mornagui, B. Bisphenol a and Human Chronic Diseases: Current Evidences, Possible Mechanisms, and Future Perspectives. *Environ. Int.* **2014**, *64*, 83–90. [CrossRef] [PubMed]
5. Uadia, P.O. A survey of the level of bisphenol A (BPA) in effluents, soil leachates, food samples, drinking water and consumer products in south-western Nigeria. *World Environ.* **2015**, *5*, 135–139.
6. Canesi, L.; Fabbri, E. Environmental effects of BPA: Focus on aquatic species. *Dose-Response* **2015**, *13*, 1559325815598304. [CrossRef] [PubMed]
7. Afzal, G.; Ahmad, H.I.; Hussain, R.; Jamal, A.; Kiran, S.; Hussain, T.; Saeed, S.; Nisa, M.u. Bisphenol a Induces Histopathological, Hematobiochemical Alterations, Oxidative Stress, and Genotoxicity in Common Carp (*Cyprinus carpio* L.). *Oxidative Med. Cell. Longev.* **2022**, *2022*, 5450421. [CrossRef]
8. Akram, R.; Iqbal, R.; Hussain, R.; Jabeen, F.; Ali, M. Evaluation of Oxidative Stress, Antioxidant Enzymes and Genotoxic Potential of Bisphenol a in Fresh Water Bighead Carp (*Aristichthys nobils*) Fish at Low Concentrations. *Environ. Pollut.* **2021**, *268*, 115896. [CrossRef]
9. Rubin, B.S. Bisphenol A: An Endocrine Disruptor with Widespread Exposure and Multiple Effects. *J. Steroid Biochem. Mol. Biol.* **2011**, *127*, 27–34. [CrossRef]

10. Crain, D.A.; Eriksen, M.; Iguchi, T.; Jobling, S.; Laufer, H.; LeBlanc, G.A.; Guillette, L.J. An Ecological Assessment of Bisphenol-A: Evidence from Comparative Biology. *Reprod. Toxicol.* **2007**, *24*, 225–239. [CrossRef]
11. Zhu, L.; Wang, L.; Fan, X.; Dong, C.; Wang, G.; Wang, Z. Chronic Exposure to Bisphenol a Resulted in Alterations of Reproductive Functions via Immune Defense, Oxidative Damage and Disruption DNA/Histone Methylation in Male Rare Minnow Gobiocypris Rarus. *Aquat. Toxicol.* **2021**, *236*, 105849. [CrossRef] [PubMed]
12. Wu, N.C.; Seebacher, F. Effect of the Plastic Pollutant Bisphenol a on the Biology of Aquatic Organisms: A Meta-Analysis. *Glob. Change Biol.* **2020**, *26*, 3821–3833. [CrossRef] [PubMed]
13. Megarani, D.V.; Hardian, A.B.; Arifianto, D.; Santosa, C.M.; Salasia, S.I.O. Comparative Morphology and Morphometry of Blood Cells in Zebrafish (Danio Rerio), Common Carp (*Cyprinus carpio* Carpio), and Tilapia (*Oreochromis niloticus*). *J. Am. Assoc. Lab. Anim. Sci.* **2020**, *59*, 673–680. [CrossRef]
14. Witeska, M. Anemia in Teleost Fishes. *Bull. Eur. Assoc. Fish Pathol.* **2015**, *35*, 148–160.
15. Kondera, E.; Witeska, M. Cadmium and Copper Reduce Hematopoietic Potential in Common Carp (*Cyprinus carpio* L.) Head Kidney. *Fish Physiol. Biochem.* **2012**, *39*, 755–764. [CrossRef] [PubMed]
16. Islam, S.M.; Rohani, M.F.; Zabed, S.A.; Islam, M.T.; Jannat, R.; Akter, Y.; Shahjahan, M. Acute Effects of Chromium on Hemato-Biochemical Parameters and Morphology of Erythrocytes in Striped Catfish Pangasianodon Hypophthalmus. *Toxicol. Rep.* **2020**, *7*, 664–670. [CrossRef] [PubMed]
17. Abdel-Aziz, E.S.H.; Abdu, S.B.; Ali, T.E.S.; Fouad, H.F. Haemopoiesis in the Head Kidney of Tilapia, *Oreochromis niloticus* (Teleostei: Cichlidae): A Morphological (Optical and Ultrastructural) Study. *Fish Physiol. Biochem.* **2010**, *36*, 323–336. [CrossRef] [PubMed]
18. Carradice, D.; Lieschke, G.J. Zebrafish in Hematology: Sushi or Science? *Blood* **2008**, *111*, 3331–3342. [CrossRef]
19. Kobayashi, I.; Sekiya, M.; Moritomo, T.; Ototake, M.; Nakanishi, T. Demonstration of Hematopoietic Stem Cells in Ginbuna Carp (*Carassius auratus* Langsdorfii) Kidney. *Dev. Comp. Immunol.* **2006**, *30*, 1034–1046. [CrossRef]
20. de Souza Anselmo, C.; Sardela, V.F.; de Sousa, V.P.; Pereira, H.M.G. Zebrafish (Danio Rerio): A Valuable Tool for Predicting the Metabolism of Xenobiotics in Humans? *Comp. Biochem. Physiol. Part C Toxicol. Pharmacol.* **2018**, *212*, 34–46. [CrossRef]
21. Gjini, E.; Jing, C.-B.; Nguyen, A.T.; Reyon, D.; Gans, E.; Kesarsing, M.; Peterson, J.; Pozdnyakova, O.; Rodig, S.J.; Mansour, M.R.; et al. Disruption of Asxl1 Results in Myeloproliferative Neoplasms in Zebrafish. *Dis. Models Mech.* **2019**, *12*, dmm035790. [CrossRef] [PubMed]
22. Yu, Y.; Wang, Q.; Huang, Z.; Ding, L.; Xu, Z. Immunoglobulins, Mucosal Immunity and Vaccination in Teleost Fish. *Front. Immunol.* **2020**, *11*, 567941. [CrossRef] [PubMed]
23. Gökalp Muranli, F.D.; Güner, U. Induction of Micronuclei and Nuclear Abnormalities in Erythrocytes of Mosquito Fish (*Gambusia affinis*) Following Exposure to the Pyrethroid Insecticide Lambda-Cyhalothrin. *Mutat. Res. Genet. Toxicol. Environ. Mutagen.* **2011**, *726*, 104–108. [CrossRef] [PubMed]
24. Bagdonas, E.; Vosylienė, M.Z. A study of toxicity and genotoxicity of copper, zinc and their mixture to rainbow trout (Oncorhynchus mykiss). *Biologija* **2006**, *52*, 8–13.
25. Canedo, A.; de Jesus, L.W.O.; Bailão, E.F.L.C.; Rocha, T.L. Micronucleus Test and Nuclear Abnormality Assay in Zebrafish (Danio Rerio): Past, Present, and Future Trends. *Environ. Pollut.* **2021**, *290*, 118019. [CrossRef] [PubMed]
26. Westerfield, M. *The Zebrafish Book: A Guide for the Laboratory Use of Zebrafish (Brachydanio rerio)*; University of Oregon Press: Eugene, OR, USA, 1995.
27. Mathieu-Denoncourt, J.; Wallace, S.J.; de Solla, S.R.; Langlois, V.S. Influence of Lipophilicity on the Toxicity of Bisphenol a and Phthalates to Aquatic Organisms. *Bull. Environ. Contam. Toxicol.* **2016**, *97*, 4–10. [CrossRef] [PubMed]
28. Chen, X.N.; Xiao, H.; Ke, W.B.; Zhan, Y.; Sun, Y. Acute and Subacute Toxicity of Bisphenol a on Zebrafish (*Danio rerio*). *Adv. Mater. Res.* **2011**, *356–360*, 138–141. [CrossRef]
29. OECD. Test no. 203: Fish, acute toxicity test. In *OECD Guidelines for the Testing of Chemicals*; OECD: Paris, France, 1992; Volume 1, pp. 1–10.
30. Di Paola, D.; Capparucci, F.; Lanteri, G.; Cordaro, M.; Crupi, R.; Siracusa, R.; D'Amico, R.; Fusco, R.; Impellizzeri, D.; Cuzzocrea, S.; et al. Combined Toxicity of Xenobiotics Bisphenol a and Heavy Metals on Zebrafish Embryos (Danio Rerio). *Toxics* **2021**, *9*, 344. [CrossRef]
31. Kochetkov, N.; Smorodinskaya, S.; Nikiforov-Nikishin, D.; Klimov, V.; Golovacheva, N.; Nikiforov-Nikishin, A.; Grozesku, Y. Evaluating possible genotoxicity of three feed additives recommended for aquaculture by using micronucleus test on danio rerio erythrocytes. *Vestn. Astrakhan State Tech. Univ. Ser. Fish. Ind.* **2022**, *3*, 48–59. [CrossRef]
32. Rusia, V.; Sood Routine, S.K. Haematological tests. In *Medical Laboratory Technology*; Kanai, L., Mukerjee, I., Eds.; McGraw Hill: New Delhi, India, 1992; pp. 252–258.
33. Ivanovski, O.; Kulkeaw, K.; Nakagawa, M.; Sasaki, T.; Mizuochi, C.; Horio, Y.; Ishitani, T.; Sugiyama, D. Characterization of Kidney Marrow in Zebrafish (*Danio rerio*) by Using a New Surgical Technique. *Contrib. Maced. Acad. Sci. Arts* **2009**, *30*, 71–80.
34. Sula, E.; Aliko, V.; Pagano, M.; Faggio, C. Digital Light Microscopy as a Tool in Toxicological Evaluation of Fish Erythrocyte Morphological Abnormalities. *Microsc. Res. Tech.* **2020**, *83*, 362–369. [CrossRef] [PubMed]
35. Ilvanova, N.T. *Atlas Kletok Krovi ryb (Sravnitel'naya Morfologiya i Klassifikatsiya Formennykh Elementov Krovi ryb) (Atlas of the Fish Blood Cells: Comparative Morphology and Classification of the Fish Blood Elements)*; Legkaya Pishch. Prom-st: Moscow, Russia, 1983; p. 325.

36. Fijan, N. Composition of Main Haematopoietic Compartments in Normal and Bled Channel Catfish. *J. Fish Biol.* **2002**, *60*, 1142–1154. [CrossRef]
37. Kondera, E. Haematopoiesis in the Head Kidney of Common Carp (*Cyprinus carpio* L.): A Morphological Study. *Fish Physiol. Biochem.* **2010**, *37*, 355–362. [CrossRef] [PubMed]
38. Wlasow, T.; Dabrowska, H. Cellular Changes in the Blood and Haemopoietic Tissues of Common Carp Exposed to Sublethal Concentration of Ammonia. *Aquat. Living Resour.* **1989**, *2*, 169–174. [CrossRef]
39. Schindelin, J.; Arganda-Carreras, I.; Frise, E.; Kaynig, V.; Longair, M.; Pietzsch, T.; Preibisch, S.; Rueden, C.; Saalfeld, S.; Schmid, B.; et al. Fiji: An open-source platform for biological-image analysis. *Nat. Methods* **2012**, *9*, 676–682. [CrossRef] [PubMed]
40. Bolognesi, C.; Hayashi, M. Micronucleus Assay in Aquatic Animals. *Mutagenesis* **2010**, *26*, 205–213. [CrossRef] [PubMed]
41. Suvarna, K.S.; Layton, C.; Bancroft, J.D. *Bancroft's Theory and Practice of Histological Techniques*, 8th ed.; Elsevier: Oxford, UK, 2019.
42. Kochetkov, N.; Smorodinskaya, S.; Vatlin, A.; Nikiforov-Nikishin, D.; Nikiforov-Nikishin, A.; Danilenko, V.; Anastasia, K.; Reznikova, D.; Grishina, Y.; Antipov, S.; et al. Ability of Lactobacillus brevis 47f to Alleviate the Toxic Effects of Imidacloprid Low Concentration on the Histological Parameters and Cytokine Profile of Zebrafish (Danio rerio). *Int. J. Mol. Sci.* **2023**, *24*, 12290. [CrossRef]
43. Bernet, D.; Schmidt, H.; Meier, W.; Burkhardt-Holm, P.; Wahli, T. Histopathology in Fish: Proposal for a Protocol to Assess Aquatic Pollution. *J. Fish Dis.* **1999**, *22*, 25–34. [CrossRef]
44. Antunes, A.M.; Rocha, T.L.; Pires, F.L.; Alves, M.; Leite, V.S.; Arana, S.; Moreira, P.; Maria, S. Gender-Specific Histopathological Response in GuppiesPoecilia ReticulataExposed to Glyphosate or Its Metabolite Aminomethylphosphonic Acid. *J. Appl. Toxicol.* **2017**, *37*, 1098–1107. [CrossRef]
45. Healy, M.J.R.; Finney, D.J. Statistical Method in Biological Assay. *J. R. Stat. Society. Ser. A Gen.* **1979**, *142*, 507. [CrossRef]
46. R Core Team. *R: A language and Environment for Statistical Computing*; R Foundation for Statistical Computing: Vienna, Austria, 2021; Available online: https://www.R-project.org/ (accessed on 2 September 2023).
47. RStudio Team. *RStudio: Integrated Development for R*; RStudio, PBC: Boston, MA, USA, 2020; Available online: http://www.rstudio.com/ (accessed on 2 August 2023).
48. Li, D.; Bi, R.; Chen, H.; Mu, L.; Zhang, L.; Chen, Q.; Xie, H.; Luo, Y.; Xie, L. The Acute Toxicity of Bisphenol a and Lignin-Derived Bisphenol in Algae, Daphnids, and Japanese Medaka. *Environ. Sci. Pollut. Res.* **2017**, *24*, 23872–23879. [CrossRef] [PubMed]
49. Asifa, K.P.; Chitra, K.C. Evaluation of LC50 and behavioural responses of bisphenol A in the cichlid fish, Etroplus maculatus. *Int. J. Curr. Res.* **2015**, *7*, 16725–16729.
50. Alexander, H.C.; Dill, D.C.; Smith, L.W.; Guiney, P.D.; Dorn, P. Bisphenol A: Acute aquatic toxicity. *Environ. Toxicol. Chem. Int. J.* **1988**, *7*, 19–26. [CrossRef]
51. Sharma, P.; Chadha, P. Bisphenol a Induced Toxicity in Blood Cells of Freshwater Fish Channa Punctatus after Acute Exposure. *Saudi J. Biol. Sci.* **2021**, *28*, 4738–4750. [CrossRef] [PubMed]
52. Khoshnood, Z. Effects of Environmental Pollution on Fish: A Short Review. *Transylv. Rev. Syst. Ecol. Res.* **2017**, *19*, 49–60. [CrossRef]
53. Soldatov, A.A.; Kukhareva, T.A. Erythropoiesis and the contents of abnormal erythroid forms in the blood of the round goby, Neogobius melanostomus Pallas, 1811 (Osteichthyes: Gobiidae). *Russ. J. Mar. Biol.* **2015**, *41*, 315–320. [CrossRef]
54. Fijan, N. Morphogenesis of Blood Cell Lineages in Channel Catfish. *J. Fish Biol.* **2002**, *60*, 999–1014. [CrossRef]
55. Liu, Y.; Xiao, Q.; Yang, S.; Zhao, L.; Fu, H.; Du, J.; Du, Z.; Yan, T.; Wu, H. Characterization of Hematopoiesis in Dabry's Sturgeon (*Acipenser dabryanus*). *Aquac. Fish.* **2017**, *2*, 262–268. [CrossRef]
56. Kondera, E.; Bojarski, B.; Ługowska, K.; Kot, B.; Witeska, M. Effects of Oxytetracycline and Gentamicin Therapeutic Doses on Hematological, Biochemical and Hematopoietic Parameters in *Cyprinus carpio* Juveniles. *Animals* **2020**, *10*, 2278. [CrossRef]
57. Kondera, E.; Teodorczuk, B.; Ługowska, K.; Witeska, M. Effect of Glyphosate-Based Herbicide on Hematological and Hemopoietic Parameters in Common Carp (*Cyprinus carpio* L.). *Fish Physiol. Biochem.* **2018**, *44*, 1011–1018. [CrossRef]
58. Suljević, D.; Islamagić, E.; Alijagić, A.; Fočak, M.; Mitrašinović-Brulić, M. Morphological identification of haematopoietic cells in pronephros of common carp (Cyprinus carpio Linnaeus, 1758). *Acta Biol. Szeged.* **2016**, *60*, 113–118.
59. Murad, A.; Houston, A.H.; Samson, L.D. Haematological Response to Reduced Oxygen-Carrying Capacity, Increased Temperature and Hypoxia in Goldfish, *Carassius auratus* L. *J. Fish Biol.* **1990**, *36*, 289–305. [CrossRef]
60. Yang, J.; Li, H.; Ran, Y.; Chan, K. Distribution and Bioconcentration of Endocrine Disrupting Chemicals in Surface Water and Fish Bile of the Pearl River Delta, South China. *Chemosphere* **2014**, *107*, 439–446. [CrossRef] [PubMed]
61. Valero, Y.; Cuesta, A.; Cammarata, M.; Cuesta, A.; Chaves-Pozo, E. Immune-Endocrine Interactions in the Fish Gonad during Infection: An Open Door to Vertical Transmission. *Fishes* **2018**, *3*, 24. [CrossRef]
62. Casanova-Nakayama, A.; Wenger, M.; Burki, R.; Eppler, E.; Krasnov, A.; Segner, H. Endocrine Disrupting Compounds: Can They Target the Immune System of Fish? *Mar. Pollut. Bull.* **2011**, *63*, 412–416. [CrossRef] [PubMed]
63. Wendelaar Bonga, S.E. The stress response in fish. *Physiol. Rev.* **1997**, *77*, 591–625. [CrossRef]
64. Dangre, A.J.; Manning, S.; Brouwer, M. Effects of Cadmium on Hypoxia-Induced Expression of Hemoglobin and Erythropoietin in Larval Sheepshead Minnow, Cyprinodon *Variegatus*. *Aquat. Toxicol.* **2010**, *99*, 168–175. [CrossRef]
65. Cui, B.; Ren, L.; Xu, Q.-H.; Yin, L.-Y.; Zhou, X.-Y.; Liu, J.-X. Silver_Nanoparticles Inhibited Erythrogenesis during Zebrafish Embryogenesis. *Aquat. Toxicol.* **2016**, *177*, 295–305. [CrossRef]

66. Belair, C.D.; Peterson, R.E.; Heideman, W. Disruption of Erythropoiesis by Dioxin in the Zebrafish. *Dev. Dyn. Off. Publ. Am. Assoc. Anat.* **2001**, *222*, 581–594. [CrossRef]
67. Lombó, M.; Getino-Álvarez, L.; Depincé, A.; Labbé, C.; Herráez, M.P. Embryonic Exposure to Bisphenol a Impairs Primordial Germ Cell Migration without Jeopardizing Male Breeding Capacity. *Biomolecules* **2019**, *9*, 307. [CrossRef]
68. Ferreira, L.; Couto, R.C.; Oliveira, P.J. Bisphenol a as Epigenetic Modulator: Setting the Stage for Carcinogenesis? *Eur. J. Clin. Investig.* **2015**, *45*, 32–36. [CrossRef] [PubMed]
69. Palhares, D.; Grisolia, C.K. Comparison between the Micronucleus Frequencies of Kidney and Gill Erythrocytes in Tilapia Fish, Following Mitomycin c Treatment. *Genet. Mol. Biol.* **2002**, *25*, 281–284. [CrossRef]
70. Cavas, T. In Vivo Genotoxicity Evaluation of Atrazine and Atrazine–Based Herbicide on Fish *Carassius auratus* Using the Micronucleus Test and the Comet Assay. *Food Chem. Toxicol.* **2011**, *49*, 1431–1435. [CrossRef] [PubMed]
71. Guo, X.; Ni, J.; Liang, Z.; Xue, J.; Fenech, M.F.; Wang, X. The molecular origins and pathophysiological consequences of micronuclei: New insights into an age-old problem. *Mutat. Res.* **2019**, *779*, 1–35. [CrossRef] [PubMed]
72. Ellwanger, J.H.; Kulmann-Leal, B.; Ziliotto, M.; Chies, J.A.B. HIV Infection, Chromosome Instability, and Micronucleus Formation. *Viruses* **2023**, *15*, 155. [CrossRef]
73. Fenech, M.; Knasmueller, S.; Bolognesi, C.; Holland, N.; Bonassi, S.; Kirsch-Volders, M. Micronuclei as biomarkers of DNA damage, aneuploidy, inducers of chromosomal hypermutation and as sources of pro-inflammatory DNA in humans. *Mutat. Res. Rev. Mutat. Res.* **2020**, *786*, 108342. [CrossRef]
74. Nwani, C.D.; Somdare, P.O.; Ogueji, E.O.; Nwani, J.C.; Ukonze, J.A.; Nwadinigwe, A.O. Genotoxicity Assessment and Oxidative Stress Responses in Freshwater African Catfish *Clarias gariepinus* Exposed to Fenthion Formulations. *Drug Chem. Toxicol.* **2016**, *40*, 273–280. [CrossRef]
75. Zaki, M.S.; Fawzi, O.M.; Shalaby, S.I. Phenol toxicity affecting hematological changes in cat fish (*Clarius lazera*). *Life Sci. J.* **2011**, *8*, 244–248.
76. Clauss, T.M.; Dove, A.D.; Arnold, J.E. Hematologic disorders of fish. *Vet. Clin. N. Am. Exot. Anim. Pract.* **2008**, *11*, 445–462. [CrossRef]
77. Minaz, M.; Er, A.; Ak, K.; Nane, İ.D.; İpek, Z.Z.; Kurtoğlu, İ.Z.; Kayış, Ş. Short-Term Exposure to Bisphenol a (BPA) as a Plastic Precursor: Hematological and Behavioral Effects on Oncorhynchus Mykiss and Vimba Vimba. *Water Air Soil Pollut.* **2022**, *233*, 122. [CrossRef]
78. Shanthanagouda Admane Holeyappa; Kaur, A.; Bansal, N.; Ansal, M.D.; Patil, J.G.; Naveenkumar Billekallu Thammegowda; Vaneet Inder Kaur; Sethi, R.S. Biomarker-Assisted Assessment of Aquatic Health Using the Cosmopolitan Common Carp, *Cyprinus carpio* (L.): A Case Study of Bisphenol-A Exposures. *Environ. Sci. Pollut. Res.* **2021**, *29*, 14206–14218. [CrossRef]
79. Witeska, M. Stress in fish-hematological and immunological effects of heavy metals. *Electron. J. Ichthyol.* **2005**, *1*, 35–41.
80. Faheem, M.; Jahan, N.; Lone, K.P. Histopathological Effects of Bisphenol-A on Liver, Kidneys and Gills of Indian Major Carp, Catla Catla (Hamilton, 1822). *J. Anim. Plant Sci.* **2016**, *26*, 514–522.
81. Lindholst, C.; Wynne, P.; Marriott, P.J.; Pedersen, S.N.; Bjerregaard, P. Metabolism of Bisphenol a in Zebrafish (Danio Rerio) and Rainbow Trout (Oncorhynchus Mykiss) in Relation to Estrogenic Response. *Comp. Biochem. Physiol. Part C Toxicol. Pharmacol.* **2003**, *135*, 169–177. [CrossRef]
82. Ike, M.; Chen, M.-Y.; Jin, C.-S.; Fujita, M. Acute Toxicity, Mutagenicity, and Estrogenicity of Biodegradation Products of Bisphenol-A. *Environ. Toxicol.* **2002**, *17*, 457–461. [CrossRef] [PubMed]
83. Hammam, A.M.; Zaki, M.S.; Yousef, R.A.; Fawzi, O. Toxicity, Mutagenicity and carcinogenicity of phenols and phenolic compounds on human and living organisms [A Review]. *Adv. Environ. Biol.* **2015**, *9*, 38–49.
84. Cengiz, E.I. Gill and Kidney Histopathology in the Freshwater Fish *Cyprinus carpio* after Acute Exposure to Deltamethrin. *Environ. Toxicol. Pharmacol.* **2006**, *22*, 200–204. [CrossRef]
85. Sanz, A.B.; Santamaría, B.; Ruiz-Ortega, M.; Egido, J.; Ortiz, A. Mechanisms of Renal Apoptosis in Health and Disease. *J. Am. Soc. Nephrol.* **2008**, *19*, 1634–1642. [CrossRef]
86. McCampbell, K.K.; Springer, K.N.; Wingert, R.A. Atlas of Cellular Dynamics during Zebrafish Adult Kidney Regeneration. *Stem Cells Int.* **2015**, *2015*, 547636. [CrossRef]
87. Mecozzi, M.; Conti, M.E. A new approach based on polynomial functions for measuring the acute toxicity of environmental samples. *Int. J. Environ. Health* **2007**, *1*, 87–97. [CrossRef]
88. Sanyal, T.; Kaviraj, A.; Saha, S. Toxicity and Bioaccumulation of Chromium in Some Freshwater Fish. *Hum. Ecol. Risk Assess. Int. J.* **2017**, *23*, 1655–1667. [CrossRef]
89. Yossa, R.; Verdegem, M. Misuse of Multiple Comparison Tests and Underuse of Contrast Procedures in Aquaculture Publications. *Aquaculture* **2015**, *437*, 344–350. [CrossRef]
90. EFSA Scientific Committee (SC); More, S.J.; Bampidis, V.; Bragard, C.; Halldorsson, T.I.; Hernández-Jerez, A.F.; Bennekou, S.H.; Koutsoumanis, K.; Lambré, C.; Machera, K.; et al. Guidance on aneugenicity assessment. *EFSA J.* **2021**, *19*, e06770.
91. Vera-Candioti, J.; Soloneski, S.; Larramendy, M.L. Evaluation of the Genotoxic and Cytotoxic Effects of Glyphosate-Based Herbicides in the Ten Spotted Live-Bearer Fish Cnesterodon Decemmaculatus (Jenyns, 1842). *Ecotoxicol. Environ. Saf.* **2013**, *89*, 166–173. [CrossRef] [PubMed]
92. Liu, J.; Zhang, L.; Lu, G.; Jiang, R.; Yan, Z.; Li, Y. Occurrence, toxicity and ecological risk of Bisphenol A analogues in aquatic environment–A review. *Ecotoxicol. Environ. Saf.* **2021**, *208*, 111481. [CrossRef]

93. Wahby, M.M.; Abdallah, Z.M.; Abdou, H.M.; Yousef, M.I.; Newairy, A.S.A. Mitigating potential of Ginkgo biloba extract and melatonin against hepatic and nephrotoxicity induced by Bisphenol A in male rats. *Egypt. J. Basic Appl. Sci.* **2017**, *4*, 350–357.
94. Sirasanagandla, S.R.; Al-Huseini, I.; Sakr, H.; Moqadass, M.; Das, S.; Juliana, N.; Abu, I.F. Natural products in mitigation of Bisphenol a toxicity: Future therapeutic use. *Molecules* **2022**, *27*, 5384. [CrossRef]
95. Giommi, C.; Habibi, H.R.; Candelma, M.; Carnevali, O.; Maradonna, F. Probiotic administration mitigates bisphenol A reproductive toxicity in zebrafish. *Int. J. Mol. Sci.* **2021**, *22*, 9314. [CrossRef]

animals

MDPI

Article

Analysis of *clasp2* Transcription Pattern in Male Germ Cells during Spermatogenesis: A Comparative Study in Zebrafish (*Danio rerio*) and Guppy (*Poecilia reticulata*)

Serena Ricci, Maurizio Lazzari, Maria Gabriella Maurizii, Valeria Franceschini, Liliana Milani and Pietro Cacialli *

Department of Biology, Geology and Environmental Sciences, University of Bologna, 40126 Bologna, Italy
* Correspondence: pietro.cacialli@unibo.it

Simple Summary: Fertility, in males and females, is linked to the correct formation of sexual line cells (spermatozoa and oocytes) through a physiological process called meiosis. In all vertebrates, meiosis is orchestrated by different molecular effectors, and among these, we focused our attention on CLASP2, a protein involved in cytoskeleton function. In several animal models, such as mice, flies, or frogs, this protein is known to participate in spermatozoa maturation. However, in fish, nothing was known about the presence of CLASP2 in sexual organs or whether it could have a role in spermatozoa formation. To investigate this point, we used two excellent vertebrate aquatic models: guppy and zebrafish. Our results open a window to knowledge about male sex cell formation effectors in fish. Considering the great affinity between zebrafish and humans, the study of CLASP2 in fish spermatogenesis could uncover a new target of male fertility in humans.

Abstract: Cytoplasmic linker-associated protein-2 (CLASP2) is a member of the CLIP-associating proteins (CLASPs) family involved in the structure and function of microtubules and Golgi apparatus. Several studies performed using different mammalian and non-mammalian model organisms reported that CLASP2 controls microtubule dynamics and the organization of microtubule networks. In *Drosophila* and mice, an important role of CLASP2 during the development of germ cell lines has been uncovered. However, no study has clearly defined its role during fish germ cell differentiation. In the present study, we used two excellent aquatic animal models among teleost fish: zebrafish (*Danio rerio*) and guppy (*Poecilia reticulata*). Using qPCR, we found that the *clasp2* transcript level is significantly high in the testis of both fish. Then, by in situ hybridization, we localized the *clasp2* transcript in the spermatozoa of zebrafish and the spermatozeugmata of guppy. Our data suggest a potential role for this gene in the last stage of spermiogenesis in fish.

Keywords: CLASP; microtubules; spermatogenesis; zebrafish; guppy

1. Introduction

Cytoplasmic linker-associated protein-2 (CLASP2) is one of the two mammalian conserved CLIP-associating proteins (CLASPs) involved in microtubule stabilization and acentrosomal nucleation [1–3]. In detail, CLASP2 strictly cooperates with its paralogue, CLASP1, to achieve functional mitotic spindle assembly [4]. In addition, several studies in epithelial [5,6] and mainly in neuronal [7,8] cell lines have defined a key role of CLASP2 in Golgi apparatus morphology, dynamics, and function. CLASP2 has received a lot of attention as a target in cell division aberration on the premise of carcinogenesis because of these important roles during cell division. *Clasp2* expression, indeed, was found to be upregulated in bladder cancer tissues, where it seems to promote endothelial-to-mesenchymal transition (EMT) and to be associated with tumor progression, gaining attention as a prognostic factor. Similarly, enrichment analysis on histological human specimens of lung cancer defined CLASP2 and its paralogue, CLASP1, as predictive markers of tumor recurrence [9–11].

The main studies that attempted to establish the role of CLASP2 in pathophysiological processes were reported in mice (*Mus musculus*). In particular, CLASP2 appears as a key factor during mouse neocortical development as it regulates neuron formation, migration, polarity, and synaptic function [7,12]. Interestingly, recent studies defined CLASP2 as a target "sensitive" to metabolic condition changes. Phospho-proteomic wide analysis of the brain from C57BL6 mice undergoing a high-fat diet showed a higher phosphorylation pattern of CLASP2 than tissues from mice on a chow diet. Further studies in vitro then clarified that CLASP2 undergoes insulin-stimulated phosphorylation, which has a crucial role in cytoskeleton modification and is necessary for GLUT4 transporter trafficking [13–15]. Also, CLASP2 was observed in mice to participate in microtubule destabilization, which is at the base of endothelial permeability and inflammation [16,17].

Although a wide body of literature is present about CLASP2 roles in mammalian cell pathways, the first evidence about CLIP protein functions was observed in *Caenorhabditis elegans*. In *C. elegans*, the synergism between the CLASP2 orthologues CLS-2, BUB-1, and KLN-1, recruited by the kinetochore, seems to be crucial during the mitotic central spindle microtubule assembly [18,19]. Similarly, during *Xenopus laevi* embryogenesis, CLASPs, mainly transcribed in cranial nerves, cooperate with microtubule plus-end tracking proteins (+TIPs) in the regulation of microtubule dynamics [20,21]. In *Drosophila melanogaster*, the gene Orbit/Mast was identified as the orthologue of the mammalian CLASP2, and it was observed to be involved in the bundling of mitotic microtubules [22,23]. However, multiple studies showed that Orbit/Mast protein plays a key role in the meiotic division of germline cells [24]. In detail, Orbit/Mast is required at several stages of oogenesis, specifically during the formation of polarized 16-cell cysts before oocyte differentiation [25]. On the other hand, studies of immunofluorescence performed in male germline cells showed a very dynamic expression of Orbit/Mast, suggesting this protein plays a role in spermatogenesis. In spermatogonia, expression of Orbit was observed to be higher in the fusomes (germline-specific cytoskeleton), where it is involved in the proper orientation of spindles during synchronous mitosis of spermatogonium cysts [26]. During cytokinesis, the accumulation of myosin was seen to be dependent on Orbit, as the myosin heavy chain localizes in proximity to Orbit in the cleavage furrow [27,28]. However, Orbit/Mast seems to also participate in the premeiotic stage of male germ cells, as it was observed to be expressed during centriole elongation [29].

As it concerns the role of CLASP2 in fish models, at present, very scant literature is present. One comparative study in mammalian cardiomyocytes and zebrafish (*Danio rerio*) embryos showed that the formation of complex CLASP2-EB1 (microtubule end-binding protein 1), as it happens in human cells, is essential for cardiac sodium channel NaV1.5 trafficking, suggesting a similar scaffold role of this protein in microtubule stabilization [30]. Moreover, interesting insights have risen on CLASP2's involvement in the regeneration processes concerning fish nervous and hemopoietic systems. Specifically, gene expression and pathway enrichment studies in the zebrafish spinal cord injury model underlined an upregulation of the CLASP2 gene associated with axon regeneration, as it acts on microtubules to favor axon extension [30,31]. Similarly, a comparative study in mice and zebrafish showed that CLASP2 deficiency affects Hematopoietic Stem Cells (HSCs) expansion and maturation during animal development as it regulates c-Kit protein levels and Golgi apparatus integrity [32]. These previous observations seem to confirm a key role of CLASP2 in fish mitotic cell division as well as in mammals and other vertebrate models. However, there are no studies that have clearly defined its role during fish germ cell formation.

In the present studies, we aim to analyze *clasp2* gene transcription levels during fish germ line cell formation. We decided to use two excellent fish models, zebrafish and guppy (*Poecilia reticulata*) teleost fishes, in which oogenesis and spermatogenesis processes are best described [33–36]. In detail, by quantitative PCR, we measured total *clasp2* transcription levels in different zebrafish and guppy tissues, observing a higher transcription in the testis, similarly in the two models. Then, we specifically localized *clasp2* mRNA by in situ hybridization, showing *clasp2* high transcription in spermatozoa in zebrafish and spermatozeugmata in guppy.

2. Materials and Methods

2.1. Phylogenetic Analysis

Sequences of the homology cluster of the *clasp* gene family were retrieved from the results of Piccinini and Milani (2023) [37] for a subset of metazoan species (namely, zebrafish, *P. reticulata*, other *Gnathostomata* species, *Drosophila melanogaster*, and *Caenorhabditis elegans*). Sequences were aligned with MAFFT [38], trimmed with BMGE [39], and a Maximum Likelihood tree was inferred with IQTREE2 [40]. *Amphimedon queenslandica* (Porifera) was used as an outgroup to root the tree.

2.2. Dissection of Animals and Organs

The study was performed by using *AB zebrafish (*Danio rerio*) and guppy (*Poecilia reticulata*) adult males and females, respectively, obtained from a local supplier and recently analyzed for other purposes, as described in our previous published reports [34,41]. Fish were housed in an aquarium under standard photoperiod conditions (14 h light and 10 h dark) and a temperature of 28 °C. The animals have not received any medical treatment before or during the experiments.

All procedures have been performed by Italian Government Decree 26/2014 and approved by the Animal Ethical Committee of the University of Bologna (protocol no. 17/79/2014). Euthanasia of animals has been performed by using specific anesthetic drugs (ethyl 3-aminobenzoate and methanesulfonate 0.1%, Sigma Chemicals Co., St. Louis, MO, USA), and organs have been dissected for the experimental procedures.

2.3. RNA Extraction

Total RNA was extracted from different organs (3 adult males and 3 adult females for zebrafish), and 3 adult males and 3 adult females for guppy were sacrificed as described above. The following organs, from different specimens: brains, eyes, kidneys, testis, and ovaries, were dissociated using RLT buffer with the RNAeasy mini kit (Qiagen, Frankfurt, Germany). To obtain purified RNA, we followed the manufacturer's protocol. This procedure was repeated in three independent experiments.

2.4. Reverse Transcriptase PCR

Reverse transcription of mRNA into cDNA was obtained by using the Superscript III First-Strand Synthesis System kit (Invitrogen, Boston, MA, USA), according to the manufacturing procedure. In detail, 0.5 µg of total RNA was incubated with buffer and enzyme mix for 10 min at 25 °C, 30 min at 50 °C, and 5 min at 85 °C. After, samples were treated with RNase-H for 30 min at 37 °C and stored at −20 °C.

2.5. Quantitative Real-Time PCR

Quantitative real-time polymerase chain reaction (PCR) was performed by using the thermocycler with the MyiQ detector (Bio-Rad, Hercules, Dallas, TX, USA). Specifically, cDNA has been mixed with specific forward and reverse primers, SYBR-Green (Bio-Rad), and RNase-free water according to the manufacturer's protocol. The above mix was incubated for 15 min at 95 °C, then 15 s at 95 °C, 30 s at 60 °C, and 30 s at 72 °C for 40 cycles. Primer sequences used for amplification of *clasp2* and *gapdh* (housekeeping) in zebrafish and guppy are listed in Tables 1 and 2:

Table 1. qPCR forward and reverse primer sequences of *clasp2* and *gapdh* in zebrafish.

	Zebrafish	(Amplicon Size)
clasp2	F: TGGAGGCACATAAAGACC R: TGACTGGATGATGGACA	(122 bp)
Gapdh	F: GTGTAGGCGTGGACTGTGG R: TGGGAGTCAACCAGGACAAA	(151 bp)

Table 2. qPCR forward and reverse primer sequences of *clasp2* and *gapdh* in guppy.

	Guppy	(Amplicon Size)
clasp2	F: GAAGGACGTTACACGTAGAC R: CCCACAGATGTCTATCCCT	(181 bp)
Gapdh	F: CTCCACTCATGGTGTCTG R: CAACATAGTCTACGGCAGC	(140 bp)

qPCR data indicate the fold change of *clasp2* transcript levels in the brain, eye, kidney, testis, and ovary of zebrafish and guppy, using *gapdh* to normalize the absolute quantification, calculated using $2-\Delta\Delta Ct$. Correct amplification and PCR efficiency were confirmed by melting curve analysis. Each qPCR experiment has been performed in biological triplicates. In the analysis of qPCR data, each n represents the average of triplicates from a single experiment. Each experiment has been repeated three times.

2.6. Synthesis of Riboprobes for clasp2 in Zebrafish and Guppy

All digoxigenin (DIG)-labeled antisense and sense riboprobes were generated using the protocol described in our previous studies [42]. In detail, *clasp2* riboprobes for zebrafish and guppy were produced by using primers listed in Tables 3 and 4.

Table 3. Forward and reverse primer sequences for *clasp2* riboprobes in zebrafish.

	Zebrafish	Length of Riboprobe
clasp2	F: GCATTGCTGGGGATCGATA R: CGTCGAAACTACGGTCTTG	(532 bp)

Table 4. Forward and reverse primer sequences for *clasp2* riboprobes in guppy.

	Guppy	Length of Riboprobe
clasp2	F: CTCAGCTCAGGCTGCTTT R: CGCAGTTGGGAATGAGGT	(494 bp)

After amplification by PCR, each insert was cloned into the TOPO-TA vector (Invitrogen). Subsequently, transformation into thermos-competent cells was performed. After applying heat shock, bacteria were plated onto Luria–Bertani (LB) agar plates containing the appropriate antibiotic to select only the transformed cells. The transformants were screened, and the white colonies containing the insert were then grown. Then, white bacterial colonies were picked and inoculated in an LB medium containing the appropriate antibiotic. Bacteria were grown for 16 h in an orbital rotator at 37 °C. Subsequently, purification of plasmid DNA was performed using the Quick Plasmid Miniprep Kit (Invitrogen, Boston, MA, USA). Next, we confirmed the antisense and sense orientation by sequencing, and the plasmids were linearized by the enzymatic restriction. Next, using T7 polymerase (Roche-Diagnostic) and SP6 polymerase (Roche-Diagnostic, Barrington, IL, USA) with DIG-RNA Labeling Mix (Roche Diagnostic, Indianapolis, IN, USA), we performed in vitro transcription. All riboprobes were purified using NucleoSpin RNA Clean-up columns (Qiagen, Frankfurt, Germany). Verification of reaction specificity was performed by hybridizing the sense and antisense riboprobes on adjacent sections.

2.7. In Situ Hybridization

Testis from zebrafish and guppy (n = 3 males per species) were quickly dissected and fixed in paraformaldehyde (PFA) at 4% overnight at 4 °C. After 24 h, tissues were processed for paraffin embedding. Sections (7 µm thick) were cut by using a rotary microtome and mounted on slides. All paraffin sections were immersed in xylene two times for 3 min and rehydrated in ethanol at 100%, 95%, 80%, 70%, 50%, and 30% (1 min each). Next, sections of zebrafish and guppy testis were immersed in PBS, adding proteinase K diluted at 2 mg/mL at 37 °C for 5 min. All slides were fixed in paraformaldehyde 4% for 20 min, washed in PBS, and then washed in standard saline sodium citrate (SSC) twice (10 min each). Then, slides were incubated at 63 °C for 24 h in a moist chamber with the probes (2 µg/mL) diluted in a specific medium (Denhart 5×; SSC 2×; 50% formamide; ethylenediamine-tetra acetic acid 4 mM; 5% dextran sulfate; yeast tRNA 50 µg/mL). After 24 h, slides were washed with SCC 2×; 50% formamide/SCC 2×; SSC 0.2×; and SSC 0.1×. After, sections were immersed in Tris-HCl/NaCl buffer (mixing 100 mM Tris-HCl pH 7.5 and 150 mM NaCl) and washed in the same buffer containing 0.5% milk powder and 0.1% Triton. For chromogenic revelation, all sections have been incubated with anti-digoxigenin alkaline phosphatase Fab fragments, dilution 1:5000 (Roche Diagnostic Company, Chicago, IL, USA), overnight at room temperature. After 24 h, all sections were washed in Tris-HCl/NaCl buffer and with 110 mM HCl-Tris (pH 8) containing 10 mM $MgCl_2$ and 100 mM NaCl. Staining was performed using NBT/BCIP buffer (Roche, Mannheim, Germany) (pH 9.5). For fluorescence revelation, all slides were immersed in anti-DIG POD antibody (Roche, Germany) at a 1:200 dilution in the above blocking solution at room temperature for 24 h. Slides were washed for 5 min in PBS 4 times. Next, the sections were visualized by the HNPP/Fast Red Fluorescent Detection set (Roche Diagnostics, Chicago, IL, USA, #11758888001) according to the kit instructions. Mounted with DAPI (Thermo Fisher, Waltham, MA, USA) and cover slide and then observed with a fluorescence microscope (Olympus Life Science, Segrate, Milan, Italy), equipped with a DP71 digital camera, or a confocal microscope (Leica SP2). The images were processed with either the Olympus (Cell), Zeiss (AxioVision4) (Zeiss, Jana, Germany), or Leica (LCS Lite) (Leica, Wetzlar, Germania) software V-10. For all reagents and kits see Table 5.

2.8. Histology (Haematoxylin and Eosin Staining)

Adult zebrafish testes were fixed in 4% paraformaldehyde in PBS for 24h at 4 °C. After washing, the tissues were stored in 70% ethanol at 4 °C. Subsequently, tissues were embedded in paraffin, sectioned using a rotary microtome, and sections were mounted on slides. To deparaffinize, sections were put in xylene and rehydrated in an ethanol series. After, paraffin sections were washed 3× in distilled water for 1 min each time. Then, to examine cell and tissue morphology, hematoxylin and eosin (HE) staining was performed.

2.9. Statistical Analysis

Data were processed for statistical analysis using GraphPad Prism version 9.4.1 by applying one-way ANOVA with Tukey–Kramer post hoc tests, adjusted for multiple comparisons. Values of p equal to or less than 0.5 were considered statistically significant.

Catalog number for all kits and reagents used:

Table 5. Catalog number of kits and reagents.

RNAeasy minikit (Qiagen)	cat # 74104
Superscript III First-Strand Synthesis kit	cat # 12574026
TOPO™ TA Cloning™	cat # K4575J10
Dig-RNA label mix (Roche)	cat # 11277073910
HNPP/Fast Red Fluorescent Detection set	cat # 11758888001
Anti-Dig antibody (Roche)	cat # 11093274910
Proteinase K	cat # 25530049
Ethanol	cat # 100983

3. Results

3.1. Phylogenetic Tree of the Clasp Gene Family

The Maximum Likelihood tree showed that the emergence of the CLASP1 and CLASP2 paralogues preceded the evolution of gnathostomes but followed at least the emergence of deuterostomes. A subsequent duplication of CLASP1 occurred in the teleost lineage (Figure 1).

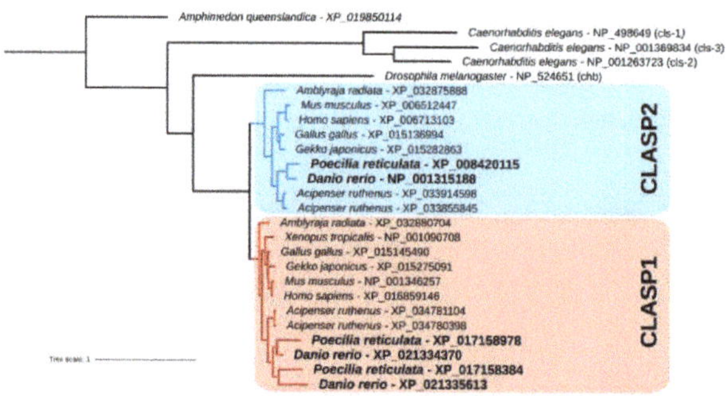

Figure 1. Phylogenetic tree of the *clasp* gene family.

3.2. Quantitative Analysis of clasp2 Transcription Level in Different Organs of Zebrafish and Guppy

For qPCR experiments, firstly, we dissected adult males and females of zebrafish (*Danio rerio*) and guppy (*Poecilia reticulata*) to remove different organs from each teleost fish (Figure 2a,b). Next, we measured the transcription level of *clasp2* in the brain, eye, kidney, testis, and ovary.

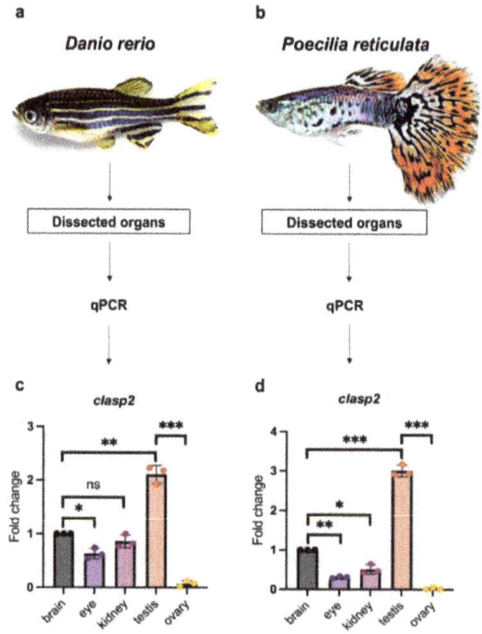

Figure 2. (**a**) Schematic experimental design outline for organs dissected from adult zebrafish (*Danio rerio*). (**b**) Schematic experimental design outline for organs dissected from an adult guppy (*Poecilia reticulata*). (**c**) qPCR analysis of *clasp2* in the brain, eye, kidney, testis, and ovary of adult zebrafish.

Clasp2 is most transcribed in the testis. Statistical analysis was completed using a one-way ANOVA with Tukey–Kramer post hoc tests, adjusted for multiple comparisons ($n = 3$ animals used). * $p < 0.01$; ** $p < 0.001$; *** $p < 0.0001$. ns—not significant. (**d**) qPCR analysis of *clasp2* in the brain, eye, kidney, testis, and ovary of an adult guppy. *Clasp2* is most transcribed in the testis. Statistical analysis was completed using a one-way ANOVA with Tukey–Kramer post hoc tests, adjusted for multiple comparisons ($n = 3$ animals used). * $p < 0.01$; ** $p < 0.001$; *** $p < 0.0001$.

In the ovary of zebrafish and guppy, *clasp2* was undetected. In the kidney and eye of both teleost fish, *clasp2* presented lower transcription compared with the brain. The most relevant data obtained from this analysis is that *clasp2* is highly transcribed in the testis of adult zebrafish and guppy compared with other organs (Figure 2c,d).

3.3. Clasp2 Transcript Is Transcribed in Spermatozoa of Adult Zebrafish Testis

We first performed hematoxylin-eosin staining of adult zebrafish testis sections to better show germ line cell morphology. In zebrafish, it is well described that spermatogenesis occurs in cysts [43]. As shown in Supplementary Figure S1, we observed numerous cysts, characterized by spermatogonia displaying heterochromatin in elongated and/or round nuclei. In the figure, it is possible to identify spermatocytes (at different meiotic division stages). Instead, spermatids, and finally, spermatozoa, are identifiable by the significant decrease in cellular and nuclear volumes.

Subsequently, to identify *clasp2* expressing cells in adult zebrafish testis, we performed fluorescence in situ hybridization. *Clasp2* mRNA was specifically transcribed in spermatozoa (Figure 3a–f).

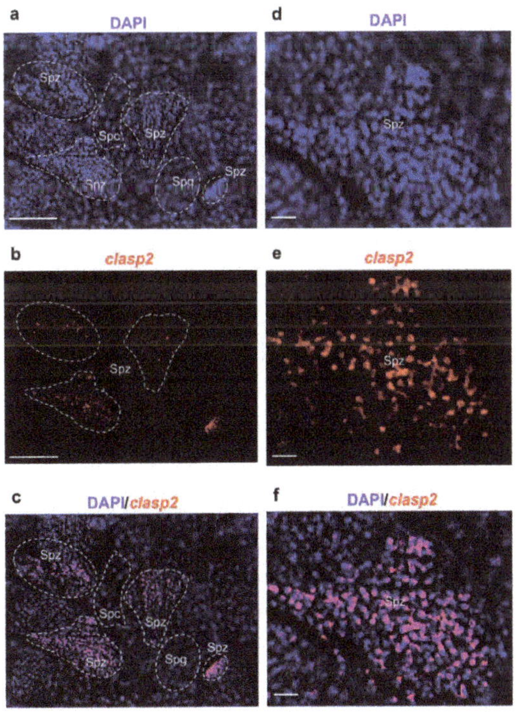

Figure 3. *Clasp2* mRNA transcription pattern in adult zebrafish testis at widefield fluorescence microscopy. (**a**) DAPI staining marks cell nuclei. (**b**) Fluorescence in situ hybridization of *clasp2*

in adult zebrafish testis. (**c**) Fluorescence in situ hybridization of *clasp2* and cell nuclei (DAPI). (**d**) High magnification of spermatozoa cell nuclei (DAPI). (**e**) High magnification of *clasp2* expressing cells. (**f**) High magnification of fluorescence in situ hybridization of *clasp2* and cell nuclei (DAPI). Abbreviations: Spz: spermatozoa; Spc: spermatocytes; Spg: spermatogonia (inside the dashed line) in adult zebrafish testis. Scale bars: 100 μ (**a**–**c**) and 25 μ (**d**–**f**).

We confirmed previous data by using confocal microscopy with higher magnification (Figure 4a–f).

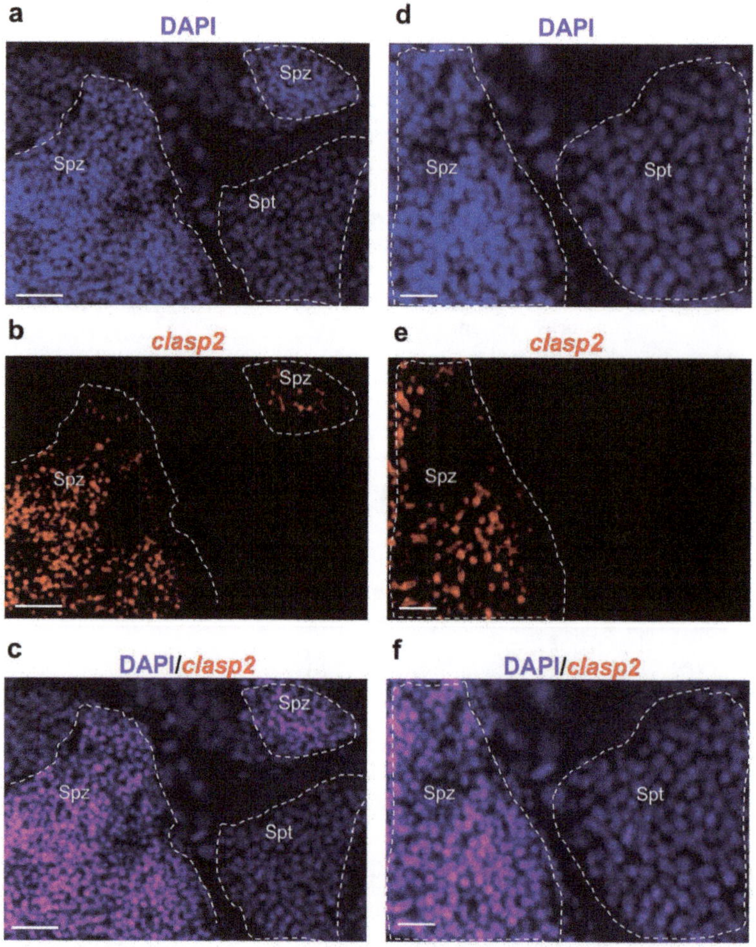

Figure 4. *Clasp2* transcription pattern in adult zebrafish testis at confocal microscopy. (**a**) DAPI staining marks cell nuclei. (**b**) Fluorescence in situ hybridization of *clasp2* in adult zebrafish testis. (**c**) Fluorescence in situ hybridization of *clasp2* and cell nuclei (DAPI) in adult zebrafish testis. (**d**) DAPI staining marks cell nuclei at high magnification. (**e**) Fluorescence in situ hybridization of *clasp2* in adult zebrafish testis at high magnification. (**f**) Fluorescence in situ hybridization of *clasp2* and cell nuclei (DAPI) in adult zebrafish testis at high magnification. Abbreviations: Spz; spermatozoa. Spt; spermatids. Scale bars: 100 μ (**a**–**c**) and 25 μ (**d**–**f**).

3.4. *Clasp2 Transcript Is Transcribed in Spermatozeugmata of Adult Guppy Testis*

As we mentioned before, the present study aimed to compare *clasp2* mRNA transcription patterns between zebrafish and guppy testis. Therefore, to identify the distribution of *clasp2* transcripts in guppy testis, we performed chromogenic and fluorescence in situ hybridization. By chromogenic revelation, we found that the *clasp2* transcript was specifically transcribed in spermatozeugmata. In detail, the staining results are dashed and present only at the periphery of spermatozeugmata where the spermatozoa heads make contact with the Sertoli cells (Figure 5a,b).

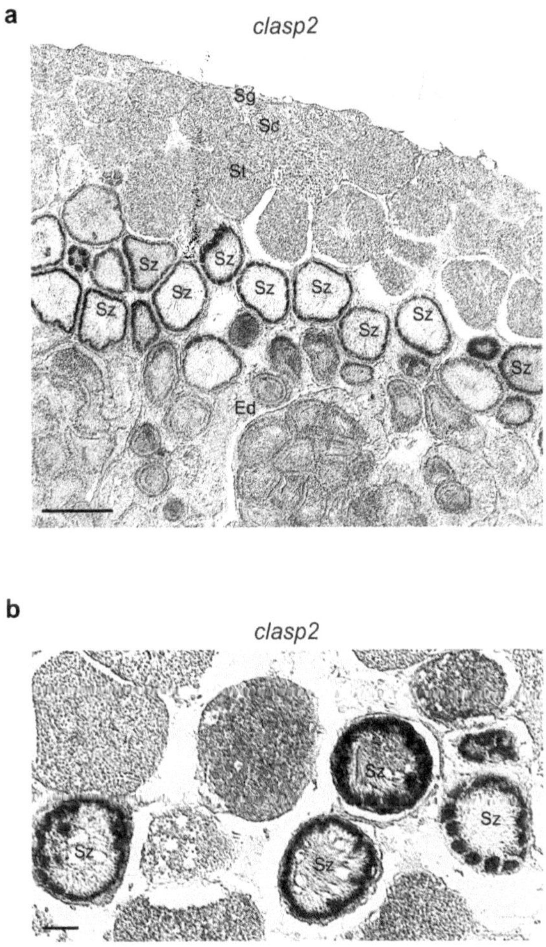

Figure 5. *Clasp2* transcript distribution in sections of adult guppy testis. (**a**) In this section, we identified the organization of the testis. Cysts containing germ cells in the early stages of spermatogenesis (spermatogonia and spermatocytes) are located near the periphery of the testis, while cysts with numerous spermatids at different stages of spermiogenesis are located deeper, near the efferent ducts. In the same region, there are present spermatozeugmata cysts, in which spermatozoa are tightly packed with the sperm heads oriented towards the Sertoli cells and flagella oriented towards the center of the cyst. Spermatozeugmata is also found inside the efferent ducts (located in the central region of the testis), where mature spermatozoa are released. Chromogenic in situ hybridization of *clasp2* in this section shows a clear marking at the periphery of the spermatozeugmata. The peripheral region of spermatozeugmata cysts located inside the efferent ducts shows weak staining (Sg: spermatogonia; Sc: spermatocytes; St: spermatids; Sz: spermatozeugmata; Ed: efferent ducts). (**b**) The strong marking of spermatozeugmata. (Sz: spermatozeugmata). Scale bars: 100 μ (**a**); 50 μ (**b**).

We confirmed these previous results by using fluorescence in situ hybridization. We found that *clasp2* transcripts are specifically distributed in spermatozoa contained in spermatozeugmata cysts (Figure 6a–e).

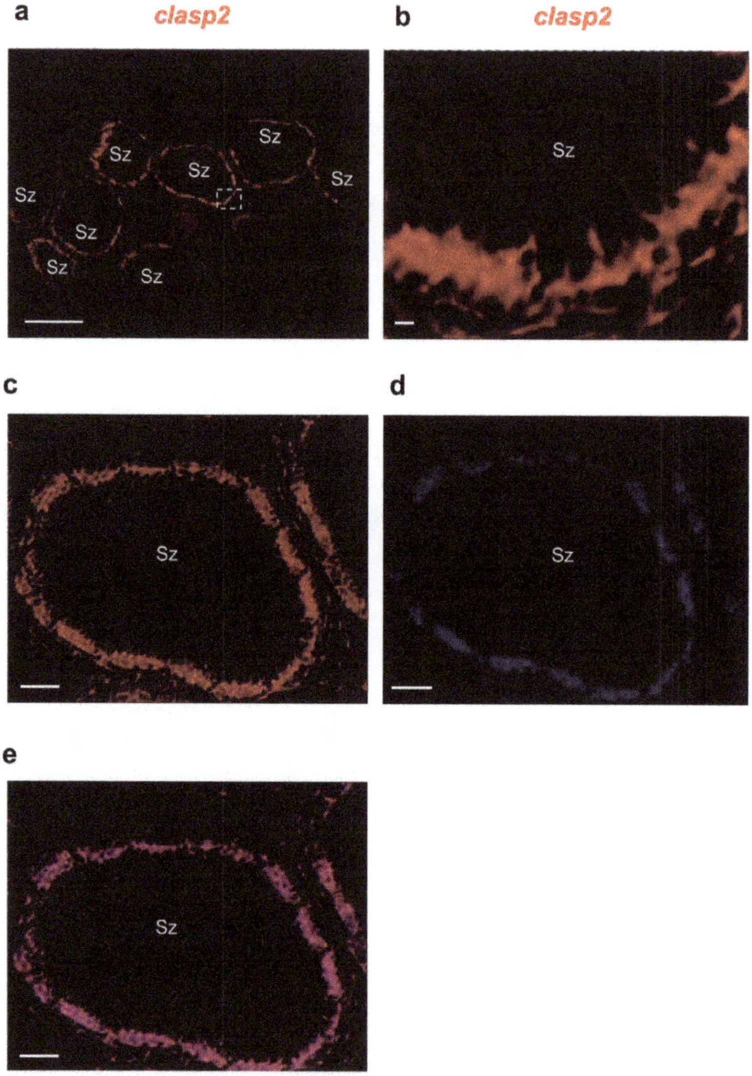

Figure 6. *Clasp2* mRNA distribution in adult guppy testis, confocal microscopy. (**a**) Fluorescence in situ hybridization of *clasp2* shows the labeling concentrated at the periphery of numerous spermatozeugmata. (**b**) High magnification of a portion of spermatozeugmata in a white rectangle is shown in panel a. (**c**) Fluorescence in situ hybridization of *clasp2* is seen at the periphery of spermatozeugmata cysts. (**d**) DAPI staining marks all the cell nuclei of the testis. (**e**) The merge of fluorescence in situ hybridization of *clasp2* and DAPI in adult guppy testis. Abbreviation: Sz; spermatozeugmata. Scale bars are 50 μ (**a**); 10 μ (**b**); and 30 μ (**c**–**e**).

4. Discussion

Microtubule dynamics and function during spindle assembly in meiotic cell divisions are crucial for the correct development of both somatic and germ cell lines in most animal species. The CLASPs protein family is a highly important regulator of the growing microtubule plus-ends, as indicated by their enrichment at these sites via tip-binding proteins such as end-binding proteins [44] (EB). Microtubule ends are also regulated by localized multimolecular ensembles of CLASPs [16,45]. For example, membrane-associated clusters of CLASP molecules capture microtubule ends at the postsynaptic membrane, whereas cortical clusters of CLASPs tether and stabilize microtubule ends near focal adhesions in motile cells or at the cell cortex [46]. Despite several animal models, such as *Drosophila*, *Xenopus*, and *C. elegans*, CLASP2 activity in meiotic divisions has been well described, and very scant knowledge is present for fish models.

For the first time in the present study, we showed the transcription pattern of *clasp2* during spermatogenesis in two small teleost fish, zebrafish and guppy. We found that *clasp2* is transcribed in several tissues, such as the brain, eye, and kidney, in a comparable fashion between zebrafish and guppy; however, higher transcription was observed in the testis. These data agree with what has been previously observed in other animal models, such as *Mus musculus* or *Drosophila* [47,48]. Thus, this might suggest a similar potential role in male germ cell differentiation. In detail, by in situ hybridization in zebrafish, we found that *clasp2* is specifically transcribed in the spermatozoa of adult testes. Interestingly, in the guppy, *clasp2* is highly transcribed in spermatozeugmata. Therefore, we could hypothesize that *clasp2* plays an important role in male germ cells during the late stage of spermatogenesis (i.e., spermiogenesis). Indeed, our results confirm previous studies performed using mouse models. In detail, breedings between homozygous CLASP2 knockout mice did not yield any pregnancies; hence, separate breedings between homozygous males and heterozygous or wild-type females did not produce any litter. This evidence led the authors to hypothesize that both male and female CLASP2$^{-/-}$ mice might be infertile. Further examination of the ovaries showed no pronounced difference in size but a difference in coloration due to hemorrhages. By contrast, testes from CLASP2 knockout mice were severely reduced in size in comparison to wild-type testes. Further histological analysis of the testes of CLASP2 knockout mice showed absent mature sperm, and the interstitium of the tubule was increased in size and filled with blood vessels [18]. In humans as well as in other animals, it is well established that spermatogenesis relies on dynamic changes in the cytoskeleton, which implies a strict regulation of microtubule dynamics [49]. Thus, microtubule dysfunction can alter the correct formation of male germ cells, and it is associated with infertility [50]. Interestingly, one recent study observed that ablation of the CLIP-170 protein (belonging to the same family of proteins as CLASP2) is associated with abnormal nucleus shapes of spermatozoa and consequent infertility [51]. However, to the best of our knowledge, no evidence has been reported about the specific role of CLASP2 in human male spermatogenesis. A possible speculation about the role of CLASP2 in male fertility could derive from preliminary data reported on The Human Protein Atlas on CLASP2 expression in Testis Cancer (https://www.proteinatlas.org/ENSG00000163539-CLASP2/pathology/testis+cancer (accessed on 12 October 2023). Indeed, it is well recognized that testicular cancer exerts profound detrimental effects on the reproductive health of men [52,53]. However, the data reported on the Protein Atlas are still preliminary, and, at present, they have not reached any prognostic value.

Thus, despite more extensive studies being needed to decipher whether and how CLASP2 is involved in fish spermatogenesis, our study represents the very first evidence of *clasp2* expression in developing male germ line cells, giving insight to further investigate this protein as a target in teleost fish meiotic cell division.

Supplementary Materials: The following supporting information can be downloaded at: https://www.mdpi.com/article/10.3390/ani13233617/s1, Supplementary Figure S1: Hematoxylin-eosin staining of adult zebrafish testis. Spg: spermatogony. Spc: spermatocyte. Spt: spermatid. Spz: spermatozoa.

Author Contributions: Conceptualization, P.C.; methodology, S.R., L.M. and P.C.; software, P.C.; validation S.R., L.M., M.L., V.F. and P.C.; formal analysis, S.R.; resources, P.C.; data curation, S.R., L.M. and M.L.; writing—original draft preparation, S.R.; writing—review and editing, P.C., L.M., S.R., M.L., M.G.M. and V.F.; visualization, S.R.; project administration, P.C.; funding acquisition, P.C. All authors have read and agreed to the published version of the manuscript.

Funding: This work was supported by national public funds grant number RFO2023-Cacialli from the Italian Ministry of University and Research (MIUR).

Institutional Review Board Statement: All procedures were performed according to the Italian Decree 26/2014 and approved by the Animal Committee of the University of Bologna (protocol no. 17/79/2014).

Informed Consent Statement: Not applicable.

Data Availability Statement: All data generated in this study are available from the corresponding author on reasonable request.

Acknowledgments: The authors thank Giovanni Piccinini for his assistance in the phylogenetic analysis.

Conflicts of Interest: The authors declare no conflict of interest.

References

1. Drabek, K.; van Ham, M.; Stepanova, T.; Draegestein, K.; van Horssen, R.; Sayas, C.L.; Akhmanova, A.; Ten Hagen, T.; Smits, R.; Fodde, R.; et al. Role of CLASP2 in microtubule stabilization and the regulation of persistent motility. *Curr. Biol.* **2006**, *16*, 2259–2264. [CrossRef]
2. Al-Bassam, J.; Chang, F. Regulation of microtubule dynamics by TOG-domain proteins XMAP215/Dis1 and CLASP. *Trends Cell Biol.* **2011**, *21*, 604–614. [CrossRef]
3. Lawrence, E.J.; Arpag, G.; Norris, S.R.; Zanic, M. Human CLASP2 specifically regulates microtubule catastrophe and rescue. *Mol. Biol. Cell* **2018**, *29*, 1168–1177. [CrossRef]
4. Pereira, A.L.; Pereira, A.J.; Maia, A.R.; Drabek, K.; Sayas, C.L.; Hergert, P.J.; Lince-Faria, M.; Matos, I.; Duque, C.; Stepanova, T.; et al. Mammalian CLASP1 and CLASP2 cooperate to ensure mitotic fidelity by regulating spindle and kinetochore function. *Mol. Biol. Cell* **2006**, *17*, 4526–4542. [CrossRef]
5. Matsui, T.; Watanabe, T.; Matsuzawa, K.; Kakeno, M.; Okumura, N.; Sugiyama, I.; Itoh, N.; Kaibuchi, K. PAR3 and aPKC regulate Golgi organization through CLASP2 phosphorylation to generate cell polarity. *Mol. Biol. Cell* **2015**, *26*, 751–761. [CrossRef]
6. Adachi, A.; Kano, F.; Tsuboi, T.; Fujita, M.; Maeda, Y.; Murata, M. Golgi-associated GSK3beta regulates the sorting process of post-Golgi membrane trafficking. *J. Cell Sci.* **2010**, *123*, 3215–3225. [CrossRef]
7. Dillon, G.M.; Tyler, W.A.; Omuro, K.C.; Kambouris, J.; Tyminski, C.; Henry, S.; Haydar, T.F.; Beffert, U.; Ho, A. CLASP2 Links Reelin to the Cytoskeleton during Neocortical Development. *Neuron* **2017**, *93*, 1344–1358.e1345. [CrossRef]
8. Sayas, C.L.; Basu, S.; van der Reijden, M.; Bustos-Moran, E.; Liz, M.; Sousa, M.; Van Ijcken, W.F.; Avila, J.; Galjart, N. Distinct Functions for Mammalian CLASP1 and -2 During Neurite and Axon Elongation. *Front. Cell Neurosci.* **2019**, *13*, 5. [CrossRef]
9. Zhu, B.; Qi, L.; Liu, S.; Liu, W.; Ou, Z.; Chen, M.; Liu, L.; Zu, X.; Wang, J.; Li, Y. CLASP2 is involved in the EMT and early progression after transurethral resection of the bladder tumor. *BMC Cancer* **2017**, *17*, 105. [CrossRef]
10. Chen, L.; Xiong, W.; Guo, W.; Su, S.; Qi, L.; Zhu, B.; Mo, M.; Jiang, H.; Li, Y. Significance of CLASP2 expression in prognosis for muscle-invasive bladder cancer patients: A propensity score-based analysis. *Urol. Oncol.* **2019**, *37*, 800–807. [CrossRef] [PubMed]
11. Valter, A.; Luhari, L.; Pisarev, H.; Truumees, B.; Planken, A.; Smolander, O.P.; Oselin, K. Genomic alterations as independent prognostic factors to predict the type of lung cancer recurrence. *Gene* **2023**, *885*, 147690. [CrossRef]
12. Beffert, U.; Dillon, G.M.; Sullivan, J.M.; Stuart, C.E.; Gilbert, J.P.; Kambouris, J.A.; Ho, A. Microtubule plus-end tracking protein CLASP2 regulates neuronal polarity and synaptic function. *J. Neurosci.* **2012**, *32*, 13906–13916. [CrossRef]
13. Siino, V.; Amato, A.; Di Salvo, F.; Caldara, G.F.; Filogamo, M.; James, P.; Vasto, S. Impact of diet-induced obesity on the mouse brain phosphoproteome. *J. Nutr. Biochem.* **2018**, *58*, 102–109. [CrossRef]
14. Langlais, P.; Dillon, J.L.; Mengos, A.; Baluch, D.P.; Ardebili, R.; Miranda, D.N.; Xie, X.; Heckmann, B.L.; Liu, J.; Mandarino, L.J. Identification of a role for CLASP2 in insulin action. *J. Biol. Chem.* **2012**, *287*, 39245–39253. [CrossRef]
15. Kruse, R.; Krantz, J.; Barker, N.; Coletta, R.L.; Rafikov, R.; Luo, M.; Hojlund, K.; Mandarino, L.J.; Langlais, P.R. Characterization of the CLASP2 Protein Interaction Network Identifies SOGA1 as a Microtubule-Associated Protein. *Mol. Cell Proteomics* **2017**, *16*, 1718–1735. [CrossRef] [PubMed]
16. Karki, P.; Ke, Y.; Zhang, C.O.; Li, Y.; Tian, Y.; Son, S.; Yoshimura, A.; Kaibuchi, K.; Birukov, K.G.; Birukova, A.A. SOCS3-microtubule interaction via CLIP-170 and CLASP2 is critical for modulation of endothelial inflammation and lung injury. *J. Biol. Chem.* **2021**, *296*, 100239. [CrossRef]

17. Karki, P.; Ke, Y.; Tian, Y.; Ohmura, T.; Sitikov, A.; Sarich, N.; Montgomery, C.P.; Birukova, A.A. Staphylococcus aureus-induced endothelial permeability and inflammation are mediated by microtubule destabilization. *J. Biol. Chem.* **2019**, *294*, 3369–3384. [CrossRef]
18. Maton, G.; Edwards, F.; Lacroix, B.; Stefanutti, M.; Laband, K.; Lieury, T.; Kim, T.; Espeut, J.; Canman, J.C.; Dumont, J. Kinetochore components are required for central spindle assembly. *Nat. Cell Biol.* **2015**, *17*, 953. [CrossRef] [PubMed]
19. Nahaboo, W.; Zouak, M.; Askjaer, P.; Delattre, M. Chromatids segregate without centrosomes during Caenorhabditis elegans mitosis in a Ran- and CLASP-dependent manner. *Mol. Biol. Cell* **2015**, *26*, 2020–2029. [CrossRef]
20. Park, E.C.; Lee, H.; Hong, Y.; Kim, M.J.; Lee, Z.W.; Kim, S.I.; Kim, S.; Kim, G.H.; Han, J.K. Analysis of the expression of microtubule plus-end tracking proteins (+TIPs) during Xenopus laevis embryogenesis. *Gene Expr. Patterns* **2012**, *12*, 204–212. [CrossRef]
21. Grimaldi, A.D.; Zanic, M.; Kaverina, I. Encoding the microtubule structure: Allosteric interactions between the microtubule +TIP complex master regulators and TOG-domain proteins. *Cell Cycle* **2015**, *14*, 1375–1378. [CrossRef]
22. Akhmanova, A.; Hoogenraad, C.C.; Drabek, K.; Stepanova, T.; Dortland, B.; Verkerk, T.; Vermeulen, W.; Burgering, B.M.; De Zeeuw, C.I.; Grosveld, F.; et al. Clasps are CLIP-115 and -170 associating proteins involved in the regional regulation of microtubule dynamics in motile fibroblasts. *Cell* **2001**, *104*, 923–935. [CrossRef]
23. Aonuma, M.; Miyamoto, M.; Inoue, Y.H.; Tamai, K.; Sakai, H.; Kamasawa, N.; Matsukage, A. Microtubule bundle formation and cell death induced by the human CLASP/Orbit N-terminal fragment. *Cell Struct. Funct.* **2005**, *30*, 7–13. [CrossRef]
24. Moriwaki, T.; Goshima, G. Five factors can reconstitute all three phases of microtubule polymerization dynamics. *J. Cell Biol.* **2016**, *215*, 357–368. [CrossRef]
25. Mathe, E.; Inoue, Y.H.; Palframan, W.; Brown, G.; Glover, D.M. Orbit/Mast, the CLASP orthologue of Drosophila, is required for asymmetric stem cell and cystocyte divisions and development of the polarised microtubule network that interconnects oocyte and nurse cells during oogenesis. *Development* **2003**, *130*, 901–915. [CrossRef]
26. Miyauchi, C.; Kitazawa, D.; Ando, I.; Hayashi, D.; Inoue, Y.H. Orbit/CLASP is required for germline cyst formation through its developmental control of fusomes and ring canals in Drosophila males. *PLoS ONE* **2013**, *8*, e58220. [CrossRef]
27. Inoue, Y.H.; Savoian, M.S.; Suzuki, T.; Mathe, E.; Yamamoto, M.T.; Glover, D.M. Mutations in orbit/mast reveal that the central spindle is comprised of two microtubule populations, those that initiate cleavage and those that propagate furrow ingression. *J. Cell Biol.* **2004**, *166*, 49–60. [CrossRef]
28. Kitazawa, D.; Matsuo, T.; Kaizuka, K.; Miyauchi, C.; Hayashi, D.; Inoue, Y.H. Orbit/CLASP is required for myosin accumulation at the cleavage furrow in Drosophila male meiosis. *PLoS ONE* **2014**, *9*, e93669. [CrossRef]
29. Shoda, T.; Yamazoe, K.; Tanaka, Y.; Asano, Y.; Inoue, Y.H. Orbit/CLASP determines centriole length by antagonising Klp10A in Drosophila spermatocytes. *J. Cell Sci.* **2021**, *134*, jcs251231. [CrossRef]
30. Marchal, G.A.; Jouni, M.; Chiang, D.Y.; Perez-Hernandez, M.; Podliesna, S.; Yu, N.; Casini, S.; Potet, F.; Veerman, C.C.; Klerk, M.; et al. Targeting the Microtubule EB1-CLASP2 Complex Modulates Na(V)1.5 at Intercalated Discs. *Circ. Res.* **2021**, *129*, 349–365. [CrossRef] [PubMed]
31. Shen, W.Y.; Fu, X.H.; Cai, J.; Li, W.C.; Fan, B.Y.; Pang, Y.L.; Zhao, C.X.; Abula, M.; Kong, X.H.; Yao, X.; et al. Identification of key genes involved in recovery from spinal cord injury in adult zebrafish. *Neural Regen. Res.* **2022**, *17*, 1334–1342. [CrossRef]
32. Klaus, A.; Clapes, T.; Yvernogeau, L.; Basu, S.; Weijts, B.; Maas, J.; Smal, I.; Galjart, N.; Robin, C. CLASP2 safeguards hematopoietic stem cell properties during mouse and fish development. *Cell Rep.* **2022**, *39*, 110957. [CrossRef]
33. Leal, M.C.; Cardoso, E.R.; Nobrega, R.H.; Batlouni, S.R.; Bogerd, J.; Franca, L.R.; Schulz, R.W. Histological and stereological evaluation of zebrafish (*Danio rerio*) spermatogenesis with an emphasis on spermatogonial generations. *Biol. Reprod.* **2009**, *81*, 177–187. [CrossRef]
34. Cacialli, P. Expression of Nerve Growth Factor and Its Receptor TrkA in the Reproductive System of Adult Zebrafish. *Vet. Sci.* **2022**, *9*, 225. [CrossRef]
35. Cacialli, P.; D'Angelo, L.; de Girolamo, P.; Avallone, L.; Lucini, C.; Pellegrini, E.; Castaldo, L. Morpho-Functional Features of the Gonads of Danio rerio: The Role of Brain-Derived Neurotrophic Factor. *Anat. Rec.* **2018**, *301*, 140–147. [CrossRef]
36. Torres-Martinez, A.; Ruiz de Dios, L.; Hernandez-Franyutti, A.; Uribe, M.C.; Sanchez, W.C. Structure of the testis and spermatogenesis of the viviparous teleost *Poecilia mexicana* (Poeciliidae) from an active sulfur spring cave in Southern Mexico. *J. Morphol.* **2019**, *280*, 1537–1547. [CrossRef] [PubMed]
37. Piccinini, G.; Milani, L. Germline-related molecular phenotype in Metazoa: Conservation and innovation highlighted by comparative transcriptomics. *Evodevo* **2023**, *14*, 2. [CrossRef]
38. Katoh, K.; Standley, D.M. MAFFT multiple sequence alignment software version 7: Improvements in performance and usability. *Mol. Biol. Evol.* **2013**, *30*, 772–780. [CrossRef]
39. Criscuolo, A.; Gribaldo, S. BMGE (Block Mapping and Gathering with Entropy): A new software for selection of phylogenetic informative regions from multiple sequence alignments. *BMC Evol. Biol.* **2010**, *10*, 210. [CrossRef]
40. Minh, B.Q.; Schmidt, H.A.; Chernomor, O.; Schrempf, D.; Woodhams, M.D.; von Haeseler, A.; Lanfear, R. IQ-TREE 2: New Models and Efficient Methods for Phylogenetic Inference in the Genomic Era. *Mol. Biol. Evol.* **2020**, *37*, 1530–1534. [CrossRef]
41. Milani, L.; Cinelli, F.; Iannello, M.; Lazzari, M.; Franceschini, V.; Maurizii, M.G. Immunolocalization of Vasa, PIWI, and TDRKH proteins in male germ cells during spermatogenesis of the teleost fish *Poecilia reticulata*. *Acta Histochem.* **2022**, *124*, 151870. [CrossRef]

42. Mahony, C.B.; Cacialli, P.; Pasche, C.; Monteiro, R.; Savvides, S.N.; Bertrand, J.Y. Hapln1b, a central organizer of the ECM, modulates kit signaling to control developmental hematopoiesis in zebrafish. *Blood Adv.* **2021**, *5*, 4935–4948. [CrossRef]
43. Menke, A.L.; Spitsbergen, J.M.; Wolterbeek, A.P.; Woutersen, R.A. Normal anatomy and histology of the adult zebrafish. *Toxicol. Pathol.* **2011**, *39*, 759–775. [CrossRef]
44. Girao, H.; Okada, N.; Rodrigues, T.A.; Silva, A.O.; Figueiredo, A.C.; Garcia, Z.; Moutinho-Santos, T.; Hayashi, I.; Azevedo, J.E.; Macedo-Ribeiro, S.; et al. CLASP2 binding to curved microtubule tips promotes flux and stabilizes kinetochore attachments. *J. Cell Biol.* **2020**, *219*, e201905080. [CrossRef]
45. Rodgers, N.C.; Lawrence, E.J.; Sawant, A.V.; Efimova, N.; Gonzalez-Vasquez, G.; Hickman, T.T.; Kaverina, I.; Zanic, M. CLASP2 facilitates dynamic actin filament organization along the microtubule lattice. *Mol. Biol. Cell* **2023**, *34*, br3. [CrossRef]
46. Mimori-Kiyosue, Y.; Grigoriev, I.; Lansbergen, G.; Sasaki, H.; Matsui, C.; Severin, F.; Galjart, N.; Grosveld, F.; Vorobjev, I.; Tsukita, S.; et al. CLASP1 and CLASP2 bind to EB1 and regulate microtubule plus-end dynamics at the cell cortex. *J. Cell Biol.* **2005**, *168*, 141–153. [CrossRef]
47. Sousa, A.; Reis, R.; Sampaio, P.; Sunkel, C.E. The Drosophila CLASP homologue, Mast/Orbit regulates the dynamic behaviour of interphase microtubules by promoting the pause state. *Cell Motil. Cytoskeleton* **2007**, *64*, 605–620. [CrossRef]
48. Drabek, K.; Gutierrez, L.; Vermeij, M.; Clapes, T.; Patel, S.R.; Boisset, J.C.; van Haren, J.; Pereira, A.L.; Liu, Z.; Akinci, U.; et al. The microtubule plus-end tracking protein CLASP2 is required for hematopoiesis and hematopoietic stem cell maintenance. *Cell Rep.* **2012**, *2*, 781–788. [CrossRef]
49. O'Donnell, L.; O'Bryan, M.K. Microtubules and spermatogenesis. *Semin. Cell Dev. Biol.* **2014**, *30*, 45–54. [CrossRef]
50. Gunes, S.; Sengupta, P.; Henkel, R.; Alguraigari, A.; Sinigaglia, M.M.; Kayal, M.; Joumah, A.; Agarwal, A. Microtubular Dysfunction and Male Infertility. *World J. Mens. Health* **2020**, *38*, 9–23. [CrossRef]
51. Lehti, M.S.; Sironen, A. Formation and function of the manchette and flagellum during spermatogenesis. *Reproduction* **2016**, *151*, R43–R54. [CrossRef]
52. Paduch, D.A. Testicular cancer and male infertility. *Curr. Opin. Urol.* **2006**, *16*, 419–427. [CrossRef]
53. Ostrowski, K.A.; Walsh, T.J. Infertility with Testicular Cancer. *Urol. Clin. N. Am.* **2015**, *42*, 409–420. [CrossRef]

animals

MDPI

Article

Influence of Transportation on Stress Response and Cellular Oxidative Stress Markers in Juvenile Meagre (*Argyrosomus regius*)

Martina Bortoletti [1,†], Elisa Fonsatti [1,†], Federico Leva [1], Lisa Maccatrozzo [1], Cristina Ballarin [1], Giuseppe Radaelli [1,*], Stefano Caberlotto [2] and Daniela Bertotto [1]

1 Department of Comparative Biomedicine and Food Science (BCA), University of Padova, Viale dell'Università 16, I-35020 Legnaro, Padova, Italy; martina.bortoletti@unipd.it (M.B.); elisa.fonsatti@phd.unipd.it (E.F.); federico.leva88@gmail.com (F.L.); lisa.maccatrozzo@unipd.it (L.M.); cristina.ballarin@unipd.it (C.B.); daniela.bertotto@unipd.it (D.B.)

2 Valle Cà Zuliani Società Agricola Srl, Via Pila 48, I-45018 Pila di Porto Tolle, Rovigo, Italy; stefano.caberlotto@vallecazuliani.it

* Correspondence: giuseppe.radaelli@unipd.it; Tel.: +39-049-8272593

† These authors contributed equally to this work.

Simple Summary: The development of aquaculture has led to an increased attention towards the welfare of farmed fish. Transportation, a routine procedure in this industry, can be stressful for fish due to handling, air exposure, confinement, and reduced oxygen levels. Therefore, it is crucial to gain insights into optimal practices for organizing transportation, ensuring the well-being of these animals and minimizing excessive stress. The objective of this study was to assess the stress response of meagre juveniles during commercial transport from Monfalcone to Stintino (Italy), which lasted 24 h. Welfare evaluation included measuring cortisol levels in muscle and assessing the potential cellular localization of oxidative stress markers in various organs and tissues. Significant increases in cortisol were observed in the fish during transfer from housing to transport tanks and in subsequent samplings at different phases of transport. This finding suggests that either the animals had insufficient time to recover or the transport conditions were suboptimal. However, in terms of oxidative stress, overall immunohistochemical analysis did not reveal variation in marker localization before and after transport, indicating that the stress experienced by the fish during transportation might have been relatively mild.

Abstract: In aquaculture, the transportation of live fish is a crucial but stress-inducing practice, necessitating a thorough understanding of its impact on fish welfare. This study aimed to assess the physiological stress response of meagre (*Argyrosomus regius*) juveniles during a 24 h commercial transport by quantifying muscle cortisol levels using a specific radioimmunoassay. Additionally, an immunohistochemical approach was used to detect and localize the cellular distribution of oxidative-stress-related biomarkers within various tissues and organs. The results demonstrated a significant increase in muscle cortisol levels following the loading procedure, remaining elevated above basal levels throughout the 24 h transport period. This effect may be attributed to either insufficient time for recovery from the loading stress or prolonged transportation-related stress. Immunostaining for all the antibodies we examined was observed in multiple tissues and organs, but we found no notable variations among the various transport phases. In conclusion, the observed stress response appears to be mainly linked to loading stress and the transport process itself, emphasizing the importance of implementing appropriate operational procedures to safeguard fish well-being during transport. Nonetheless, the unaltered distribution of oxidative stress markers between the control and transported groups suggests that the experienced stress might be within tolerable limits.

Keywords: 24 h transport; fish welfare; HSP70; HNE; NT; 8-OHdG

1. Introduction

In aquaculture, transportation is a crucial process for transferring juveniles from the hatchery to the commercial farm, where they are reared before being sent to the food market. Furthermore, in farms located far from the slaughterhouse, transportation becomes essential to convey the fish to the facility where they are sacrificed and prepared for sale [1]. Changes in water quality can adversely affect the physiology of fish [2] as well as confinement inside transport tanks, especially when fish are transported at high densities, exacerbates the situation, causing skin abrasions, loss of scales and mucus, and creating entry routes for pathogens, thereby increasing the likelihood of disease incidence [3,4].

Any stressor elicits a physiological response in animals known as the stress response [5], which unfolds in three distinct stages: primary, secondary and tertiary response. In fish, the primary response encompasses the triggering of two neuroendocrine axes. On one hand, the hypothalamus–sympathetic–chromaffin axis stimulates the release of catecholamines, primarily adrenaline and noradrenaline, from chromaffin cells. On the other hand, the hypothalamus–pituitary–interrenal axis prompts the secretion of corticosteroids, with cortisol as the primary hormone, from the interrenal tissue. In fish, as in mammals, these coordinated responses work in tandem to mobilize energy reserves, addressing the increased metabolic demands arising from the stressful situation [6].

Following the primary response, a secondary one develops, characterized by the effects produced by hormones once they reach their target tissues through the bloodstream. At the cellular level, stress frequently induces an elevation in Heat Shock Proteins' (HSPs) synthesis [7]. HSPs play a crucial role in cellular protection and stress response mechanisms [8]. The synthesis of HSP70, a member of the HSP family, is induced in fishes under stressful conditions, such as transport-related stress [9,10].

The secondary response to stress may also lead to a decline in the antioxidant defense capacity. This decrease arises from an imbalance between reactive oxygen species (ROS), which are byproducts of normal metabolic processes, and antioxidants that typically neutralize ROS [11]. Consequently, the imbalance gives rise to oxidative stress, resulting in lipid oxidation, the formation of protein carbonyls, genetic damage, and ultimately culminating in cellular demise [12,13]. Lipid peroxides are utilized as unstable indicators of oxidative stress, undergoing decomposition to produce more complex and reactive compounds, among which 4-Hydroxy-2-nonenal (HNE) is commonly employed as a biomarker for lipid peroxidation [14]. Another noteworthy ROS is the superoxide radical, which, when combined with nitric oxide, gives rise to peroxynitrite, a potent oxidant with the capacity to impact DNA, lipids, and proteins. 3-Nitrotyrosine (NT), arising from the oxidative transformation of the amino acid tyrosine, is employed as a relatively stable marker for peroxynitrite production [15]. DNA damage induced by oxidative stress can encompass the occurrence of single- and double-strand breaks, as well as base modifications such as the oxidation of deoxyguanosine to form 8-Hydroxy-2′-deoxyguanosine (8-OHdG) [16]. 8-OHdG represents a prominent biomarker of DNA damage, and it is associated with a well-established mutagenic potential, particularly involving G to T transversions [17]. The presence of all these markers has been previously observed in various fish species exposed to diverse stressors, including transportation [18], handling, temperature fluctuations [19], ammonia elevation [20], and environmental contaminants [21,22].

If the stress response continues for an extended period, a tertiary response, or the so-called chronic stress response, takes place. This response involves comprehensive changes at the organism level, including reduced growth rates and reproductive capacities, behavioral modifications, and immunosuppression. Ultimately, the chronic stress response may even lead to the animal's death [6].

In this study, we investigated the effects of a 24 h transport on the stress response in meagre (*Argyrosomus regius*) juveniles by using radioimmunoassay to measure muscle cortisol levels and immunohistochemistry to detect the presence and intracellular localization of HSP70, HNE, NT, and 8-OHdG in various organs and tissues. *Argyrosomus regius* has recently been introduced into aquaculture, gaining popularity due to its notably high growth rates and meat quality, leading to its widespread adoption [23]. Consequently, there is currently limited information available regarding the stress response of this species following commercial transport. To our knowledge, the sole available study on this subject originates from our research group [18]. Our findings will hopefully contribute to a deeper understanding of the challenges posed by live fish transportation in aquaculture and offer insights into potential strategies to enhance fish welfare during this critical phase.

2. Materials and Methods

The animals included in this study were sampled at different time intervals during a commercial transport from the hatchery to the fattening facility. All animals used in this study were handled in accordance with the guidelines of the Council Regulation (EC) No. 1/2005 which establishes transport conditions to be fulfilled in order to ensure the welfare of animals, and Directive 2010/63/EU, which regulates the use of animals for experimental and other scientific purposes.

2.1. Transport Conditions, Animals and Sampling

Argyrosomus regius juveniles were transported from Monfalcone (Gorizia, Italy) to Stintino (Sassari, Italy) following a 72 h fasting period. The fish were gently crowded using nets and then rapidly transferred by hand-nets to the tanks of the vehicle for transportation by trained personnel. The entire transportation process lasted for 24 h and was conducted using a commercial truck equipped with ten specialized tanks (2.7 m^3), insulated and equipped with aeration and temperature control systems. The fish weighed 7.5 g, and the tank density was maintained at 20 kg/m^3. During the transport, which took place in the summer season (mid-July), the temperature and level of oxygen dissolved in water were, respectively, 19–24 °C and 16–20 ppm.

Sampling occurred at four phases: before transport (control), after loading (which lasted 4 h), during the travel (16 h after departure), and at the end of the travel (24 h after departure). At each sampling point, 25 fish were sacrificed using an overdose of MS222 (1 g/L Sandoz, Milan, Italy).

2.2. Cortisol Measurement

Cortisol was measured in muscle samples obtained from the caudal peduncle of 20 fish, for each sampling time. Muscle samples were frozen by immersion in liquid nitrogen and pulverized. Then, 100 mg of the resulting powder was resuspended in 1 mL of phosphate buffer, pH 7.2 and extracted in diethyl ether. Cortisol was measured with a microtitre RIA (Radioimmunoassay), as described by Bertotto et al. [24]. The procedure was validated for the meagre and the sensitivity of the assay was 3.125 pg $well^{-1}$ [18].

2.3. Immunohistochemistry (IHC)

Immunohistochemical staining was evaluated within several organs (skin, gills, stomach, intestine, liver, pancreas, muscle) obtained from 5 fish for each sampling time. Tissues were fixed in 4% paraformaldehyde in phosphate-buffered saline (PBS, 0.1 M, pH 7.4) at 4 °C overnight, rinsed in PBS, dehydrated in a rising alcohol series and embedded in paraffin. Serial sections were cut at 4 μm for the immunohistochemical analysis.

The list of primary antibodies used in this study is consistent with that described in Bortoletti et al. [18] and is reported in Table 1. The Elite ABC KIT system (Vector Laboratories, Inc., Burlingame, CA, USA) was used for IHC, as described in Pascoli et al. [21]. Briefly, blocking of endogenous peroxidase activity was performed by incubating the sections in 3% H_2O_2 in PBS. Non-specific binding sites were blocked by incubation with normal goat serum (Dakocytomation, Milano, Italy). Following the incubation in primary antibodies (see Table 1) overnight at 4 °C, sections were washed and incubated with biotin-conjugated anti-mouse Ig antibodies (Dakocytomation, Milano, Italy). The immunoreactive sites were detected using the avidin-biotin peroxidase complex (Vector Laboratories, Inc., Burlingame, CA, USA), adding a solution of 10 mg of 3.3′-diaminobenzidine tetrahydrochloride (DAB,

Sigma, Milano, Italy) in 15 mL of 0.5 M Tris buffer at pH 7.6, containing 1.5 mL of 0.03% H_2O_2. After IHC, a nuclear counterstaining with Mayer's haematoxylin was performed. To verify the immunostaining specificity sections were incubated with: (i) PBS instead of the specific primary antibodies; (ii) pre-immune sera instead of the primary antisera; (iii) PBS instead of the secondary antibodies. The results of these controls were negative.

Table 1. Details of the primary antibodies (data from [18]).

Target Protein	Antibody	Species	Dilution	Characterization	Immunizing Antigen/Source
Heat Shock Protein 70 (HSP70)	Anti-HSP70	mouse monoclonal	1:600 (IHC)	Immunohistochemistry	Recombinant fragment from human Hsp70 (Abcam, Cambridge, UK)
4-Hydroxy-2-nonenal (HNE)	Anti-HNE	mouse monoclonal	1:50 (IHC)	Immunohistochemistry	4-Hydroxy-2-nonenal modified KLH (Abcam, UK)
3-Nitrotyrosine (NT)	Anti-NT	mouse monoclonal	1:1000 (IHC)	Immunohistochemistry	3-(4-Hydroxy-3-nitrophenylacetamido) propionic acid conjugated to bovine serum albumin (BSA) (GeneTex, Inc., Irvine, CA, USA)
8-Hydroxy-2′-guanosine (8-OHdG)	Anti-8-OHdG	mouse monoclonal	1:3000 (IHC)	Immunohistochemistry	8-Hydroxy-2′-deoxyguanosine conjugated Keyhole Limpet Hemocyanin (Abcam, UK)

2.4. Statistical Analysis

Statistical analysis was carried out by means of STATISTICA 8.0 (StatSoft, Tulsa, OK, USA). After transformation in a logarithmic scale (with a logarithmic function), the cortisol concentrations were compared by means of one-way analysis of variance (ANOVA). In the presence of significant differences between the means, a post hoc comparison of Honestly Significant Difference (HSD) for unbalanced designs was carried out.

3. Results

3.1. Stress Response

The average cortisol values detected in the muscle of the sampled fish before transport (control), after loading, during transport, and at the end of the travel are presented in Figure 1. A significant increase in cortisol levels was observed when comparing fish from the facility's holding tanks (2.28 ± 0.47 ng/g) to those in the transport tanks of the truck (6.97 ± 1.37 ng/g; $p < 0.0001$). However, no significant differences in cortisol values were found among the different transport phases ($p > 0.05$), although higher values were noted in fish at the conclusion of the travel (11.16 ± 2.15 ng/g).

3.2. HSP70 and Oxidative Stress Markers

Immunopositivity was observed in various tissues and organs for all antibodies, including the skin, stomach, intestine, hepatopancreas, and gills, while the skeletal muscle tested negative. The summary of the immunohistochemical analysis results is presented in Table 2, indicating the positivity and signal intensity levels in various tissues and organs of the animals sampled before and at the end of transportation. The data presented represent the average positivity observed in five animals per experimental group.

Overall, the immunohistochemical analysis revealed the cellular localization of all tested antibodies. However, no significant differences in signal intensity were detected between the control and transported animals, except for HNE intestinal distribution, which was more pronounced in transported animals.

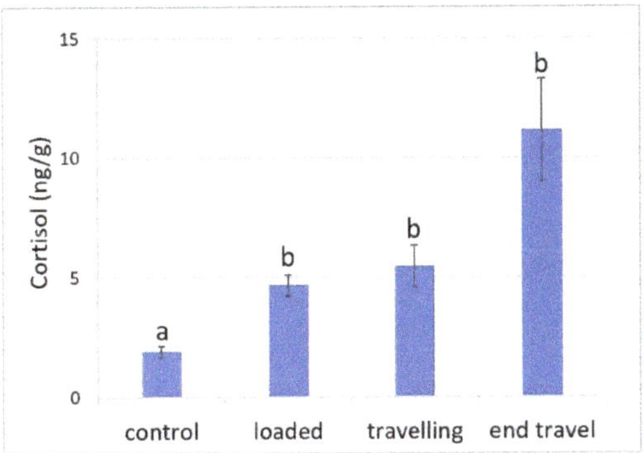

Figure 1. Cortisol concentrations detected in meagre muscle at four phases: before transport (control); after loading; during travel (after 16 h from departure) and at the end of travel (after 24 h). Data are expressed as mean ± SE (n = 20). Different letters denote statistically significant differences at the different sampling points ($p < 0.0001$).

Table 2. Immunohistochemical detection of Heat Shock Protein 70 (HSP70), 4-hydroxy-2-nonenal (HNE), 3-nitrotyrosine (NT) and 8-hydroxy-2′-deoxyguanosine (8-OHdG) in tissues of meagre (control animals and after transport): -, not detectable; +/-, slight signal; +, moderate signal; ++ marked signal.

Tissue	Control				Transported			
	HSP70	**NT**	**HNE**	**8-OHdG**	**HSP70**	**NT**	**HNE**	**8-OHdG**
Skin	+	-	-	+	+	-		+
Stomach	++	+	++	++	++	+	++	++
Intestine	++	+	+	++	++	+	++	++
Hepatopancreas	++	+/-	++	+	++	+/-	++	+
Gills	++	+/-	++	++	++	+/-	++	++
Muscle	-		-	-	-	-	-	-

For all antibodies, positive immunostaining was evident in the epithelial cells of the skin, in gill epithelium lining primary and secondary lamellae and in the cells lining the gastric mucosa and gastric glands of the stomach. Additionally, positive staining was found in the enterocytes of the intestinal mucosa as well as in the hepatocytes and pancreatic glands within the hepatic parenchyma.

In all cases, the positive immunostaining was observed at the cytoplasmic level, except for anti-8-OHdG, which exhibited nuclear positivity.

3.2.1. HSP70 Immunohistochemistry

Figure 2 illustrates a series of panels depicting the immunohistochemical localization of HSP70 in various tissues and organs of both control (A, C, E) and transported animals (B, D, F). No significant difference in terms of immunopositivity was observed between the two groups (see also Table 2 for comparison). In the hepatopancreas (Figure 2A,B), immunopositivity was evident in the cytoplasm of hepatocytes comprising the hepatic parenchyma and the acini (glands) of the pancreas. In the stomach (Figure 2C,D), immunopositivity was primarily observed in the gastric glands, while being less pronounced in the pits and epithelium lining the gastric mucosa. Similarly, in the intestine, immunopositivity was noticeable in the cytoplasm of enterocytes within the intestinal mucosa, with positive results also observed in rodlet cells (Figure 2E,F).

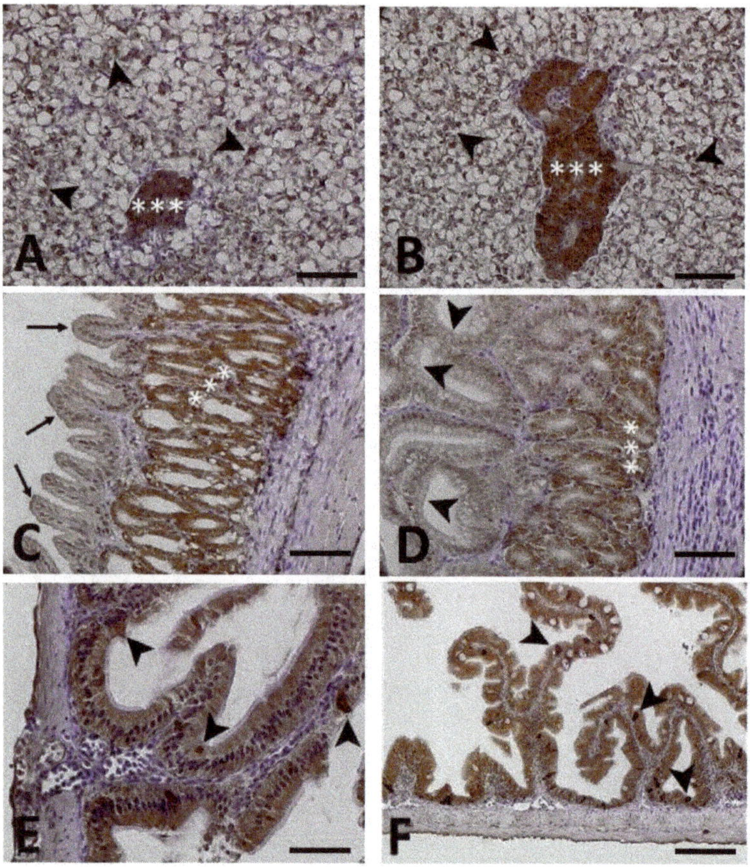

Figure 2. Immunohistochemical localization of Heat Shock Protein 70 (HSP70) in different tissues of control (**A**,**C**,**E**) and transported animals (**B**,**D**,**F**). All sections are counterstained with Mayer's hematoxylin. No notable variations in immunoreactivity were evident when comparing control and transported animals (refer also to Table 2). (**A**,**B**) Hepatopancreas: marked HSP70 immunostaining is evident in the parenchyma within the cytoplasm of hepatocytes (arrowheads) and in the acini (glands) (asterisks). (**C**,**D**) Stomach: marked immunopositivity is observed in the gastric glands (asterisks), and it is less pronounced in the pits (arrowheads) and epithelium lining the gastric mucosa (arrows). (**E**,**F**) Intestine: marked HSP70 immunoreactivity is detected in the cytoplasm of enterocytes and in the rodlet cells (arrowheads). Scale bar (**A**–**E**): 200 μm; scale bar (**F**): 100 μm.

3.2.2. HNE, NT and 8-OHdG Immunohistochemistry

Figure 3 illustrates a series of panels depicting the immunohistochemical localization of HNE (A–D), NT (E,F) and 8-OHdG (G,H) in various tissues and organs of both control (A,C,E,G) and transported animals (B,D,F,H). Except for intestinal HNE immunopositivity, no significant differences in terms of immunopositivity were observed between the two groups (see also Table 2 for comparison).

In relation to HNE, positive immunostaining was detected in the cytoplasm of hepatocytes of both control (Figure 3A) and transported animals (Figure 3B), with no discernible differences observed between the two groups. However, in the intestine, a more pronounced positivity was evident in the enterocytes of transported animals (Figure 3C) compared to the controls (Figure 3D), indicating a higher level of lipid peroxidation.

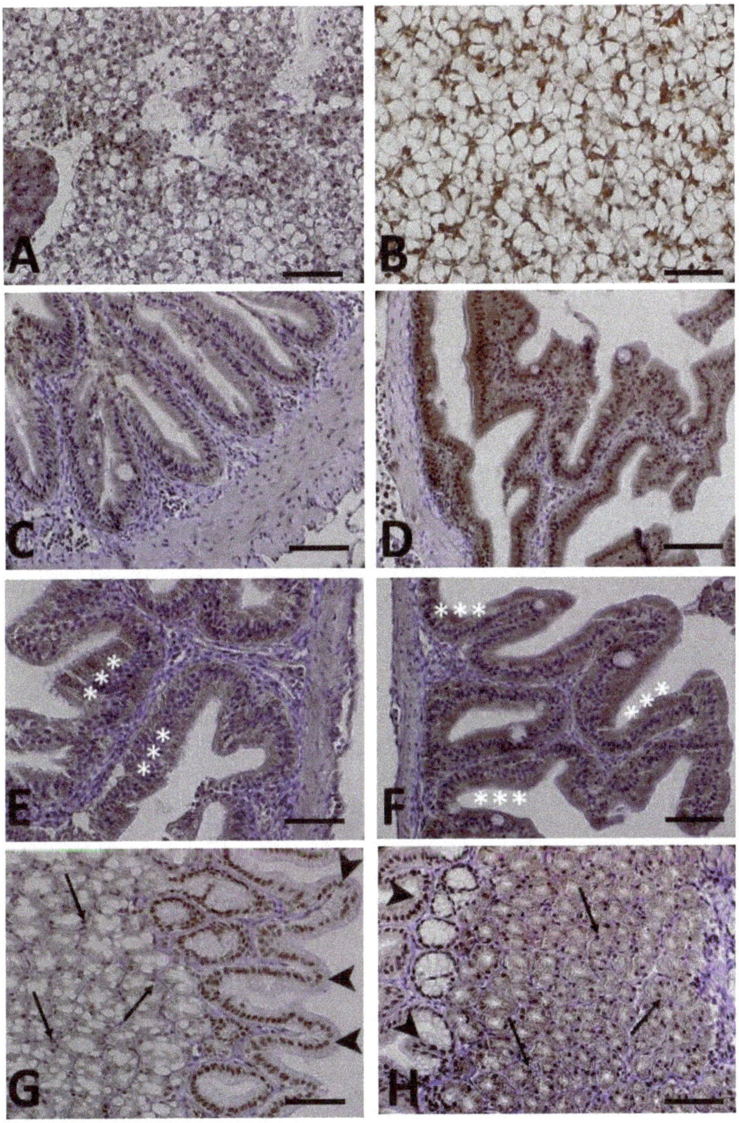

Figure 3. Immunohistochemical localization of 4-Hydroxy-2-nonenal (HNE) in panels (**A–D**), 3-Nitrotyrosine (NT) in panels (**E**,**F**) and 8-Hydroxy-2′-deoxyguanosine (8-OHdG) in panel (**G**,**H**) in meagre (*Argyrosomus regius*). Panels (**A**,**C**,**E**,**G**) show control animals, while panels (**B**,**D**,**F**,**H**) depict animals exposed to transport stress. All sections are counterstained with Mayer's hematoxylin. No notable variations in immunoreactivity were evident when comparing control and transported animals, except for HNE (refer also to Table 2). (**A**,**B**) Hepatopancreas: slight HNE immunostaining is evident in the parenchyma within the cytoplasm of hepatocytes and in the acini (glands). (**C**,**D**) Intestine: moderate (**C**) and marked (**D**) HNE immunopositivity is observed in the enterocytes of control and transported animals, respectively. (**E**,**F**) Intestine: moderate NT immunostaining in the enterocytes (asterisks). (**G**,**H**) Stomach: marked 8-OHdG immunostaining is visible in the nuclei of the secretory epithelium lining the mucosa (arrowheads), as well as in some of the nuclei of the underlying glands (arrows). Scale bar: 200 μm.

Positive immunostaining to the 3-Nitrotyrosine (NT) antibody was observed in the intestine at the level of the enterocytes (Figure 3E,F).

Figure 3G,H display a stomach positive to the 8-OHdG antibody. The immunoreactivity was observed in the nuclei of the secretory epithelium lining the mucosa, as well as in some of the nuclei of the underlying glands.

4. Discussion

The objective of this study was to assess the stress response in meagre juveniles during transportation. The assessment of physiological parameters during stressful operations, such as transportation, yields valuable insights for establishing appropriate management practices [25]. In the present study, we assessed stress parameters, including cortisol levels and the cellular expression of specific oxidative stress indicators.

The muscle cortisol levels in the fish transferred to the transport tanks were significantly higher than those of the controls. Elevations in plasma cortisol levels as a result of transport-induced stress have previously been reported in various marine and freshwater species, such as European seabass (*Dicentrarchus labrax*) [9], Pacific bluefin tuna (*Thunnus orientalis*) [26], Nile tilapia (*Oreochromis niloticus*), channel catfish (*Ictalurus punctatus*) and rainbow trout (*Oncorhynchus mykiss*) [27–30]. Consistent with other studies, fishing and loading operations onto transport tanks are commonly identified as the most stressful procedures for fish [31–34]. Before the loading procedure, fish confined in fishing nets tend to exhibit elevated cortisol levels [35]. However, in contrast to our previous study [18], cortisol levels did not show a significant decrease during transportation; instead, they exhibited a non-significant upward trend until the end of the travel. One plausible hypothesis to account for this difference is the variation in transportation duration (48 h vs. 24 h), indicating that the fish transported in the present study might not have had sufficient time for cortisol levels to return to baseline. The limited time available for recovery could be the reason explaining the maintenance of elevated levels in the end-of-travel sampling.

In our prior investigation [18], we observed a noteworthy decrease in muscle cortisol levels after 16 h from departure. However, our current study yielded contrasting results despite employing an equivalent sampling timeframe (16 h) for both transportation methods. Hence, we speculate that additional factors, such as transportation conditions, pre-transportation fish condition, and potential disparities between fish stocks, might have influenced the observed variation. In our analysis, we meticulously assessed essential transportation parameters that influence animal welfare, encompassing water quality and stocking density, and found them to be consistent between both trips.

Another hypothesis is that differences in the fish stocks transported in the two travels led to varied initial conditions. It is possible that some fish within the transported groups may be more susceptible to stressors, such as disturbances caused by inefficient personnel or suboptimal water conditions. It is important to acknowledge individual differences in stress response, with varying degrees of "stress responsiveness" [6]. These individual variations could have contributed to the observed differences in cortisol levels and stress response between the two transportation scenarios.

Furthermore, variations in cortisol levels during the two transports may result from differences in the loading procedure and/or transportation process due to different operators involved. It is possible some fish might have experienced more manipulations or impactful events, such as air exposure, during loading compared to others. The observed differences in transported animals could be influenced by a combination of these factors. This underscores the necessity for precise study, planning, and execution of practices during complex and impactful procedures like transportation. Implementing standardized protocols and minimizing potential stressors can help reduce variability and enhance the accuracy and reliability of experimental results.

In addition to evaluating the primary stress response in the present study, we also investigated the secondary stress response by assessing indicators such as Heat Shock Protein 70 (HSP70) and oxidative stress products (4-Hydroxynonenal (HNE), 3-Nitrotyrosine

(NT), and 8-Hydroxy-2′-deoxyguanosine (8-OHdG)) through immunohistochemistry. This technique facilitated the examination of the presence and cellular localization of these molecules of interest using specific antibodies.

The immunohistochemical investigation provided valuable insights into the distribution of different markers across various tissues and organs in both the control and transported animals. Positive results for the various markers were observed in the epidermis of the skin, intestine, gill lamellae, as well as the gastric glands and hepatopancreas. However, no substantial variations in signal intensity were observed when comparing control and transported animals, in accordance with previous studies on the effect of transport stress in marine fish species using immunohistochemical techniques [9,18]. Nevertheless, a notable exception was observed in the case of HNE. In the present study, most tissues exhibited comparable levels of HNE immunopositivity between transported and control animals, except for the intestine, where the signal appeared more pronounced in the transported animals. This observation aligns with the understanding that lipid peroxides, sensitive markers of cellular oxidative stress, can undergo subsequent reactions, leading to the formation of more complex and reactive compounds, including MDA (malondialdehyde) and HNE (4-Hydroxynonenal) [16]. Recent research conducted by Fiocchi et al. [19] provided further evidence of increased HNE immunopositivity in various tissues of European seabass subjected to handling and temperature increase, reinforcing the role of HNE as an indicator of oxidative stress in fish. Additionally, a separate investigation on the effects of transportation on rainbow trout reported a significant increase in MDA content during the transportation process [36], underscoring MDA as a pertinent indicator of oxidative damage [37]. These findings collectively suggest that the transportation process can impact oxidative stress markers in fish, with HNE showing a pronounced response in the intestine. Understanding these oxidative stress indicators contributes to a deeper comprehension of the physiological responses of fish during transportation, thereby enhancing knowledge of fish welfare and health aspects in the context of transportation-related stress.

To summarize, the detected cortisol levels confirmed the presence of a stress response due to transportation, with the pre-transport procedures, particularly the loading process, being a significant stress factor. Following this operation, cortisol levels increased and remained elevated throughout transportation, indicating that the animals did not fully recover from the stress experienced. There are two possible explanations for this: the limited available time or the highly stressful transportation conditions.

With respect to the cellular localization of oxidative stress markers, immunohistochemistry analysis showed, overall, no significant variation before and after transportation, suggesting that the stress experienced by the fish during transportation might have been relatively mild.

Lastly, it is essential to highlight that cortisol continues to be a valuable indicator of well-being in farmed species, and muscle serves as an intriguing matrix for evaluating cortisol levels and monitoring the stress response. Muscle is less influenced by the sampling effect while remaining highly comparable to a blood matrix, making it applicable even in situations where blood sampling is not possible, such as in *post-mortem* examinations.

5. Conclusions

This study focused on describing and analyzing the stress response to transportation in meagre juveniles. The observed stress response appeared to be primarily associated with loading stress and the transport process itself, underscoring the significance of implementing appropriate operational protocols to preserve fish welfare. Nevertheless, the overall unaltered distribution of oxidative stress markers between control and transported fish suggests that the experienced stress might have been relatively mild. The use of muscle instead of blood for cortisol evaluation once again proved to be an excellent tool in assessing *post-mortem* stress, particularly when blood sampling is not feasible. These findings contribute to our understanding of the stress factors involved in fish transportation and emphasize the need for further research to enhance fish well-being during transport operations.

Author Contributions: Conceptualization, M.B., G.R. and D.B.; formal analysis, D.B.; funding acquisition, G.R. and D.B.; investigation, M.B., E.F., F.L., C.B. and S.C.; project administration, G.R. and D.B.; resources, G.R., S.C. and D.B.; supervision, L.M., G.R. and D.B.; validation, M.B., G.R. and D.B.; visualization, M.B., E.F. and F.L.; writing—original draft, M.B., C.B., G.R. and D.B.; writing—review and editing, E.F., F.L. and L.M. All authors have read and agreed to the published version of the manuscript.

Funding: This research was funded by project Fondi di Ateneo of the University of Padova to Daniela Bertotto (60A08-1811/13) and Giuseppe Radaelli (CPDA123738).

Institutional Review Board Statement: The study was conducted in accordance with the principles outlined in Regulation (EC) No. 1/2005 for the protection of animals during transport and Directive 2010/63/EU for the protection of animals used for experimental and other scientific purposes.

Data Availability Statement: The data presented in this study are available within the article.

Acknowledgments: The authors wish to thank Elena Negrato, Giovanni Caporale and Carlo Poltronieri for their technical assistance.

Conflicts of Interest: The authors declare no conflict of interest. The funders had no role in the design of the study; in the collection, analyses, or interpretation of data; in the writing of the manuscript; or in the decision to publish the results.

References

1. Bergqvist, J.; Gunnarsson, S. Finfish Aquaculture: Animal Welfare, the Environment, and Ethical Implications. *J. Agric. Environ. Ethics* **2013**, *26*, 75–99. [CrossRef]
2. Sampaio, F.D.F.; Freire, C.A. An overview of stress physiology of fish transport: Changes in water quality as a function of transport duration. *Fish Fish.* **2016**, *17*, 1055–1072. [CrossRef]
3. King, H.R. Fish transport in the aquaculture sector: An overview of the road transport of Atlantic salmon in Tasmania. *J. Vet. Behav.* **2009**, *4*, 163–168. [CrossRef]
4. Urbinati, E.C.; Abreu, J.S.; Camargo, A.C.; Parra, M.A. Loading and transport stress of juvenile matrinxã (*Brycon cephalus*, Characidae) at various densities. *Aquaculture* **2004**, *229*, 389–400. [CrossRef]
5. Helmut, S.; Henrik, S.; Kurt, B.; Jessica, D.; Kristina, S.S.; Cédric, M.; Neil, R.; Fredrik, J.; Hilde, T.; Lloyd, V. Health of farmed fish: Its relation to fish welfare and its utility as welfare indicator. *Fish Physiol. Biochem.* **2011**, *38*, 85–105. [CrossRef]
6. Wendelaar Bonga, S.E. The stress response in fish. *Physiol. Rev.* **1997**, *77*, 591–625. [CrossRef]
7. Iwama, G.K.; Thomas, P.T.; Forsyth, R.B.; Vijayan, M.M. Heat shock protein expression in fish. *Rev. Fish Biol. Fish.* **1998**, *8*, 35–56. [CrossRef]
8. Yamashita, M.; Yabu, T.; Ojima, N. Stress protein HSP70 in fish. *Aqua-Biosci. Monogr.* **2010**, *3*, 111–141. [CrossRef]
9. Poltronieri, C.; Maccatrozzo, L.; Simontacchi, C.; Bertotto, D.; Funkenstein, B.; Patruno, M.; Radaelli, G. Quantitative RT-PCR analysis and immunohistochemical localization of HSP70 in sea bass *Dicentrarchus labrax* exposed to transport stress. *Eur. J. Histochem.* **2007**, *51*, 125–135.
10. Poltronieri, C.; Negrato, E.; Bertotto, D.; Majolini, D.; Simontacchi, C.; Radaelli, G. Immunohistochemical localization of constitutive and inducible heat shock protein 70 in carp (*Cyprinus carpio*) and trout (*Oncorhynchus mykiss*) exposed to transport stress. *Eur. J. Histochem.* **2008**, *52*, 191–198. [CrossRef]
11. Halliwell, B.; Gutteridge, J.M.C. *Free Radicals in Biology and Medicine*, 3rd ed.; Oxford University Press: Oxford, UK, 1999; pp. 1–905. [CrossRef]
12. Davies, K.J.A. Oxidative stress, the paradox of aerobic life. In *Free Radical and Oxidative Stress: Environment, Drugs and Food Additives*, 2nd ed.; Rice Evans, C., Halliwell, B., Land, G.G., Eds.; Portland Press: London, UK, 1995; Volume 40, pp. 1–31.
13. Winston, G.W.; Di Giulio, R.T. Prooxidant and antioxidant mechanisms in aquatic organisms. *Aquat. Toxicol.* **1991**, *19*, 137–161. [CrossRef]
14. Aldini, G.; Dalle-donne, I.; Facino, R.M.; Milzani, A.; Carini, M. Intervention Strategies to Inhibit Protein Carbonylation by Reactive Carbonyls. *Med. Res. Rev.* **2006**, *27*, 817–868. [CrossRef]
15. Ischiropoulos, H.; Zhu, L.; Chen, J.; Tsai, M.; Martin, J.C.; Smith, C.D.; Beckman, J.S. Peroxynitrite-mediated tyrosine nitration catalyzed by superoxide dismutase. *Arch. Biochem. Biophys.* **2002**, *298*, 431–437. [CrossRef]
16. Ploch, S.A.; Lee, Y.; MacLean, E.; Di Giulio, R.T. Oxidative stress in liver of brown bullhead and channel catfish following exposure to tertbutyl hydroperoxide. *Aquat. Toxicol.* **1999**, *46*, 231–240. [CrossRef]
17. Pilger, A.; Rüdiger, H.W. 8-Hydroxy-2′-deoxyguanosine as a marker of oxidative DNA damage related to occupational and environmental exposure. *Int. Arch. Occup. Environ. Health* **2006**, *80*, 1–15. [CrossRef]
18. Bortoletti, M.; Maccatrozzo, L.; Radaelli, G.; Caberlotto, S.; Bertotto, D. Muscle Cortisol Levels, Expression of Glucocorticoid Receptor and Oxidative Stress Markers in the Teleost Fish *Argyrosomus regius* Exposed to Transport Stress. *Animals* **2021**, *11*, 1160. [CrossRef]

19. Fiocchi, E.; Civettini, M.; Carbonara, P.; Zupa, W.; Lembo, G.; Manfrin, A. Development of molecular and histological methods to evaluate stress oxidative biomarkers in sea bass (*Dicentrarchus labrax*). *Fish Physiol. Biochem.* **2020**, *46*, 1577–1588. [CrossRef]
20. Zarantoniello, M.; Bortoletti, M.; Olivotto, I.; Ratti, S.; Poltronieri, C.; Negrato, E.; Caberlotto, S.; Radaelli, G.; Bertotto, D. Salinity, Temperature and Ammonia Acute Stress Response in Seabream (*Sparus aurata*) Juveniles: A Multidisciplinary Study. *Animals* **2021**, *11*, 97. [CrossRef]
21. Pascoli, F.; Negrato, E.; Di Giancamillo, A.; Bertotto, D.; Domeneghini, C.; Simontacchi, C.; Mutinelli, F.; Radaelli, G. Evaluation of oxidative stress biomarkers in *Zosterisessor ophiocephalus* from the Venice Lagoon, Italy. *Aquat. Toxicol.* **2011**, *101*, 512–520. [CrossRef]
22. Topal, A.; Gergit, A.; Özkaraca, M. Assessment of oxidative DNA damage, oxidative stress responses and histopathological alterations in gill and liver tissues of *Oncorhynchus mykiss* treated with linuron. *Hum. Exp. Toxicol.* **2020**, *40*, 1112–1121. [CrossRef]
23. Parisi, G.; Terova, G.; Gasco, L.; Piccolo, G.; Roncarati, A.; Moretti, V.M.; Centoducati, G.; Gatta, P.P.; Pais, A. Current status and future perspectives of Italian finfish aquaculture. *Rev. Fish Biol. Fish.* **2014**, *24*, 15–73. [CrossRef]
24. Bertotto, D.; Poltronieri, C.; Negrato, E.; Majolini, D.; Radaelli, G.; Simontacchi, C. Alternative matrices for cortisol measurement in fish. *Aquac. Res.* **2010**, *41*, 1261–1267. [CrossRef]
25. Gomes, L.C.; Edsandra, C.C.; Richard, P.B.; Rodrigo, R.; Carlos, E.C.; Bernardo, B. Use of salt during transportation of air breathing pirarucu juveniles (*Arapaima gigas*) in plastic bags. *Aquaculture* **2006**, *256*, 521–528. [CrossRef]
26. Honryo, T.; Oakada, T.; Kawahara, M.; Kurata, M.; Agawa, Y.; Sawada, Y.; Miyashita, S.; Takii, K.; Ishibashi, Y. Estimated time for recovery from transportation stress and starvation in juvenile Pacific bluefin tuna *Thunnus orientalis*. *Aquaculture* **2018**, *484*, 175–183. [CrossRef]
27. Morteza Yousefi, M.; Hoseini, S.M.; Weber, R.A.; Silva, E.; Rajabiesterabadi, H.; Arghideh, M.; Delavar, F.H. Alleviation of transportation-induced stress in Nile tilapia, *Oreochromis niloticus*, using brackish water. *Aquac. Rep.* **2022**, *27*, 101378. [CrossRef]
28. Refaey, M.M.; Tian, X.; Tang, R.; Li, D. Changes in physiological responses, muscular composition and flesh quality of channel catfish *Ictalurus punctatus* suffering from transport stress. *Aquaculture* **2017**, *478*, 9–15. [CrossRef]
29. Refaey, M.M.; Li, D. Transport stress changes blood biochemistry, antioxidant defense system, and hepatic HSPs mRNA expressions of channel catfish *Ictalurus punctatus*. *Front. Physiol.* **2018**, *9*, 1628. [CrossRef]
30. Tacchi, L.; Lowrey, L.; Musharrafieh, R.; Crossey, K.; Larragoite, E.T.; Salinas, I. Effects of transportation stress and addition of salt to transport water on the skin mucosal homeostasis of rainbow trout (*Oncorhynchus mykiss*). *Aquaculture* **2015**, *435*, 120–127. [CrossRef]
31. Schreck, C.; Solazzi, M.; Johnson, S.; Nickelson, T. Transportation stress affects performance of coho salmon, *Oncorhynchus kisutch*. *Aquaculture* **1989**, *82*, 15–20. [CrossRef]
32. Weirich, C.; Tomasso, J. Confinement- and Transport-Induced Stress on Red Drum Juveniles: Effect of Salinity. *Progress. Fish-Cult.* **1991**, *53*, 146–149. [CrossRef]
33. Erikson, U.; Sigholt, T.; Rustad, T.; Einarsdóttir, I.E.; Jørgensen, L. Contribution of Bleeding to Total Handling Stress During Slaughter of Atlantic Salmon. *Aquac. Int.* **1999**, *7*, 101–115. [CrossRef]
34. Dobšíková, R.; Svobodová, Z.; Blahova, J.; Modra, H.; Velisek, J. The effect of transport on biochemical and haematological indices of common carp (*Cyprinus carpio* L.). *Czech J. Anim. Sci.* **2009**, *54*, 510–518. [CrossRef]
35. Skjervold, P.O.; Fjæra, S.O.; Østby, P.B.; Einen, O. Live-chilling and crowding stress before slaughter of Atlantic salmon (*Salmo salar*). *Aquaculture* **2001**, *192*, 265–280. [CrossRef]
36. Ren, Y.; Men, X.; Yu, Y.; Li, B.; Zhou, Y.; Zhao, C. Effects of transportation stress on antioxidation, immunity capacity and hypoxia tolerance of rainbow trout (*Oncorhynchus mykiss*). *Aquac. Rep.* **2022**, *22*, 100940. [CrossRef]
37. Liu, X.L.; Xi, Q.Y.; Yang, L.; Li, H.Y.; Jiang, Q.Y.; Shu, G.; Wang, S.B.; Gao, P.; Zhu, X.T.; Zhang, Y.L. The effect of dietary Panax ginseng polysaccharide extract on the immune responses in white shrimp, *Litopenaeus vannamei*. *Fish Shellfish Immunol.* **2011**, *30*, 495–500. [CrossRef] [PubMed]

animals

MDPI

Article

Morphology, Histology, and Histochemistry of the Digestive Tract of the Marbled Flounder *Pseudopleuronectes yokohamae*

Jeong-Hyeon Cho [1], Jin Woo Park [1], Yong-Woon Ryu [1], Kang-Woong Kim [2] and Sang-Woo Hur [2,*]

[1] Jeju Fisheries Research Institute, National Institute of Fisheries Science, Jeju 63610, Republic of Korea
[2] Aquafeed Research Center, National Institute of Fisheries Science, Pohang 37517, Republic of Korea
* Correspondence: maverickhur@korea.kr; Tel.: +82-54-230-3630

Simple Summary: The marbled flounder (*Pseudopleuronectes yokohamae*) is a fish species with high commercial value in Korea and has been gaining attention as a new aquaculture target. To clarify the digestive physiology and feeding habits of the marbled flounder, this study conducted morphological, histological, and histochemical analyses of its digestive tract. In conclusion, the feeding habits of marbled flounder are close to those of carnivorous fish. In addition, we found that the marbled flounder has a stomach with a structure suitable for eating small amounts of hard food, and that most of its digestion takes place in the anterior intestine portion. Our study presents the histological and histochemical characteristics of the digestive tract of the marbled flounder, providing basic knowledge for both physiological and nutritional studies.

Abstract: This study investigated the morphological, histological, and histochemical characteristics of the digestive tract of the marbled flounder (*Pseudopleuronectes yokohamae*). The relative length of the gut of the marbled flounder digestive tract was 1.54 ± 0.10 ($n = 20$), and it had a simple stomach and 6–9 pyloric caeca. The mucosal folds of the marbled flounder digestive tract exhibited a general branched morphology. The thickness and mucosal fold length of the intestinal muscularis externa showed similar aspects in all areas. The thickness of the intestinal muscularis externa was the thickest in the posterior intestine portion, and the length of mucosal folds was the longest in the anterior intestine portion. It was indicated that food digested by gastric acid in the stomach moves to the anterior portion (including pyloric caeca) and mid portion of the intestine, ensuring effective stimulation of cholecystokinin (CCK)-producing cells. In addition, the distribution pattern of CCK-producing cells in the intestine was very similar to that of mucus-secreting goblet cells. The CCK-producing cells and goblet cells in the marbled flounder were well-adapted to promote optimal control of the digestive process. Based on the morphological and histochemical studies, it was concluded that the marbled flounder displays a digestive tract comparable to that of fish species with carnivorous habits.

Keywords: marbled flounder; digestive tract; digestive physiology; immunohistochemistry; endocrine cells

1. Introduction

The digestive tract of fish is anatomically composed of a long hollow tube, and as in other vertebrates, it is functionally in charge of important processes such as digestion, absorption, and transportation of nutrients necessary for body growth from food [1,2]. While the digestive tract of most fish is generally composed of a series of tubes including the mouth, pharynx, esophagus, stomach, intestine, rectum, and anus, the shape and structure vary depending on the species and feeding habits [3]. In particular, due to the diversity of feeding and phylogenetic characteristics of the digestive tract of teleosts, many studies related to the morphological and structural characteristics and ontogenetic development of the digestive tract have been reported [4–8]. The diversity of morphological and structural characteristics of fish digestive tracts indicates that each species has different dietary and digestive physiology characteristics [9,10]. Therefore, to understand the diet and digestive physiology of fish, histological studies on digestive tract morphology and internal structure must be performed; and since the understanding of the fish digestive tract is related to fish nutrition, such studies can also contribute to fish management and conservation [10].

The digestive system is the largest endocrine organ in vertebrates [2]. Hormones and a wide range of signaling molecules secreted by various types of endocrine cells can rapidly and reversibly change the properties of the digestive system and other organ systems [11]. Energy and nutrients are provided in a sequential process, where complex polymers are absorbed across the apical membrane of epithelial cells, transported to the systemic circulation, and hydrolyzed into smaller molecules [12–14]. Digestive system function is regulated by various digestive hormones and substances. Among them, cholecystokinin (CCK) and mucus-secreting goblet cells play an important physiological role in the intestine of vertebrates, including fish. CCK, a gastrointestinal hormone, is a neurotransmitter peptide abundant in the brain and gut [15]. CCK plays an important role in pancreatic enzyme secretion, gallbladder contraction, amino acid and sugar transport regulation, and intestinal peristalsis regulation [15]. In addition, mucus-secreting goblet cells in the fish digestive tract produce lubricant on the surface of the mucous membrane, protecting the mucous membrane from damage caused by physical or chemical substances and digestive enzymes [16]. In addition, mucus secreted from goblet cells of vertebrates, including fish, plays an important role in the digestion and absorption of substances [16].

The marbled flounder (*Pseudopleuronectes yokohamae*), found throughout Korea, southern Hokkaido in Japan, and the East China Sea, is a fish species with high commercial value in both Korea and Japan [17,18] and has been gaining attention as a new aquaculture target [19]. Previous studies determined the marbled flounder to be a benthic fish that feeds on macrozoobenthos, such as Pelecypoda, Amphipoda, and Polychaete [19]; however, there have been few histological and histochemical studies concerning the digestive tract of species of the *Pseudopleuronectes* genus, including the marbled flounder. Interest in the digestive physiology of cultured fish is based on the expectation that the application of histological and histochemical knowledge to the development of aquaculture and husbandry prevents digestive disorders and helps in understanding the regulatory aspects of fish nutrition [20]. Therefore, this study aims to provide basic knowledge for physiological and nutritional studies by describing the histological and histochemical characteristics of the marbled flounder digestive tract.

2. Materials and Methods

Sample collection was carried out at the Aquafeed Research Center (Pohang, Korea) of the National Institute of Fisheries Science (NIFS) of Korea in accordance with the Guidelines for Experimental Animals of NIFS Institutional Animal Care Use Committee (2019-NIFS-IACUC-13).

2.1. Specimens

Twenty marbled flounder juveniles, with average body length and body weight of 14.6 ± 0.5 cm and 56.6 ± 5.1 g, respectively, were obtained from the Aquafeed Research Center, Pohang, Korea. The first broodstock generation had been hatched in the Fisheries Resources Institute, Gyeongsangbuk-do, located near our research facility (ca. 45 km). Fish were kept in natural photoperiod and water temperature conditions. The collected specimens were anaesthetized with an overdose of 2-phenoxyethanol (150 ppm; Sigma-Aldrich, St. Louis, MO, USA) and their body lengths were measured. The body cavity was cut open through a mid-ventral incision and the digestive system, from the esophagus to the anus, along with the accessory organs, was removed through dissection from the body cavity. The length of the digestive tract was measured, and relative length of the gut (RLG) was calculated using the equation RLG = DL/SL, where DL is the digestive tract length (cm) and SL is the standard length (cm) (Table 1).

Table 1. Body length, body weight, digestive tract length, and relative length of gut of the marbled flounder *Pseudopleuronectes yokohamae*.

	Marbled Flounder
Sample number	20
Body length (cm)	14.6 ± 0.5
Body weight (g)	56.6 ± 5.1
Digestive tract length (cm)	22.5 ± 2.0
Relative length of the gut	1.54 ± 0.10

Values are the means ± standard deviation (n = 20).

2.2. Histochemistry

Samples from nine different regions of the digestive tract (esophagus, cardiac stomach portion, fundic stomach portion, pyloric stomach portion, pyloric caeca, anterior intestine portion, mid intestine portion, posterior intestine portion, and rectum) (Figure 1) were fixed in Bouin's solution (Sigma-Aldrich, St. Louis, MO, USA), dehydrated through an ethanol–xylene series, and embedded in paraffin. Sagittal sections were cut at a thickness of 4 µm, and 10 sections per region of the digestive tract were prepared for staining. Sections were stained with Hansen's hematoxylin and 0.5% eosin (HE) for histological observation and were stained with Alcian blue (AB) at pH 2.5 and periodic acid–Schiff (PAS) for observations of mucus-secreting goblet cells [21,22]. Microscopy of the length of the mucosal folds, thickness of the muscularis externa, and characteristics of the goblet cells from different regions of the digestive tract was carried out using a light microscope (Leica, DM3000 LED, Wetzlar, Germany) attached to a camera (Dhyana 400DC, Tucsen Photonics, Fuzhou, China) with a capture program (Tucsen Mosaic version 2.2, Fuzhou, China).

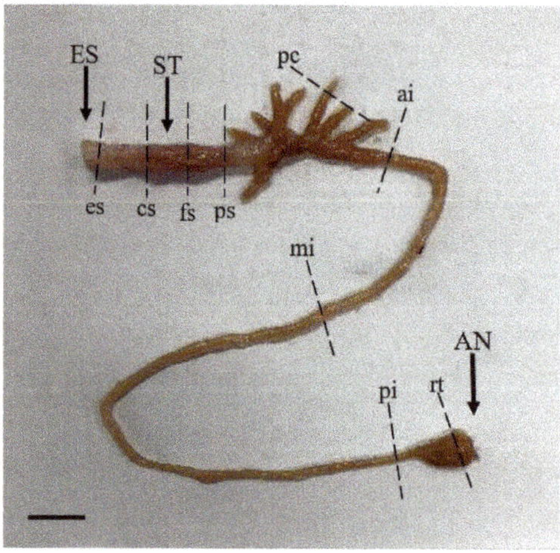

Figure 1. Anatomical features of the marbled flounder *Pseudopleuronectes yokohamae* digestive tract divided into nine segments. Abbreviations: ES, esophagus; ST, stomach; AN, anus; es, esophagus portion; cs, cardiac stomach portion; fs, fundic stomach portion; ps, pyloric stomach portion; pc, pyloric caeca portion; ai, anterior intestine portion; mi, mid intestine portion; pi, posterior intestine portion; rt, rectum portion. Scale bar: 10 mm.

2.3. Immunohistochemistry

Formalin-fixed tissue samples were embedded in paraffin and 4-µm sections were cut. For paraffin-embedded tissue immunohistochemistry staining, deparaffinized and rehydrated sections were incubated with 3% hydrogen peroxidase solution for 10 min at room temperature and stained with primary antibodies against cholecystokinin-8 (CCK-8) (1:20,000, Sigma-Aldrich, St. Louis, MO, USA). The stained sections were further incubated with peroxidase-conjugated secondary antibodies (EnVision+Rabbit, Dako) for 30 min, and Mayer's Hematoxylin was applied for counterstaining. The CCK-producing cells were observed using a light microscope (Leica, DM3000 LED, Wetzlar, Germany) attached to a camera (Dhyana 400DC, Tucsen Photonics, Fuzhou, China) with a capture program (Tucsen Mosaic version 2.2, Fuzhou, China).

2.4. Statistical Analysis

All data were analyzed using the IBM SPSS 19 software package for Windows (SPSS Inc., Chicago, IL, USA). Data were evaluated for assumptions, including normality and homogeneity of variance, using the Shapiro–Wilk and Levene's tests, respectively, and no violation was detected ($p > 0.05$). Statistical analyses of data were conducted using one-way analysis of variance with a 95% significance level ($p < 0.05$). When a significant treatment effect was detected, the Tukey's honestly significant difference (HSD) test was used to assess significant differences among means.

3. Results

3.1. Histological Characteristics of the Digestive Tract

The tissue of the digestive tract can be largely divided into four layers consisting of the serosa, muscularis externa, submucosa, and muscularis mucosae from outside to inside. The internal structure characteristics of each part are described as follows.

Esophagus: The serosa of the marbled flounder esophagus is thin, and the muscularis externa is widely and well-developed. The submucosa between the muscularis externa and muscularis mucosae is thin, and the lamina propria is well-developed. The mucosal folds of the esophagus are thin, long, and are arranged to form regular branches towards the lumen. The average thickness of the muscularis externa and length of the mucosal fold were 485 and 1031 μm, respectively (Table 2 and Figure 2).

Table 2. Histological features of the digestive tract of the marbled flounder *Pseudopleuronectes yokohamae*.

	Muscularis Externa Thickness (μm)	Mucosal Fold Length (μm)
Esophagus	483.5 ± 23.1	1031.7 ± 26.0
Cardiac stomach portion	245.3 ± 31.8	1044.3 ± 90.4
Fundic stomach portion	150.1 ± 3.0	1383.3 ± 51.3
Pyloric stomach portion	603.5 ± 40.3	906.8 ± 45.0
Pyloric caeca	32.2 ± 1.2	669.4 ± 34.0
Anterior intestine portion	57.4 ± 8.6	630.7 ± 21.9
Mid intestine portion	50.9 ± 6.2	486.5 ± 20.2
Posterior intestine portion	70.4 ± 4.1	458.6 ± 34.1
Rectum	198.0 ± 10.8	859.5 ± 49.6

Values are the mean ± standard error (n = 20)

Stomach: The stomach was investigated by dividing it into cardiac, fundic, and pyloric portions, and the mucosal folds of the stomach showed a regular shape without forming branches. The serosa is thin, similar to that in the esophagus, and the muscularis externa is widely and well-developed. In the center, the gastric glands with clear lumen formation are well-developed and are distributed primarily in the fundic portion. The thickness of the muscularis externa sequentially decreases along the pyloric portion, cardiac portion, and fundic portion (603, 245, and 150 μm, respectively); while the length of mucosal folds is reversed, sequentially increasing along the fundic portion, cardiac portion, and pyloric portion (1383, 1044, and 906 μm, respectively) (Table 2 and Figure 2).

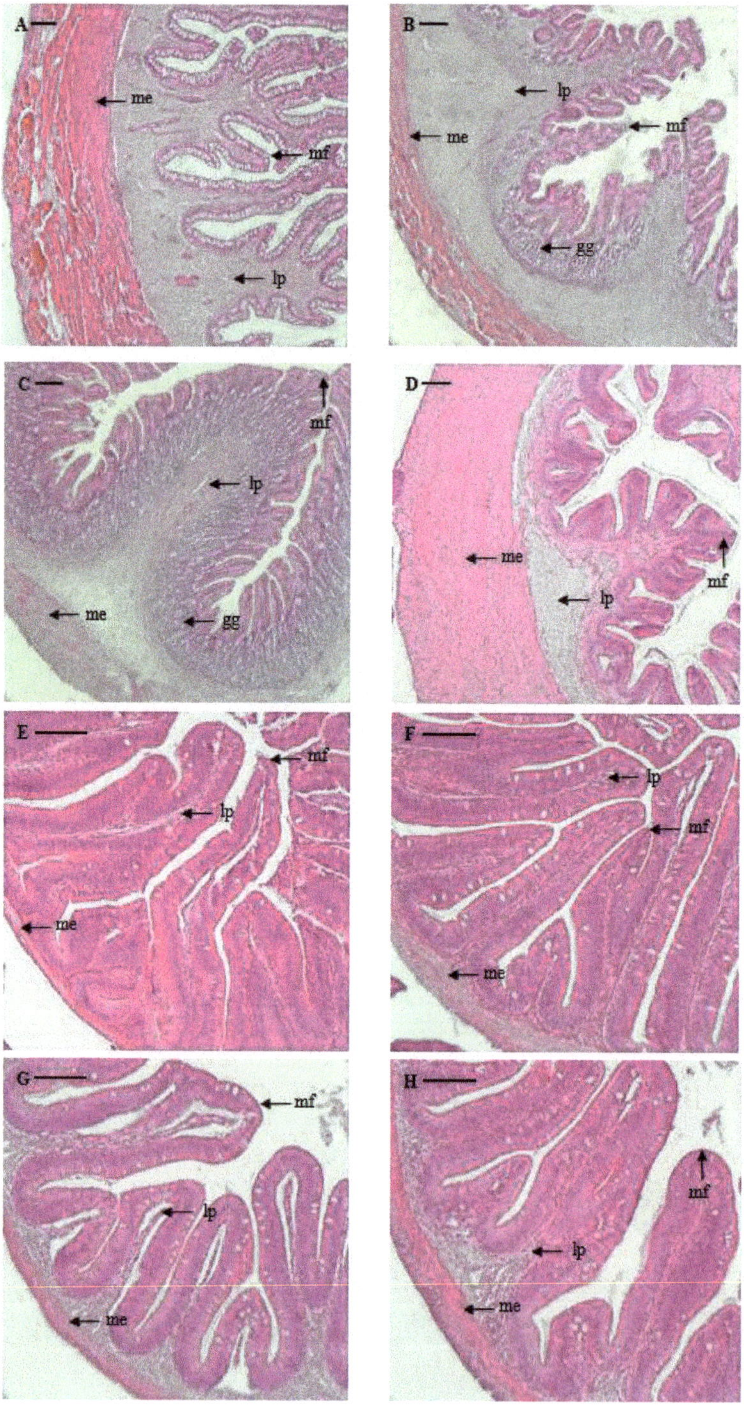

Figure 2. *Cont.*

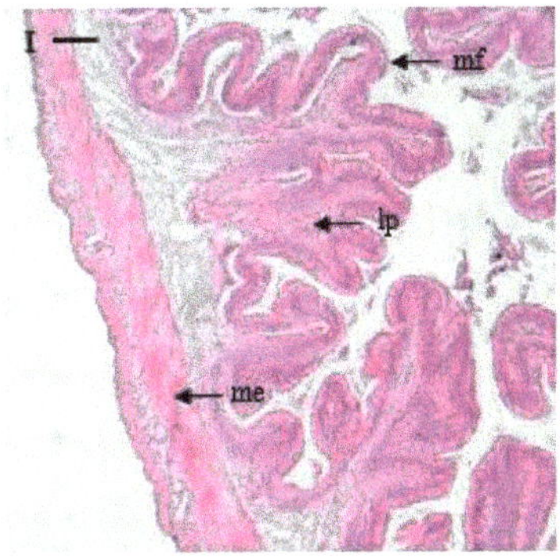

Figure 2. Microphotographs of cross sections of the digestive tract of the marbled flounder *Pseudopleuronectes yokohamae*: (**A**) esophagus; (**B**) cardiac stomach portion; (**C**) fundic stomach portion; (**D**) pyloric stomach portion; (**E**) pyloric caeca; (**F**) anterior intestine portion; (**G**) mid intestine portion; (**H**) posterior intestine portion; (**I**) rectum. Abbreviations: lp, lamina propria; me, muscularis externa; mf, mucosal fold; gg, gastric gland. Scale bars indicate 100 μm.

Pyloric caeca: There were six to nine pyloric caeca in marbled flounder, depending on the individual. The serosae of the pyloric caeca are so thin that it is difficult to differentiate them. The mucosal folds are irregular in shape and form branches. The degree of development of the muscularis externa is very weak compared to that of the esophagus and stomach (32 μm). The mucosal folds are shorter compared to those of the esophagus and stomach but longer than that of the intestine (669 μm) (Table 2 and Figure 2).

Intestine: The intestine could be divided into the anterior portion, mid portion, and posterior portion according to the internal structures such as external shape, shape and length of mucosal folds, thickness of the muscularis externa, and degree of development of the epithelial layer. In the anterior portion, mucosal folds are irregular in shape and show clear formation of branches. Although the muscularis externa is very thin (57 μm), it is thicker than that of the pyloric caeca. The mucosal folds in the anterior portion are the longest (630 μm) among all intestine portions and are similar in length to those in the pyloric caeca. The mucosal folds of the mid portion are also irregular in shape, with clear formation of branches. While the muscularis externa is of a similar thickness (50.9 μm) to that of the anterior portion, the mucosal folds are shorter (486.6 μm). The mucosal folds of the posterior portion are irregular in shape and form branches. The muscularis externa is similar to those in other parts of the intestine but is thicker (70.4 μm). The mucosal folds are a similar length to those of the mid portion (458 μm) (Table 2 and Figure 2).

Rectum: The rectum is well-developed, with distinct divisions in the serosa, muscularis externa, and submucosa, and the mucosal folds are irregular and complicated in shape compared to those in the intestine, forming branches. The muscularis externa is thicker (198 μm) and the mucosal folds are longer (859 μm) compared to those in the intestine and pyloric caeca (Table 2 and Figure 2).

3.2. Distribution and Characteristics of Goblet Cells

The distribution and relative frequency of goblet cells differed according to their location in the digestive tract, as shown in Table 3 and Figure 3. Notably, no goblet cells were identified in the stomach of the marbled flounder. Mucus-secreting goblet cells were identified in the esophagus (5743 cells), pyloric caeca (300 cells), anterior intestine portion (650 cells), mid intestine portion (528 cells), posterior intestine portion (467 cells), and rectum (1943 cells). Among the digestive tract regions, goblet cells had significantly higher distribution in the esophagus, followed by the anus ($p < 0.05$). Goblet cells were more highly distributed in the intestine than in the pyloric caeca, and the number of goblet cells decreased when moving from the anterior to the posterior intestine portion. However, there was no significant difference in the number of goblet cells between the intestine and pyloric caeca ($p > 0.05$). Esophageal goblet cells were densely distributed and filled the lumen, portraying a spherical or oval shape. Goblet cells from the anterior intestine portion to the anus were distributed throughout the mucosal folds in a spherical or oval shape.

Table 3. Numbers of goblet cells and CCK-producing cells in different digestive tract regions of the marbled flounder *Pseudopleuronectes yokohamae*.

Tissue Section	Numbers/Tissue Section	
	Goblet Cells	**CCK-Producing Cells**
Esophagus	5743 ± 107 [a]	None
Cardiac stomach portion	None	None
Fundic stomach portion	None	None
Pyloric stomach portion	None	None
Pyloric caeca	300 ± 32 [c]	65 ± 6 [a]
Anterior intestine portion	650 ± 86 [c]	62 ± 3 [a]
Mid intestine portion	528 ± 59 [c]	13 ± 1 [b]
Posterior intestine portion	467 ± 72 [c]	8 ± 2 [b]
Rectum	1943 ± 144 [b]	4 ± 1 [b]

Values are the mean ± standard error (n = 20). Values followed by differing superscript letters are significantly different (Tukey's HSD test, $p < 0.05$).

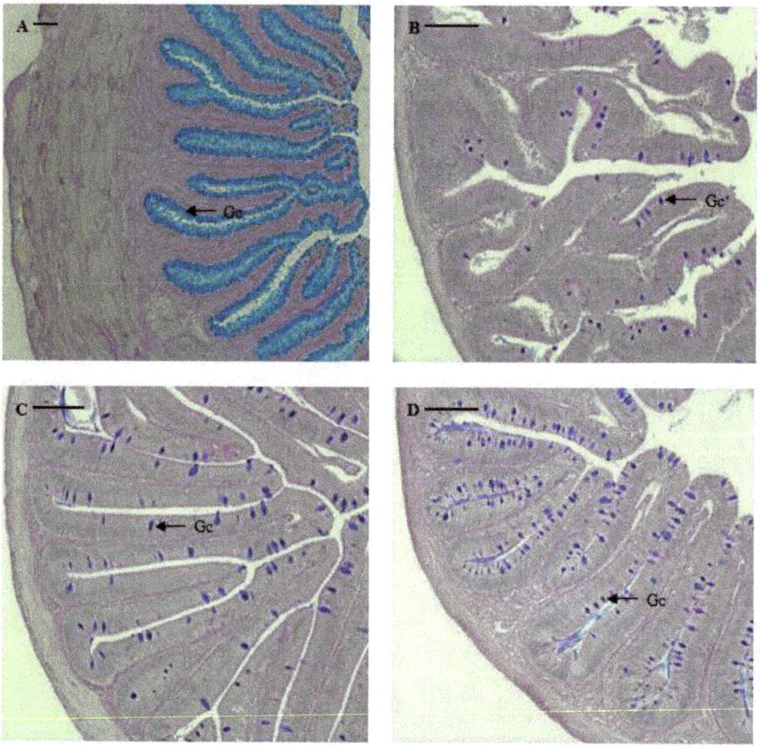

Figure 3. *Cont.*

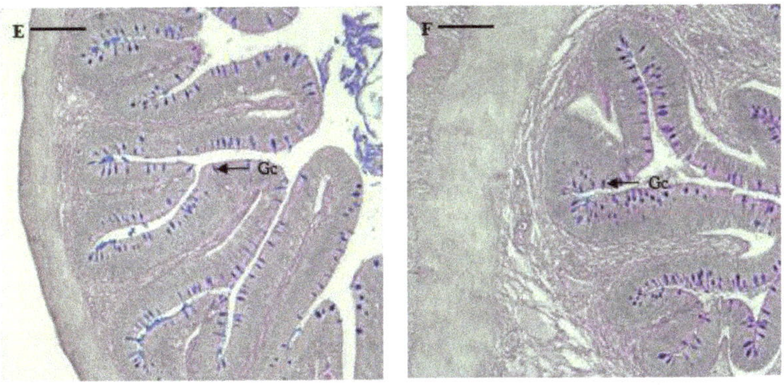

Figure 3. Microphotographs of mucus-secreting goblet cells in the digestive tract of marbled flounder *Pseudopleuronectes yokohamae*: (**A**) esophagus; (**B**) pyloric caeca; (**C**) anterior intestine portion; (**D**) mid intestine portion; (**E**) posterior intestine portion; (**F**) rectum. Abbreviation: Gc, goblet cell. Scale bars indicate 100 μm.

3.3. Distribution and Characteristics of CCK-Producing Cells

CCK-producing cells are typical endocrine-like cells, characterized by a long spindle shape with a narrow apex pointing towards the intestinal lumen. Although marbled flounder CCK-producing cells were not found in the esophagus or stomach, they were found at various frequencies in the pyloric caeca and intestine, extending to the rectum. CCK-producing cells were identified in the pyloric caeca (65 cells), anterior intestine portion (62 cells), mid intestine portion (13 cells), posterior intestine portion (8 cells), and rectum (4 cells). They were primarily distributed in the pyloric caeca and anterior intestine portion, showing a lower frequency towards the rectum (Table 3 and Figure 4).

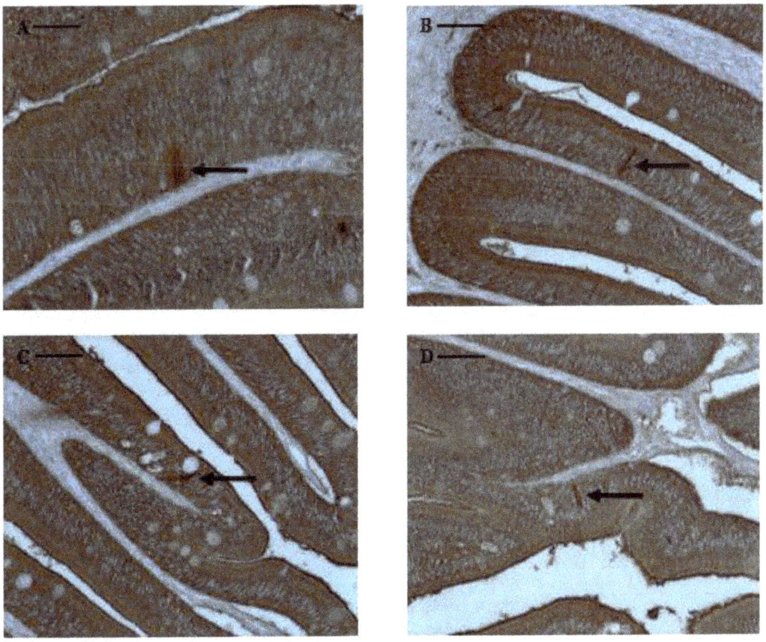

Figure 4. *Cont.*

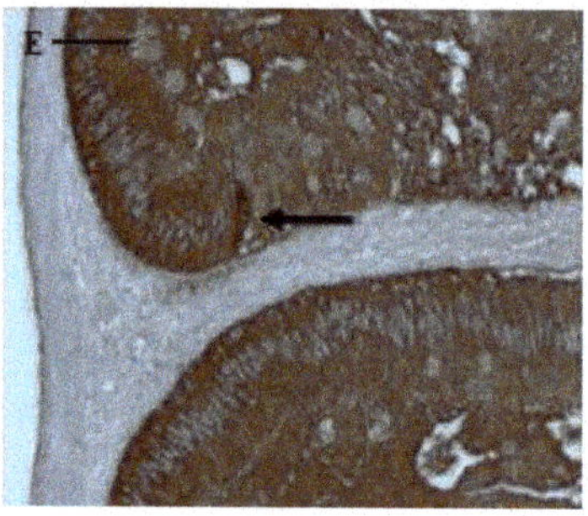

Figure 4. Microphotographs of CCK-producing cells in the digestive tract of marbled flounder *Pseudopleuronectes yokohamae*. (**A**) pyloric caeca; (**B**) anterior intestine portion; (**C**) mid intestine portion; (**D**) posterior intestine portion; (**E**) rectum. Arrows indicate CCK-producing cells. Scale bars indicate 40 μm.

4. Discussion

To clarify the physical and chemical characteristics of the marbled flounder digestive tract, this study conducted morphological, histological, and histochemical analyses. The length of the intestine in the digestive tract of teleosts is an important parameter to distinguish the dietary patterns of fish. Al-Hussaini (1947) [23] divided fish by their feeding habit based on their RLG values: carnivorous fish, 0.5~2.4; omnivorous fish, 1.3~4.3; and herbivorous fish, 3.7~6.0. In addition, Dasgupta (2004) [24] observed that the mean RLG values were 0.7 in carnivorous fish, 1.4 in omnivorous fish, 4.8 in herbivorous fish, and 3.7 in plankton feeder fish. The RLG value of marbled flounder in this study was found to be 1.54 ± 0.10, which was similar to that of omnivorous fish; however, the shape and anatomical structure of the digestive tract showed similar characteristics to those of carnivorous fish. According to the results of a recent study, the feeding habits of marbled flounder are in fact close to those of carnivorous fish, as their main prey is Polychaeta (45.3%) and Gastropoda (23.2%) [25].

The esophagus of the marbled flounder showed a branched shape with mucosal folds along the longitudinal axis, similar to the esophagus of many teleosts [26], and a stratified epithelial layer consisting mainly of epithelial and secretory cells [27,28]. This structure suggests that the esophagus of the marbled flounder can be expanded for the digestion and transportation of food. In the esophagus, the number of goblet cells was significantly higher than in other digestive tract tissues. Goblet cells secrete mucus to protect the mucosal epithelial layer and facilitate the transport of food substances and excretion into the digestive tract [29]. Therefore, it is suggested that the presence of many goblet cells in the esophagus of the marbled flounder is for lubricating and protecting the esophagus.

Digestion in the stomachs of teleosts can be divided into physical and chemical digestion. Physical digestion varies depending on the development of the muscularis externa and muscularis mucosa, and chemical digestion varies depending on the degree of gastric gland development [30,31]. Fish with poorly developed muscularis mucosa but well-developed muscularis externa are suited for digesting a large amount of soft food. In contrast, fish with poorly developed muscularis externa but well-developed muscularis mucosa are suited for ingesting a small amount of hard food [32]. In the case of the marbled flounder, the muscularis externa of the cardiac and fundic portions of the stomach has weak muscularis externa development compared to that of the pyloric portion, but the muscularis mucosa of the cardiac and fundic portions has well-developed gastric glands for chemical digestion. In addition, although there are no gastric glands in the pyloric portion, the muscularis externa is very well-developed, suggesting that the marbled flounder has a stomach with a structure suitable for eating a small amount of hard food.

The pyloric caeca of fish are an evolutionary strategy to increase the total surface area of the digestive epithelium, without increasing the length or thickness of the intestine [33]. Pyloric caeca are found in approximately 60% of teleosts, especially in carnivorous species [34]. The number of pyloric caeca in the marbled flounder was found to be six to nine, and the morphological structure of the digestive tract was similar to those of carnivorous fish. However, since little research has been conducted on the correlation between the number of pyloric caeca and size of the teleost intestine and the relationship between the number of pyloric caeca and digestive physiology, additional research is needed.

The shape of the mucosal folds of the intestine is more complex than that of the esophagus or stomach, and these structural characteristics increase the dwell time of food in the intestine and increase its surface area, improving digestion function and absorption of nutrients [35]. In other words, the mucosal folds of the fish intestine are well developed (more complex and more branched) and have higher digestive activity [31]. Mucosal folds in the anterior, mid, and posterior intestine portions of the marbled flounder showed similar branched patterns; however, the mucosal folds became shorter moving from the anterior intestine portion to the posterior intestine portion, suggesting that more digestion takes place in the anterior intestine portion than in the posterior intestine portion.

Previous studies investigating the histochemical characteristics of mucus-secreting cells found that the shape, size, and distribution of goblet cells in the digestive tract differ depending on the fish species and digestive tract structure [35–38]. The goblet cells in the digestive tract of marbled flounder showed a different shape and distribution compared to those of other fish. For marbled flounder, goblet cells were irregularly distributed without a regular arrangement from the upper part to the lower part of the mucosal folds, and the shape of goblet cells appeared to vary from spindle to oval. The number of goblet cells in the intestine of the marbled flounder decreased from the anterior to the posterior portion. Lee and Chin (1995) [32] reported that the distribution of goblet cells in the gut of stomachless fish increased towards the posterior intestine portion, and Hur et al. (2013) [39] reported that the distribution of goblet cells decreased towards the posterior intestine portion in stomach fish. Judging from the results of previous studies, marbled flounder displayed a digestive characteristic typical of stomach fish.

The final absorption of water, ions, and proteins takes place in the teleost rectum [9] and many mucin-containing epithelial cells exist in the rectum to contribute to the passage of feces, protection of the epithelium, and final absorption of substances [35]. The histological and histochemical analysis results revealed that the marbled flounder rectum was composed of a complex columnar shape, similar to that of other teleost fish. In the epithelial lining, epithelial cells coated in secretions of neutral and acidic mucus and a number of goblet cells were observed, which were judged to contribute to the absorption of substances generally performed in the rectum.

In fish, CCK cells are present among the endocrine cells of digestive tract and within the central and peripheral nervous system [40,41]. These cells influence digestive processes by regulating food intake through peripheral satiety signals [42]. Marbled flounder CCK cells were distributed throughout the digestive tract, including the pyloric caeca, intestine, and rectum, and CCK cell frequency was highest in the pyloric caeca and anterior portion of the intestine. CCK cells are found in various fish species, including flat fish species [43,44], and distribution patterns similar to that of the marbled flounder CCK cells were observed in the digestive tracts of hairychin goby *Sagamia geneionema* [30], blacktip grouper *Epinephelus fasciatus* [39], silver catfish *Rhamdia quelen* [45], dorado *Salminus brasiliensis* [46], and Neotropical freshwater fish *Prochilodus lineatus* [47]. Considering the biological roles of CCK cells, their distribution pattern in marbled flounder suggests that the pyloric caeca and anterior intestine portion of the intestine play an important role in gallbladder contraction, pancreatic enzyme secretion, stimulation of gastrointestinal motility, and inhibition of gastric emptying functions [48]. However, the distribution characteristics of neuroendocrine cells, including those that secrete CCK, in fish digestive tracts do not show common distribution characteristics among fish species and dietary patterns [29,49,50].

5. Conclusions

The anatomical and functional structures of the marbled flounder digestive tract were investigated and considered to closely represent the morphology of the carnivorous fish digestive tract. Through investigating the physical and chemical characteristics of the digestive tract, basic information regarding the physical characteristics and nutritional studies of food suitable for the marbled flounder were provided. However, studies on the characteristics of goblet cells according to dyeing properties, absorptive cells, and endocrine cells distributed in the digestive tract should be conducted in more detail at the ultrastructural level.

Author Contributions: J.-H.C.: conceptualization, methodology, data curation, writing—original draft, writing—review and editing, visualization. J.W.P.: investigation, validation, writing—review and editing. Y.-W.R.: investigation, validation, writing—review and editing. K.-W.K.: conceptualization, supervision. S.-W.H.: conceptualization, supervision, project administration, writing—review and editing. All authors have read and agreed to the published version of the manuscript.

Funding: This research did not receive any specific grant from funding agencies in the public, commercial, or not-for-profit sectors.

Institutional Review Board Statement: All activities related to animal ethical considerations, such as euthanasia, were approved by the Institutional Animal Care and Use Committee of National Institute of Fisheries Science, Republic of Korea (2019-NIFS-IACUC-13) on 1 February 2019.

Informed Consent Statement: Not applicable.

Data Availability Statement: Not applicable.

Acknowledgments: This study was supported by grants from the National Institute of Fisheries Science, Republic of Korea (R2023025 and R2023029).

Conflicts of Interest: We declare that we have no financial and personal relationship with other people or organizations that might inappropriately influence our work, and there is no professional or other personal interest of any nature or kind in any product, service, and/or company that could be construed as influencing the content of this paper.

References

1. Ray, A.K.; Ringø, E. Chapter 1: The gastrointestinal tract of fish. In *Aquaculture Nutrition: Gut Health, Probiotics and Prebiotics*; John Wiley and Sons: Hoboken, NJ, USA, 2014; pp. 1–13.
2. Kirschbaum, F.; Formicki, K. *Digestive System. The Histology of Fishes*; CRC Press LLC: Boca Raton, FL, USA, 2019; pp. 88–115.
3. Rønnestad, I.; Yúfera, M.; Ueberschär, B.; Ribeiro, L.; Sæle, Ø.; Boglione, C. Feeding behaviour and digestive physiology in larvae fish: Current knowledge, and gaps and bottlenecks in research. *Rev. Aquac.* **2013**, *5*, S59–S98. [CrossRef]
4. Pena, R.; Dumas, S.; Villalejo-Fuerte, M.; Ortiz-Galindo, J.L. Ontogenetic development of the digestive tract in reared spotted sand bass *Paralabrax maculatofasciatus* larvae. *Aquaculture* **2003**, *219*, 633–644. [CrossRef]
5. Wegner, A.; Ostaszewska, T.; Rożek, W. The ontogenetic development of the digestive tract and accessory glands of sterlet (*Acipenser ruthenus* L.) larvae during endogenous feeding. *Rev. Fish Biol. Fish.* **2009**, *19*, 431–444. [CrossRef]
6. Ostaszewska, T.; Krajnik, K.; Adamek-Urbanska, D.; Kasprzak, R.; Rzepkowska, M.; Luczynski, M.; Karczewska, A.T.; Dabrowski, K. Effect of feeding strategy on digestive tract morphology and physiology of lake whitefish (*Coregonus lavaretus*). *Aquaculture* **2018**, *497*, 32–41. [CrossRef]
7. Lipscomb, T.N.; Yanong, R.P.; Ramee, S.W.; DiMaggio, M.A. Histological, histochemical and biochemical characterization of larval digestive system ontogeny in black tetra *Gymnocorymbus ternetzi* to inform aquaculture weaning protocols. *Aquaculture* **2020**, *520*, 734957. [CrossRef]
8. Alves, A.P.C.; Pereira, R.T.; Rosa, P.V. Morphology of the digestive system in carnivorous freshwater dourado *Salminus brasiliensis*. *J. Fish Biol.* **2021**, *99*, 1222–1235. [CrossRef]
9. Carrassón, M.; Grau, A.; Dopazo, L.R.; Crespo, S. A histological, histochemical and ultrastructural study of the digestive tract of *Dextex dextex* (Pisces, Sparidae). *Histol. Histopathol.* **2006**, *21*, 579–593.
10. Wilson, J.M.; Castro, L.F.C. Morphological diversity of the gastrointestinal tract in fishes. In *Fish Physiology: The Multifunctional Gut of Fish*; Grosell, M., Farrell, A.P., Brauner, C.J., Eds.; Academic Press: San Diego, CA, USA, 2010; pp. 1–55.

11. Buddington, R.K.; Krogdahl, Å. Hormonal regulation of the fish gastrointestinal tract. *Comp. Biochem. Physiol. A* **2004**, *139*, 261–271. [CrossRef]
12. Collie, N.L.; Ferraris, R.P. Nutrient fluxes and regulation in fish intestine. In *Biochemistry and Molecular Biology of Fishes, Metabolic Biochemistry*; Hochachka, P.W., Mommsen, T., Eds.; Elsevier Science: New York, NY, USA, 1995; Volume IV, pp. 221–239.
13. Takei, Y.; Loretz, C.A. The gastrointestinal tract as an endocrine/neuroendocrine/paracrine organ: Organization, chemical messengers and physiological targets. *Fish Physiol.* **2010**, *30*, 261–317.
14. Won, E.T.; Borski, R.J. Endocrine regulation of compensatory growth in fish. *Front. Endocrinol.* **2013**, *4*, 74. [CrossRef]
15. Miller, L.J.; Harikumar, K.G.; Wootten, D.; Sexton, P.M. Roles of cholecystokinin in the nutritional continuum. Physiology and potential therapeutics. *Front. Endocrinol.* **2021**, *12*, 684656. [CrossRef]
16. Gustafsson, J.K.; Johansson, M.E.V. The role of goblet cells and mucus in intestinal homeostasis. *Net. Rev. Gastroenterol. Hepatol.* **2022**, *19*, 785–803. [CrossRef]
17. Kim, I.S.; Youn, C.H. Taxonomic revision of the flounders (pisces: Pleuronectiformens) from Korea. *Korean J. Ichthyol.* **1994**, *6*, 99–131. (In Korean with English Abstract)
18. Ji, H.S.; Kim, J.K.; Kim, B.J. Molecular phylogeny of the families Pleuronectidae and Poecilopsettidae (PISCES, Pleuronectiformes) from Korea, with a proposal for a new classification. *Ocean Sci. J.* **2016**, *51*, 299–304. [CrossRef]
19. Hur, S.H.; Nam, K.M.; Park, J.M.; Jeong, J.M.; Baeck, G.W. Feeding habits of the mabled flounder, *Peluronectes yokohamae* in the Coastal waters off Tong-yeong. *Korean J. Ichthyol.* **2012**, *24*, 77–83. (In Korean with English Abstract)
20. Hernández, D.R.; Pérez Gianeselli, M.; Domitrovic, H.A. Morphology, histology and histochemistry of the digestive system of south American catfish (*Rhamdia quelen*). *Int. J. Morphol.* **2009**, *27*, 105–111. [CrossRef]
21. Cao, X.J.; Wang, W.M. Histology and mucin histochemistry of the digestive tract of yellow catfish, *Pelteobagrus fulvidraco*. *Anat. Histol. Embryol.* **2009**, *38*, 254–261. [CrossRef]
22. Suvarva, S.K.; Layton, C.; Bancroft, J.D. *Bancroft's Theory and Practice of Histological Techniques*, 7th ed.; Churchill Livingstone: London, UK, 2012.
23. Al-Hussaini, A.H. *The Feeding Habits and the Morphology of the Alimentary Tract of Some Teleosts Living in the Neighbourhood of the Marine Biological Station, Ghardaqa, Red Sea*; Publications of the Marine Biology Station Ghardaga (Red Sea); Fouad I Univ. Press: Cairo, Egypt, 1947; p. 61.
24. Dasgupta, M. Relative length of the gut of some freshwater fishes of West Benga in relation to food and feeding habits. *Indian J. Fish.* **2004**, *51*, 381–384.
25. Roh, T.H.; Choi, D.H.; Lee, S.H.; Kwon, D.H.; Han, K.H.; Kim, M. J Feeding habits of the marbled flounder *Pseudopleuronectes yokohamae* in the West Sea in Korea. *Korean J. Fish. Aquat. Sci.* **2022**, *55*, 583–589. (In Korean with English Abstract)
26. Park, J.Y.; Kim, I.S. Histology and mucin histochemistry of the gastrointestinal tract of the mud loach, in relation to respiration. *J. Fish Biol.* **2001**, *58*, 861–872. [CrossRef]
27. Germano, R.M.; Stabille, S.R.; Mari, R.B., Pereira, J.N.B.; Faglioni, J.R.S.; Miranda Neto, M.H. Morphological characteristics of the *Pterodoras granulosus* digestive tuve (Valenciennes, 1821) (Osteichthyes, Doradidae). *Acta Zool.* **2014**, *95*, 166–175. [CrossRef]
28. Matheus, V.A.; Faccioli, C.K.; Chedid, R.A.; Senhorini, J.A.; Franceschini-Vicentini, I.B.; Vicentini, C.A. Morphological and histochemical features of the digestive tract of *Leiarius marmoratus* (Gill, 1870). *J. Fish Biol.* **2021**, *99*, 1622–1631. [CrossRef]
29. Løkka, G.; Austbø, L.; Falk, K.; Bjerkås, I.; Koppang, E.O. Intestinal morphology of the wild Atlantic salmon (*Salmo salar*). *J. Morphol.* **2013**, *274*, 859–876. [CrossRef]
30. Hur, S.W.; Kim, S.K.; Kim, D.J.; Lee, B.I.; Park, S.J.; Hwang, H.G.; Jun, J.C.; Myeong, J.I.; Leem, C.H.; Leem, Y.D. Digestive physiological characteristics of the Gobiidae—Characteristics of CCK-producing cells and mucus-secreting goblet cells of stomach fish and stomachless fish. *Dev. Reprod.* **2016**, *20*, 207–217. [CrossRef]
31. Cho, J.H.; Park, J.W.; Lee, B.J.; Kim, K.W.; Hur, S.W. Low extrusion pressure and small feed particle size improve the growth performance and digestive physiology of rockfish (*Sebastes shclegeli*). *Aquaculture* **2023**, *566*, 739199. [CrossRef]
32. Lee, J.S.; Chin, P. Morphology and histology of the digestive tract of the black sea bream, *Acanthopagrus shcelegeli*. *J. Korean Fish. Soc.* **1999**, *32*, 642–648. (In Korean with English Abstract)
33. Banan Khojasteh, S.M. The morphology of the post-gastric alimentary canal in teleost fishes: A brief review. *Int. J. Aquat. Sci.* **2012**, *3*, 71–88.
34. Hossain, A.M.; Dutta, H.M. Phylogeny, Ontogeny, Structure and Function of Digestive Tract Appendages (Caeca) in Teleost Fish. In *Fish Morphology: Horizons of New Research*; Datta-Munshi, J.S., Dutta, M.H., Eds.; Taylor & Francis Inc.: London, UK, 1996.
35. Faccioli, C.K.; Chedid, R.A.; do Amaral, A.C.; Vicentini, I.B.F.; Vicentini, C.A. Morphology and histochemistry of the digestive tract in carnivorous freshwater *Hemisorubim platyrhynchos* (Siluriformes: Pimelodidae). *Micron* **2014**, *64*, 10–19. [CrossRef]
36. Diaz, A.O.; Escalante, A.H.; Garcia, A.M.; Goldemberg, A.L. Histology and histochemistry of the pharyngeal cavity and oesophagus of the silverside *Odontesthes bonariensis* (Cuvier and Valencienns). *Ant. Histol. Embryol.* **2006**, *35*, 42–46. [CrossRef]
37. Kalhoro, H.; Tong, S.; Wang, L.; Hua, Y.; Volatiana, J.A.; Shao, Q. Morphological study of the gastrointestinal tract of *Larimichthys crocea* (Acanthopterygil: Perciformes). *Zoologia* **2018**, *35*, 1–9. [CrossRef]
38. Okuthe, G.E.; Bhomela, B. Morphology, histology and histochemistry of the digestive tract of the Banded tilapia, *Tilapia sparrmanii* (Perciformes: Cichlidae). *Zoologia* **2021**, *37*, 1–14. [CrossRef]
39. Hur, S.W.; Lee, C.H.; Lee, S.H.; Kim, B.H.; Kim, H.B.; Baek, H.J.; Lee, Y.D. Characterization of cholecystokinin-producing cells and mucus-secreting goblet cells in the blacktip grouper, *Epinephelus fasciatus*. *Tissue Cell* **2013**, *45*, 153–157. [CrossRef]

40. Ku, S.K.; Lee, J.H.; Lee, H.S. Immunohistochemical study on the endocrine cells in gut of the stomachless teleost, *Zacco platypus* (Cyprinidae). *Anat. Histol. Embtyol.* **2004**, *33*, 212–219. [CrossRef]
41. Vigliano, F.A.; Muñoz, L.; Hernández, D.; Cerutti, P.; Bermúdez, R.; Quiroga, M.I. Immunohistochemical study on the gut neuroendocrine system of juvenile pejerrey (*Odontesthes bonariensis*). *J. Fish Biol.* **2011**, *78*, 901–911. [CrossRef]
42. Murashita, K.; Fukada, H.; Hosokawa, H.; Matsumoto, T. Changes in cholecystokinin and peptide Y gene expression with feeding in yellowtail (*Seriola quinqueradiata*): Relation to pancreatic exocrine regulation. *Comp. Biochem. Physiol. B Biochem. Mol. Biol.* **2007**, *146*, 318–325. [CrossRef]
43. Kurokawa, T.; Suzuki, T.; Andoh, T. Development of cholecystokinin and pancreatic polypeptide endocrine systems during the larval stage of Japanese flounder, *Paralichthys olivaceus*. *Gen. Comp. Endocrinol.* **2000**, *120*, 8–16. [CrossRef]
44. Kamisaka, Y.; Totland, G.K.; Tagawa, M.; Kurokawa, T.; Suzuki, T.; Tanaka, M.; Ronnestad, I. Ontogeny of cholecystokinin-immunoreactive cells in the digestive tract of Atlantic halibut, *Hippoglossus hippoglossus*, larvae. *Gen. Comp. Endocrinol.* **2001**, *123*, 31–37. [CrossRef]
45. Hernández, D.R.; Vigliano, F.A.; Sánchez, S.; Bermúdez, R.; Domitrovic, H.A.; Quiroga, M.I. Neuroendocrine system of the digestive tract in *Rhamdia quelen* juvenile: An immunohistochemical study. *Tissue Cell* **2012**, *44*, 220–226. [CrossRef]
46. Pereira, R.T.; Costa, L.S.; Oliveira, I.R.C.; Araújo, J.C.; Aerts, M.; Vigliano, F.A. Relative distribution of gastrin-, CCK-8-, NPY- and CGRP-immunoreactive cells in the digestive tract of dorado (*Salminus brasiliensis*). *Tissue Cell* **2015**, *47*, 123–131. [CrossRef]
47. Barrios, C.E.; Santinón, J.J.; Domitrovic, H.A.; Sánchez, S.; Hernández, D.R. Localization and distribution of CCK-8, NPY, Leu-ENK-, and ghrelin- in the digestive tract of *Prochilodus lineatus* (Valenciennes, 1836). *An. Acad. Bras. Cienc.* **2020**, *92*, e20181165. [CrossRef]
48. Jensen, H.; Rourke, I.J.; Moller, M.; Jonson, L.; Johnsen, A.H. Identification and distribution of CCK-related pepteides and mRNAs in the rainbow trout, *Oncorhynchus mykiss*. *BBA-Mol. Cell Res.* **2001**, *1517*, 190–201.
49. Lin, X.; Wang, P.; Ou, Y.; Li, J.; Wen, J. An immunohistochemical study on endocrine cells in the neuroendocrine system of the digestive stract of milkfish *Chanos chanos* (Forsskal, 1775). *Aquac. Res.* **2017**, *48*, 1439–1449. [CrossRef]
50. Volkoff, H. The neuroendocrine regulation of food intake in fish: A review of current knowledge. *Front. Neurosci.* **2016**, *10*, 540. [CrossRef]

animals

MDPI

Article

Physicochemical and Biochemical Properties of Trypsin-like Enzyme from Two Sturgeon Species

Abbas Zamani [1,2,*], Maryam Khajavi [1], Abdolmohammad Abedian Kenari [3], Masoumeh Haghbin Nazarpak [2], Atefeh Solouk [4], Mina Esmaeili [5] and Enric Gisbert [6]

1 Fisheries Department, Faculty of Natural Resources and Environment, Malayer University, 4th km of Arak Road, Malayer 6574184621, Iran
2 New Technologies Research Center, Amirkabir University of Technology, Tehran 1591634653, Iran
3 Department of Aquaculture, Faculty of Natural Resources and Marine Sciences, Tarbiat Modares University, Noor P.O. Box 46414-356, Iran
4 Department of Biomaterial and Tissue Engineering, Medical Engineering Faculty, Amirkabir University of Technology, Tehran 1591634311, Iran
5 Department of Fisheries, Faculty of Animal Sciences and Fisheries, Sari Agricultural Sciences and Natural Resources University, Sari 4818168984, Iran
6 IRTA, Centre de la Rápita, Aquaculture Program, Crta. del Poble Nou Km 5.5, 43540 la Rápita, Spain
* Correspondence: a.zamani@malayeru.ac.ir; Tel./Fax: +98-81-32355330

Simple Summary: Beluga and sevruga are *two* highly valuable sturgeon species from the *Acipenseride* family in Iran. In recent years, research has been focused on commercial rearing of these species. A very important aspect in the sturgeon farming industry is the development of formulated compound diets for promoting growth. However, the ability of fish to digest compound diets is mostly related to the existence of the digestive enzymes in different parts of the *gastrointestinal* tract. In gastric species, protein digestion is conducted along the gastrointestinal tract by several proteases such as pepsin, trypsin, and chymotrypsin. Trypsin, as an alkaline protease, is able to hydrolyze protein residues and peptides to release free amino acids and small peptides for intestinal absorption; therefore, the activity of trypsin has been widely used as a valuable indicator of digestive capacity in fish. In this work, we aimed to characterize trypsin from beluga and sevruga for the first time. The results of our study show that the physicochemical and biochemical properties of trypsin from beluga and sevruga were in agreement with data reported in bony fish and may be considered a preliminary step to design in vitro tests for the assessment of protein digestibility in these primitive species.

Abstract: This work aimed to determine the physicochemical and biochemical properties of trypsin from beluga *Huso huso* and sevruga *Acipenser stellatus*, *two* highly valuable sturgeon species. According to the results obtained from the methods of casein-zymogram and inhibitory activity staining, the molecular weight of trypsin for sevruga and beluga was 27.5 and 29.5 kDa, respectively. Optimum pH and temperature values for both trypsins were recorded at 8.5 and 55 °C by BAPNA (a specific substrate), respectively. The stability of both trypsins was well-preserved at pH values from 6.0 to 11.0 and temperatures up to 50 °C. TLCK and SBTI, two specific trypsin inhibitors, showed a significant inhibitory effect on the enzymatic activity of both trypsins ($p < 0.05$). The enzyme activity was significantly increased in the presence of Ca^{+2} and surfactants and decreased by oxidizing agents, Cu^{+2}, Zn^{+2}, and Co^{+2} ($p < 0.05$). However, univalent ions Na^{+} and K^{+} did not show any significant effect on the activity of both trypsins ($p > 0.05$). The results of our study show that the properties of trypsin from beluga and sevruga are in agreement with data reported in bony fish and can contribute to the clear understanding of trypsin activity in these primitive species.

Keywords: beluga; physicochemical properties; *sevruga*; trypsin; digestive physiology

1. Introduction

Beluga (*Huso huso*) and sevruga (*Acipenser stellatus)* are among the most important species of sturgeon fish (*Acipenseridae*) inhabiting the Caspian Sea with a high demand for products such as caviar, meat, skin, and cartilage [1,2]. Today, sturgeons are considered a vulnerable group of fish species for different reasons such as overfishing for production of meat and caviar, water pollution, and destruction of their natural habitats [3,4]. Therefore, in recent years, researchers have focused their studies on restocking and commercial rearing purposes. According to the Iranian Fisheries Organization report, the aquaculture production of sturgeon has increased from 363 t in 2009 to 2516 t in 2020 [5]. A very important aspect in sturgeon farming industry, affecting its production efficiency and long-term sustainability, is the development of formulated compound diets for promoting growth and product quality. However, the ability of fish to digest compound diets is mostly related to the existence of the digestive enzymes in different parts of the *gastrointestinal* tract [6]. Digestive enzymes reflect the capability of digestion in the organism under study and thus indicate the nutritional status at different stages of growth [7,8]. Herein, analysis of digestive enzymes activity is regarded as a biochemical procedure which may contribute to generate valuable information for understanding the physiology of digestion in fish [9,10]. This important issue can also help to define the requirements of fish for essential nutrients such as proteins, lipids, or carbohydrates [11]. Among macronutrients, dietary proteins are key nutrients for fish growth, since proteins are the building blocks of muscle cells and organs. They occur in a great array of forms, and their nutritional value depends on their amino acid composition. As Moraes and Almeida, 2020 [12] reviewed, the use of dietary proteins depends on a wide array of functional, biochemical, and genetic species-specific characteristics such as the age of organisms; the range of environmental factors (pH, dissolved oxygen, and ammonia levels); the amino acid profile of dietary protein; the digestible usable dietary energy levels; and the presence of antinutritional factors, among others.

Protein digestion is conducted along the gastrointestinal tract by several proteases, with specific actions on the polypeptide chain. In gastric species, pepsin, trypsin, and chymotrypsin are the most important proteolytic enzymes in fish [13–15]. As shown by Nolasco-Soria, 2021 [16], trypsin in combination with other alkaline proteases and peptidases such as chymotrypsin, aminopeptidases, and carboxypeptidases complete the acid predigestion conducted in the stomach, hydrolyzing protein residues and peptides to release free amino acids and small peptides for intestinal absorption; therefore, the activity of trypsin has been widely used as a valuable indicator of digestive capacity in fish, as well as a useful biomarker for its nutritional and physiological condition [17]. According to surveys conducted in various species of fish, trypsin participates in activating trypsinogen and other zymogens in the intestine and plays an effective role in protein degradation of the consumed diet in the carnivorous fish up to 40–50% [14,18–21]. Furthermore, trypsin quantification is essential for the design of in vitro digestibility protocols of feed ingredients and for the formulation of high digestible compound feeds for aquaculture fish species [22].

Hence, a better understanding of the properties of trypsin is necessary to generate valuable information for protein degradation in the fish digestive tract. The characterization of trypsin, especially its physicochemical and biochemical properties, has been thoroughly studied from the intestine of various fish including grass carp (*Ctenopharyngodon idellus*), spotted goatfish (*Pseudupeneus maculatus*), grey triggerfish (*Balistes capriscus*), skipjack tuna (*Katsuwonus pelamis*), smooth hound (*Mustelus mustelus*), and Brazilian flounder (*Paralichthys orbignyanus*) [23–28]. The activity of trypsin among several sturgeon species has been mostly studied during larval ontogeny in the members of the genus *Acipenser* such as *A. transmontanus* [29], *A. fulvescens* [30], *A. oxyrinchus* [31,32], *A. baerii* [33], *A. persicus* [34], *A. nacarii* [35,36], *A. stellatus* [37], and genus *Huso* such as *H. huso* [38]. However, there are no studies evaluating the physicochemical and biochemical characteristics of trypsin in sturgeons; thus, this study attempts to characterize trypsin from beluga and sevruga, as two of the main sturgeon species from the Caspian Sea, for the first time.

2. Materials and Methods

2.1. Fish Samples

Viscera from five specimens of beluga and sevruga (8.0 ± 0.4 kg; 95 ± 8 cm) were obtained from Saei sturgeon rearing center, Sari, Mazandaran, Iran. Fish were fed a commercial diet (crude protein 46%, crude lipid 16%, ash 8.5%, crude fiber 2.5%, and moisture 9%, Mazandaran Animal & Aquatic Feed Company, Semeskandeh Olya, Iran) and kept in fasting condition for 72 h before sampling. Those samples were packed in polyethylene bags, placed in ice with the sample/ice ratio of approximately 1:3 (*w*/*w*), and directly transported to the laboratory. Upon arrival, the intestine was separated from the rest of the collected viscera, washed with cold distilled water (4 °C), pooled, and stored at −80 °C for further analysis.

2.2. Preparation of Intestinal Crude Extracts for Trypsin Characterization

The frozen intestine of beluga and sevruga was partially thawed in the refrigerator at 4 °C for 2 h. The samples were then cut into small pieces and homogenized in 50 volumes of 50 mM Tris–HCl buffer (pH 7.5, 10 mM $CaCl_2$, 0.5 M NaCl) by a tissue homogenizer (Heidolph Diax 900, Sigma Co., St. Louis, MO, USA) at 4 °C for 2 min. The homogenate was then filtered with a cheese cloth to separate the floating fat phase and centrifuged for 45 min at 4 °C at 14,000× *g* by a refrigerated centrifuge (Hettich Benchtop Centrifuge Rotina 420R, Berlin, Germany). The resulting supernatant from each sample was collected, defined as intestinal crude extract (ICE), and then used throughout this study.

2.3. Reagents

EDTA (Ethylenediaminetetraacetic acid), Pepstatin A, PMSF (phenylmethanesulfonyl fluoride), and sodium cholate were obtained from Molekula Co (Gillingham, UK). BAPNA (Nα-benzoyl-DL-arginine-ρ-nitroanilide hydrochloride), ß-mercaptoethanol, BSA (bovine serum albumin), iodoacetic acid, saponin, SBTI (soybean trypsin inhibitor), TLCK (N-ρ-tosyl-L lysine-chloromethyleketone), and TPCK (N-tosyl-L-phenylalanine chlorom ethyleketone) were purchased from Sigma Chemical Co (St. Louis, MO, USA). Molecular weight marker (PM 2700) was obtained from SMOBIO Technology, Inc. (Hsinchu, Taiwan).

2.4. Trypsin Assay

To measure the enzyme activity in ICE, BAPNA was used as a substrate at a concentration of 1 mM in 50 mM Tris–HCl, 20 mM $CaCl_2$ (pH 8.5) according to the method of Erlanger et al., 1961 [39]. Each ICE (25 μL) was mixed with the prepared substrate (1250 μL) and incubated at 55 °C for 20 min. The reaction was terminated by adding 30% (*v*/*v*) acetic acid (250 μL) to the mixture and followed by measuring the trypsin activity at an absorbance of λ = 410 nm using a spectrophotometer (UV-1601, Shimadzu, Kyoto, Japan). One unit of activity was defined as 1 μmol of ρ-nitroaniline released per min and calculated with the following equation [40]:

$$\text{Trypsin activity (unit/mL)} = \frac{\text{Absorbance 410 nm} \times 1000 \times \text{mixture volume (mL)}}{8800 \times \text{reaction time (min)} \times 0.025}$$

where 8800 (cm^{-1} M^{-1}) is the molar extinction coefficient of ρ-nitroaniline measured at λ = 410 nm.

2.5. Protein Assay

The concentration of protein in both ICEs was determined at λ = 750 nm by using BSA (1 mg mL^{-1} as a standard) and Folin–Ciocalteau reagent according to the Lowry et al., 1951 [41] method.

2.6. Characterization of Trypsin by SDS-PAGE Electrophoresis

SDS-PAGE electrophoresis was performed to determine the protein pattern in ICEs from both sturgeon species [42]. Each ICE was mixed at 2:1 (*v*/*v*) ratio with sample buffer (62.5 mM Tris–HCl pH 6.8, 2% SDS (*w*/*v*), 10% (*v*/*v*) glycerol, 0.3% (*w*/*v*) bromophenol blue and 5% (*v*/*v*) ß-mercaptoethanol) and boiled for 10 min. Thereafter, the ICEs (with protein concentration of 15 µg) were loaded onto the gel made of 4% stacking gel and 12% separating gel and the electrophoresis was run at a constant current of 15 mA using a vertical electrophoresis system (Bio-Rad Laboratories, Inc., Hercules, CA, USA). After the run, protein bands present in the gel were stained with 0.1% Coomassie Brilliant Blue (G-250) in methanol (35%) and acetic acid (7.5%) and unstained in methanol (35%) and acetic acid (7.5%).

Casein-zymography was performed after electrophoresis for detection of proteases in both ICEs as described by Garcia-Carreno et al., 1993 [43]. Both ICEs were submitted to native-PAGE electrophoresis in a same manner of SDS-PAGE where samples were not boiled, and SDS and reducing agent were removed. After the run, the gel was immersed in 50 mL of a casein solution (20 mg mL^{-1} in 50 mM Tris–HCl, pH 7.5) for 1 h at 4 °C with gentle agitation to allow diffusion of the casein into the gel. Thereafter, the gel was transferred to another solution (50 mL) containing casein (20 mg mL^{-1} in 50 mM Tris–HCl, pH 8.5, 10 mM $CaCl_2$) for 20 min at 55 °C with continuous agitation. The gel was then stained with 0.1% Coomassie Brilliant Blue (R-250) in methanol (35%) and acetic acid (7.5%) and unstained in methanol (35%) and acetic acid (7.5%). The presence of proteolytic activities in both ICEs was indicated by the appearance of clear zones on the blue background of the gel, which meant that casein was digested by the targeted protease in these areas.

To reveal the trypsin present in both ICEs, the inhibitory activity staining was used after the submission of both ICEs to native-PAGE electrophoresis as described by Ahmad and Benjakul [44] with a slight modification. After the run, the gel was immersed in 30 mL of an SBTI solution (1 mg mL^{-1} in 50 mM Tris–HCl, pH 8.5, 10 mM $CaCl_2$) for 30 min at 4 °C to allow diffusion of SBTI into the gel. Thereafter, the incubation of gel was performed for 40 min at 55 °C and followed by washing in cold distilled water and staining with 0.05% Coomassie Brilliant Blue (R-250) to appear inhibitory zones, indicating the presence of the trypsin in both ICEs. The molecular weight of the trypsin that appeared in both ICEs was estimated using wide-range molecular weight markers (PM2700, SMOBIO, Hsinchu, Taiwan) by calculating the trypsin Rf in comparison with those of protein markers.

2.7. Optimum Temperature and Thermostability

To determine the optimum temperature for trypsin activity, the activity of this alkaline protease was measured in both ICEs at different temperatures, including 10, 25, 35, 45, 50, 55, 60, 65, and 70 °C after 20 min of incubation at pH 8.5, using 1 mM BAPNA as a substrate. For the thermostability test, both ICEs were incubated at the above-mentioned temperatures for 30 min and then cooled in an ice bath for assay of residual activity of the enzyme at pH 8.5 as described by Zamani et al., 2014 [40].

2.8. Optimum pH and Stability

Different buffers in the pH range of 4.0 to 11.0 (50 mM acetic acid–sodium acetate for pHs 4–6; 50 mM Tris–HCl for pHs 7–9 and 50 mM glycine–NaOH for pHs 10–11) were used for determining the optimum pH for trypsin activity. Both ICEs were used, and they were incubated using1 mM BAPNA as a substrate after 20 min of incubation at 55 °C at different pHs. For the pH stability test, the remaining activity of the trypsin from each ICE was measured using 1 mM BAPNA as a substrate at 55 °C after being incubated at the above-mentioned pHs for 30 min [40].

2.9. Effect of Inhibitors

Several protease inhibitors (0.01 mM pepstatin A, 0.05 mM SBTI, 1 mM iodoacetic acid, 2 mM EDTA, 5 mM TLCK, 5 mM TPCK, 5 mM ß-mercaptoethanol, and 10 mM PMSF) were prepared in the relevant solvents and incubated with an equal volume of each ICE at room temperature for 15 min. The remaining activity of the enzyme was then measured by 1 mM BAPNA as a substrate (at 55 °C, pH 8.5) and the percent inhibition was calculated according to the method of Khantaphant and Benjakul, 2010 [45]. The trypsin activity of control was measured in the same manner without the presence of inhibitors and scored to 100%.

2.10. Effect of Metal Ions

To investigate the effect of metal ions (5 mM) on the trypsin activity of both ICEs, univalent (K^+, Na^+) and divalent (Ca^{2+}, Cu^{2+}, Zn^{2+} and Co^{2+}) cations were dissolved in 50 mM Tris–HCl (pH 8.5) and then incubated with an equal volume of each ICE for 30 min at room temperature. The residual activity of the enzyme was determined using 1 mM BAPNA as a substrate at 55 °C and pH 8.5 [40]. The enzymatic activity of control was assayed without the presence of metal ions and taken as 100%.

2.11. Effect of Surfactants and Oxidizing Agents

The effect of surfactants (anionic: SDS and sodium cholate; non-ionic: saponin and Triton X-100, all at 1%) and oxidizing agents (sodium perborate at a concentration of 1% and H_2O_2 at three concentrations of 5%, 10%, and 15%) on the trypsin activity was measured by incubation of the above-mentioned surfactants and oxidizing agents with an equal volume of each ICE for 1 h at 40 °C. The residual activity of the enzyme was then determined at 55 °C and pH 8.5 using 1 mM BAPNA as a substrate. The assessment of control enzymatic activity was conducted in a similar condition in the absence of chemicals and scored to 100% [25].

2.12. Statistical Analysis

This study was conducted on the basis of a completely randomized design, and a one-way ANOVA was used for data analysis using SPSS package 22.0 (SPSS Inc., Chicago, IL, USA). All experimental assessments were performed in triplicate, and data was expressed as the mean ± standard deviation (SD). The comparison of means was carried out by Duncan's multiple range tests with a statistical significance at $p < 0.05$.

3. Results and Discussion

Fish proteases such as trypsin have been the main objective of many of studies, but it is difficult to compare the results obtained in different species because the data are affected by the use of many different methodologies, the state of feeding of experimental animals (fed vs. starved fish), and the type of enzyme preparation (intestinal tissues alone vs. intestinal extracts with the intestinal content) [46].

3.1. Protein Pattern, Casein-Zymographyand Inhibitory Activity

Protein pattern of ICE from beluga and sevruga is depicted in Figure 1a. Based on results obtained from the SDS-PAGE, a number of proteins with different molecular weights were shown in the ICE of both sturgeon species. The major bands of each ICE appeared between molecular weights comprised between 10 and 60 kDa.

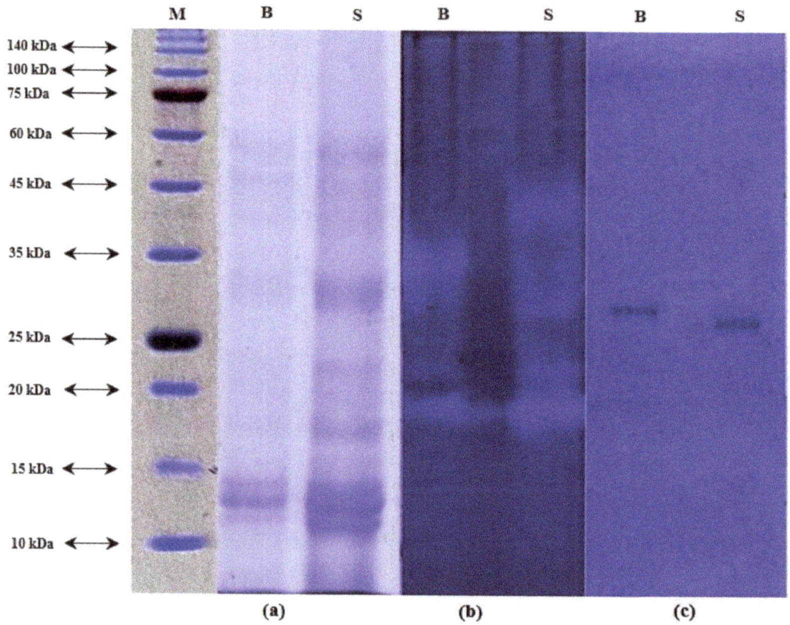

Figure 1. SDS–PAGE (**a**), zymography (**b**), and inhibitory activity staining (**c**) of the intestinal crude extract (ICE) of *Huso huso* and *Acipenser stellatus*. (**a**): Molecular weight marker (M); ICE from *Huso huso* intestine (B); ICE from *Acipenser stellatus* intestine (S). Coomassie blue G-250 (0.1%) was used for staining proteins from SDS-PAGE. Zymogram activity and inhibitory activity staining were stained by using 0.1% Coomassie blue R-250.

Casein-zymogram can be used as a highly sensitive and fast assay method for detecting nanograms of protein. The protease activity of both ICEs was demonstrated by casein-zymography as illustrated in Figure 1b. The clear bands, showing the presence of protease, appeared on the gel with different molecular weights. Based on the zymogram pattern, proteases present in ICE from beluga were observed in the range of molecular weights between 19 to 35 kDa, while those in the ICE of sevruga ranged from molecular weights of 19 to 45 kDa.

Inhibitory activity staining for detection of trypsin in ICEs is depicted in Figure 1c. Results showed that a single band for each ICE clearly appeared on the gel with a molecular weight of 27.5 and 29.5 kDa for sevruga and beluga, respectively. In general, trypsins have molecular weights in the range of 20–30 kDa [47]. In particular, different molecular weights for trypsins have been reported in various fish species such as 21.7 kDa for mrigal carp [48], 23.2 kDa for common kilka [40], 23.5 kDa for pirarucu [49], 24 kDa for small red scorpion fish [50], 21 and 24 kDa for liver of albacore tuna [51], 24 kDa for catfish [52], 24.4 kDa for gulf corvina [53], 25 kDa for Monterey sardine [54], 26 kDa for common dolphinfish [55], 27 kDa for zebra blenny [56], 28.8 kDa for sardinelle [57], 29 kDa for Atlantic bonito [58], 38.5 kDa for tambaqui [59], and 42 kDa for skipjack tuna [60]. However, several reasons such as different habitat and climate, autolytic degradation, and genetic variation among fish species may explain why trypsins from various sources have different molecular weights [60,61].

3.2. *Optimum Temperature and Thermostability*

Enzymes are one of the main biological macromolecules and their maximum activity depends on an optimum temperature to make them functional. Figure 2a revealed that optimum temperature of the trypsin in ICE prepared from beluga and sevruga was found to be 55 °C, although 92.70% of the maximum activity of the enzyme was still maintained at 60 °C for both sturgeon species. However, an obvious decrease in the trypsin activity of both ICEs was observed at temperatures above 60 °C, probably due to thermal inactivation of this enzyme caused by protein unfolding [40]. Similar optimum temperature (55 °C) was recorded for trypsins in skipjack tuna [26], gilthead seabream [46], sardinelle [57], and silver mojarra [62]. Optimum temperature of trypsin for both sturgeon species was higher than values reported for cold-water fish such as Atlantic cod (*Gadus morhua*) [63], grey triggerfish [25], lane snapper (*Lutjanus synagris*) [64], and Japanese sea bass (*L. japonicus*) [65] indicating optimum temperatures over a range of 40–45 °C. These differences could be attributed to the temperature of the fish habitat or experimental conditions used in assessments [66]. The optimum temperature for enzyme maximum activity may be interesting for comparative physiological studies, even though such data offer limited information on enzyme activity under normal rearing conditions. Although fish trypsins are mostly unstable at temperatures higher than 40–50 °C, their thermal stability is well known to be at temperatures below 40 °C [67]. Trypsin thermal stability from ICE of beluga and sevruga is displayed in Figure 2b. As can be observed in this figure, the stability of both trypsins was highly maintained up to 50 °C with a remaining activity of 90.2% and 91.7% for sevruga and beluga, respectively. A gradual decrease in the activity of both trypsins was recorded at 55 °C, whereas enzymatic activity sharply decreased at 60 °C. After heating the ICEs at 70 °C, the relative activities for both trypsins were only about 0.9% and 1.6% of their initial activity for sevruga and beluga, respectively. These results were in accordance with those of sardinelle, common kilka, mrigal carp, and pirarucu, which were exhibited to be stable up to 50 °C [40,48,49,57]. The trypsins from beluga and sevruga showed to be more stable at high temperatures in comparison with those reported for the Monterey sardine, chinook salmon, bluefish, Tunisian barbell, and common dolphinfish that the enzymatic activity was rapidly lost at temperatures above 40 °C [54,55,58,68,69]. In general, thermostability of the trypsin enzyme might vary by some factors such as fish species and experimental conditions [23,70]. From a biological point of view, it is difficult to deduce any advantage for beluga and sevruga in possessing proteases showing different resistances to heating, since the normal temperature of water rarely exceeds 21–24 °C. Nevertheless, from a biotechnological perspective, it may be interesting to have information about active and easily denaturalizable proteases potentially useful in the feed industry [71].

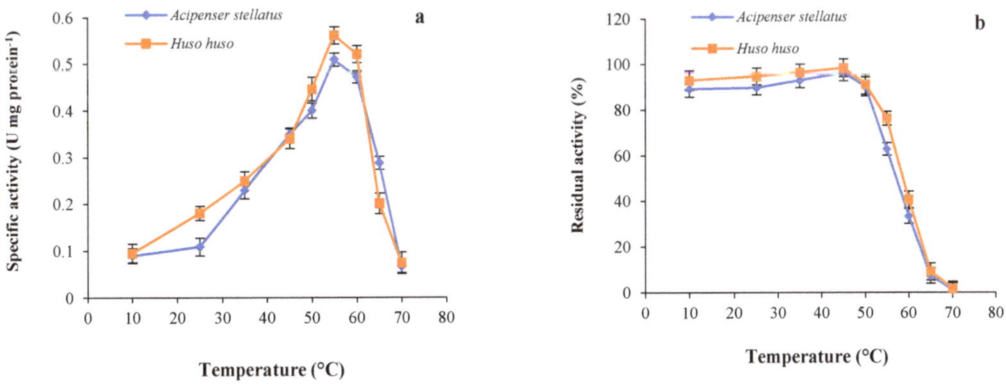

Figure 2. (**a**) Optimum temperature: the enzymatic activity of both trypsins was measured at different temperatures at pH 8.5 using BAPNA as a substrate; (**b**) thermostability: residual activity of both enzymes was determined at 55 °C and pH 8.5 after incubating at different temperatures for 30 min using BAPNA as a substrate.

3.3. *Effect of pH on Trypsin Activity and Stability*

The results observed from the effect of pH on the activity and stability of trypsin from beluga and sevruga are illustrated in Figure 3. Trypsins from both species had a maximal activity at pH 8.5 (Figure 3a). Trypsin activity was dramatically reduced at pH values ranging from 4.0 to 5.0. Our results showed that the stability of trypsin from both sturgeon species was highly preserved at pH values comprised between 5.0 and 11.0 with activity values of 75% for beluga and 80% for sevruga. The high ranges of pH may change the net charge and conformation of an enzyme and inhibit to bind to the substrate properly, resulting in the abrupt loss of enzymatic activity [60,72]. Trypsins are mainly known to have more activity within a range of pH values comprised between 7.5 and 10.5 [46,73]. The optimum pH (8.5) recorded for trypsin in both sturgeon was similar with results reported for trypsins from the brownstripe red snapper viscera and the albacore tuna hepatopancreas [45,51], whereas both trypsins indicated the lower optimum pH than those recorded for the Japanese sea bass, gilthead seabream and common dentex, and the pirarucu [46,49,65]. However, optimum pH may differ depending upon the experimental conditions such as concentration and type of substrate, temperature, and type of metal ions [60]. For instance, Martinez et al., 1988 [74] showed that the trypsin from pyloric caeca of the anchovy had the optimum pHs of 8.0 and 9.5 for the hydrolysis of BAPNA and casein, respectively.

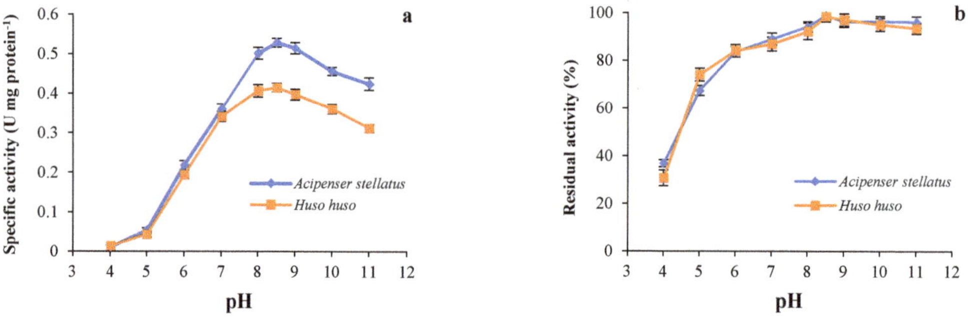

Figure 3. (**a**) Optimum pH: the enzymatic activity of both trypsins was determined at different pHs at 55 °C using BAPNA as a substrate; (**b**) pH stability: residual activity of both enzymes was measured at 55 °C and pH 8.5 after incubating at different pHs for 30 min using BAPNA as substrate.

The effect of pH on the trypsin stability in both ICEs is displayed in Figure 3b. The stability of both enzymes was considerably retained between pH 6.0 to 11.0. Trypsin activity was lost about 63.15% and 69.26% at pH 4.0 in *sevruga* and beluga, respectively, while the loss of the enzyme activity at pH 5.0 was recorded by 34.13% and 35.84% for *sevruga* and beluga, respectively. However, trypsin in ICE of *sevruga* and beluga lost only 1.19% and 1.51% of its activity at pH 8.5, respectively. Similar behavior was reported for trypsins from common kilka [40], common dolphinfish [55], and zebra blenny [56] which retained 80–100% of the activity at pH ranges of neutral and alkaline. The high catalytic activity of the trypsin is observed in alkaline pHs, and its stability at a particular pH may be linked to the net charge of the enzyme at that pH [75].

3.4. *Effect of Inhibitors on Trypsin Activity*

The sensitivity of protease enzymes to various inhibitors is a valuable tool for their proper functional characterization [55]. Based on their nature, inhibitors can be classified into two classes: chemical inhibitors and protein inhibitors [76]. A trypsin inhibitor is a type of serine protease inhibitor that reduces the biological activity of trypsin thereby rendering it unavailable to bind with proteins for the digestion process [77,78]. Therefore, it can be considered important to characterize the effect of different inhibitors on the trypsin activity.

Table 1 shows the effect of different inhibitors on the activity of trypsin from beluga and sevruga. As it is shown in this table, a serine protease inhibitor such as PMSF inhibited 39.11% and 36.29% of the trypsin activity in *sevruga* and beluga ICE samples, respectively. Both enzymes were completely inhibited by trypsin specific inhibitors such as SBTI and TLCK, while a chymotrypsin-specific inhibitor (TPCK) did not show any inhibitory effect on their enzymatic activity ($p > 0.05$). Furthermore, a metalloproteinase inhibitor (EDTA) and a disulfide bond reducing agent (ß-mercaptoethanol) had a partial inhibitory effect on trypsin activity in both sturgeon species, although the inhibition rates varied between both species. In particular, trypsin from *sevruga* was inhibited by ß-mercaptoethanol (25.33%) and EDTA (23.55%) more than those in beluga (22.84 and 21.06%, respectively) where no significant difference was shown between both species ($p > 0.05$). Results from EDTA indicate the high dependence of trypsin activity from both sturgeon species on divalent cations [46]. However, an aspartic proteinase inhibitor (Pepstatin A) and a cysteine proteinase inhibitor (iodoacetic acid) exhibited a negligible inhibitory effect on the trypsin activity of both species. Similar results have been observed in other fish species [40,51,56]. For instance, Khangembam and Chakrabarti, 2015 [48] reported that trypsin activity from the digestive system of mrigal carp was inhibited by SBTI and TLCK. SBTI is a single polypeptide chain that acts as a reversible competitive inhibitor of trypsin and forms a stable, enzymatically inactive complex with trypsin, resulting in reduction of the enzyme availability [79]. TLCK is an irreversible inhibitor of trypsin and trypsin-like serine protease that deactivates these enzymes through the formation of a covalent bond with histidine residue in the catalytic site of the enzyme and blocks the active center of the enzyme for binding to substrate [80]. As reported by several authors [40,49,56], PMSF strongly inhibited the activity of trypsin from viscera of common kilka, pirarucu, and zebra blenny, respectively, whereas TPCK had no effect on the enzyme activity from common kilka [40]. The trypsin activity from the intestine of common dolphinfish was partially inhibited by β-mercaptoethanol and EDTA [54], while the trypsin activity from liver of albacore tuna was not reduced in the presence of pepstatin A and iodoacetic acid [51].

Table 1. Effect of various inhibitors on the activity of trypsin from *Huso huso* and *Acipenser stellatus*.

Inhibitors	**Concentration**	**Inhibition (%)**	
		Huso huso	***Acipenser stellatus***
Control	-	0.0 ± 0.00 [a]	0.0 ± 0.00 [a]
PMSF	10 mM	36.29 ± 0.39 [c]	39.111 ± 0.45 [c]
SBTI	0.05 mM	99.81 ± 0.54 [d]	99.29 ± 0.49 [d]
TLCK	5 mM	99.19 ± 0.57 [d]	99.51 ± 0.47 [d]
TPCK	5 mM	0.00 ± 0.00 [a]	0.00 ± 0.00 [a]
Pepstatin A	0.01 mM	0.51 ± 0.003 [a]	0.89 ± 0.002 [a]
Iodoacetic acid	1 mM	0.25 ± 0.002 [a]	0.44 ± 0.003 [a]
EDTA	2 mM	21.06 ± 0.36 [b]	23.55 ± 0.51 [b]
ß-Mercaptoethanol	5 mM	22.84 ± 0.41 [b]	25.33 ± 0.37 [b]

Each intestinal crude extract (ICE) was incubated with same volume of inhibitor at room temperature for 15 min and the enzyme activity was determined at pH 8.5 and 55 °C. The control was prepared with similar conditions in absence of inhibitors. Data were represented by mean ± standard deviation and compared by Duncan's multiple-range test. Different superscripts in the same column show statistical difference while there is no significant difference in the same row ($p < 0.05$).

3.5. *Effect of Metal Ions*

Metal ions have a key role in the activity regulation of many enzyme-catalyzed reactions [81]. Our results on the effect of metal ions on the trypsin activity in *beluga* and sevruga are detailed in Table 2. No significant effect on the activity of *both enzymes* was found in the presence of univalent cations Na^+ and K^+ ($p > 0.05$). The enzymatic activity of trypsin in both species was significantly reduced by divalent cations Cu^{2+}, Zn^{2+} and Co^{2+}, whereas Ca^{2+} significantly enhanced the activity of both trypsins ($p < 0.05$). Similar results on the effect of Ca^{2+} on the trypsin activity were also observed in common kilka, common dolphinfish, and zebra blenny [40,55,56]. The attachment of Ca^{2+} to the active site of serine proteases such as trypsin not only increases the stability of the enzyme structure, but it also protects the enzyme from self-digestion [66,69,82]. The enzymatic activity of trypsin in common dolphinfish was reduced by 82% and 81% by Zn^{2+} and Cu^{2+}, respectively [55], while 100% of enzymatic activity of tryspin from zebra blenny was lost in the presence of Zn^{2+} and Cu^{2+} [56]. In common kilka [40], no inhibition was observed in the trypsin activity in presence of Na^{+} and K^{+}. Differences in percent inhibition might be linked to species diversity, environmental adaptations and feeding habits of fish [83].

Table 2. Effect of various metal ions on activity of trypsin from *Huso huso* and *Acipenser stellatus*.

Metal Ions	Concentration	Residual Activity (%)	
		Huso huso	*Acipenser stellatus*
Control	-	100 ± 0.00 ^{d}	100 ± 0.00 ^{d}
KCl	5 mM	99.6 ± 1.05 ^{d}	99.2 ± 0.92 ^{d}
NaCl	5 mM	99.3 ± 1.03 ^{d}	99.5 ± 1.01 ^{d}
$CaCl_2$	5 mM	107.21 ± 0.81 ^{e}	105.32 ± 0.98 ^{e}
$CuCl_2$	5 mM	48.31 ± 0.88 ^{a}	44.58 ± 0.96 ^{a}
$ZnCl_2$	5 mM	67.79 ± 1.15 ^{b}	64.83 ± 1.23 ^{b}
$CoCl_2$	5 mM	76.69 ± 0.85 ^{c}	74.77 ± 1.05 ^{c}

Univalent and divalent metal ions were incubated with same volume of each intestinal crude extract (ICE) at room temperature for 30 min and residual activity of the enzyme was measured at pH 8.5 and 55 °C. The control was prepared with similar conditions without metal ions and scored as 100%. Data were represented by mean ± standard deviation and compared by Duncan's multiple-range test. Different superscripts in the same column show statistical difference while there is no significant difference in the same row ($p < 0.05$).

3.6. Effect of Surfactants and Oxidizing Agents

Surfactants are the most widely used groups of compounds today, with wide application in industry and household. These are unique substances that contain hydrophobic and hydrophilic moieties within their molecule and find enormous applications in biology. Oxidizing agents such as sodium perborate and H_2O_2 are also used in the detergent industry as bleaching agent. Surfactants and oxidizing agents may be harmful or even toxic to aquatic organisms. These compounds can penetrate through tissues and bind to biomolecules, such as enzymes, causing changes in cellular activity [84]. As reviewed by Rubingh, 1996 [85], surfactants can influence the activity of enzymes in two ways. Firstly, by binding to the enzyme, surfactants can influence intrinsic enzyme properties such as the secondary and tertiary structure or flexibility, and thereby, affect its ability to serve as a catalyst. A less direct, but equally important, way in which surfactants affect enzyme activity is by changing the environment in which the enzyme functions. It is well-known that SDS disrupts non-covalent bonds within and between enzymes, denaturing them, and resulting in the loss of their native conformation and function [86], whereas saponin, Triton X-100, and sodium cholate are the non-denaturing surfactants [87–89]. Hence, characterizing the effect of these chemical compounds on trypsin activity is of relevance for proper characterizing its activity.

The results on the effect of various surfactants and oxidizing agents on the trypsin activity in *sevruga* and beluga are shown in Table 3. A significant increase in the activity of both trypsins was observed after incubation for 1 h at 40 °C in the presence of surfactants tested, including saponin, sodium cholate, and Triton X-100 at final concentrations of 1% ($p < 0.05$). Both trypsins were highly unstable against sodium dodecyl sulfate (SDS), in which trypsins from *sevruga* and beluga significantly lost about 94% and 97% of their activity in the presence of 0.1% SDS, respectively ($p < 0.05$). Similar results were found in trypsins of other fish species in the presence of saponin, sodium cholate, Triton X-100 and SDS [40,56]. The obtained results on the effect of oxidizing agents on both trypsins showed that the enzymatic activity was reduced in the presence of sodium perborate (1%) in sevruga and beluga by 22.23% and 24.37%, respectively. The activity of both enzymes was also decreased significantly with an increase in H_2O_2 concentrations from 5% to 15%, as described in Table 3 ($p < 0.05$). Trypsin from sevruga showed significantly higher activity than trypsin from beluga in the presence of H_2O_2 ranging from 5% to 15%, indicating that trypsin from sevruga was more tolerant to H_2O_2 than trypsin from beluga. The biochemical and structural properties of enzyme can affect its ability as a catalyst in presence of oxidizing agents [85]. These results showed that trypsins from *sevruga* and beluga were more stable against H_2O_2 than trypsins from grey triggerfish and zebra blenny [25,56], whereas most proteases have shown to be unstable in the presence of oxidizing agents like hydrogen peroxide [25].

Table 3. Trypsin activity from *Huso huso* and *Acipenser stellatus* in presence of surfactants and oxidizing agents.

Chemicals		Concentration	Residual Activity %	
			Huso huso	*Acipenser stellatus*
Surfactants	Control	-	100 ± 0.00 A ^{d}	100 ± 0.00 A ^{d}
	Triton X-100	1%	108.81 ± 1.62 A ^{e}	112.55 ± 1.11 A ^{e}
	SDS	1%	2.79 ± 0.21 A ^{a}	5.77 ± 0.33 B ^{a}
	Sodium cholate	1%	132.23 ± 1.12 A ^{j}	197.77 ± 0.99 B ^{j}
	Saponin	1%	126.14 ± 1.06 A ^{f}	128.44 ± 1.46 A ^{f}
Oxidising agents	Sodium perborate	1%	75.63 ± 1.07 A ^{c}	77.77 ± 0.97 A ^{b}
	H_2O_2	5%	70.30 ± 1.02 A ^{c}	87.55 ± 1.02 B ^{c}
		10%	64.97 ± 1.19 A ^{c}	84.55 ± 0.73 B ^{c}
		15%	40.35 ± 0.65 A ^{b}	73.77 ± 0.91 B ^{b}

Each intestinal crude extract (ICE) was incubated with same volume of the chemicals for 1 h at 40 °C and residual activity of the enzyme was measured at pH 8.5 and 55 °C. The control was prepared with similar conditions without chemicals and scored as 100%. Data were represented by mean ± standard deviation and compared by Duncan's multiple-range test. Different capital letters in the same row and different superscripts in the same column show statistical difference ($p < 0.05$).

4. Conclusions

The results of our study indicated that trypsin from intestine of beluga and sevruga had similar properties to trypsins from bony fish. The enzyme had an optimum temperature of 55 °C and thermal stability was maintained over 90% up to 55 °C. This alkaline protease had an optimum pH of 8.5 and showed to be tolerant in the pH range of 6.0 to 11.0 in both studied sturgeon species. The molecular weight of trypsin for sevruga and beluga was estimated to be 27.5 and 29.5 kDa, respectively, as data from inhibitory activity staining indicated. Both trypsins were inhibited by main specific inhibitors, SBTI and TLCK. Additionally, the enzymatic activity of trypsins was still detected after 1 h in the presence of surfactants and oxidative agents. Information provided in this manuscript related to trypsin activity for beluga and sevruga based on changes in the activity of this alkaline protease based on a tested range of pH and temperature values, and presence of potential inhibitors, ions and cations may be considered a preliminary step to design in vitro tests for the assessment of protein digestibility in these species.

Author Contributions: Conceptualization and methodology, A.Z.; writing—review and editing, A.Z. and E.G.; investigation and writing—original draft, M.K.; data analysis and measurement of physicochemical properties of the enzyme, A.A.K., M.H.N., A.S. and M.E. All authors have read and agreed to the published version of the manuscript.

Funding: This work was funded by the Vice-Presidency of Research and Technology from Malayer University, Iran (grant number 84/9-1-310).

Institutional Review Board Statement: This study conformed to the current Iranian law regarding the care and use of laboratory animals and was approved by the Animal Ethical Committee of Malayer University (Code: IR.UMSHA.REC.1397.994).

Informed Consent Statement: Informed consent was obtained from all subjects involved in the study.

Data Availability Statement: Not applicable.

Acknowledgments: We would like to express our sincere gratitude to laboratory staffs of Fisheries Department of Natural Resources and Environment Faculty of Malayer University for helpful assistance.

Conflicts of Interest: The authors declare no conflict of interest.

References

1. Kalbassi, M.R.; Abdollahzadeh, E.; Salari-Joo, H. A review on aquaculture development in Iran. *Ecopersia* **2013**, *1*, 159–178.
2. Vasilyeva, L.M.; Elhetawy, A.I.G.; Sudakova, N.V.; Astafyeva, S.S. History, current status and prospects of sturgeon aquaculture in Russia. *Aquac. Res.* **2019**, *50*, 979–993. [CrossRef]
3. Aghilinejhad, S.M.; Gorgin, S.; van Uhm, D.; Joolaie, R.; Ghorbani, R.; Paighambari, S.Y.; Mohammadi, J.; Jalali, A. What are the drivers of the occurrence of illegal fishing and conservation barriers of sturgeons in the Caspian Sea? *Aquat. Conserv. Mar. Freshw. Ecosyst.* **2018**, *28*, 690–701. [CrossRef]
4. Mirrasooli, E.; Ghorbani, R.; Gorgin, S.; Aghilinejhad, S.M.; Jalali, A. Factors associated with illegal fishing and fisher attitudes toward sturgeon conservation in the southern Caspian Sea. *Mar. Policy* **2019**, *100*, 107–115. [CrossRef]
5. Iranian Fisheries Organization Statistical Yearbook. *Budget and Planning Deputy;* Iranian Fisheries Organization Statistical Yearbook: Tehran, Iran, 2022; p. 64. (In Persian)
6. Furne, M.; García-Gallego, M.; Hidalgo, M.C.; Morales, A.E.; Domezain, A.; Domezain, J.; Sanz, A. Effect of starvation and refeeding on digestive enzyme activities in sturgeon (*Acipenser naccarii*) and trout (*Oncorhynchus mykiss*). *Comp. Biochem. Physiol.* **2008**, *149A*, 420–425. [CrossRef] [PubMed]
7. Kolkovski, S. Digestive enzymes in fish larvae and juveniles-Implications and applications to formulated diets. *Aquaculture* **2001**, *200*, 181–201. [CrossRef]
8. Yufera, M.; Darias, M.J. The onset of exogenous feeding in marine fish larvae. *Aquaculture* **2007**, *268*, 53–63. [CrossRef]
9. Twining, S.S.; Alexander, P.A.; Huibregste, K.; Glick, D.M. A pepsinogen from rainbow trout. *Comp. Biochem. Physiol.* **1983**, *75B*, 109–112. [CrossRef]
10. Navarro-Guillen, C.; Yufera, M.; Perera, E. Biochemical features and modulation of digestive enzymes by environmental temperature in the greater amberjack, *Seriola dumerili*. *Front. Mar. Sci.* **2022**, *28*, 1391. [CrossRef]
11. Bolasina, S.; Perez, A.; Yamashita, Y. Digestive enzymes activity during ontogenetic development and effect of starvation in Japanese flounder, *Paralichthys olivaceus*. *Aquaculture* **2006**, *252*, 503–515. [CrossRef]
12. Moraes, G.; de Almeida, L.C. Nutrition and functional aspects of digestion in fish. In *Biology and Physiology of Freshwater Neotropical Fish;* Academic Press: Cambridge, MA, USA, 2020; pp. 251–271.
13. Glass, H.J.; MacDonald, N.L.; Moran, R.M.; Stark, J.R. Digestion of protein in different marine species. *Comp. Biochem. Physiol.* **1989**, *94B*, 607–611. [CrossRef]
14. Eshel, A.; Lindner, P.; Smirnoff, P.; Newton, S.; Harpaz, S. Comparative study of proteolytic enzymes in the digestive tracts of the European sea bass and hybrid striped bass reared in freshwater. *Comp. Biochem. Physiol.* **1993**, *106A*, 627–634. [CrossRef]
15. Rungruangsak-Torrissen, K.; Moss, R.; Andresen, L.H.; Berg, A.; Waagbo, R. Different expressions of trypsin and chymotrypsin in relation to growth in Atlantic salmon (*Salmo salar* L.). *Fish Physiol. Biochem.* **2006**, *32*, 7–23. [CrossRef] [PubMed]
16. Nolasco-Soria, H. Improving and standardizing protocols for alkaline protease quantification in fish. *Rev. Aquac.* **2021**, *13*, 43–65. [CrossRef]
17. Nazdar, N.; Imani, A.; Noori, F.; Sarvi Moghanlou, K. Effect of silymarin supplementation on nickel oxide nanoparticle toxicity to rainbow trout (*Oncorhynchus mykiss*) fingerlings: Pancreas tissue histopathology and alkaline protease activity. *Iran. J. Sci. Technol. Trans. A Sci.* **2018**, *42*, 353–361. [CrossRef]
18. Krogdahl, A.; Lea, T.B.; Olli, J.J. Soybean proteinase inhibitors affect intestinal trypsin activities and amino acid digestibilities in rainbow trout (*Oncorhyncus mykiss*). *Comp. Biochem. Physiol.* **1994**, *107A*, 215–219. [CrossRef]
19. Haard, N.F.; Dimes, L.E.; Arndt, R.E.; Dong, F.M. Estimation of protein digestibility: IV. Digestive proteinases from the pyloric caeca of coho salmon (*Oncorhynchus kisutch*) fed diets containing soybean meal. *Comp. Biochem. Physiol.* **1996**, *115B*, 533–540. [CrossRef]
20. Chong, A.S.C.; Hashim, R.; Chow-Yang, L.; Ali, A.B. Partial characterization and activities of proteases from the digestive tract of discus fish, *Symphysodon aeguifasciata*. *Aquaculture* **2002**, *203*, 321–333. [CrossRef]
21. Hedstrom, L. Serine protease mechanism and specificity. *Chem. Rev.* **2002**, *102*, 4501–4524. [CrossRef]
22. Gilannejad, N.; Martínez-Rodríguez, G.; Yúfera, M.; Moyano, F.J. Estimating the effect of different factors on the digestive bioaccessibility of protein by the Senegalese sole (*Solea senegalensis*); combination of response surface methodology and in vitro assays. *Aquaculture* **2017**, *477*, 28–34. [CrossRef]
23. Liu, Z.Y.; Wang, Z.; Xu, S.Y.; Xu, L.N. Two trypsin isoforms from the intestine of the grass carp (*Ctenopharyngodon idellus*). *J. Comp. Physiol.* **2007**, *177B*, 655–666. [CrossRef] [PubMed]
24. Souza, A.A.; Amaral, I.P.; Santo, A.R.E.; Carvalho, J.R.L.B.; Bezerra, R.S. Trypsin-like enzyme from intestine and pyloric caeca of spotted goatfish (*Pseudupeneus maculatus*). *Food Chem.* **2007**, *100*, 1429–1434. [CrossRef]
25. Jellouli, K.; Bougatef, A.; Daassi, D.; Balti, R.; Barkia, A.; Nasri, M. New alkaline trypsin from the intestine of grey triggerfish (*Balistes capriscus*) with high activity at low temperature: Purification and characterisation. *Food Chem.* **2009**, *116*, 644–650. [CrossRef]

26. Klomklao, S.; Kishimura, H.; Nonami, Y.; Benjakul, S. Biochemical properties of two isoforms of trypsin purified from the intestine of skipjack tuna (*Katsuwonus pelamis*). *Food Chem.* **2009**, *115*, 155–162. [CrossRef]
27. Bougatef, A.; Balti, R.; Nasri, R.; Jellouli, K.; Souissi, N.; Nasri, M. Biochemical properties of anionic trypsin acting at high concentration of NaCl purified from the intestine of a carnivorous fish: Smooth hound (*Mustelus mustelus*). *J. Agric. Food Chem.* **2010**, *58*, 5763–5769. [CrossRef] [PubMed]
28. Candiotto, F.B.; Freitas-Junior, A.C.V.; Neri, R.C.A.; Bezerra, R.S.; Rodrigues, R.V.; Sampaio, L.A.; Tesser, M.B. Characterization of digestive enzymes from captive Brazilian flounder *Paralichthys orbignyanus*. *Braz. J. Biol.* **2017**, *78*, 281–288. [CrossRef] [PubMed]
29. Buddington, R.K.; Doroshov, S.I. Feeding trials with hatchery produced white sturgeon juveniles (*Acipenser transmontanus*). *Aquaculture* **1984**, *36*, 237–243. [CrossRef]
30. Buddington, R.K. Digestive secretions of lake sturgeon, *Acipenser fulvescens*, during early development. *J. Fish Biol.* **1985**, *26*, 715–723. [CrossRef]
31. Dabrowski, K.; Kaushik, S.J.; Fauconneau, B. Rearing of sturgeon (*Acipenser baeri* Brandt) larvae: I. Feeding trial. *Aquaculture* **1985**, *47*, 185–192. [CrossRef]
32. Ostaszewska, T.; Kolman, R.; Kamaszewski, M.; Wiszniewski, G.; Adamek, D.; Duda, A. Morphological changes in digestive tract of Atlantic sturgeon (*Acipenser oxyrinchus*) during organogenesis. *Int. Aquat. Res.* **2011**, *3*, 101–105.
33. Bardi, R.W.; Chapman, F.A.; Barrows, F.T. Feeding trials with hatchery-produced Gulf of Mexico sturgeon larvae. *Prog. Fish Cult.* **1998**, *60*, 25–31. [CrossRef]
34. Babaei, S.S.; Abedian Kenari, A.; Rajabmohammad Nazari, R.M.; Gisbert, E. Developmental changes of digestive enzymes in Persian sturgeon (*Acipenser persicus*) during larval ontogeny. *Aquaculture* **2011**, *318*, 138–144. [CrossRef]
35. Camacho, S.; Carmona, R.; Llorente, J.I.; Sanz, A.; Garcia-Gallego, M.; Domezain, A.; Dominguez, N.; Ostos-Garrido, M.V. Stomach development in the sturgeon *Acipenser naccarii*: Histoenzymatic and ultrastructural analysis. *J. Appl. Ichthyol.* **2011**, *27*, 693–700. [CrossRef]
36. Sanz, A.; Llorente, J.I.; Furne, M.; Ostos-Garrido, M.V.; Carmona, R.; Domezain, A.; Hidalgo, M.C. Digestive enzymes during ontogeny of the sturgeon *Acipenser naccarii*: Intestine and pancreas development. *J. Appl. Ichthyol.* **2011**, *27*, 1139–1146. [CrossRef]
37. Ghasemi, N.; Imani, A.; Noori, F.; Shahrooz, R. Ontogeny of digestive tract of stellate sturgeon (*Acipenser stellatus*) from hatching to juvenile stage: Digestive enzymes activity, stomach and proximal intestine. *Aquaculture* **2020**, *519*, 734751. [CrossRef]
38. Asgari, R.; Rafiee, G.; Eagderi, S.; Noori, F.; Agh, N.; Poorbagher, H.; Gisbert, E. Ontogeny of the digestive enzyme activities in hatchery produced beluga (*Huso huso*). *Aquaculture* **2013**, *416*, 33–40. [CrossRef]
39. Erlanger, B.F.; Kokowski, N.; Cohen, W. The preparation and properties of two new chromogenic substrates of trypsin. *Arch. Biochem. Biophys.* **1961**, *95*, 271–278. [CrossRef]
40. Zamani, A.; Rezaei, M.; Madani, R.; Habibi Rezaie, M. Trypsin enzyme from viscera of common kilka (*Clupeonella cultriventris caspia*): Purification, characterization, and its compatibility with oxidants and surfactants. *J. Aquat. Food Prod. Technol.* **2014**, *23*, 237–252. [CrossRef]
41. Lowry, O.H.; Rosebrough, N.J.; Farr, A.L.; Randall, R.J. Protein measurement with the folin phenol reagent. *J. Biol. Chem.* **1951**, *193*, 265–275. [CrossRef]
42. Laemmli, U.K. Cleavage of structural proteins during assembly of the head bacteriophague T4. *Nature* **1970**, *227*, 680–685. [CrossRef]
43. Garcia-Carreno, F.L.; Dimes, L.E.; Haard, N.F. Substrate-gel electrophoresis for composition and molecular weight of proteinases or proteinaceous proteinase inhibitors. *Anal. Biochem.* **1993**, *214*, 65–69. [CrossRef]
44. Ahmad, M.; Benjakul, S. Impact of legume seed extracts on degradation and functional properties of gelatin from unicorn leatherjacket skin. *Process. Biochem.* **2011**, *46*, 2021–2029. [CrossRef]
45. Khantaphant, S.; Benjakul, S. Purification and characterization of trypsin from the pyloric caeca of brownstripe red snapper (*Lutjanus vitta*). *Food Chem.* **2010**, *120*, 658–664. [CrossRef]
46. Alarcón, F.J.; Díaz, M.; Moyano, F.J. Characterization and functional properties of digestive proteases in two sparids; gilthead seabream (*Sparus aurata*) and common dentex (*Dentex dentex*). *Fish Physiol. Biochem.* **1998**, *19*, 257–267. [CrossRef]
47. Gendry, P.; Launay, J.F. Pancreatic anionic trypsin: Evidence for the existence of a 30 kDa form. *Comp. Biochem. Physiol.* **1992**, *102B*, 37–41. [CrossRef] [PubMed]
48. Khangembam, B.K.; Chakrabarti, R. Trypsin from the digestive system of carp *Cirrhinus mrigala*: Purification, characterization and its potential application. *Food Chem.* **2015**, *175*, 386–394. [CrossRef] [PubMed]
49. De Freitas-Junior, A.C.V.; da Costa, H.M.S.; Marcuschi, M.; Icimoto, M.Y.; Machado, M.F.; Machado, M.F.; Ferreira, J.C.; de Oliveira, V.M.; Buarque, D.S.; Bezerra, R.S. Substrate specificity, physicochemical and kinetic properties of a trypsin from the giant Amazonian fish pirarucu (*Arapaima gigas*). *Biocatal. Agric. Biotechnol.* **2021**, *35*, 102073. [CrossRef]
50. Aissaoui, N.; Marzouki, M.N.; Abidi, F. Purification and biochemical characterization of a novel intestinal protease from *Scorpaena notata*. *Int. J. Food Prop.* **2017**, *20*, 2151–2165. [CrossRef]
51. Klomklao, S.; Benjakul, S. Two trypsin isoforms from albacore tuna (*Thunnus alalunga*) liver: Purification and physicochemical and biochemical characterization. *Int. J. Biol. Macromol.* **2018**, *107*, 1864–1870. [CrossRef]
52. Dos Santos, C.W.V.; da Costa Marques, M.E.; de AraujoTenorio, H.; de Miranda, E.C.; Pereira, H.J.V. Purification and characterization of trypsin from *Luphiosilurus alexandri* pyloric cecum. *Biochem. Biophys. Rep.* **2016**, *8*, 29–33. [CrossRef]

53. Gonzalez-Felix, M.L.; De La Ree-Rodriguez, C.; Perez-Velazquez, M. Partial characterization, quantification and optimum activity of trypsin and lipase from the sciaenids *Cynoscion othonopterus*, *Cynoscion parvipinnis* and *Cynoscion xanthulus*. *Arch. Biol. Sci.* **2020**, *72*, 81–93. [CrossRef]
54. Castillo-Yanez, F.J.; Pacheco-Aguilar, R.; Garcia-Carreno, F.L.; Del Toro, M.A.N. Isolation and characterization of trypsin from pyloric caeca of Monterey sardine (*Sardinops sagax caerulea*). *Comp. Biochem. Physiol.* **2005**, *140B*, 91–98. [CrossRef] [PubMed]
55. Dos Santos, D.M.R.C.; dos Santos, C.W.V.; de Souza, C.B.; de Albuquerque, F.S.; dos Santos Oliveira, J.M.; Pereira, H.J.V. Trypsin purified from *Coryphaena hippurus* (common dolphinfish): Purification, characterization, and application in commercial detergents. *Biocatal. Agric. Biotechnol.* **2020**, *25*, 101584. [CrossRef]
56. Ktari, N.; Ben Khaled, H.; Nasri, R.; Jellouli, K.; Ghorbel, S.; Nasri, M. Trypsin from zebra blenny (*Salaria basilisca*) viscera: Purification, characterisation and potential application as a detergent additive. *Food Chem.* **2012**, *130*, 467–474. [CrossRef]
57. Ben Khaled, H.; Jellouli, K.; Souissi, N.; Ghorbel, S.; Nasri, M. Purification and characterisation of three trypsin isoforms from viscera of sardinelle (*Sardinella aurita*). *Fish Physiol. Biochem.* **2011**, *37*, 123–133. [CrossRef] [PubMed]
58. Klomklao, S.; Benjakul, S.; Visessanguan, W.; Kishimura, H.; Simpson, B.K. 29 kDa trypsin from the pyloric ceca of Atlantic bonito (*Sarda sarda*): Recovery and characterization. *J. Agric. Food Chem.* **2007**, *55*, 4548–4553. [CrossRef]
59. Bezerra, R.S.; Santos, J.F.; Paiva, P.M.; Correia, M.T.S.; Coelho, L.C.B.B.; Vieira, V.L.L.; Carvalho, J.R.L.B. Partial purification and characterization of a thermostable trypsin from pyloric caeca of tambaqui (*Colossoma macropomum*). *J. Food Biochem.* **2001**, *25*, 199–210. [CrossRef]
60. Klomklao, S.; Benjakul, S.; Visessanguan, W.; Kishimura, H.; Simpson, B.K. Proteolytic degradation of sardine (*Sardinella gibbosa*) proteins by trypsin from skipjack tuna (*Katsuwonus pelamis*) spleen. *Food Chem.* **2006**, *98*, 14–22. [CrossRef]
61. Lu, B.J.; Zhou, L.G.; Cai, Q.F.; Hara, K.; Maeda, A.; Su, W.J.; Cao, M.J. Purification and characterisation of trypsins from the pyloric caeca of Mandarin fish (*Siniperca chuatsi*). *Food Chem.* **2008**, *110*, 352–360. [CrossRef]
62. Silva, J.F.; Esposito, T.S.; Marcuschi, M.; Ribeiro, K.; Cavalli, R.O.; Oliveira, V.; Bezerra, R.S. Purification and partial characterisation of a trypsin from the processing waste of the silver mojarra (*Diapterus rhombeus*). *Food Chem.* **2011**, *129*, 777–782. [CrossRef]
63. Stefansson, B.; Sandholt, G.B.; Gudmundsdottir, Á. Elucidation of different cold-adapted Atlantic cod (*Gadus morhua*) trypsin X isoenzymes. *Biochim. Biophys. Acta-Proteins Proteom.* **2017**, *1865*, 11–19. [CrossRef] [PubMed]
64. Esposito, T.S.; Marcuschi, M.; Amaral, I.P.; Carvalho, L.B., Jr.; Bezerra, R.S. Trypsin from the processing waste of the lane snapper (*Lutjanus synagris*) and its compatibility with oxidants, surfactants and commercial detergents. *J. Agric. Food Chem.* **2010**, *58*, 6433–6439. [CrossRef] [PubMed]
65. Cai, Q.F.; Jiang, Y.K.; Zhou, L.G.; Sun, L.C.; Li, G.M.; Osatomi, K.; Cao, M.J. Biochemical characterization of trypsins from the hepatopancreas of Japanese sea bass (*Lateolabrax japonicus*). *Comp. Biochem. Physiol.* **2011**, *159B*, 183–189. [CrossRef]
66. Kishimura, H.; Klomklao, S.; Benjakul, S.; Chun, B.S. Characteristics of trypsin from the pyloric ceca of walleye pollock (*Theragra chalcogramma*). *Food Chem.* **2008**, *106*, 194–199. [CrossRef]
67. Klomklao, S.; Kishimura, H.; Benjakul, S. Anionic trypsin from the pyloric ceca of Pacific saury (*Cololabis saira*): Purification and biochemical characteristics. *J. Aquat. Food. Prod. Technol.* **2014**, *23*, 186–200. [CrossRef]
68. Kurtovic, I.; Marshall, S.N.; Simpson, B.K. Isolation and characterization of a trypsin fraction from the pyloric ceca of chinook salmon (*Oncorhynchus tshawytscha*). *Comp. Biochem. Physiol.* **2006**, *143B*, 432–440. [CrossRef] [PubMed]
69. Sila, A.; Nasri, R.; Jridi, M.; Balti, R.; Nasri, M.; Bougatef, A. Characterisation of trypsin purified from the viscera of Tunisian barbel (*Barbus callensis*) and its application for recovery of carotenoproteins from shrimp wastes. *Food Chem.* **2012**, *132*, 1287–1295. [CrossRef]
70. Kanno, G.; Kishimura, H.; Ando, S.; Nalinanon, S.; Klomklao, S.; Benjakul, S.; Chun, B.S.; Saeki, H. Structural properties of trypsin from cold-adapted fish, arabesque greenling (*Pleurogrammus azonus*). *Eur. Food Res. Technol.* **2011**, *232*, 381–388. [CrossRef]
71. Haard, N.F. A review of proteolytic enzymes from marine organisms and their application in the food industry. *J. Aquatic Food Product. Technol.* **1992**, *1*, 17–35. [CrossRef]
72. Wu, X.; Wang, Y.; Tong, B.; Chen, X.; Chen, J. Purification and biochemical characterization of a thermostable and acid-stable alpha-amylase from *Bacillus licheniformis* B4-423. *Int. J. Biol. Macromol.* **2018**, *109*, 329–337. [CrossRef]
73. Simpson, B.K. Digestive proteinases from marine animals. In *Seafood Enzymes: Utilization and Influence on Postharvest Seafood Quality*; Haard, N.F., Simpson, B.K., Eds.; Marcel Dekker: New York, NY, USA, 2000; pp. 531–540.
74. Martinez, A.; Olsen, R.L.; Serra, J.L. Purification and characterization of two trypsin like enzymes from the digestive tract of anchovy *Engraulis encrasicholus*. *Comp. Biochem. Physiol.* **1988**, *91B*, 677–684. [CrossRef]
75. Mamimin, C.; O-thong, S.; Nitipan, S. Application of metagenomics technique in gene encoding thermostable enzyme discovery from hot spring. *Thaksin Univ. J.* **2016**, *19*, 96–105.
76. Cristina Oliveira de Lima, V.; Piuvezam, G.; Leal Lima Maciel, B.; Heloneida de Araújo Morais, A. Trypsin inhibitors: Promising candidate satietogenic proteins as complementary treatment for obesity and metabolic disorders? *J. Enzyme. Inhib. Med. Chem.* **2019**, *34*, 405–419. [CrossRef] [PubMed]
77. Farady, C.J.; Craik, C.S. Mechanisms of macromolecular protease inhibitors. *Chembiochem* **2010**, *11*, 2341–2346. [CrossRef] [PubMed]
78. Avilés-Gaxiola, S.; Chuck-Hernández, C.; Serna Saldívar, S.O. Inactivation methods of trypsin inhibitor in legumes: A review. *J. Food Sci.* **2018**, *83*, 17–29. [CrossRef] [PubMed]

79. Senphan, T.; Benjakul, S.; Kishimura, H. Purification and characterization of trypsin from hepatopancreas of Pacific white shrimp. *J. Food Biochem.* **2015**, *39*, 388–397. [CrossRef]
80. Choi, S.M.; Oh, E.S.; Kim, D.S.; Pyeun, J.H.; Cho, D.M.; Ahn, C.B.; Kim, H.R. Comparative Biochemical Properties of Proteinases from the Hepatopancreas of Shrimp.-I. Purification of Protease from the Hepatopancreas of *Penaeus japonicus*. *Fish Aquatic. Sci.* **1998**, *1*, 201–208.
81. Page, M.J.; Di Cera, E. Role of Na+ and K+ in enzyme function. *Physiol. Rev.* **2006**, *86*, 1049–1092. [CrossRef]
82. Fu, X.Y.; Xue, C.H.; Miao, B.C.; Li, Z.J.; Gao, X.; Yang, W.G. Characterization of proteases from the digestive tract of sea cucumber (*Stichopus japonicus*): High alkaline protease activity. *Aquaculture* **2005**, *246*, 321–329. [CrossRef]
83. Bougatef, A. Trypsins from fish processing waste: Characteristics and biotechnological applications—Comprehensive review. *J. Clean. Prod.* **2013**, *57*, 257–265. [CrossRef]
84. Mandal, R.; Mandal, D.; Mishra, N.; Bahadur, A. Effect of surfactants on phosphatase level of fresh water fish *Labeo rohita*. *J. Environ. Biol.* **2010**, *31*, 395–398. [PubMed]
85. Rubingh, D.N. The influence of surfactants on enzyme activity. *Curr. Opin. Colloid Interface Sci.* **1996**, *1*, 598–603. [CrossRef]
86. Jelińska, A.; Zagożdżon, A.; Górecki, M.; Wisniewska, A.; Frelek, J.; Holyst, R. Denaturation of proteins by surfactants studied by the Taylor dispersion analysis. *PLoS ONE* **2017**, *12*, e0175838. [CrossRef] [PubMed]
87. Lin, S.; Blankschtein, D. Role of the bile salt surfactant sodium cholate in enhancing the aqueous dispersion stability of single-walled carbon nanotubes: A molecular dynamics simulation study. *J. Phys. Chem.* **2010**, *114B*, 15616–15625. [CrossRef] [PubMed]
88. Parra, J.G.; Iza, P.; Dominguez, H.; Schott, E.; Zarate, X. Effect of Triton X-100 surfactant on the interfacial activity of ionic surfactants SDS, CTAB and SDBS at the air/water interface: A study using molecular dynamic simulations. *Colloids Surf. A Physicochem. Eng. Asp.* **2020**, *603*, 125284. [CrossRef]
89. Rai, S.; Acharya-Siwakoti, E.; Kafle, A.; Devkota, H.P.; Bhattarai, A. Plant-derived saponins: A review of their surfactant properties and applications. *Science* **2021**, *3*, 44. [CrossRef]

animals

MDPI

Article

A Morphological and Ultrastructural Study of the Anterior Digestive Tract of Adult Nile Tilapia *Oreochromis niloticus*

Antonio Palladino [1,†], Elena De Felice [2,*,†], Chiara Attanasio [3], Carmela M. A. Barone [1], Antonio Crasto [1], Livia D'Angelo [3], Daniela Giaquinto [3], Claudia Lambiase [1], Paola Scocco [2], Francesco Serrapica [1] and Lucianna Maruccio [3]

1 Department of Agricultural Sciences, University of Naples Federico II, Via Università 100, 80055 Portici, Italy
2 School of Biosciences and Veterinary Medicine, University of Camerino, Via Gentile III da Varano, 62032 Camerino, Italy
3 Department of Veterinary Medicine and Animal Production, University of Naples Federico II, Via F. Delpino 1, 80137 Naples, Italy
* Correspondence: elena.defelice@unicam.it
† These authors contribute equally to the work.

Simple Summary: Aquaculture is currently one of the fastest growing food-producing systems due to its relevance as source of livelihood worldwide. As fishes live in very diversified environmental conditions and this diversification also impacts their diet, it is not surprising that a considerable variability exists in the organization of the digestive tract. Nile tilapia is among the most-used fish species in aquaculture due to its growth and ability to adapt to a wide range of growing conditions. The aim of this work is an integrated description of esophagus and stomach morphology. We identified five distinct zones of the stomach via light and scanning microscopy. Histochemical investigation showed the presence of carboxylated and sulphated mucins along the esophagus and stomach capable of attracting a great amount of water and enhancing the protective function of the mucosa mucous gel, especially in the highly acidic stomach environment that digests algae and detrital bacteria. Therefore, our study provides new insights concerning the morphological structure of the anterior digestive tract of Nile tilapia, relevant to better understand the several aspects related to physiological growth and the health status of this species, while at the same time comparing different or pathological conditions.

Abstract: Among the most-used fish species in aquaculture is the Nile tilapia, due to its rapid growth rate and its adaptation to a wide range of farming conditions. A careful description of the morphology of the digestive tract, particularly the esophagus and stomach, allows a better understanding of the relationship between structure and function. Combining scanning and light microscopy we highlighted the presence of five different zones in the stomach (1: esophagus-gastric lumen passage; 2: descending glandular portion; 3: fundic portion; 4: ascending glandular portion; 5: gastric-pyloric transition portion). Histochemical investigation showed a secretion of carboxylates mucopolysaccharides along the esophagus and sulphated complex carbohydrates in the stomach. These results suggest that mucins play a protective role of the epithelial lining, which is essential for a correct digestive process. Finally, the characterization of the main cellular structures may be inspiring for more advanced studies aiming to decipher the role of specific molecules, such as neuropeptides, involved in the physiological digestive process.

Keywords: *Oreochromis niloticus*; esophagus; stomach; morphometry; conventional glycohistochemistry; SEM

1. Introduction

Aquaculture is currently one of the fastest growing food-producing systems due to its relevance as a source of livelihood worldwide. In commercial farming, fish exposed to suboptimal environmental conditions can alter feeding behavior and growth rate as well as energy metabolism [1]. In addition, fish living in highly diversified aquatic environments due to the different types of food available will display morpho-functional variations of the digestive tract, mainly related to the stomach [2,3]. Indeed, this organ may be even absent; when present, it may show different shapes [4,5] or be modified as a swelling structure (called intestinal bulb) [6–11]. The extreme variability of stomach shape and structure in Teleosts is mainly related to phylogeny and ontogenesis, but also to functional adaptation [12–16]. Compliant with different feeding habits, carnivorous fishes feature large, extensible stomachs with a short intestine [9–11,17,18], while omnivorous fishes display a large stomach and a long intestine [19,20]. Interestingly, strictly herbivorous fish feature a small stomach with a long and complex intestine, but in some cases the stomach may be completely absent [14,18,19,21–27]. As the morphology of the digestive system of fishes is strictly related to specific diet, correct management of feeding habits is mandatory to improve production rates of fish.

Of the fish species used in aquaculture, Tilapia is the third most-used after carp and salmon [28,29]. It is an omnivorous fish of the Cichlid family which lives in Africa, South America and Asia. Nile tilapia (*Oreochromis niloticus*) is one of the most farmed tilapia species, due to its rapid growth rate and ability to adapt to a wide range of environmental conditions [28]. Tilapia is appreciated by consumers for the softness of its muscles, its low fat content and good flavour [29]. This is a tropical species living in shallow and sweet waters (lakes and lacustrine areas) with temperatures ranging from 31 to 36 °C; however, Tilapia is resistant to variations of these environmental parameters. It can live up to 10 years and reach a weight of about 5 kg [28].

Very few studies describing the ultrastructure of the digestive tract of Nile tilapia are reported in literature [30–32] and only one shows a figure of the entire stomach stained by Masson's trichrome [33]. Although the basic histological structure of the digestive tract is similar in the different fish species [34], significant differences can be observed macroscopically in the different regions of the stomach. In particular, the differences highlighted in wall architecture and mucosa ultrastructure of the digestive tract of Nile tilapia [30–32] determine different secreting behavior and mucosa architecture of each one of the different regions of the stomach.

The shape of the stomach and the topographical proximity between the cardia and pylorus probably influences food transit; therefore, also influencing the morphological and functional features, such as enzyme secretion.

The aim of this research is an integrated description of esophagus and stomach morphology of Nile tilapia by combining scanning electron and optical microscopy, highlighting the different possible distributions of mucopolysaccharides to better understand the relationship between structure and function.

2. Materials and Methods

2.1. Sample Collection

For this study, ten adult female Nile tilapia *O. niloticus*, with an average weight of 200 g and 20 cm long, were used. These fish, intended for human consumption, were obtained from a local farmer in the north of Italy. The animal care procedures were in accordance with legislative decree No. 146, which implements directive 98/58/EC, 20 July 1998, concerning the protection of farmed animals. After a 24-hour fast, the fish were suppressed at lower temperature. Then, samples of the esophagus and stomach were removed through a ventral incision of the body from the anal opening to the operculum, washed in phosphate buffer (PBS) pH 7.4 and fixed as follows:

1. in Bouin's liquid for 24 hours (h) at room temperature (RT) for histological and histochemical analyses;
2. in 2.5% glutaraldehyde solution in PBS pH 7.4 for 24 h at 4 °C for scanning electron microscopy (SEM).

2.2. Morphological and Conventional Glycohistochemical Analysis

Whole stomachs with the respective esophagus were fixed and then dehydrated in graded ethanol concentrations (70% to absolute), clarified through two baths in xylene and embedded in paraffin wax at 56–58 °C under a thermostat-vacuum paraffin-embedding bath for 1.5 h. Sagittal sections of 7 µm thick were obtained by using a Reichert–Jung 2050 microtome, then mounted on clean and dry glass slides. Sections were stained with Harris's hematoxylin-eosin (HE) for morphological evaluation. For histochemical analysis, samples were treated with periodic acid Schiff (PAS), thereby visualizing the vicinal hydroxyls groups; Alcian blue (AB) pH 2.5, thereby visualizing acidic groups pertaining to carboxylated and sulphated complex carbohydrates (sialylated glycoproteins, hyaluronic acid, chondroitin, chondroitin-sulphates A/B/C, heparin, heparansulphate glycosaminoglycan-like material); AB pH 1, thereby visualizing acidic groups pertaining to sulphated complex carbohydrates (chondroitin-sulphates A/B/C, heparin, heparansulphate glycosaminoglycan-like material); and periodic acid Schiff in combination with the Alcian blue (AB/PAS), thereby co-visualizing acidic groups and vicinal hydroxyls [35–37]. Staining solutions were purchased from Bio-Optica Milano SPA: HE (code 05-M06004); AB pH 2.5 (code 05-M26003); AB pH 1 (code 05-M26005); PAS (code 04-130802A); AB/PAS (code 04-163802). All reactions were carried out according to the manufacturer's instructions.

All stained sections were observed and photographed (ten for 20× for the mosaic and 40× for the particulars) under Leica DM 1000 light microscope. The digital raw images were optimized for image resolution, contrast, evenness of illumination and background using Adobe Photoshop CS5 (Adobe Systems, San Jose, CA, USA).

2.3. Morphometry

The thickness of the superficial and glandular epithelium, submucosa, and circular and longitudinal musculature of the stomach were measured by digital imaging and Image J software. For each sample, a total of 3 sections have been processed and 10 random microscopic fields were analyzed at 20 × magnification. In particular, the thickness was calculated as the average of four measurements. In order to highlight statistically significant differences among the same layers in different zones, we performed the statistical analysis described below. The analysis consisted in a one-way ANOVA test on each mucosal layer. This test was conducted in multiple comparison within each group, posing the values of Zone 1 as "internal control". We evaluated measurements of the same zone from 13 different slides. Through this approach, 4 p-values were calculated corresponding to Zone 1 vs. Zone 2; Zone 1 vs. Zone 3; Zone 1 vs. Zone 4; and Zone 1 vs. Zone 5. The differences were considered significant when the p-value was <0.05.

2.4. Scanning Electron Microscopy (SEM)

Specimens of 5 × 5 mm were postfixed with 1% osmium tetroxide in PBS for 6 h at 4 °C in the dark. After three washes in PBS, the specimens were dehydrated through a graded series of ethanol and subsequently in CO_2 atmosphere, using a critical point dried EMITECH K850 (Quorum, East Sussex, UK). They were attached to aluminum stubs facing upwards, covered with carbon tabs, and then the samples were sputtered with gold palladium (AGARIB 7340, Agar Scientific Ltd, Stansed, UK). The sections were examined with a scanning electron microscope ZEISS EVO40 (20.000KU EHT, 10.0 mm < ωD < 13.0 and 90× < magnification < 8k).

3. Results

3.1. Gross Morphology

The esophagus appeared as a short and dilatable membranous tube connecting the pharyngeal cavity to the initial part of the stomach, displaying the same diameter along its entire length (Figure 1A).

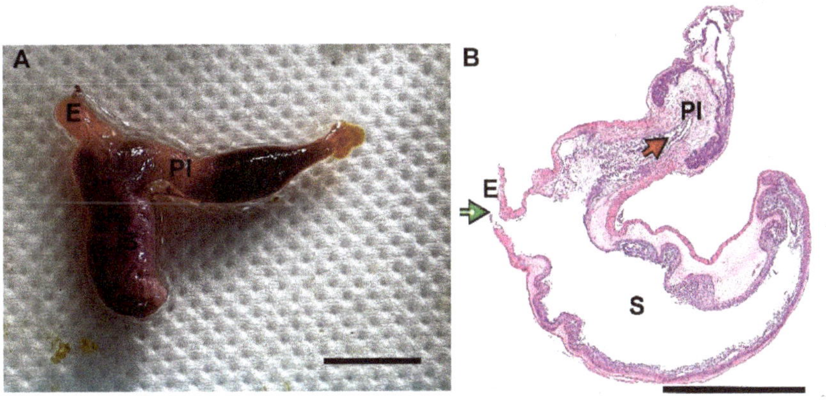

Figure 1. Gross morphology and microscopical organization of esophagus and stomach of Nile tilapia. (**A**): esophagus (E), stomach (S) and proximal portion of the intestine (PI); (**B**): Hematoxylin/Eosin staining of a para-sagittal section of the two organs. In the upper left is visible the esophagus outlet (green arrow), while in the upper right is the pyloric sphincter (red arrow). The figure is representative of a reconstructed mosaic of 180 individual photos taken using a 20× objective to visualize in detail the stratigraphy of the organs. Scale bars = 1 cm.

The stomach was similar to a U-shaped pouch, with a descending portion and an ascending portion positioned in the abdominal cavity ventrally and dorsally, respectively (Figure 1A).

3.2. Microscopical Organization

The esophagus and the stomach showed the typical stratigraphy of hollow organs consisting of four layers or tunics; from the external layer to the lumen these are: the serous; the double muscle layer (inner circular and outer longitudinal), composed of striated fibers; the submucosal and the mucosal. In particular, the muscle layer displayed a different thickness along the stomach, allowing us to identify up to three layers in the regions featured by the thickest wall. These layers were arranged as follows: inner circular, middle oblique and outer longitudinal (Figure 1B).

The serous tunic of the esophagus was composed of two layers: the subserous lamina, made of loose connective tissue and highly vascularized; and the serous epithelial lamina. The inner longitudinal muscle layer was quite thick, while the outer circular layer was characterized by bundle of muscle fibers separated by loose connective tissue. The lamina propria of the mucosa was organized in a thick stratum of collagen fibers. The mucosa showed numerous longitudinal folds along the entire length of the organ, giving the lumen of the esophagus a star-shaped appearance, mainly when observed in a transversal section. These mucosa folds were covered with a layer of squamous epithelium, including mucous secreting cells (Figure 2A) featuring an elongated or a rounded shape. The rounded cells that appeared larger compared with the elongated cells were concentrated basally, while the small elongated cells were localized on the epithelium surface. Interestingly, we didn't detect taste buds within the esophagus mucosa and the component known as *muscularis mucosae*. The transition from the esophagus to the stomach, corresponding to the cardial zone, was characterized by a cylindrical glandular epithelium (Figure 3C).

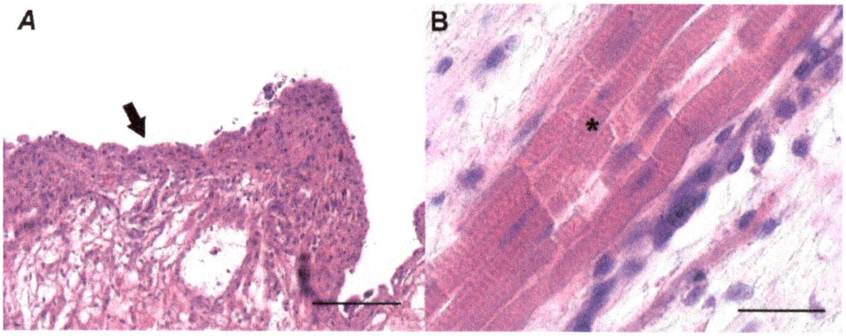

Figure 2. Histological characteristics of sagittal sections of esophagus walls of Nile tilapia. (**A**): mucosa folds of the esophagus covered with layer of squamous epithelium (arrow), (**B**): pyloric zone with striated muscle layers (asterisk). Scale bar = (**A**): 100 µm; (**B**): 25 µm.

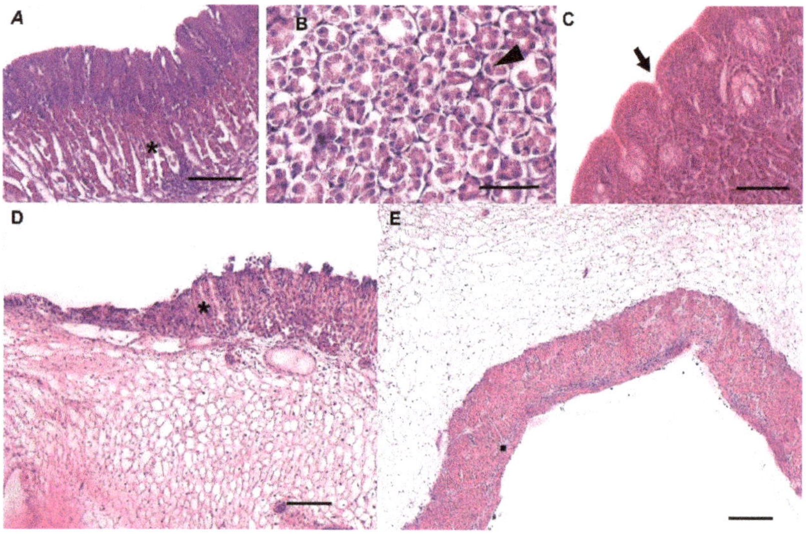

Figure 3. Histological characteristics of sagittal sections of stomach wall of Nile tilapia. (**A**): epithelial layer that coats the tubular glands (asterisk); (**B**) cross section of the tubular glands with prismatic cells (arrow head); (**C**): monolayer of cylindrical cells displaying the typical folds and delimiting the gastric pits (arrow); (**D**): transition from the esophageal to the gastric mucosa, showing tubular glands at the beginning (asterisk); (**E**): fundic zone (square putative third transversal muscle layer). Scale bars = (**A**,**C**,**D**,**E**): 100 µm; (**B**): 50 µm.

The stomach serosa was formed by a vascularized and innervated tissue composed of a simple epithelium and a thin layer of connective tissue exerting mainly a mechanical support function. In the muscular layer, we observed a smooth musculature organized in fibers displaying a circular and a longitudinal course in the inner and outer layer, respectively. In addition, in the fundic zone, towards the lumen, we identified a third internal layer formed by diagonal fibers (Figure 3E). The submucosa architecture displayed a stratum of loose connective tissue joining the mucosa to the muscle tunica and including large blood and lymphatic vessels. We also highlighted the presence of circular structures in the submucosal layer that could be a nervous net comparable with those described in the mucosal wall of rainbow trout stomach [18]. The connective tissue allows the epithelium to slide over the underlying layers of the wall. The mucosa was characterized by two portions:

one external, including the gastric tubular glands, and one internal (luminal), featuring elongated cylindrical cells with the nucleus located in the basal portion (Figure 3A–C). This layer of the mucosa was organized in folds showing basal "gastric areoles" featured by apical holes known as "gastric pits". The distal portion of the stomach, namely the pyloric zone, was characterized by a muscular layer of striated muscle, as well as a reduction, up to the disappearance, of the glandular component (Figure 2B).

Based on morphological analyses, we identified five distinct zones with different morpho-functional features in the stomach (Figure 4):

- ZONE 1: esophagus–gastric lumen passage
- ZONE 2: descending glandular portion
- ZONE 3: fundic portion
- ZONE 4: ascending glandular portion
- ZONE 5: gastric–pyloric transition portion

Zone 1 corresponded to the transition area from the esophagus to the stomach. The first segment of the esophagus presented a squamous epithelium rich in mucous cells, while towards the stomach we observed an initial reduction in the number of mucous cells, then a progressive replacement of the squamous epithelium with that typical of the gastric region. A gradual increase of gastric glands was evident immediately beneath this superficial epithelium. The submucosa was made up of connective tissue, while the muscle fibers in this area were exclusively smooth. Further, in correspondence with the appearance of the population of the tubular glands, the organization of the muscular tunic included two layers, one (internal) with circular fibers and one (external) with longitudinal fibers.

Zone 2, descending glandular portion, showed the overlapping of the tunics of the gastric glandular portion. Here it was evident the presence of folds, including columnar cells, that covered an inner layer of glands reaching the organ lumen through the gastric pits. The submucosal layer was made up of dense connective tissue; thicker and less compact in the section corresponding to the fold, and thinner and denser in the portion at the base of the folds. In addition, circular corpuscles were observed in this layer. The smooth muscle tunic was composed of fibers arranged to form the two patterns previously described.

Zone 3, fundic area, was located between the descending (ventral) and the ascending (dorsal) tracts of the organ and showed an architecture intermediate to those of the two tracts. This area retained a well-developed mucosa lying superficially to the tubular glands on both sides. In contrast, the submucosal layer was different in the two tracts, appearing thin and compact in the descending tract, and thick and loose in the ascending one. At the apex, moving towards the ascending tract, there was a thickening of the muscular circular layer, while the longitudinal layer remained constant. In the muscle stratum, some transversal fibers still appeared.

Zone 4, ascending tract, featured both the epithelial layer and the glands. An increased height of the connective layer of the submucosa was observed in this area. At the same time, the muscular tunic continued the progressive thickening that was observed starting from the apex of the fundus. In addition, in this area the thickening was due to the internal fibers with a circular pattern, as reported for Zone 3.

Zone 5, gastric pyloric, the transition area between the proper gastric and the pyloric portions. The various layers appeared highly modified in this area. Starting from the epithelium, we witnessed a gradual change towards the lining of the intestinal mucosa. The tubular glands gradually disappeared, and the submucosa changed towards the pyloric sphincter. The muscular layer increased considerably in thickness, becoming the predominant layer in the proximity of the pyloric sphincter. In the area immediately preceding the pylorus, striated muscles appeared.

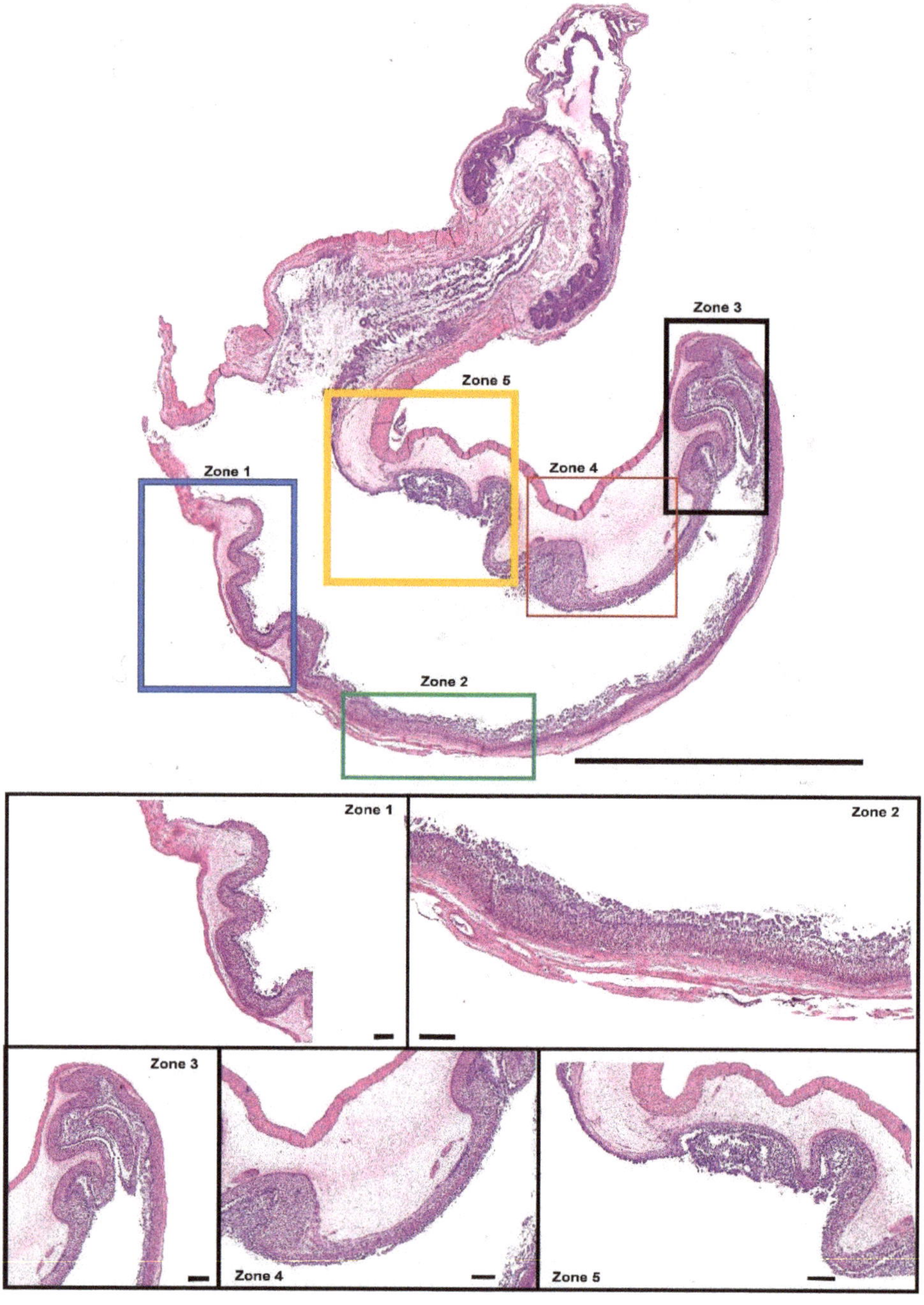

Figure 4. Morpho-functional organization of the stomach. Five zones based on morphological differences we highlighted: Zone 1: esophagus–gastric lumen passage, Zone 2: descending glandular portion, Zone 3: fundic portion, Zone 4: ascending glandular portion, Zone 5: gastric-pyloric transition portion. Scale bars = complete reconstruction: 1 cm; Zone 1, Zone 2, Zone 3, Zone 4, Zone 5: 200 μm.

3.3. Morphometrical Analysis

We measured the thicknesses of the different layers composing tilapia stomach by evaluating superficial and glandular epithelia, epithelium, submucosal, circular and longitudinal muscle components. The results are summarized in the diagrams shown in Figure 5.

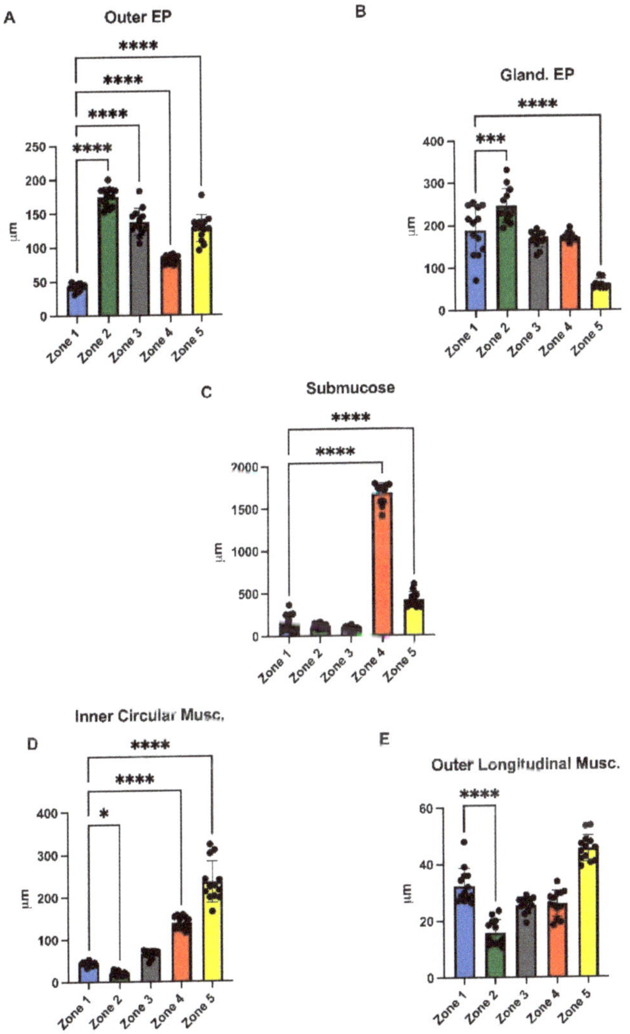

Figure 5. Diagrams representing the results of the morphometric analysis. (**A**): the thickness of the coating epithelium was reduced in Zone 1 and significantly more developed in Zones 2, 3 and 5, with a slight decrease in Zone 4. (**B**): the glandular epithelium, uniform in Zones 3 and 4, showed a significant increase in Zone 2 and decrease in the region of gastric–pyloric transition. (**C**): the submucosa showed a constant thickness in all zones except in Zone 4, where it increased. (**D**): the inner circular muscle layer showed a gradual increase from Zone 1 to Zone 5, except in Zone 2. (**E**): the outer longitudinal muscle layer displayed a uniform thickness, except in Zone 2 where it decreased. * *p*-values < 0.05; *** *p*-values < 0.001; **** *p*-values < 0.0001.

The mean thickness of the lining epithelium in zone 1 was 42 μm; it was significantly higher in the other zones: Zone 2, 175 μm; Zone 3, 137 μm; Zone 4, 81 μm; and Zone 5, 127 μm (*p*-values: <0.0001, <0.0001, <0.0001 and <0.0001, respectively) (Figure 5A). The glandular epithelium was uniform in Zones 1, 3 and 4 (188 μm, 165 μm, 172 μm, respectively) (*p*-values: 0.2320 and 0.5095). We measured a significant increase in Zone 2 (246 μm) (*p*-value: 0.0001), and a significant decrease in the gastric–pyloric transition region (60 μm) (*p*-value: 0.0001) (Figure 5B). The submucosa showed a constant thickness in the first three zones: 1, 2 and 3; 58 μm, 117 μm, and 87 μm, respectively, and a significant increase in Zones 4 and 5 (1676 μm and 417 μm, respectively) (*p*-values: <0.0001, <0.0001, respectively) (Figure 5C). The circular muscular layer in Zone 1 was 43 μm thick, showing a gradual increase in Zones 3, 4 and 5 (65 μm, 128 μm and 235 μm, respectively); however, this increase is significant only in zones 4 and 5 (*p*-values: <0.0001, <0.0001). Conversely, we detected a slight reduction in Zone 2 (20 μm) that significantly differs from Zone 1 (*p*-value: 0.0374) (Figure 5D). By contrast, the thickness of the longitudinal layer of the muscular tunic was generally uniform (32 μm, 15 μm, 25 μm, 26μm and 45μm) (*p*-values: 0.2900, 0.6870 and 0.3614), except in Zone 2, where it significantly decreased (*p*-value: 0.0001) (Figure 5E).

3.4. *Histochemical Analysis*

The histochemical investigation showed differentiated results among the treatments carried out. Moderate positivity was evidenced in all the histochemical reactions at mucous cell level in the esophagus epithelial lining, with the exception of AB pH 1.0, which displayed a negative reaction. In the stomach, the epithelial cell coat showed an intense staining to PAS, and AB/PAS (Figure 6A,D and Figure 6G, respectively), while moderately positive to AB pH 2.5 and negative to AB 1.0 (Figure 6B,E and Figure 6C,F, respectively). The gastric pits proved to be the most reactive stomach structures; in fact, they were strongly positive to PAS, AB pH 2.5 and 1.0 (Figure 6D,E,F), and prevalently AB positive to AB/PAS treatment, showing a blue staining (Figure 6G). Gastric glands showed a strong reactivity to both PAS (Figure 6A,D) and AB/PAS (Figure 6G), to which they were prevalently PAS positive, showing a pink staining. Gastric glands showed a slight AB pH 2.5 reactivity (Figure 6B,E), while they were negative to AB pH 1.0 (Figure 6C,F).

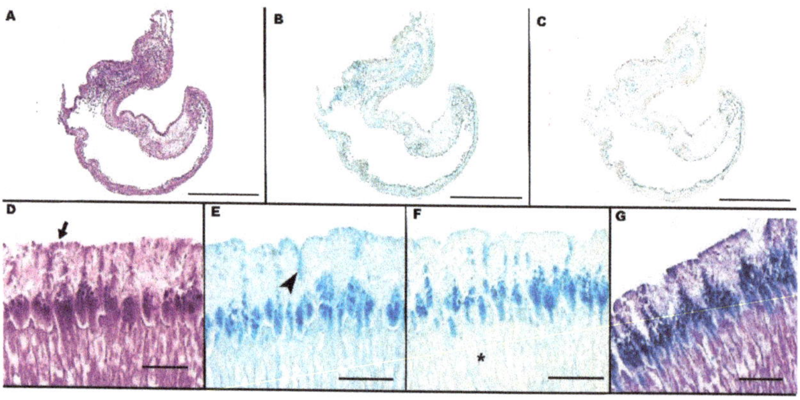

Figure 6. Histochemical investigation of the stomach of Nile tilapia. Figure shows periodic acid Schiff (PAS) (**A,D**); Alcian blue at pH 2.5 (**B,E**); Alcian blue at pH 1.0 (**C,F**); and Alcian blue/PAS (**G**) binding sites on mosaics of the whole stomach. Epithelial cell coat (arrow); Gastric pits (asterisk); Gastric glands (arrow head). D, E and F represent magnifications (40×) of PAS, Alcian blue at pH 2.5, and Alcian blue at pH 1.0, respectively. Scale bars = (**A–C**): 1 cm; (**D–G**): 100 μm.

3.5. Scanning Electron Microscopy Obsevations

Through scanning electron microscopy (SEM) we observed the esophagus and stomach in three dimensions.

In cross-section, the esophagus was tubular (Figure 7A), and the four layers were distinguishable from outside to inside as follows: serous, muscular with longitudinal (outer) and circular (inner) fibers, a very thick submucosa, and a mucosa featuring folds of different sizes.

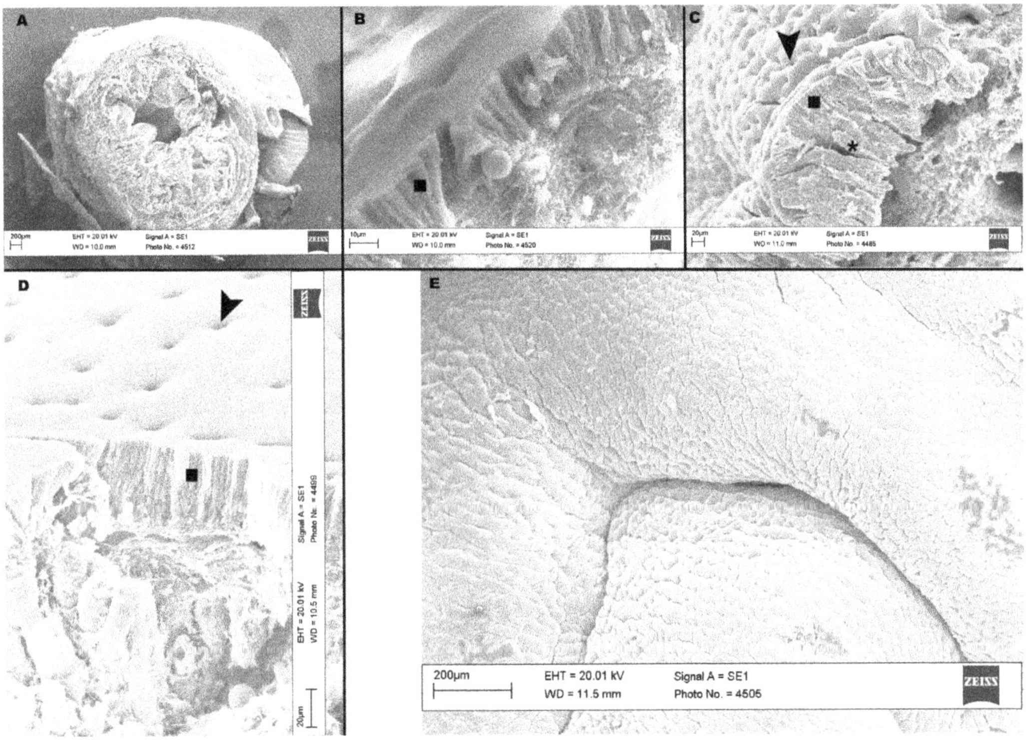

Figure 7. Scanning electron microscopy (SEM) images of esophagus and stomach of Nile tilapia. (**A**): cross section of esophagus. The four layers are clearly evident: serous, muscular (inner circular and outer longitudinal), submucosa and mucosa, featuring folds of different heights; (**B**): esophagus–gastric lumen transition. This section shows a single layer of columnar epithelial cells (square) and gastric glands. Vessels are evident in the submucosa; (**C**): descending glandular portion of the stomach. The outlets of the gastric pits (arrow heads) are noted, as well as the epithelial columns lining the mucosa (square) and a gastric gland (asterisk); (**D**): fundic portion of the stomach. Evident columnar epithelium (square) and gastric pits (arrow heads) laying on periglandular connective tissue (arrow); (**E**): ascending glandular portion of the stomach. Scale bars are indicated in each picture.

We further identified five different areas of the stomach, consistent with what we observed via light microscopy. The transition area between the esophagus and the first part (Zone 1) of the stomach was characterized by the appearance of a columnar epithelium laying on a basal lamina, below which the submucosa was also evident. In addition, we noticed gastric pits with gland outlets (Figure 7B). Also, through this technique we observed that the mucosa of the descending glandular portion (Zone 2) featured numerous folds. The columnar epithelium laying on the submucosa was very evident. In this region, the glands with their outlets appeared surrounded by epithelial cells (Figure 7C). The fundic portion (Zone 3) did not differ from Zone 2. Therefore, this portion was characterized by an increase of the glands highlighted on the epithelial surface by numerous gastric pits (Figure 7D). The ascending glandular portion (Zone 4) began to show significant changes, mainly due the increase of connective tissue and, therefore, to the presence of very large mucous folds (Figure 7E). Finally, the last portion, transitioning between the properly glandular gastric portion and the non-glandular pyloric portion (Zone 5), was characterized by morphological modifications typical of the intestinal epithelium. A mucosa that is no longer glandular was clearly visible, as well as an increased muscular component, typical of the pyloric sphincter.

4. Discussion

In this study we described the esophagus and stomach of Nile tilapia to highlight their different cellular distribution, their morphological structure through scanning electron microscopy, and to characterize their complex carbohydrates through conventional glyco-histochemical stainings. Our observations allow us to relate each morphological difference to a specific function.

Regarding the esophagus, our SEM data highlighted a mucosa architecture characterized by thin longitudinal folds of different heights, extending along the whole organ length. Microscopically, these folds were deep and included the mucosa, thus increasing the distension capacity of the esophagus during food transport [38,39]. Further, our results corroborate the hypothesis reported in other studies [30,33], according to which the typical protective function of the esophagus from mechanical insults is played by the squamous epithelium. Histochemical treatments evidenced the production by the esophagus mucous cells of carboxylated complex carbohydrates, represented by hyaluronic acid, chondroitin, chondroitin-sulphates A/B/C and glycosaminoglycan-like material that form a viscoelastic barrier that, covering the outer epithelial layer, protects the esophageal mucosa. Intriguingly, the presence of a great number of muciparous cells (both elongated and rounded) suggests a pre-gastric digestion, postulating an additional function of tilapia esophagus in digestion, as is already reported in other fish [2].

Literature data show a subdivision in the stomach of this fish. Al-Hussaini and Kholy [40] described two zones in tilapia stomach, namely one large caecal and one pyloric; in contrast, Caceci and collaborators [32] highlighted three regions, an initial or cardiac, a middle or fundic, and a terminal or pyloric. Further studies [41,42] report that the stomach of poly-hybrid Tilapia displays an identical structure in all the regions.

However, in our work we identified for the first time, via scanning microscopy and histochemical techniques, five morphologically distinct zones in the stomach, based on the composition of the gastric wall and the architecture of the superficial epithelium. The first zone (Zone 1) showed a replacement of the stratified squamous epithelium, including mucous cells of the esophagus, with a simple columnar epithelium typically displayed in the stomach. This organization is due to the transition from a non-glandular to a glandular structure, very common in Teleosts [43–47]. The columnar epithelium lined the folds which basis is featured by the outlet of the gastric glands, which secrete acidic and neutral mucins, as reported by several authors [30,31,33]. Different studies showed that the active secretion of mucins and the structure of the epithelium play fundamental roles in the maintenance of the physiological microenvironment, which is essential for digestive function [18]. To further confirm this, in Zone 5, corresponding to the gastric–pyloric transition portion, epithelium morphology changes, gradually displaying similarities with that of the first tract of the intestine.

In this study we also described a progressive increase in the glandular component of the gastric mucosa, which confirms that already found in the first tract (Zone 1); subsequently, in the descending glandular portion (Zone 2), the digestion starts continuing up to the fundic (Zone 3) as well as in the ascending glandular portion (Zone 4). The disappearance of the tubular glands in the gastric–pyloric passage portion (Zone 5) is indicative of a cessation of digestive function typical of this zone. In fact, it is reported that the glandular region of the stomach exerts digestive functions while the non-glandular region is mainly devoted to the transit of food to the intestine due to its morphology (muscular layer) and function (epithelial secretions) [40]. In addition, the folds of the mucosa delimiting all the gastric lumen in the first four areas could also play a role in delaying the passage of food through the stomach. This function maximizes the mixing of food with digestive enzymes, enhancing the digestive process and nutrient absorption, as confirmed in other studies [20,31].

Folds are closely related to an increase in the submucosal connective layer. In thicker areas, the connective tissue is much less dense, thus ensuring proper mechanical resistance to the gastric wall. The three muscular layers contribute to mixing the food and to ensure its progression towards the pylori. The smaller quantity of muciparous cells in the pyloric region, compared with other regions of the stomach, suggests a role in food retention rather than a proper digestive function. Indeed, the preservation of food in the pyloric region is probably relevant for better digestion, as already observed in other fish [48].

Furthermore, the presence of a crest-shaped sphincter (valve) characterized by the striated muscles between the stomach and the first tract of the intestine could avoid the transit to the intestine of food that has not undergone chemical digestion, as reported by Morrison and Wright [33].

From a morphological standpoint, the stomach of Nile tilapia displays the same cellular organization of most teleost fishes, including other cichlids [10,11,49]. The glandular cells promote digestion [42] by secreting hydrochloric acid through the oxyntic cells. Therefore, their secretion plays a key role in regulating gastric pH since, unlike what happens in mammals, fish gastric glands produce only hydrochloric acid. As reported in the literature, it seems that to promote acid lysis, as well as digestion of algae and detrital bacteria, tilapia needs an acidic gastric environment during the early stage of digestion due to the lack of adaptation to trituration [31,42,50,51]. From the histochemical point of view, the tilapia stomach shows the production of a great variety of glycoconjugates. The gastric pits secrete high-acid complex carbohydrates, represented by chondroitin-sulphates A/B/C, heparin, and heparan-sulphate glycosaminoglycan-like material, with numerous negative charges protecting the gastric mucosa from the acidic environment of the stomach. The protective action is probably carried out in two ways: by repelling the negative charges and by attracting large quantities of water to form a thick layer of gel. The cell coat is also involved in the protective function; in fact, the presence of sialylated glycoproteins in this site is evidenced by histochemical treatments. Gastric glands show the presence of neutral, and in minor measure, sialylated glycoproteins [18,32,41,52]. This finding causes us to hypothesize a role for each class of mucins that, on one hand, preserve the cellular epithelial layer, and on the other, help to ensure the right conditions for correct digestive processes.

In addition, the morphometric analysis of the muscle layers confirms a rather peculiar organization of Nile tilapia stomach. The external longitudinal layer, displaying a constant thickness throughout the lumen of the organ, exerts a key role in food stirring. Conversely, the circular layer, which increases significantly from the apex of the stomach, ensures the progression of almost completely digested food towards the pyloric sphincter. While in the descending portions of the stomach, the food advances due to gravity as well, in the ascending tracts it must proceed against gravity, therefore requiring a force able to push it up. The increased thickness of the circular muscle layer could then be attributed to this specific function.

5. Conclusions

In view of the key role played by fish both in translational research [53,54] and aquaculture, our work may be of interest in different research settings. This preliminary study provides in-depth morphological data of the stomach of Nile tilapia, inspiring us to formulate a functional hypothesis. The identification of the main cellular structures within the epithelium may contribute to more advanced studies aiming to unravel the basic mechanisms of digestion in this species. In addition, our dataset may be useful for studies focused on understanding the morphological changes induced by different dietary plans and aimed at improving production rates while also increasing animal welfare.

Author Contributions: Conceptualization, L.M., A.P. and E.D.F.; methodology, P.S.; software, A.P.; validation, C.L. and F.S.; formal analysis, A.P.; investigation, L.M., A.P. and E.D.F.; resources, A.C.; data curation, A.P. and D.G.; writing—original draft preparation, A.P. and L.M.; writing—review and editing, E.D.F.; visualization, C.A.; supervision, L.D.; project administration, A.C. and C.M.A.B.; funding acquisition, A.C. All authors have read and agreed to the published version of the manuscript.

Funding: This research received no external funding.

Institutional Review Board Statement: Not applicable.

Data Availability Statement: Datasets used in the analyses are stored at the authors' home institution and will be provided upon request from the corresponding author.

Conflicts of Interest: The authors declare no conflict of interest.

References

1. Ishibashi, Y.; Ekawa, H.; Hirata, H.; Kumai, H. Stress response and energy metabolism in various tissues of Nile tilapia (*Oreochromis niloticus*) exposed to hypoxic conditions. *Fish. Sci.* **2002**, *68*, 1374–1383. [CrossRef]
2. Díaz, A.O.; García, A.M.; Devincenti, C.V.; Goldemberg, A.L. Morphological and Histochemical Characterization of the Mucosa of the Digestive Tract in *Engraulis anchoita* (Hubbs and Marini, 1935). *Anat. Histol. Embryol.* **2003**, *32*, 341–346. [CrossRef] [PubMed]
3. Carrassón, M.; Grau, A.; Dopazo, L.R.; Crespo, S. A histological, histochemical and ultrastructural study of the digestive tract of Dentex dentex (Pisces, Sparidae). *Histol. Histopathol.* **2006**, *21*, 579–593. [CrossRef] [PubMed]
4. Klumpp, D.W.; Nichols, P.D. Nutrition of the Southern Sea garfish Hyporhamphus rnelanoehir: Gut passage rate and daily consumption of two food types and assimilation of seagrass components. *Mar. Ecol. Prog. Ser.* **1983**, *12*, 207–216. [CrossRef]
5. Unal, G.; Cetinkaya, O.; Kankaya, E.; Elp, M. Histological study of the organogenesis of the digestive system and swim bladder of the *Chalcalburnus tarichi* Pallas, 1811 (Cyprinidae). *Turk. J. Zool.* **2001**, *25*, 217–228.
6. Sinha, B.M.; Moitra, S.K. Morpho-histology of the intestine in a freshwater major carp, Cirrhinus mrigala (Hamilton) during the different life-history stages in relation to food and feeding habits. *Anat. Anz.* **1974**, *137*, 395–407.
7. Dasgupta, M. Morphological adaptation of the alimentary canal of four Labeo species in relation to their food and feeding habits. *Indian J. Fish* **2001**, *48*, 255–257.
8. Naguib, S.A.A.; EI-Shabaka, H.A.; Ashour, F. Comparative histological and ultrastructural studies on the stomach of *Schilbe mystus* and the intestinal swelling of *Labeo niloticus*. *J. Am. Sci.* **2011**, *7*, 251–263.
9. Musa, M.; Yanuhar, U.; Susilo, E.; Soemarno. Stomach histological decay of milkfish, Chanos Chanos (Forsskål, 1775): Ontogeny, environmental stress, shifting food composition, and disease infection. *J. Nat. Sci. Res.* **2013**, *3*, 194–200.
10. Ghosh, S.K.; Chakrabarti, P. Histological and histochemical characterization on the stomach of *Mystus cavasius* (Hamilton), *Oreochromis niloticus* (Linnaeus) and *Gambusia chapra* (Hamilton): Comparative study. *J. Basic Appl. Zool.* **2015**, *70*, 16–24. [CrossRef]
11. Ghosh, S.K.; Chakrabarti, P. Histological, surface ultrastructural, and histochemical study of the stomach of red piranha, Pygocentrus nattereri (Kner). *Arch. Pol. Fish.* **2015**, *23*, 205–215. [CrossRef]
12. Reifel, C.W.; Travill, A.A. Structure and carbohydrate histochemistry of the stomach in eight species of teleosts. *J. Morphol.* **1978**, *158*, 155–167. [CrossRef] [PubMed]
13. Kozarić, Z.; Kužir, S.; Petrinec, Z.; Gjurčević, E.; Baturina, N. Histochemistry of complex glycoproteins in the digestive tract mucosa of Atlantic Bluefin tuna (*Thynnus* L.). *Vet. Arch.* **2007**, *77*, 441–452.
14. Hernándes, D.R.; Pérez Gianeselli, M.; Domitrovic, H.A. Morphology and Histochemistry of the Digestive system of South American Catfish (*Rhamdia quelen*). *Internat. J. Morphol.* **2009**, *27*, 105–111. [CrossRef]
15. Santos, M.L.D.; Arantes, F.P.; Pessali, T.C.; Santos, J.E.D. Morphological, histological and histochemical analysis of the digestive tract of *Trachelyopterus striatulus* (Siluriformes: Auchenipteridae). *Zoologia* **2015**, *32*, 296–305. [CrossRef]
16. Xiong, D.; Zhang, L.; YU, H.; Xie, C.; Kong, Y.; Zeng, Y.; Huo, B.; Liu, Z. A study of morphology and histology of the alimentary tract of *Glyptosternum maculatum* (Sisoridae, Siluriformes). *Acta Zool.* **2011**, *92*, 161–169. [CrossRef]
17. Qu, M.; Ding, S.; Xu, X.; Shen, M.; You, Y.; Su, Y. Ontogenetic development of the digestive system and growth in coral trout (*Plectropomus leopardus*). *Aquaculture* **2012**, *334*, 132–141. [CrossRef]
18. De Felice, E.; Palladino, A.; Tardella, F.M.; Giaquinto, D.; Barone, C.M.A.; Crasto, A.; Scocco, P. A morphological, glycohistochemical and ultrastructural study on the stomach of adult Rainbow trout *Oncorhynchus mykiss*. *Eur. Zool. J.* **2021**, *88*, 269–278. [CrossRef]
19. Smith, L.S. Digestive functions in teleost fishes. In *Fish Nutrition*, 2nd ed.; Halver, J.E., Ed.; Academic Press: London, UK, 1989; pp. 331–421.

20. Okuthe, G.E.; Bhomela, B. Morphology, histology and histochemistry of the digestive tract of the Banded tilapia, *Tilapia sparrmanii* (Perciformes: Cichlidae). *Zoologia* **2020**, *37*, e51043. [CrossRef]
21. Moraes, M.F.P.G.; Barbola, I.F.; Guedes, E.A.C. Alimentac¸ão e relac¸ões morfológicas com o aparelho digestivo do "curimbatá", *Prochilodus lineatus* (Valenciennes) (Osteichthyes, Prochilodontidae), de uma lagoa do sul do Brasil. *Rev. Bras. Zool.* **1997**, *14*, 169–180. [CrossRef]
22. Seixas Filho, J.T.; Brás, J.M.; Gomide, A.T.M.; Oliveira, M.G.A.; Donzele, J.L.; Menin, E. Anatomia funcional e morfometria dos intestinos e dos cecos pilóricos do *Teleostei* (Pisces) de água doce *Brycon orbignyanus* (Valenciennes 1849) *Rev. Bras. Zootec.* **2000**, *29*, 313–324. [CrossRef]
23. Seixas Filho, J.T.; Brás, J.M.; Gomide, A.T.M.; Oliveira, M.G.A.; Donzele, J.L.; Menin, E. Anatomia funcional e morfométrica do intestino no *Teleostei* (Pisces), de água doce surubim (*Pseudoplatystoma corruscans*—Agassiz 1829). *Rev. Bras. Zootec.* **2001**, *30*, 1670–1680. [CrossRef]
24. Peretti, D.; Andrian, I.F. Feeding and morphological analysis of the digestive tract of four species of fish (*Astyanax altiparanae, Parauchenipterus galeatus, Serrasalmus marginatus* and *Hoplias aff. malabaricus*) from the upper Paraná River floodplain, Brazil. *Braz. J. Biol.* **2008**, *68*, 671–679. [CrossRef] [PubMed]
25. Rodrigues, S.S.; Menin, E. Anatomia do tubo digestivo de *Salminus brasiliensis* (Cuvier, 1817) (Pisces, Characidae, Salmininae). *Biotemas* **2008**, *21*, 65–75. [CrossRef]
26. Sanchez-Amaya, M.I.; Ortiz-Delgado, J.B.; Garcia-Lopez, A.; Cardenas, S.; Sarasquet, C. Larval ontogeny of red banded seabream Pagrus auriga valencieness, 1843 with special reference to the digestive system. A histological and histochemical approach. *Aquaculture* **2007**, *263*, 259–279. [CrossRef]
27. Smith, L.S. Introduction to Fish Physiology. In *An Atlas of Fish Histology: Normal and Pathological Features*; Takashima, F., Hibiya, T., Eds.; Argent Laboratories Press: Redmond, WA, USA; Gustav Fisher Verlag: Kodansha, Tokyo, 1995; p. 192.
28. Yan, B.; Wang, Z.H.; Zhao, J.L. Mechanism of osmoregulatory adaptation in tilapia. *Mol. Biol. Rep.* **2013**, *40*, 925–931. [CrossRef] [PubMed]
29. Akalu, B. The main factors affecting growth performance of *Oreochromis niloticus* L. (1758) in aquaculture system. *J. Fish. Livest. Prod.* **2021**, *9*, 1–9. [CrossRef]
30. Awaad, A.S.; Moawad, U.K.; Tawfiek, M.G. Comparative Histomorphological and histochemical studies on the oesophagus of Nile Tilapia *Oreochromis niloticus* and African Catfish *Clarias Gariepinus*. *J. Histol.* **2014**, *2014*, 1–10. [CrossRef]
31. Osmand, A.H.K.; Caceci, T. Histology of the stomach of *Tilapia nilotica* (Linnaeus 1758) from the River Nile. *J. Fish Biol.* **1991**, *38*, 211–223. [CrossRef]
32. Caceci, T.; EL-Habback, H.A.; Smith, S.A.; Smith, B.J. The stomach of *Oreochromis niloticus* has three regions. *J. Fish Biol.* **1997**, *50*, 939–952. [CrossRef]
33. Morrison, C.M.; Wright, J.R., Jr. A study of the histology of the digestive tract of the Nile tilapia. *J. Fish Biol.* **1999**, *54*, 597–606. [CrossRef]
34. Wilson, J.M.; Castro, L.F. Morphological diversity of the gastrointestinal tract in fishes. *Fish Physiol.* **2010**, *30*, 1–55. [CrossRef]
35. Pearse, A.G.E. *Histochemistry: Theoretical and Applied*, 4th ed.; Churchill Livingstone: Edinburgh, UK; London, UK; Melbourne, Australia; New York, NY, USA, 1985.
36. Mercati, F.; Dall'Aglio, C.; Acuti, G.; Faeti, V.; Tardella, F.M.; Pirino, C.; De Felice, E.; Scocco, P. Oregano feed supplementation affects glycoconjugates production in swine gut. *Animals* **2020**, *10*, 149. [CrossRef] [PubMed]
37. Bancroft, J.D.; Gamble, A. *Theory and Practice of Histological Techniques*, 6th ed.; Churchill-Livingstone: London, UK, 2008; pp. 1–744.
38. Vieira-Lopes, D.A.; Pinheiro, N.L.; Sales, A.; Ventura, A.; Araújo, F.G.; Gomes, I.D.; Nascimento, A.A. Immunohistochemical study of the digestive tract of *Oligosarcus hepsetus*. *World J. Gastroenterol.* **2013**, *19*, 1919–1929. [CrossRef]
39. Zug, G.R.; Vittand, L.J.; Caldwell, J.P. *Introduction Biology of Amphibian and Reptiles*, 2nd ed.; Academic Press: San Diego, CA, USA, 2001; pp. 1–776.
40. Al-Hussaini, A.H. On the functional morphology of the alimentary tract of some omnivorous fish. *Egypt. Acad. Sci.* **1953**, *4*, 17–39.
41. Scocco, P.; Ceccarelli, P.; Menghi, G. Glycohistochemistry of the *Tilapia* spp. stomach. *J. Fish Biol.* **1996**, *49*, 584–593. [CrossRef]
42. Gargiulo, A.M.; Ceccarelli, P.; Dall'Aglio, C.; Pedini, V. Ultrastructure study on the stomach of *Tilapia* spp. (Teleostei). *Anat. Histol. Embryol.* **1997**, *26*, 331–336. [CrossRef]
43. Grau, A.; Crespo, S.; Sarasquete, M.C.; De Canales, M.G. The digestive tract of the amberjack *Seriola dumerili*, Risso: A light and scanning electron microscope study. *J. Fish Biol.* **1992**, *41*, 287–303. [CrossRef]
44. Murray, H.M.; Wright, G.M.; Goff, G.P. A comparative histological and histochemical study of the stomach from three species of pleuronectid, the Atlantic halibut, *Hippoglossus hippoglossus*, the yellowtail flounder, *Pleuronectes ferruginea*, and the winter flounder, *Pleuronectes americanus*. *Canad. J. Zool.* **1994**, *72*, 1199–1210. [CrossRef]
45. Oliveira-Ribeiro, C.A.D.; Fanta, E. Microscopic morphology and histochemistry of the digestive system of a tropical freshwater fish *Trichomycterus brasiliensis* (Lütken) (Siluroidei, Trichomycteridae). *Rev. Brasil. Zool.* **2000**, *17*, 953–971. [CrossRef]
46. Dai, X.; Shu, M.; Fang, W. Histological and ultrastructural study of the digestive tract of rice field eel, *Monopterus albus*. *J. Appl. Ichthyol.* **2007**, *23*, 177–183. [CrossRef]
47. Ikpegbu, E.; Ezeasor, D.N.; Nlebedum, U.C.; Nnadozie, O. Morphological and histochemical observations on the oesogaster of the domesticated African catfish (*Clarias gariepinus Burchell*,1822). *Bulg. J. Vet. Med.* **2013**, *16*, 88–95.

48. Raji, A.R.; Norouzi, E. Histological and histochemical study on the alimentary canal in walking catfish (*Clarius batrachus*) and Piranha (*Serrasalmus nattereri*). *Iran. J. Vet. Res.* **2010**, *11*, 255–261.
49. Caceci, T.; Hrubec, T.C. Histology and ultrastructure of the gut of the black mollie (*Poecilia* spp.), a hybrid teleost. *J. Morphol.* **1990**, *204*, 265–280. [CrossRef] [PubMed]
50. Bowen, S.H. Detrital nonprotein amino acids are the key to rapid growth of Tilapia in Lake Valencia, Venezuela. *Science* **1980**, *207*, 1216–1218. [CrossRef]
51. Lobel, P.S. Trophic biology of herbivorous reef fishes: Alimentary pH and digestive capabilities. *J. Fish Biol.* **1981**, *19*, 365–397. [CrossRef]
52. Domeneghini, C.; Arrighi, S.; Radaelli, G.; Bosi, G.; Mascarello, F. Morphological and histochemical peculiarities of the gut in the white sturgeon, *Acipenser transmontanus*. *Eur. J. Histochem.* **1999**, *43*, 135–145.
53. Giaquinto, D.; De Felice, E.; Attanasio, C.; Palladino, A.; Schiano, V.; Mollo, E.; Lucini, C.; de Girolamo, P.; D'Angelo, L. Central and peripheral NPY age-related regulation: A comparative analysis in fish translational models. *Int. J. Mol. Sci.* **2022**, *23*, 3839. [CrossRef]
54. Montesano, A.; De Felice, E.; Leggieri, A.; Palladino, A.; Lucini, C.; Scocco, P.; de Girolamo, P.; Baugmart, M.; D'Angelo, L. Ontogenetic pattern changes of Nucleobindin-2/Nesfatin-1 in the brain and intestinal bulb of the short lived African turquoise killifish. *J. Clin. Med.* **2019**, *9*, 103. [CrossRef]

animals MDPI

Article

Comparative Research on Intestinal Functions of Wild and Cultured *Hemibarbus maculatus* in Jialing River

Bangyuan Wu [1,2], Hong Lei [2,3], Jie Zhen [2,4], Limin Zhao [2], Baolin Song [2,5] and Yu Zeng [1,2,*]

1 Key Laboratory of Southwest China Wildlife Resources Conservation, Ministry of Education, Nanchong 637000, China
2 College of Life Science, China West Normal University, Nanchong 637000, China
3 College of Life Science and Technology, Huazhong Agriculture University, Shizishan 1#, Wuhan 430070, China
4 School of Medicine, Kunming University of Science and Technology, Kunming 650500, China
5 Department of Infectious Diseases and Public Health, Jockey Club College of Veterinary Medicine and Life Sciences, City University of Hong Kong, Hong Kong SAR, China
* Correspondence: zengyu@cwnu.edu.cn; Tel.: +86-15908370168

Simple Summary: *Hemibarbus maculatus* is a common economic fish in Jialing River. We tested the intestinal and liver digestive function of wild and cultured *Hemibarbus maculatus*. We found that the absorptive function and intestinal lymphocytes number are weaker or lower in the cultured *Hemibarbus maculatus*, and the decreased digestive function of the cultured *Hemibarbus maculatus* was also found. In conclusion, the intestinal digestion, absorption and lymphocytes of the wild are generally better than those of the cultured. Future aquacultural activities should consider these changes when facing pragmatic problems in order to resolve the difficulties in aquacultura cultivation.

Abstract: *Hemibarbus maculatus* is a common economic fish in the midstream and downstream of the Jialing River. In order to resolve the difficulties in aquacultural cultivation, we tested the intestinal and liver digestive function of wild and cultured *Hemibarbus maculatus*. Histological methods and special biochemical staining methods were used to compare the differences of morphological structure, goblet cells, argyrophil cells, lymphocytes and Na^+/K^+ATPase in the intestine, and the morphological structure, glycogen and lipid in the liver between the two kinds of *Hemibarbus maculatus*. The results showed that higher amount of fat was found to attached to the gut, lower Na^+/K^+ATPase vitality in the foregut and hidgut ($p < 0.01$) and lower number of goblet cells in the hindgut ($p < 0.01$) of the cultured *Hemibarbus maculatus* when compared to the wild ones. The number of the argyrophilic cells did not show significant differences between the two kinds, but the number of lymphocytes was significantly lower in the segments of gut in cultured. This suggests the absorptive function and intestinal immunity are weaker in the cultured *Hemibarbus maculatus*. In addition, more glycogen and lipid were found in the liver of cultured fishes, which indicates the decreased digestive function of the cultured *Hemibarbus maculatus*. In conclusion, the intestinal digestion, absorption and lymphocytes level of the wild are generally better than those of the cultured, and more hepatic lipopexia and glycogen are present in the cultured ones. Future aquacultural activities should consider these changes when facing pragmatic problems.

Keywords: intestinal functions; *Hemibarbus maculatus*; wild and cultured

1. Introduction

Hemibarbus maculatus, in the Cyprinidae and Gobioninae subfamily, is distributed in the Jialing River in China, and these fish are commonly found in the Nanchong part where there is a subtropical humid monsoon climate zone with an annual average water temperature of 16.9 °C [1]. This species is benthic omnivorous and mainly feeds on aquatic insects and algae. Due to its high nutritional value and delicate meat, *Hemibarbus maculatus* is a popular food among people. However, in aquaculture, the traditional pond culture fish usually die quickly and do not grow the same size as the wild ones, which requires us to explore the functional changes of the digestive system and its limitations in artificial culture. As gut plays the key roles in preventing harmful substances from entering the tissue, digesting and absorbing the nutrients at the same time, it was proved to be the cause of such kinds of problems in some species [2,3]. We also targeted the gut of *Hemibarbus maculatus* living in different environments and aimed to find out the physiological changes within this cultivation issue.

The gut of fish relates closely with the growth, development and reproduction of the fish [4]. The gut anatomical structure of fish is nearly the same as that of mammals, but it mainly depends on various cells to secrete mucus and enzymes for the functions of digestion, absorption and immunity [5]. For example, goblet cells not only secrete mucus to lubricate the gut to help with digestion and absorption, but also form a barrier to protect the gut mucosa from harmful microorganisms [6]. In addition, lymphocytes and argyrophil cells also exercise immunity and secrete mucus, respectively. Thus, the cells undertake an important function in fish's gut, and combining the number and distribution collected from the gut morphology data will truly reflect the causes of susceptibility and poor capacity of digestion and absorption. Liver, as the largest digestive gland, plays very key roles in the digestion or metabolism, and the fatty change in the liver and glycogen storage level are very important to test their ability to metabolize the fat.

Previous studies on *Hemibarbus maculatus* mainly focused on its mitochondrial genome and DNA [7,8], and in aquaculture, some researchers tried to propose better culture strategies to improve its growth [9–11]. In addition, more researchers have realized the importance of the gut recently; they are paying attention to gut microorganisms and have conducted experiments on how to cure fish with enteropathogenic bacterial infection [12,13]. Despite many studies that have been conducted, the main cause of the low levels of digestion and absorption remains unknown; this study aims at studying the gut changes of wild and cultured fish in different growth environments intuitively using histology and biochemistry methods. To clarify the ingestion, absorption function and lymphocyte levels of the gut, methods of comparing the changes of secretory cells' number and the activities of the digestive enzyme are used, and based on these, we propose solutions to the current culture problems.

2. Materials and Methods

2.1. Animals

There were 60 healthy *Hemibarbus maculatus* used in this study, and all were 15–20 cm long (about 180 g); half of them were wild, captured from the midstream of Jialing River, and the remaining half were cultured and purchased from Nanchong Aquatic Market (Nanchong, China). The diets of the cultured fish are shown in Table 1. *Hemibarbus maculatus* live in the middle or lower layers of the water and prefer benthic drilling holes, often inhabited or haunted by moss along the coast near the stone cracks, or deadman. This fish is gentle and sensitive to water flow, especially during the spring flood breeding period, when it is excited to swim when it encounters a small stream, and even jumps out of the water. In the juvenile stage, it feeds on zooplankton, and concurrently eats some algae or hydrophytes. Li et al., 2013 and Xia et al., 2011 [14,15] captured and dissected *Hemibarbus maculatus* separately to examine its digestive tract; combined with their feeding studies of *Hemibarbus maculatus*, the fish mainly feeds on benthic invertebrates, shrimps and insect larvae, and it is a partial carnivorous fish (as shown in Table 2). In our study, the *Hemibarbus maculatus* fish were captured from Jialing River, Nanchong, Sichuan Province, which possesses abundant plankton populations, thus meeting the ingestion demand of *Hemibarbus maculatus* adequately. Besides, Song et al., 2021 [16] set five sampling point in Jialing River in Nanchong and obtained the physical and chemical indicators that the water transparency is 113 ± 5.40 cm, the dissolved oxygen is 6.70 ± 0.10 mg/L, the water temperature is 7–31 °C and the pH value is 6.60 ± 0.07, which basically suitable for the growth and reproduction of *Hemibarbus maculatus*. The valley landform, riverbed, sediment and hydrological grade provide a complex environment for the survival of a variety of aquatic organisms including of wild *Hemibarbus maculatus*. In artificial culture, the artificial feed is the main food for *Hemibarbus maculatus* adult fish, and plankton cultured in the water is supplemented as complementary natural feed. The protein content in the artificial feed is from 36% to 40%, which is determined by the weather, water temperature, fish growth speed and food intake of fish. The feeding time was about 1 h, and fish predation is not obvious, with 70–80% full considered appropriate. The artificial feed formula was as follows: bean cake 30%, rapeseed cake 40%, sesame or peanut cake 15%, fish meal 5% and flour 10%. Feed ingredients were fully mixed into powder or mixed with water for feeding. The phytoplankton's culture was as follows: ammonium bicarbonate and phosphate fertilizer were added to the pond regularly, which is conducive to plankton reproduction. Pond water was mainly yellow-green, and water was injected or changed once every 10–15 days. The water transparency was 30–35 cm, which is good in the state of "nutrient rich, live and fresh". The ammonia nitrogen content was controlled at 0.2 mg/L, and the pH value of water quality was maintained at 7.2–7.8 and no less than 5 mg/L dissolved oxygen. The provided diet was determined according to the weather, water temperature, fish growth and feeding conditions. The feeding time was about 1 h, the fish do not obviously scramble for food and the appropriate amount is 70 or 80% full. In addition, water quality, temperature and the stocking density of fish species were executed according to the feeding standard of the fish.

Table 1. Ingredient composition of the basal diet in cultured fish.

Ingredient	Content (%)	Ingredient	Content (%)
Entrance fish meal	25	Rapeseed oil	1.5
Soybean meal	15	Soybean oil	1.5
Rapeseed meal	16	Vitamin	0.5
Corn	10	$Ca(H_2PO_4)_2$	2.2
Wheat middling	17	Econazole	0.1
Soybean meal fermenting	6	Barren rock powder	2
Diggested tankage	2.7	mineral	0.5

Table 2. Composition and occurrence rate of food in *Hemibarbus maculatus* Bleeker in the wild.

Food Species	Representative Food	Occurrence Rates (%)
Aquatic insects	Tendipes larva	72.24
Alga	-	58.47
Hydrophyte	-	32.62
Mussels	-	27.26
oligochaete	aquatic worms	33.80
protozoan	Vorticellidae	26.98
Gastropoda	-	16.78
decapod	-	6.53

2.2. Gross Observation of the Hemibarbus Maculatus

Firstly, the fish were immersed in the MS-222 anesthetic of 55 mg/L to anesthetize them (AQUI-S, 5 mg/kg.), and then routine anatomy was carried out. Secondly, the gut tract, liver and other organs were removed and washed with normal saline. Thirdly, the length of small gut was measured quickly, and the total weight of gut tract was measured by electronic balance. Then the gut was divided into foregut, midgut and hindgut. The weight and length of each part of the gut was measured using the same method, and the unit weight index was calculated (Unit weight (g/cm) = total gut weight (g)/total length (cm)). Lastly, the liver and intestine samples were fixed in 4% neutral paraformaldehyde fixative.

2.3. Histology Preparation of the Tissues for Microscopic Observation

After fixation for more than 24 h, the histological method (paraffin embedding tissue section technique) was used for the production of the liver and the intestines (foregut, midgut and hindgut) tissue slide, which includes washing dehydration, transparency, dipping wax and encapsulation, sliding for 5 μm thick with microtome (HM340E, Thermofisher, Waltham, MA, USA) for the staining.

The method was consistent with that used by Fischer et al., 2008 [17], which involves xylene dewaxing; 100% alcohol debenzene; 95%, 90%, 80% and 70% alcohol internal gradient hydration; washing with distilled water; Hematoxylin solution (ST2001, SAINT BIO) impregnation; washing with distilled water; differentiation with 1% hydrochloric acid; washing with water stain in Eosin (ST2001, SAINT BIO) for 30 s; washing with distilled water; concentration gradient alcohol dehydration (70–100%); Xylene transparency; sealing piece; and eventually drying for observation.

2.4. Morphological Measurement of the Gut

Six sections and measurements were performed per fish and per group for each of the parameters. Tissue slices (5 visual fields) were observed under optical microscope (BA410, Olympus, Tokyo, Japan) and photographed or analyzed with image-pro software under 100 times. The villus height (VH), crypt depth (CD) (according to the method of Wang et al., 2008) [18], villus width, intestinal wall thickness and diameter were measured, and the ratio of VH/CD was calculated. Absorption surface area was also r measured according to Kisielinski et al., 2002 [19].

2.5. Measurement of Lymphocytes in the Villi

For the measurement of lymphocytes in the villi, intensely stained basophilic nuclei with hematoxyline sections of Hematoxylin-eosin (H.E) staining were counted from above 1.2. At least 1000 intraepithelial nuclei located above the root of villi were counted. All mononuclear (roundness, the chromatin is compact and uniform, without nucleolus, cytoplasmic minima, dark blue) and non-epithelial cells were considered as lymphocytes. The calculation results were expressed by the number of lymphocytes per 100 epithelial cells of each villus.

2.6. Goblet Cells Staining by AB-PAS

The steps of kit usage were consistent with that of Chen et al., 2017 [20] and Miller and Zachary, 2017 [21], which include indirect differentiation of goblet cells by coloration of alexin blue and Schiff reagents with different mucin proteins in combination with acid and alkali, routine section dewaxing water washing, AB-PAS dye staining for 5 min, clear water washing for 2 min for 3 times, periodate staining for 5 min, washing, Schiff staining for 25 min, water washing for about 10 min, alum hematoxylin dyeing for about 10 s, hydrochloric acid alcohol differentiation, gradient alcohol dehydration after clear water washing, xylene transparent seal sheet and natural drying. The results showed that acid mucin goblet cells were stained blue, neutral mucin goblet cells were red and acid and neutral mucin mixed goblet cells were generally purplish red or blue-purple, respectively. The number of goblet cells was calculated on each section and adopted the unit by number, at X200 magnification.

2.7. Observation of Argyrophil Cells: Silver Leaching Method of Longguikai

The description of this method was consistent with Zhang et al., 1999 [22]. They were immersed into distilled water first. Then, silver solution was preheated to 60 °C and sections were immersed into it for 3 h. Then, the sample was rinse once using distilled water, poured into 45 °C reduction solution for 1 h, rinsed with running water and dehydrate, and the piece was transparentized and sealed and dried naturally. The results of staining showed that the particles of argyrophil cells were brown and black. The counting method for the number of argyrophil cells is the same as goblet.

2.8. Determination of Gut Absorption Function

The Na+/K+-ATP enzyme vitality was detected according to the instructions on the Na+/K+-ATP Kit (BC0056, Solarbio, Beijing, China). Briefly, prepare of sample enzyme solution was performed by instructions, followed by ice bath homogenization according to the volume (mL) of tissue mass (g) (extract at 1:10), centrifuging with 3000× *g* at 4 °C for 10 min, clearing and placing on ice to be tested. Whereafter the activity of the Na+/K+-ATP enzyme was determined, and the formula was used to calculate the results: Na+/K+-ATP enzyme activity (μmol/h/g) = [standardC × V total] × (standard A)/(standard blank)/(total number of W × V samples)/T ≤ 7.5 × (test-control)/(standard-blank)/W. (C: Standard tube concentration, 0.5 μmol/mL; V total: Total volume of enzymatic reaction, 0.5 mL; V samples: sample volume, 0.2 mL; T: Reaction time, 1/6 h; W: Sample weight, g).

2.9. Observation of the Liver Morphology, Glycogen and Lipid: H.E, Periodic Acid-Schiff Stain (PAS) and Oil Red O Staining

For observation of liver tissue, we directly made paraffin sections of liver tissue, and then the H.E, PAS and Oil Red O (ORO) staining were used for morphology, glycogen and lipid observation. HE and PAS staining was made according to intestinal morphology and goblet cells staining. The lipid droplets were made with Oil Red O staining according to the Oil Red O Staining Kit (Servicebio, G1016, Wuhan, China). Briefly, the slices were fixed in the fixative solution for 15 min, washed with tap water, and dried; then the sections were stained with Oil Red solution for 10 min in the dark, and then immersed in 60% isopropanol for differentiation and immersed in pure water, hematoxylin, and 60% alcohol; at last, they were observed with microscope inspection, image acquisition and analysis.

2.10. Data Analysis

All the data were formatted as mean values ± standard deviation (M ± SD); data were all consistent with normal distribution, and Student's *t*-test was used. All data analysis was performed using SPSS 16.0 (SPSS Inc., Chicago, IL, USA). $p < 0.05$ means statistically significant.

3. Results

3.1. Intestinal Anatomy and Morphology

From the anatomical results, there was a large amount of fat on the gut surface of the cultured fish, as shown in Figure 1A,B. No significant difference was observed in the weight of fish; weight of the total gut; or weight of the foregut, midgut and hindgut; no significant difference was observed in the length of the full gut or foregut, midgut or hindgut (Table 3).

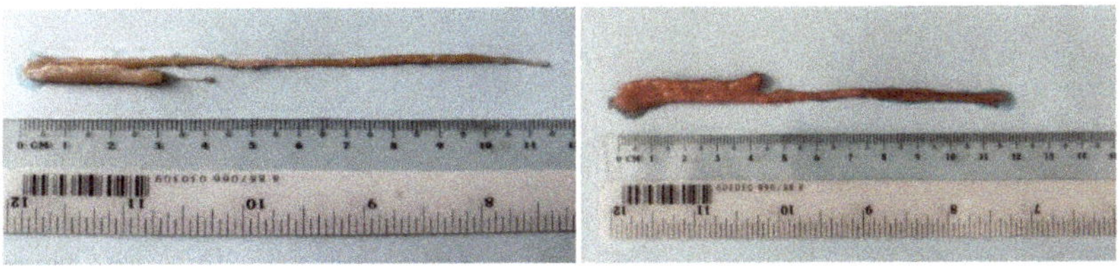

(A) (B)

Figure 1. Intestinal anatomical structure of wild (**A**) and cultured (**B**) *Hemibarbus maculatus*. A large amount of fat on the gut surface of the cultured, but not wild fish, was found.

Table 3. Morphological parameters of wild and cultured *Hemibarbus maculatus*.

Variety	Weight of Fish (g)	Weight of Total Gut (g)	Weight of Foregut (g)	Weight of Midgut (g)	Weight of Hindgut (g)
Wild	43.544 ± 1.221	0.824 ± 0.1196	0.398 ± 0.0652	0.168 ± 0.0287	0.182 ± 0.020
Cultured	44.822 ± 1.358	0.896 ± 0.1771	0.382 ± 0.0719	0.202 ± 0.0455	0.246 ± 0.038
Variety	**Length of full gut (cm)**	**Length of foregut (cm)**	**Length of Midgut (cm)**	**Length of Hindgut (cm)**	**Unit weight**
Wild	13.26 ± 1.212	3.67 ± 0.2048	2.984 ± 0.149	6.408 ± 0.279	0.062 ± 0.010
Cultured	13.134 ± 1.552	3.224 ± 0.3606	3.234 ± 0.244	6.524 ± 0.470	0.066 ± 0.013

Note: n = 30, data are presented as mean ± standard deviation.

3.2. Intestinal Histological Structure

The results showed that the width of foregut villus of the cultured was wider than the wild group (Figure 2A,B and Table 4, $p < 0.05$). The absorption area of the cultured hindgut was larger than the wild hindgut (Figure 2C, $p < 0.01$). No significant difference was observed for villus height, crypt depth, intestinal wall thickness and diameter, VH/CD.

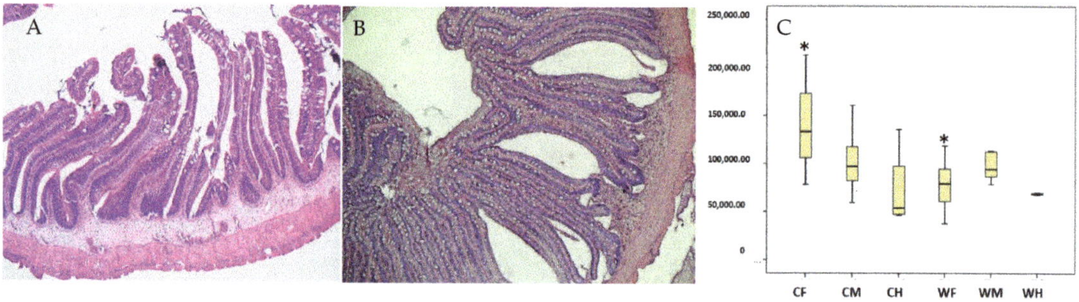

Figure 2. Histological structure the foregut of cultured (**A**) (H.E., bar = 500 μm) and wild *Hemibarbus maculatus* (**B**) (H.E., bar = 500 μm), and the absorption area (**C**) of gut segments in cultured and wild *Hemibarbus maculatus*. The width of foregut villus of the cultured was wider than the wild group. The absorption area of the cultured hindgut was larger than the wild group. Note: n = 30, the data were misrepresented by average ± standard deviation, * indicates that the data are significantly dif-ferent from those of cultured (wild) ($p < 0.05$). CF: Cultured foregut; CM: Cultured midgut; CH: Cultured hindgut; WF: Wild foregut; WM: Wild midgut; WH: Wild hindgut. (means ± SD, n = 10).

Table 4. Indexes of gut tract in Cultured and wild *Hemibarbus maculatus*.

		Villus Height (μm)	Villus Width (μm)	Crypt Depth (μm)	Thickness of Muscle Layer (μm)	Tube Radius (μm)
Cultured	foregut	386.51 ± 53.51	110.24 ± 21.58 *	58.26 ± 6.93	51.47 ± 9.61	995.38 ± 31.56
	midgut	426.79 ± 69.06	62.70 ± 9.20	62.70 ± 9.54	86.16 ± 14.73	803.24 ± 77.24
	hindgut	293.51 ± 34.35	63.84 ± 11.32	47.54 ± 6.55	36.95 ± 5.64	559.60 ± 20.37
Wild	foregut	445.05 ± 48.17	50.47 ± 4.86 *	66.01 ± 7.89	92.05 ± 22.36	1004.43 ± 126.35
	midgut	541.01 ± 31.21	56.43 ± 4.34	61.08 ± 6.42	67.42 ± 5.67	969.88 ± 75.23
	hindgut	305.86 ± 80.26	52.52 ± 1.68	49.23 ± 3.59	38.78 ± 4.53	565.26 ± 64.50

Note: n = 30, the data were misrepresented by average ± standard deviation, * indicates that the data were significantly different from those of cultured (wild) ($p < 0.05$).

Na+/K+-ATPase vitality is the main factor to monitor the gut digestion and absorption function; it could be used to reflect the differences between the wild and the cultured guts in this study. The vitality of Na+/K+-ATPase in the wild foregut and hindgut was significantly higher than that of the cultured group, while it was significantly higher in the midgut of the cultured than that of the wild group (Figure 3).

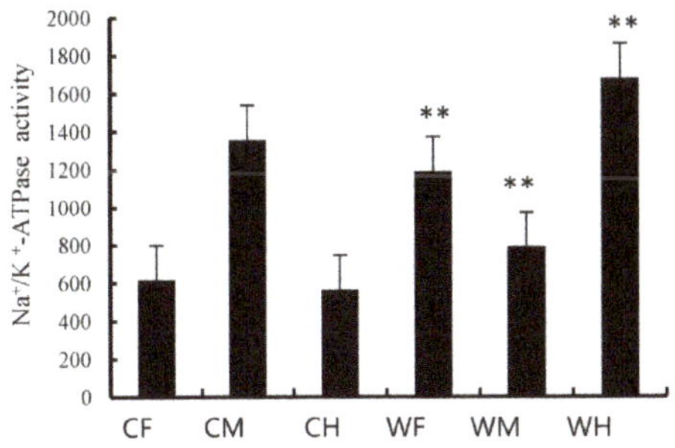

Figure 3. The vitality of Na^+/K^+-ATPase in cultured and wild *Hemibarbus maculatus*. The activity of Na^+/K^+-ATPase in the wild foregut and hindgut was significantly higher than that of the cultured group, while was significantly higher in the midgut of the cultured than that of the wild group. Note: $n = 30$, the data were misrepresented by average ± standard deviation ** indicates that the data were significantly different from those of cultured (wild) ($p < 0.01$).

3.3. Intestinal Secretory Cells Argyrophil Cells and Goblet Cells

Two of the main secretory cells in the gut-argyrophil cells and goblet cells were also used to determine the secretory function of the gut. Three types of argyrophil cells, such as fusiform, round and tadpole, were found (Figure 4A,B). Different shapes of cells determined different secretory functions. In this study, we found that the number of argyrophil cells between the same segments of the wild and the cultured was different, and the number of argyrophil cells in foregut was generally higher (Figure 4C). This suggested that the foregut secretory function of *Hemibarbus maculatus* was better than that of other gut segments. In addition, there was no significant difference in the number of goblet cells in the foregut and midgut between the wild and cultured groups, but the number of goblet cells in the wild hindgut was more than the cultured group ($p < 0.01$) (Figure 5A–C).

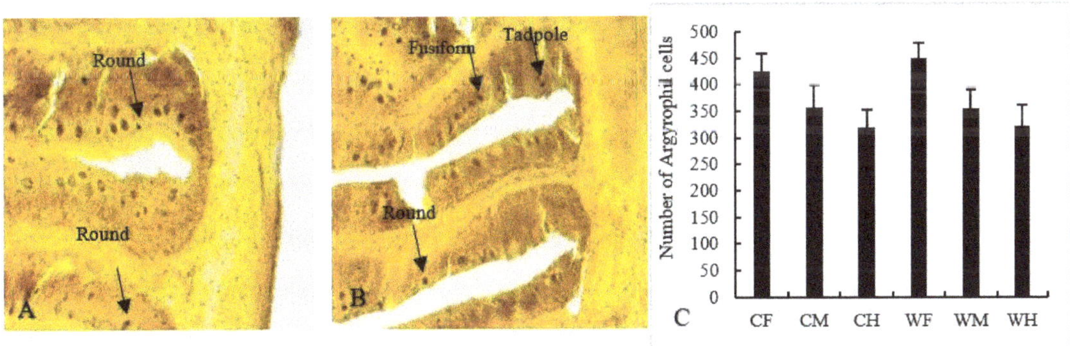

Figure 4. Distribution (**A**,**B**) and the number (**C**) of argyrophil cells in wild (**A**) and cultured (**B**) *Hemibarbus maculatus*. Three types of argyrophil cells were found—fusiform, round and tadpole—and the number of argyrophil cells in foregut was generally higher. Note: $n = 30$, the data were misrepresented by average ± standard deviation. (Longguikai, bar = 50 μm) Note: $n = 30$, the data were misrepresented by average ± standard deviation.

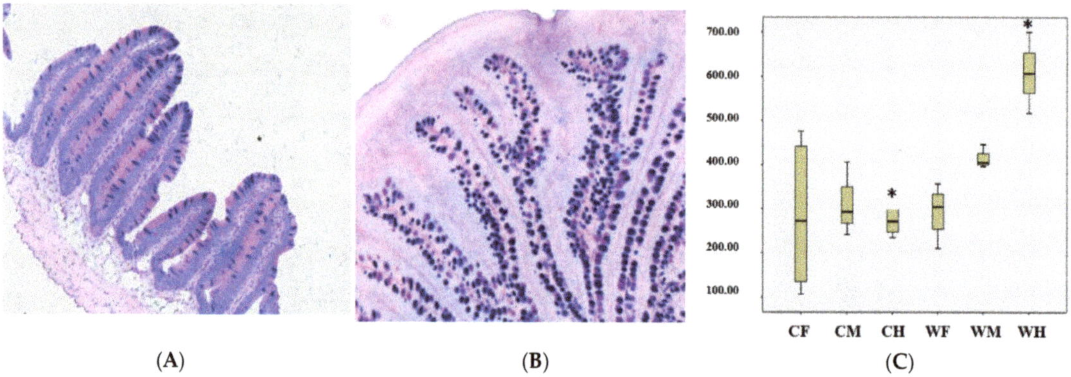

Figure 5. Distribution (**A**,**B**) and comparison of goblet cells in different intestinal segments (**C**) between cultured (**A**) and wild (**B**) *Hemibarbus maculatus*. There was no significant difference in the number of goblet cells in foregut and midgut between the wild and the cultured groups, but the wild hindgut was more than the cultured group. (AB-PAS, bar = 50 μm) Note: $n = 30$, the data were misrepresented by average ± standard deviation, * indicates that the data were different from those of cultured (wild) ($p < 0.05$).

3.4. Lymphocyte Number

Lymphocytes are indispensable cellular components of immune response function. They are the main executors of almost all immune function of lymphoid system, while they are also the primary index to detect immune function. Therefore, the number of lymphocytes in different gut segments of the cultured and the wild groups were measured. We found that the number of lymphocytes in the gut tract of the wild group was larger, while the difference between the foregut and hindgut was significant ($p < 0.05$), and that in the midgut was very significant ($p < 0.01$). The results indicate that the wild fish have better intestinal immune function (Figure 6).

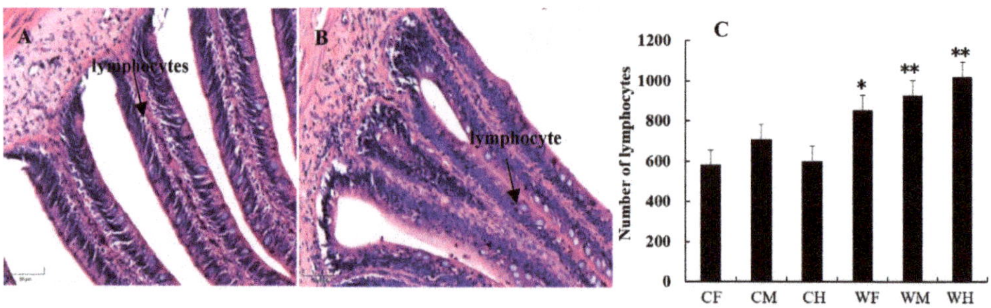

Figure 6. The distribution (**A**,**B**) and number (**C**) of lymphocytes in different gut segments between cultured (**A**) and wild (**B**) *Hemibarbus maculatus*. The number of lymphocytes in the gut tract of the wild group was larger, while the difference between foregut and hindgut was significant, and that in the midgut was very significant. Note: $n = 30$, the data were misrepresented by average ± standard, * indicates significantly different from those of the cultured (wild) group ($p < 0.05$), ** indicates significantly different from those of cultured (wild) group ($p < 0.01$). (H.E., bar = 50 μm).

3.5. Hepatic Morphology, Glycogen and Lipid

The result showed that in the wild, normal structures were observed (Figure 7A); however, in the cultured fish, rounded vacuole was observed (Figure 7B); they were confirmed by Oil Red O Staining, lipid vacuole degeneration was observed, and many lipid oil droplets were found in the cells but fewer in the wild group (Figure 7C,D). In addition, the area of glycogen deposition was hardly observed in the wild group, while in the cultured group, the deposition of glycogen was present (Figure 7E,F).

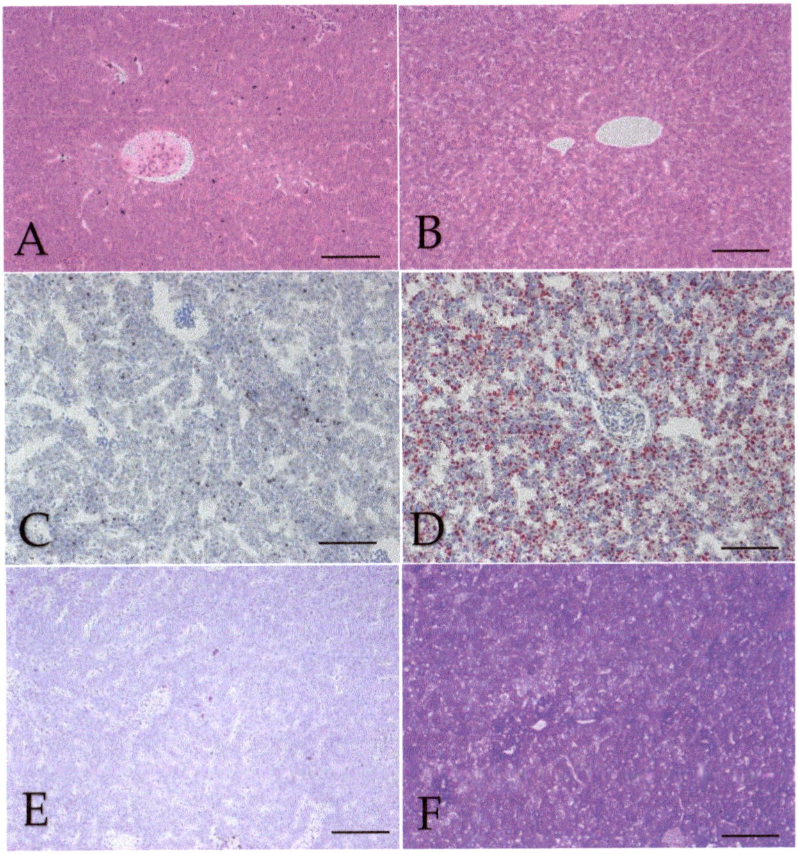

Figure 7. Liver morphology (**A**,**B**) (H.E., bar = 200 μm), lipid (**C**,**D**) (O.R.O., bar = 200 μm) and glycogen (**E**, **F**) (PAS, bar = 200 μm) stained by H.E, Oil Red O and PAS between wild and cultured *Hemibarbus maculatus*. Rounded vacuole was observed in the cultured group (Figure 7B) but very little in wild group (Figure 7A), and many lipid droplets and deposition of glycogen were found in the cells, but fewer in the wild group (Figure 2C,D). More deposition of glycogen was presented in the cultured group (Figure 7F), but little was found in the wild group (Figure 7E).

4. Discussion

Gross observation results indicated that there was a large amount of fat attached on the intestinal surface of the cultured group, which might be related to the limited vitality space, low energy consumption, rich diet nutrition and low conversion rate for cultured fish. As Table 1 shows, the feeds of the cultured group contain a lot of fat; the high-fat diet (HFD) has posed a considerable threat to the fish aquaculture industry due to the ability to induced fatty liver [23], which is caused by the unbalanced nutrition, improper feeding and unscientific management of culture [24]. Fatty liver fish will have a reduced grow rate and weaken disease resistance [25,26]; in addition, excessive fats that are contained in commercial aquatic feeds may lead to increased oxidative stress result and chaos in liver of the glucose and lipid balance, ultimately leading to disease and death [27]. In this experiment, via H.E. and Oil Red Staining, the liver of the cultured group was observed to have steatosis and intercellular fat infiltration, suggesting that fatty liver occurs, and according to the result of glycogen accumulation in the liver, the oxidative stress of the liver may also increase [28]; they may adversely affect the digestion and absorption of *Hemibarbus maculatus*. In addition, the length and weight of gut did not differ significantly between the wild and cultured groups, but the absorption area of the foregut of the cultured group was significantly larger than that of the wild group. We speculate that the foregut lumen enlarges in order to store more food to adapt to the large number of dry pellets fed to them on a fixed time in artificial breeding. *Hemibarbus maculatus* is a kind of stomachless fish, whose foregut has the function of digestion and storage [29]. However, will this change of the foregut affect its digestion function?

Wang et al. (2018) [30] pointed out that the change of intestinal structure was related to food habit. The intestinal length and digestive enzymes could be affected by animal habitat and food type [31]. In order to investigate whether the digestion and absorption function of foregut has changed Na^+/K^+-ATPase vitality, we chose to compare the digestion and absorption function of cultured and wild *Hemibarbus maculatus*. Na^+/K^+-ATPase, which is an integral membrane protein, could be found in all higher-class eukaryotes, and is used to drive numerous transport processes [30] and also acts as a marker enzyme of the basolateral membrane in the gut [32]. In this study, the enzyme vitality of each gut segment was different between the wild and cultured fish. The enzyme vitality of the foregut and hindgut of the wild group was significantly higher than that of the cultured group, while in the midgut the observation was the opposite. This result confirmed the above-mentioned fact that the absorption area of foregut of the cultured group was increased. Therefore, we concluded that the foregut of the cultured fish acts as a temporary storage for food and weakens the digestion and absorption function, which coincide with the intestinal function of the stomachless fish [33]. The food began to be digested and absorbed in the midgut, which corresponds with the fact that enzyme vitality of the midgut of the cultured fish was significantly higher than that of the wild fish. On the contrary, the foregut of the wild fish had a fixed digestion and absorption function, and food started to be digested and absorbed in the foregut, and then continued to be further digested and absorbed through the midgut and hindgut. For the enzyme vitality, the activity of Na^+/K^+-ATPase in the wild foregut and hindgut was significantly higher than that of the cultured group, while in the midgut it was significantly lower; this is because the foregut of the cultured fish primarily acts as a store site [29], and the midgut is the main site to digest and absorb, while in the wild, the foregut has a degree of digestion and absorption and the hindgut has the strongest digestion and absorption function. However, digestion and absorption functions could interact with many factors, especially the secretion in the gut.

The digestion and absorption function of the gut is closely related to the secretion function [34]. Secretion can provide digestive enzymes, solubilize nutrients to ensure optimal digestion performance, protect the gut mucosa from harsh dietary component as well as from the alimentary tract's own acid and protect the entire organism against microbes and chemicals that may be detrimental to the animals' health [35]. There are a large number of mucilage cells in the intestinal villus, which include two cell types: larger ovoid saccular cells and elongated oval goblet cells. In the gut, the mucus cells were mainly goblet cells [32]. In our study, the results showed that the number of goblet cells in the hindgut of the cultured group was significantly smaller than that of the wild group. This conformed to the conclusion that the activities of Na+/K+-ATPase are secreted by goblet cells, which in the hindgut of the cultured group was significantly lower than that of the wild group. In addition, argyrophilic cells in the digestive tract have both endocrine and exocrine functions [28].

The intestinal mucosa is in contact with microbial flora, pathogens, etc., and plays a key role in the immune response against infection. The gut mucosal barrier represents a complex environment consisting of different physical barriers and immune cells (mainly the lymphocytes), which prevent potentially harmful pathogens from entering the intestine while maintaining tolerance to food antigens and commensal bacteria [36–38]. In this study, the result showed the counts of lymphocytes in each section of the gut of the wild fish were significantly greater than that of the cultured fish, among which the largest number of lymphocytes reside in the midgut. Based on this result, we thought that the wild species' immune system could adapt to the living environment and improve autoimmunity to defend against adverse factors in the environment, which was consistent with the fish study in the laboratory and the wild fish [39].

5. Conclusions

In conclusion, the intestinal digestion, absorption and lymphocytes level of the wild fish are generally better than those of the cultured fish, and more hepatic lipopexia and glycogen are present in the cultured ones. Through the intestinal morphological data and the studies of the intestinal functions of fish, we will reflect the aquaculture ecology and formulate appropriate strategies in aquaculture.

Author Contributions: H.L., J.Z. and L.Z. completed the experiment and prepared the manuscript; B.W., B.S. and H.L. carried out the growth experiment and analyzed the samples; B.W. and Y.Z. designed the experiment and revised the manuscript. All authors have read and agreed to the published version of the manuscript.

Funding: This research was funded by the Fundamental Research Funds of Sichuan Provincial Department of Science and Technology (Project No. 2022NSFSC0129) and the program for the Fundamental Research Funds of China West Normal University (Project No. 20A003).

Institutional Review Board Statement: The animal study protocol was approved by the Ethics Committee of China West Normal University (protocol code cwnu2022D008, 2022.03.04). for studies involving animals.

Informed Consent Statement: Informed consent was obtained from all subjects involved in the study.

Data Availability Statement: Not applicable.

Acknowledgments: The authors would like to thank the co-workers of China West Normal University for their assistance in performing the experiment and the analysis.

Conflicts of Interest: The authors declare that the research was conducted without any commercial or financial relationships that could be construed as a potential conflict of interest.

References

1. Jiang, C.M. *The Differential Comparison of Biologicalcharacteristics of Saurogobio Dabryi in the Jialing River*; China West Normal University: Nanchong, China, 2017.
2. Bakke, A.M.; Glover, C.; Krogdahl, A. Feeding, digestion and absorption of nutrients. *Fish Physiol.* **2010**, *30*, 57–110. [CrossRef]
3. De Schryver, P.; Vadstein, O. Ecological theory as a foundation to control pathogenic invasion in aquaculture. *ISME J.* **2014**, *8*, 2360–2368. [CrossRef] [PubMed]
4. Wilson, J.M. Morphological diversity of the gastrointestinal tract in fishes. *Fish Physiol.* **2010**, *30*, 1–55. [CrossRef]
5. Banan Khojasteh, S.M. The morphology of the post-gastric alimentary canal in teleost fishes: A brief review. *Int. J. Aqua. Sci.* **2012**, *3*, 71–88. [CrossRef]
6. Cinar, K.; Senol, N. Histological and histochemical characterization of the mucosa of the digestive tract in Flower Fish (Pseudophoxinus antalyae). *Anat. Histol. Embryol.* **2006**, *35*, 147–151. [CrossRef]
7. Li, P.; Xu, D.D.; Peng, Z.G.; Zhang, Y.G. The complete mitochondrial genome of the spotted steed, *Hemibarbus maculatus* (Teleostei, Cypriniformes). *Mitochondr. DNA* **2012**, *23*, 34–36. [CrossRef]
8. Han, M.; Wang, C.L.; Shen, C.N.; Shen, I.S. The complete mitochondrial genome of half-spined barbel *Acrossocheilus hemispinus* (Nichols) (Teleostei, Cyprinidae, Barbinae). *Mitochondr. DNA* **2015**, *26*, 133–134. [CrossRef]
9. Lian, Q.P.; Mi, G.Q.; Hu, T.J.; Wang, Y.C.; Yao, Z.L.; Qin, D. Study on embryo development of crossbreed F1 by *Hemibarbus maculates* (♀) × *Hemibarbus labeo* (♂). *Hubei Agric. Sci.* **2012**, *51*, 131–135.
10. Yun, T. Study on the morphological biology of *Hemibarbus maculatus* Bleeker in Yuanhe river. *J. Anhui Agric. Sci.* **2012**, *40*, 17605–17607. [CrossRef]

11. Tuo, Y.; Xiao, T.Y. Analysis and discrimination of sexual dimorphism of *Hemibarbus maculatus* Bleeker by morphometry. *Fisheries Sci.* **2015**, *34*, 603–608.
12. Zhu, Z.; Tianle, X.U.; Deng, S.; Chen, J.; Chen, R.; Xie, L.; Huang, Z.; Wu, S.H.; Wu, Y.Z. Isolation, identification and antimicrobial test in vitro of Aeromonas from *Hemibarbus maculatus* Bleeker. *Agric. Sci. Tech.* **2017**, *18*, 156–160. [CrossRef]
13. Luo, L.; Zhao, Z.G.; Du, X.; Xu, Q.Y. Relationship between intestinal flora and immunity in fish. *Chin. J. Fish.* **2018**, *31*, 51–54. [CrossRef]
14. Li, X.X.; Liu, W.; Tang, F.J. Analysis of feeding and nutritional level of three species of subfamilies in Harbin section of Songhua River. *J. Fish.* **2013**, *26*, 29–33. [CrossRef]
15. Xia, X.; Guan, S.F.; Gong, S.Y. Fengxi river flower eating study. *Jiangxi Aquat. Prod. Sci. Tech.* **2011**, *2*, 24–29. [CrossRef]
16. Song, L.; Wang, J.; Yang, K. Phytoplankton community structure and water quality assessment in Nanchong outbound section of Jialing River in autumn. *J. China West Norm. Univ.* **2021**, *42*, 215–224. [CrossRef]
17. Fischer, A.H.; Jacobson, K.A.; Rose, J.; Zeller, R. Hematoxylin and eosin staining of tissue and cell sections. *Csh. Protoc.* **2008**, *6*, 415–417. [CrossRef]
18. Wang, J.X.; Peng, K.M. Developmental morphology of the small intestine of African Ostrich chicks. *Poult. Sci.* **2008**, *87*, 2629–2635. [CrossRef]
19. Kisielinski, K.; Willi, S.; Prescher, A.; Klosterhalfen, B.; Schumpelick, V. A simple new method to calculate small intestine absorptive surface in the rat. *Clin. Exp. Med.* **2002**, *2*, 131–135. [CrossRef]
20. Chen, J.W.; Zhang, W.; Wang, T.W.; Cheng, J.; Luo, J.F. Application of AB-PAS staining after digestion in mucus staining. *Int. J. Clin. Exp. Patho.* **2017**, *7*, 112–113. [CrossRef]
21. Miller, M.A.; Zachary, J.F. Mechanisms and morphology of cellular injury, adaptation, and death. *Pathol. Basis Vet. Dis.* **2017**, *6*, 2–43. [CrossRef]
22. Zhang, S.Z.; Wu, X.B.; Chen, B.H.; Wang, R.P.; Wang, C.L.; Xie, W.S. Distribution and morphological observation of argyrophil cells in digestive tract of alligator sinensis. *Chin. J. Zool.* **1999**, *34*, 31–34.
23. Norambuena, F.; Estevez, A.; Bell, G.; Carazo, I.; Duncan, N. Proximate and fatty acid compositions in muscle, liver and gonads of wild versus cultured broodstock of Senegalese sole (Solea senegalensis). *Aquaculture* **2012**, *356*, 176–185. [CrossRef]
24. Yu, K.; Huang, K.; Tang, Z.; Huang, X.; Sun, L.; Pang, L.; Mo, C. Metabolism and antioxidation regulation of total flavanones from Sedum sarmentosum Bunge against high-fat diet-induced fatty liver disease in Nile tilapia (Oreochromis niloticus). *Fish. Physiol. Biochem.* **2021**, *47*, 1149–1164. [CrossRef]
25. Goessling, W.; Sadler, K.C. Zebrafish: An important tool for liver disease research. *Gastroenterology* **2015**, *149*, 1361–1377. [CrossRef] [PubMed]
26. Prisingkorn, W.; Prathomya, P.; Jakovlic, I.; Liu, H.; Zhao, Y.H.; Wang, W.M. Transcriptomics, metabolomics and histology indicate that high-carbohydrate diet negatively affects the liver health of blunt snout bream (Megalobrama amblycephala). *BMC Genom.* **2017**, *18*, 856. [CrossRef] [PubMed]
27. Layman, J.I.; Pereira, D.L.; Chellan, N.; Huisamen, B.; Kotzé, S.H. A histomorphometric study on the hepatoprotective effects of a green rooibos extract in a diet-induced obese rat model. *Acta Histochem.* **2019**, *121*, 646–656. [CrossRef] [PubMed]
28. Berdja, S.; Boudarene, L.; Smail, L.; Neggazi, S.; Boumaza, S.; Sahraoui, A.; Haffaf, E.I.; Kacimi, G.; Bouguerra, S.A. Scolymus hispanicus (golden thistle) ameliorates hepatic steatosis and metabolic syndrome by reducing lipid accumulation, oxidative stress, and inflammation in rats under hyperfatty diet. *eCAM* **2021**, *14*, 5588382–5588395. [CrossRef]
29. Grosell, M.; Farrell, A.P.; Brauner, C.J. (Eds.) *Fish Physiology: The Multifunctional Gut of Fish*, 1st ed.; Academic Press: Cambridge, MA, USA, 2011; Volume 30. [CrossRef]
30. Wang, Y.Z.; Sun, J.F.; Lv, A.J.; Zhang, S.L.; Sung, Y.Y.; Shi, H.Y.; Hu, X.C.; Chen, S.J.; Xing, K.Z. Histochemical distribution of four types of enzymes and mucous cells in the gastrointestinal tract of reared half-smooth tongue sole Cynoglossus semilaevis. *J. Fish. Biol.* **2018**, *92*, 3–16. [CrossRef]
31. Day, R.D.; German, D.P.; Manjakasy, J.M.; Farr, I.; Hansen, M.J.; Tibbetts, I.R. Enzymatic digestion in stomachless fishes: How a simple gut accommodates both herbivory and carnivory. *J. Comp. Physiol. B.* **2011**, *181*, 603–613. [CrossRef]
32. Lingrel, J.B.; Kuntzweiler, T. Na+-K+-ATPase. *J. Biol. Chem.* **1994**, *269*, 19659–19662. [CrossRef]
33. Le, H.T.M.D.; Shao, X.; Krogdahl, A.; Kortner, T.M. Intestinal function of the stomachless fish, Ballan Wrasse (Labrus bergylta). *Front. Mar. Sci.* **2019**, *6*, 1–15. [CrossRef]
34. Van der Klis, J.; Jansman, A. Optimising nutrient digestion, absorption and gut barrier function in monogastrics: Reality or illusion. In *Nutrition and Health of the Gastrointestinal Tract*; Wageningen Academic Publishers: Wageningen, The Netherlands, 2002; pp. 15–36.
35. Madara, J.L. *Functional Morphology of Epithelium of the Small Intestine. Comprehensive Physiology*; John Wiley & Sons, Inc.: Hoboken, NJ, USA, 2011. [CrossRef]
36. Israeli, E. Goblet cells deliver luminal antigen to CD103+ dendritic cells in the small intestine. *Isr. Med. Assoc. J.* **2012**, *483*, 345–349. [CrossRef]
37. Melo-Gonzalez, G.; Felipe, T.M. *Investigating the Role of the Intestinal Barrier in Regulation of Immune Homeostasis in the Gut*; The University of Manchester: Manchester, UK, 2016.

38. Kober, O. *The Role of Mucus in the Cross Talk between Gut Bacteria and the Host*; University of East Anglia: Norwich, UK, 2013.
39. Robertson, S.; Bradley, J.E.; MacColl, A.D.C. Measuring the immune system of the three-spined stickleback—Investigating natural variation by quantifying immune expression in the laboratory and the wild. *Mol. Ecol. Resour.* **2015**, *16*, 701–713. [CrossRef] [PubMed]

animals

MDPI

Review

A Pretty Kettle of Fish: A Review on the Current Challenges in Mediterranean Teleost Reproduction

Marta Lombó [1,2,3,†], Christian Giommi [1,2,†], Matteo Zarantoniello [1,*] and Giulia Chemello [1,2,*]

1 Department of Life and Environmental Sciences (DiSVA), Università Politecnica delle Marche, 60131 Ancona, Italy; mloma@unileon.es (M.L.); c.giommi@pm.univpm.it (C.G.)
2 INBB—Consorzio Interuniversitario di Biosistemi e Biostrutture, 00136 Roma, Italy
3 Department of Molecular Biology, Faculty of Biology and Environmental Sciences, Universidad de León, 24071 León, Spain
* Correspondence: m.zarantoniello@staff.univpm.it (M.Z.); g.chemello@staff.univpm.it (G.C.)
† These authors contributed equally to this work.

Simple Summary: In the Mediterranean region, teleost species are facing diverse challenges resulting from the combined effects of climate change, pollution, and overfishing. The survival of these species and the overall stability of the ecosystem are closely linked to their reproductive success and their capacity to produce healthy offspring. Teleost reproduction is a crucial aspect in aquaculture as well since it ensures the productivity of high-quality animal proteins as an alternative to overexploited wild stocks. Therefore, deep knowledge on the factors influencing the reproduction of species devoted to aquaculture at different levels is necessary to propose efficient management approaches, aiming to identify sustainable farming techniques. To attain this goal, it is initially necessary to comprehend the molecular and physiological mechanisms that regulate various levels of the reproductive process in a changing scenario. The use of model organisms allows scientists to test specific stressors without external interference, thanks to their availability, easier handling, and biological characteristics. However, caution is needed to avoid overestimating their applicability to other species. The present review aims to explore the current factors affecting reproduction in wild, farmed, and model teleosts.

Abstract: The Mediterranean region is facing several environmental changes and pollution issues. Teleosts are particularly sensitive to these challenges due to their intricate reproductive biology and reliance on specific environmental cues for successful reproduction. Wild populations struggle with the triad of climate change, environmental contamination, and overfishing, which can deeply affect reproductive success and population dynamics. In farmed species, abiotic factors affecting reproduction are easier to control, whereas finding alternatives to conventional diets for farmed teleosts is crucial for enhancing broodstock health, reproductive success, and the sustainability of the aquaculture sector. Addressing these challenges involves ongoing research into formulating specialized diets, optimizing feeding strategies, and developing alternative and sustainable feed ingredients. To achieve a deeper comprehension of these challenges, studies employing model species have emerged as pivotal tools. These models offer advantages in understanding reproductive mechanisms due to their well-defined physiology, genetic tractability, and ease of manipulation. Yet, while providing invaluable insights, their applicability to diverse species remains constrained by inherent variations across taxa and oversimplification of complex environmental interactions, thus limiting the extrapolation of the scientific findings. Bridging these gaps necessitates multidisciplinary approaches, emphasizing conservation efforts for wild species and tailored nutritional strategies for aquaculture, thereby fostering sustainable teleost reproduction in the Mediterranean.

Keywords: reproductive performance; aquaculture; climate change; water pollution; fish diet; overfishing

1. Introduction

The Mediterranean Sea is a marginal and semi-enclosed sea situated at the crossroads of Africa, Europe, and Asia. Since ancient times, it has been an important route for merchants and travelers and, currently, it represents an essential hub for global economy and trade [1]. The Mediterranean Sea is characterized by its semi-enclosed nature, limited connection to the Atlantic Ocean, and high variability in temperature and salinity. These environmental factors shape the unique ecosystems found within its waters [2]. As a result of that, it has long been considered a focal point of scientific inquiry and environmental concern [3,4]. Although it represents only 0.82% of the sea water surface, the Mediterranean Sea harbors nearly 17,000 known marine species, representing 4–18% of global marine biodiversity. Regarding fish, teleost species are the most abundant, representing 85% of Mediterranean fish [5]. These bony fish, characterized by their advanced reproductive systems, are vital components of the Mediterranean ecosystem, playing key roles in nutrient cycling, trophic interactions, and maintaining ecological balance [6]. In the last decades, the reproduction of teleosts has been threatened by the ongoing environmental changes and the increasing anthropogenic impacts [7].

1.1. Mediterranean Teleosts in a Changing Environment

Climate change can exert an influence on the marine environment through various processes, directly and indirectly altering the stability of environmental conditions. Rising sea temperatures and declining oceanic pH levels can induce ocean acidification, driven by the increased absorption of atmospheric carbon dioxide (CO_2) by the oceans. Additionally, climate change has the potential to disturb ocean circulation patterns, resulting in alterations in currents and upwelling systems. Finally, climate fluctuations could induce hypoxia, shifts in precipitation patterns, and an increase in the frequency and intensity of extreme weather events, such as hurricanes and typhoons [8–10].

These variations can alter the reproductive success of teleosts by affecting the proper function of the brain–pituitary–gonadal axis (HPG) at different levels [11]. Temperature is one of the predominant factors influencing the reproductive processes of teleosts. Mediterranean surface waters are warming up at a rate > 2 °C/100 year along the whole water column [12]. The sensitivity of fish to thermal fluctuations varies, depending not only on the species and life stages but also on the range of thermal tolerance fish display [13,14]. Considering that most teleost species are poikilotherms, changes in temperature can impact different vital parameters, such as metabolic rate and growth, natural death, and reproductive success [7]. Another factor with outstanding contribution is salinity, which is often linked to climate-induced alterations in precipitation patterns. Slight shifts in salinity levels can affect the osmoregulatory mechanisms of teleosts, potentially leading to reduced hatching success and larval survival [15]. Altogether, these environmental stressors collectively contribute to the vulnerability of teleost reproductive success in the Mediterranean.

1.2. Impact of Anthropic Activities

The Mediterranean region is currently home to over 525 million people and this figure is expected to increase by an additional 130 million by 2050, particularly in the eastern and southern countries. Additionally, the Mediterranean countries remain the most important global tourist destination, welcoming over 360 million international tourists annually [16]. The concentration of population along the coastal trips has led to an increase in waste production, discharges of wastewater, and pressure on natural resources. Consequently, human activities exacerbate the challenges faced by teleosts in the Mediterranean. A major threat specific to the Mediterranean concerns fishery-related activities, with 88% of stocks being overfished [16]. The impacts of fisheries go beyond stocks; they affect seafloor habitats, including benthic fauna damage, and contribute to marine litter and microplastic generation [12]. Overfishing has led to a decline in teleost populations, affecting both their abundance and age structure. This can disrupt the intricate balance of reproductive dynamics, with potential consequences for the long-term sustainability of teleost populations [17].

The introduction of non-native species, whether intentionally or inadvertently, can also disrupt the reproductive ecology of native teleosts. Competition for resources, predation pressure, and disease transmission associated with invasive species can all interfere with successful teleost reproduction and survival [12]. In this regard, compared to the other regional seas, the Mediterranean has the highest number of invasive species that can jeopardize both the survival and reproduction of native species [16]. Moreover, pollutants, including heavy metals, (micro)plastics, agricultural runoffs, pharmaceuticals, and personal care products, have become pervasive in the Mediterranean. These contaminants can bioaccumulate in fish tissues [18] and can even be maternally transmitted to the offspring through the yolk, as demonstrated in model species [19]. The consequences range from impaired gametogenesis to skewed sex ratios, posing serious threats to the future of teleost populations in the region [20]. Nowadays, new approaches have been developed to monitor the levels of pollutants in sea water and in invertebrates and fish tissues. The derived results could provide useful information about the type and concentrations of different chemicals and will enable the correlation of fish survival and welfare with specific polluted conditions [21,22].

1.3. Implications for Teleost Farming

The aquaculture sector in the Mediterranean Sea represents the fastest-growing industry in food production, especially finfish farming [23]. Farmed teleosts help alleviate pressure on wild fish populations by offering an alternative source of seafood. However, as the challenges intensify for wild teleost populations, the aquaculture industry, which mostly relies on the breeding success of farmed species, faces its own set of obstacles [24]. Understanding how environmental changes and anthropogenic impacts influence the reproduction of teleosts is also crucial for developing sustainable aquaculture practices. In farmed species, it has been proven that adjustments in photoperiod, sea temperature, density, and salinity may ensure optimal reproductive conditions, as reviewed by Mylonas and Zohar [25]. Dietary factors can also influence reproductive performance and progeny survival of farmed species since the nutritional composition of feed provided in aquaculture settings may differ significantly from the natural diet of teleosts [26,27]. In fact, the dietary requirements of broodstock as well as those of larvae remain one of the critical points in aquaculture since they depend on the availability of nutrients and raw materials, and they can imbalance the ratio of inputs to outputs [28].

1.4. Model Species and Predictive Studies

To gain insights into the effects of environmental changes on teleost reproduction, researchers often turn to model species, such as zebrafish (*Danio rerio*) and medaka (*Oryzias latipes*). These species, with their rapid reproduction, transparency in early developmental stages, and genetic tractability, offer valuable tools for studying the impacts of temperature, salinity, and pollutants on teleost reproduction [29]. In addition, zebrafish can represent a useful model organism to conduct preliminary evaluations of innovative dietary formulations. In fact, due to their features, zebrafish can (i) simplify the evaluation of a wide range of alternative ingredients, tested at different inclusion levels, compared to the costly and more time-consuming feeding trials conducted on farmed fish, and (ii) enable a deeper understanding of the molecular mechanisms involved in nutrient metabolism [30,31]. Despite not being native to the Mediterranean, the findings from zebrafish and medaka studies can provide valuable insights into the potential challenges faced by native teleosts. It is noteworthy that studies on model species may oversimplify the complex interactions among different factors, neglecting the intricate ecological relationships that impact reproductive processes.

In this review, we have compiled information about the different factors affecting the reproductive performance of wild and farmed teleost species, focusing on those of the Mediterranean Sea. Moreover, we describe the most outstanding results obtained using model species to shed light on the impact of environmental pollution on teleost reproduction.

2. Reproductive Challenges in Wild Teleost Populations

In the present global context, the significant and irreversible alterations in marine conditions result from the combined impact of three primary stressors: climate change, environmental pollution, and overfishing [32]. The influence of these events on marine ecosystems depends on their close correlation. Indeed, climate change has the potential to alter both the toxicity and fate of marine contaminants as well as the extent of fishing im-pact [8,33,34]. The environmental distribution of persistent organic pollutants (POPs) is closely tied to sea temperature, organic carbon content, and their recycling among environmental reservoirs [34]. Alterations in environmental parameters affect the distribution, species composition, seasonality, and reproduction in both marine and freshwater ecosystems, thereby influencing the availability of fisheries' resources. Consequently, to meet market demand, fishing activities may shift to other targets, intensifying efforts on new species or smaller specimens, often during their vulnerable life stages [33]. However, marine teleosts, which exhibit the most diverse reproductive strategies among vertebrates, may display varied responses when subjected to the same conditions. This complexity makes it challenging to generalize and predict the potential consequences of specific stressors. Therefore, research has focused on investigating aspects of the reproductive cycle shared by different species, regardless of their reproductive strategy, such as the development and growth of gametes and their quality and the preparation for the subsequent reproductive season (e.g., proliferation of germ cells in iteroparous species) [35].

2.1. Climate Change

Several reviews have detailed the impact of climate change on teleost reproductive physiology, focusing mainly on global warming, which has exhibited the most concerning effects [7,11,36–38]. For the record, in temperate teleosts, including those inhabiting the Mediterranean area, the photoperiod has been considered the primary factor influencing the onset of the reproductive cycle [39]. While the photoperiod itself is not directly affected by climate change, the migration of many marine species towards the poles in response to global warming has resulted in significant shifts in their photoperiodic reproductive periods [40]. This phenomenon may lead to anticipated delays and a shorter reproductive season [37].

In this review, we refrained from discussing the intricate molecular mechanisms governing reproduction impacted by variations in environmental temperature or other parameters. This decision has been made to avoid presenting redundant information already extensively covered in the afore-mentioned literature, although we briefly summarized the main effects observed in laboratory trials. The observed results highlighted that temperature fluctuations possess the capability to influence the HPG axis at multiple levels, thereby affecting hormone synthesis, action, and structure [11]. Departure from the optimal thermal range also exerts an impact on gamete development, spawning timing, and egg quality in female specimens [7]. Increasing temperature and hypoxia reduce sperm motility and increase the sperm oxygen consumption, affecting the quality of spermatozoa during fertilization [41]. Hypoxic events also have the potential to influence gonadal steroidogenesis, thereby affecting the availability of cholesterol precursors for steroid production or directly impacting aromatase activity, as reported by Servilli et al. [37]. Notable correlations were observed between changes in salinity or water acidification and a reduction in sperm quality and reproductive output [42,43].

2.2. Environmental Contaminations Move to Next Page

Numerous chemical substances released into the aquatic environment are categorized as persistent organic pollutants (POPs) due to their demonstrated high toxicity, capacity for long-range transport, persistence, and bioaccumulation potential [44]. Within this classification, certain pesticides like dichlorodiphenyltrichloroethane (DDT) are included, while other compounds, such as pharmaceutical products, display POP-like characteristics by persisting in aquatic environments and exhibiting the potential for environmental accumulation [44,45].

Some POPs (polychlorinated biphenyls (PCBs), dichlorodiphenyltrichloroethane (DDT)) express their toxicity by interfering with the endocrine pathway (synthesis, secretion, transport, or elimination) or interacting with receptors changing hormonal responses during fish reproduction and therefore have been named endocrine disruptors (EDCs) [46]. General mechanisms by which contaminants affect the normal reproductive process in fish are described in Section 4.

As a result of extensive anthropization and industrialization, Mediterranean coastal areas are major recipients of pollutants [47]. Indeed, the literature has documented the presence of various pesticides, pharmaceuticals products, and EDCs in sediments, seawater, and fish species within the Mediterranean regions [48–54]. The first study investigating the occurrence of various contaminants, in open surface water, concerned the western Mediterranean Sea [54]. From the 70 total samples collected in 10 different sampling areas, 3 pesticides and 11 pharmaceuticals and personal care products were detected. Other studies investigated the presence of various contaminants in fish from different areas of the Mediterranean Sea, as stated below.

Organochlorine pesticides including DDT and PCBs were detected in different teleost species from different areas of the Adriatic Sea [52,53]. Bisphenol A (BPA), 4-nonylphenol, and 4-t-octylphenol were found in red mullet (*Mullus barbatus*) from the Tyrrhenian Sea [48]. High levels of BPA were also found in the muscle of different fish species in the Cabrera Marine Protected Area (Balearic Islands) [49]. BPA and 4-alkylphenols were detected in sediments and seawater of two coastal areas of the Iberian Peninsula [50]. Different PCBs and organochlorines were detected in European anchovies from the Adriatic Sea [51].

Finally, the presence of heavy metals in the waters and fish species of the Mediterranean Sea has also been a matter of concern for a considerable period, and investigations into their presence persist to the present day, as reported in different recent publications [55–57].

2.3. *Overfishing*

Over the last twenty years in the Mediterranean area, different wild stocks have experienced a decline, while their exploitation levels have increased [58]. Generally, the overall impact of overfishing surpasses, to varying degrees, the direct impact of global warming on species and ecosystem diversity, as well as ecosystem functioning. This trend is likely to intensify in the coming decades [59]. Both the direct and indirect effects of fishing activity on the survival of a stock highly depend on the species' reproductive strategy. Adult males and females and immature and mature specimens exhibit different behaviors during the reproductive season. These behavioral differences may lead to the selective removal of specific subgroups within the population, consequently altering the sex ratio and size distribution. This reduction in potential mate encounters can thereby decrease reproductive output [60]. Independently of the seasonal period, specific fishing techniques (such as bottom trawling and dredging) may alter the spawning habitat of certain demersal and benthic species. An interesting example of the aforementioned discussion is represented by the case study on the red mullet (*Mullus barbatus*) in the Gulf of Castellammare (northwestern Sicily, central Mediterranean). This study reported a significant increase in the spawning stock and recruitment pattern of red mullet after a 14-year ban on trawling in that area [61].

It is important to emphasize that the impacts of fishing activities on the reproduction of wild populations did not only concern large naval fleets; artisanal fisheries can also influence the survival of Mediterranean coastal species [62]. This issue is particularly significant, given that artisanal fisheries represent approximately 85% of the Mediterranean fishing fleet [63].

Among the consequences of fishing overexploitation, fluctuations in life history traits of marine fish, such as changes in the reproductive period and in size or age at first maturity, are the most common [64]. This phenomenon has been observed in species with long life cycles, such as the European hake (*Merluccius merluccius*) in the western Mediterranean Sea, although, in this study, the effects on the long-term dynamics of European hake population have been attributed to both fishing and hydroclimatic variability [65].

This example demonstrates how the impact of fishing on wild stocks depends on their interaction with other factors, including environmental conditions. Indeed, a species already stressed by adverse climate changes or contamination may not be able to thrive under exploitation. Conversely, fishing activity has the potential to disrupt the ability of a wild stock to endure or adapt to climate changes [8,63]. Given the critical nature of the current situation, immediate and effective management is essential to safeguard biodiversity and ensure the sustainability of fisheries' resources for future generations.

2.4. Reproductive Challenges in Mediterranean Area

Due to the variable nature of all the forces influencing the intricate process of fish reproduction, it is evident that neither laboratory trials nor predictive models can comprehensively incorporate all factors impacting the reproductive cycle of wild teleost stocks. Our current level of knowledge is still inadequate for making precise predictions beyond general assumptions, except for the relatively few species that have garnered research interest. The main deficiency lies in the limited understanding of reproductive plasticity among teleost species when adapting over multiple generations, for instance in scenarios characterized by gradual temperature increases [37].

Though it is challenging to identify the predominant stressor driving changes in the reproductive cycle of a species, different populations have experienced alterations in specific reproductive traits in the Mediterranean area. These changes are attributed to variations in environmental conditions associated with climate change, contamination, and high exploitation over the years. An exemplary demonstration of population dynamics influenced by the combination of these three stressors is observed in the case of small pelagic species such as European sardines (*Sardina pilchardus*) and European anchovies (*Engraulis encrasicolus*) [66].

These two small pelagic species stand out as the most significant in terms of abundance, ecological relevance, and economic value, being the principal targets for purse seiners and pelagic trawlers in the Mediterranean Sea [67]. European sardines and European anchovies typically experience substantial fluctuations in their abundance and biomass due to their strong reliance on environmental variability [68]. As a result, a growing focus has been developed on exploring the correlation with ongoing environmental changes to effectively implement conservation programs for these species. The observed decline in the abundance of European anchovy eggs and larvae in the north-west Mediterranean Sea during surveys conducted in the 1980s, 2000s, and 2010s was attributed to the gradual increase in seawater temperature and salinity, paired with a concurrent decrease in chlorophyll a levels [69]. Basilone and collaborators (2021) during a 9-year-long time series obtained within the European Data Collection Framework (DCF; EC 665/2008) provided basic knowledge on the reproductive biology of European sardines in the central Mediterranean Sea [70].

Along the analyzed period, the size at first maturity ranged from 108 to 124 mm (total length) for females and from 102 to 122 mm for males, with significant differences among years linked to sardines' flexibility to environmental variability [70].

An evident decline in stock population with possible consequences on the reproductive parameters was observed for sardines from the Adriatic population. Indeed, a general reduction in the mean size (15.5 cm the highest size class observed) and age (greater than 2 years old but less than 3 years) of the population was observed, together with a predominance of male specimens in the population [71]. These data differed from those obtained in previous surveys, where sardines' total length was observed up to 17.0 cm and 21.0 cm [72–74]. The situation is likely to impact the reproductive success and overall stability of the species, particularly due to the reduction in female specimens. A higher number of females is considered advantageous for the population, as it correlates with a greater reproductive rate [75].

An alarming effect of endocrine disruptors was observed in the European anchovies from the western Adriatic Sea, showing the effects of overexposure to contaminants. In male specimens where PCBs and organochlorines were detected, transcription of vitellogenin, vitellogenin receptors, and genes encoding zona radiata proteins was observed in gonad and liver samples. Additionally, histological analysis identified intersex characteristics in 13% of the analysed testes [51].

A similar scenario was observed in two species of top predators inhabiting the Mediterranean Sea: bluefin tuna (*Thunnus thynnus*) and swordfish (*Xiphias gladius*), for which high induction of vitellogenin and zona radiata proteins was detected in the serum of male specimens of these species [76,77]. In *Xiphias gladius*, evidence of intersex in male gonads was also observed [78,79].

Although presently there are limited records of reproductive alterations in teleost that can be unequivocally linked to specific factors, the consequences of environmental changes resulting from climate change, contamination, and overfishing may manifest across various species in the years to come. Therefore, anticipating the future threats to the marine ecosystem is expected to pose a significant challenge.

3. Impact of Broodstock Nutrition on Reproduction

Broodstock nutrition represents the main factor affecting reproductive performance in farmed fish, with further consequences on the development and survival of embryos and larvae [80,81]. During oocyte maturation, maternal reserves, built up through proper nutrition, are invested in the yolk sac formation that, in turn, will constitute the first energy and nutrient substrates for the developing embryo [81]. Consequently, the early nutritional environment of fish larvae (from fertilization to yolk absorption) is fixed before fertilization and will sustain the newly hatched fish until the occurrence of exogenous feeding [82]. For that reason, diet formulation and feeding frequency must be considered as a crucial point in farmed broodstock nutrition. In this regard, particular emphasis must be given to the essential fatty acids fraction of aquafeeds, especially polyunsaturated fatty acids (PUFAs), being essential to sustain highly energy-demanding processes like reproduction, proper gamete maturation, and correct larval development [83,84]. In fact, long-chain PUFAs, such as eicosapentaenoic acid (20:5n-3; EPA), docosahexaenoic acid (22:6n-3; DHA), and arachidonic acid (20:4n-6; ARA) are able to positively modulate the (i) gamete quality of male fish, including sperm velocity and motility; (ii) reproductive processes of female fish, including oocyte maturation, ovulation, fertility, and spawning rate; and (iii) offspring development, including hatching success and larval quality [85–87].

3.1. *Plant-Derived Ingredients*

Traditionally, the main source of long-chain PUFAs in dietary formulations intended for farmed fish is constituted by marine-derived ingredients like fish meal (FM) and, particularly, fish oil (FO) [83,88]. These raw materials represent the gold standard in the aquafeed industry but, in the last decades, their use has been severely restricted due to economic and environmental concerns [89]. In this context, the partial or total replacement of marine-derived ingredients in aquafeeds with vegetable meal (VM) and/or oil (VO) has gained increasing attention, mainly due to their higher abundance and lower price compared to FM and FO [90,91]. Particularly, VOs have been reported to be good candidates since they are characterized by increased sustainable production and price stability [92,93]. However, VO sources are often scarce in long-chain PUFAs (mainly EPA and DHA), despite showing a significant amount of shorter-chain precursors like α-linolenic acid (18:3n-3; ALA) and linoleic acid (18:2n-6; LA). Despite the fact that FM and FO inclusion rates in aquafeeds have shown a clear downward trend over the last decades [94], their total replacement by plant sources is still not optimal, with potential consequences on fish reproductive performances.

In this regard, due to the suboptimal long-chain PUFA profile of plant-derived ingredients, several studies have been conducted on zebrafish, as a widely recognized model for aquaculture nutrition [95], and on both freshwater and marine fish of commercial interest to investigate how the reproductive performance can be affected by replacing marine-derived ingredients with plant-derived ones. In this context, promising results were obtained in rainbow trout (*Oncorhynchus mykiss*) fed diets in which FO was totally replaced by linseed oil (LO) or sesame oil (SO), over a short-term period [96]. Particularly in rainbow trout males, semen volume, pH, sperm motility, sperm density, and kinematic parameters were not affected after a 7-month feeding trial, leading to the production of functional gametes able to guarantee proper fertilization and consequent eyeing and hatching rates of embryos [96]. Similarly, no differences in egg indices, fertilization, embryos' survival, and hatching rates were evident in rainbow trout fed diets characterized by the total replacement of dietary FO with LO or SO for about 5 months prior to spawning [97]. In addition, the contents of ARA, EPA, and DHA of eggs from rainbow trout females fed VO-based diets were greater than the dietary concentrations, highlighting the efficient ALA and LA bioconversion to long-chain PUFA in this species to alleviate the dietary deficiencies [97]. The reproductive performance and fatty acid dynamics of embryos of rainbow trout were also evaluated in response to FO's replacement with a mixture of VOs (LO, canola, olive, sunflower, corn, and coconut oils) in plant protein-rich broodstock diets, during a 3-month feeding trial prior to spawning [98]. Particularly, the use of an FO:VO ratio of 25:75 led to positive results in terms of fertilization rate, survival at eyed-eggs stage, eggs weight and diameter, and hatching rate; however, the complete replacement of FO with VO blends resulted in a pronounced reduction in the long-chain PUFA content of eggs which may have consequently affected their quality parameters [98]. Longer-term studies (3 years) revealed that rainbow trout reared entirely on plant-based diets, totally deprived of marine raw materials, can successfully produce eggs in which neo-synthesized n-3 long-chain PUFAs are accumulated, leading to viable offspring, despite a reduction in egg size [99]. In addition, complete FM replacement with cottonseed over a 3-year trial did not affect the rainbow trout males' reproductive performance in terms of steroid hormones, the sperm concentration, and motility but resulted in reduced fertility in females [100].

An additional drawback related to the use of high inclusion levels of vegetable ingredients is the presence of toxic metabolites like zearalenone, a mycotoxin produced by several species of *Fusarium* moulds that are commonly found in plant products used as raw materials for aquafeed production [101]. In this regard, Woźny et al. [102] observed that feed-borne exposure to zearalenone negatively affected gonadal development in rainbow trout, with occurrence of intersex fish and sex-reversed (feminized) males, and offspring survival. These results bring attention to the potential contamination of plant-derived ingredients that can be detrimental to different phases of the reproductive cycle, from gonad maturation to reproductive events.

3.2. Nutritional Programming

Freshwater fish are generally able to elongate shorter-chain precursors to long-chain PUFAs, even if supplying them preformed is more advantageous due to their pivotal role in several physiological processes, including reproduction [103]. Conversely, marine fish have had no evolutionary pressure to retain the ability to endogenously produce lc-PUFAs and thus they must be provided through the diet [104]. In this context, an interesting strategy to optimize the use of plant-derived ingredients in marine fish is represented by the production of specimens characterized by a higher capacity of long-chain PUFAs.

Environmental factors in the periconceptional and early developmental stages, including nutrition, can deeply influence the offspring metabolic pathways to prepare them for nutritional challenges through the nutritional programming [105]. In this context, research has been focused on (i) challenge nutrition during very early life history that can improve the ability of fish to effectively rely on suboptimal diets later in life [106–110] and (ii) the maternal role in progeny programming, producing offspring better adapted to being fed suboptimal diets and expanding the existent knowledge of how broodstock nutrition can deeply influence the fish's reproductive performance, as well as the embryo and larval development.

The effect of a broodstock diet on nutritionally programmed progeny has been widely explored in gilthead seabream (*Sparus aurata*), in which the optimal development of eggs and, then, embryos relies on a proper provision of sufficient quantities of essential nutrients, particularly long-chain PUFA, due to its scarce bioconversion ability [111]. Gilthead seabream broodstock challenged with a 60% replacement of FO by LO resulted in offspring with a greater ability to cope with low FM and FO diets and this nutritional programming persisted with a modulation of selected genes in the offspring, which regulated energy metabolism in the liver, even after 16 months, when fish reached the reproductive age [112,113]. However, higher LO levels (80% of FO replacement) markedly reduced spawning quality, larval survival, and larval and juvenile growth [112,113] as a consequence of the deleterious effects of a too drastic n-3 long-chain PUFA reduction in broodstock diets [114]. Similarly, Xu et al. [115] highlighted that, in gilthead seabream broodstock, the provision of a diet in which FM (15%) was replaced by VM, despite not altering the fish's reproductive performance, negatively affected the offsprings growth and their ability to use 18C-PUFA through a downregulation of the fatty acyl desaturase 2 (*fads2*). Conversely, the same FM-VM ratio combined with an FO replacement with a mixture of VO (LO, palm oil, and rapeseed oil) in broodstock diets resulted in a reduced female fecundity and egg EPA and DHA contents. However, an upregulation of hepatic *fads2* and elongase of long-chain PUFAs (*elovl6*) was detected in juvenile progenies, highlighting an enhanced biosynthesis of these fatty acids in the offspring related to the FO replacement with VO [115].

In light of these results, optimizing broodstock diets to obtain a nutritionally programmed progeny without compromising spawning quality represents a crucial point. In this regard, a recent study proposed to challenge gilthead seabream broodstock for three months during the spawning season with a diet characterized by mixture of 20% FO and 80% rapeseed oil, after being fed for three months before the spawning season with a low FM (5%) and FO (3%) diet, high in LA and ALA and low in long-chain PUFAs to induce a *fads2* upregulation [116]. Interestingly, broodstock with high blood levels of fads2 showed better reproductive performance, in terms of fecundity and sperm and egg quality. This study consequently identified a feasible percentage reduction in long-chain PUFAs (ARA, EPA and DHA reduced to 0.4, 6.6 and 8.4% of total fatty acids) for a diet designed to induce nutritional programming in the offspring that, on the other hand, preserved reproductive performance in broodstock [116].

3.3. *Alternative Ingredients*

The aquaculture sector has experienced the introduction of functional ingredients that are able to replace unsustainable marine-derived raw materials or fill the gaps originated from the introduction of suboptimal plant-derived ingredients, even at low concentrations, in fish diets [117]. These solutions can represent an alternative dietary-derived solution to positively modulate the overall welfare and reproductive performance of fish fed diets with reduced levels of marine-derived ingredients. Particularly, potential solutions of both animal and microbial origin have been tested in order to evaluate the effects of alternative ingredients on fish reproduction.

Considering the ingredients of animal origin, the use of crustacean biomass as a suitable and cost-effective choice to improve the fish's reproductive performance was tested [118]. Particularly, it has been evidenced that supplementing *Artemia salina* biomass (2.5%) or *Palaemon elegans* (7.5%) in a plant-based diets low in FM (17%) for European seabass (*Dicentrarchus labrax*) broodstock resulted in improved egg quality (egg diameter and oil droplet size), fecundity, viability, fertilization, and hatching percentages, with a positive modulation of hormones involved in reproduction [118]. Furthermore, partial FM replacement with *Rapana venosa* meal up to 50 g/kg in diets for rainbow trout broodstock has been shown to not significantly affect fish reproductive parameters and did not alter the morphological architecture of different target organs, including the intestine, liver, and muscle [119], usually analysed for determining the overall welfare of farmed fish in response to alternative ingredients [120,121].

Dried microbial biomasses from protist, cyanobacteria, and microalgae have gained interest for their content in proteins, lipids (particularly PUFAs), vitamins, and pigments with antioxidant properties and for their consequent beneficial roles on fish welfare, even at low dietary concentrations [122,123]. Considering microalgae, a recent study on zebrafish evidenced that dietary FM replacement with *Chlorella* sp. meal was able to enhance the fish's reproductive performance [124]. Particularly, the highest inclusions of *Chlorella* sp. meal tested (40 and 50 g/kg) ensured the highest egg production, hatching rate, and larval survival at 6 days post-fertilization (dpf) [124]. The cyanobacterium spirulina (*Arthrospira platensis*) has gained notable attention for aquafeed formulation due to its antioxidant properties and the abundance of ALA and LA, which are precursors of arachidonic acid, fundamental for prostaglandin synthesis involved in the modulation of steroidogenesis, oocyte maturation, and ovulation [125–127]. However, the positive effect of spirulina on fish reproduction has been widely explored only in ornamental fish, including the three-spot gourami (*Trichopodus trichopterus*) [128] and the yellowtail cichlid (*Pseudotropheus acei*) [129]. Considering species of commercial interest, Lu and Takeuchi [130] investigated the reproductive performance of Nile tilapia (*Oreochromis niloticus*) fed solely on raw spirulina, concluding that this dietary solution was able to maintain normal reproduction throughout three generations.

Finally, in light of their fatty acids profile, microbial biomass from certain protists species has gained attention in the aquafeed production sector [131]. In this regard, thraustochytrids are protists that have lost photosynthesis, and they are able to store large amounts of long-chain PUFAs in heterotrophic conditions [132]. Particularly, *Schizochytrium* sp. is characterized by a high lipid content (55–75% in dry matter) and up to 49% DHA of total lipids [133]. Interestingly, the inclusion of *Schizochytrium* sp. oil (6.9%) in a plant-based diet intended for rainbow trout broodstock resulted in reproductive performance and egg quality comparable to those observed in fish fed a conventional commercial FM- and FO-rich feed and led to a significant increase in fry survival [134]. Additionally, the 4-month-old offspring obtained by the broodstock fed the plant-based diet enriched with *Schizochytrium* sp. oil were characterized by a notable growth improvement and a positive modulation of long-chain PUFA biosynthesis and β-oxidation pathways when challenged with the same diet over a 1-month feeding trial [134]. Interestingly, the *Schizochytrium* sp. was also proposed as a solution to improve the nutritional profile of black soldier fly (*Hermetia illucens*) prepupae meal (HM), subsequently used as feed ingredients in diets for zebrafish (*Danio rerio*) [135,136].

The full-fat HM represents a widely explored raw material for diet formulations in aquaculture, characterized by a proper protein content but also by an unbalanced fatty acid profile, high in saturated fatty acid (SFA) and low in long-chain PUFA [137]. This fatty acid profile was previously addressed as the main factor affecting the reproductive performance of female zebrafish fed, over their whole life cycle, 50% of an FM replacement with not-enriched full-fat HM, resulting in a reduction in spawning rate and in an increased number of previtellogenic and atretic oocytes [138]. However, in that case, no alteration in the hatching rate was evident, highlighting the effort that female fish put in the reproductive event, choosing quality rather than quantity of spawned eggs [138]. Interestingly, using *Schizochytrium*-enriched HM, 50% inclusion with respect to FM did not impair zebrafish female reproduction, ensuring performances similar to those observed in fish fed a commercial FM- and FO-based diet; an adverse effect on spawning, hatching, and oocyte classes' frequency was only evident at 75% or total FM replacement [136]. Furthermore, the long-term provision of the HM-based diets (characterized by 25, 50, 75, and 100% of FM replacement with *Schizochytrium*-enriched HM) during the whole life cycle of zebrafish (from larvae to adult) resulted in offspring better adapted to cope with low long-chain PUFA dietary formulations that did not show adverse effects on their growth and welfare [135].

In general, the main drawbacks regarding microbial biomasses as ingredients for fish diets are represented by the high production costs that still limit their use in the aquaculture sector [139,140] and by the presence, in certain species, of a thick cell wall that can impair nutrient absorption, especially in carnivorous fish [141]. However, the use of a combination of different microalgal species may represent a solution to enable higher dietary inclusion levels respect to those achievable using a single species [142].

3.4. Functional Feed Additives

Several studies have been conducted to improve the broodstock diet quality by enrichment with functional additives. Al-Feky et al. [143] evaluated the effects of different supplementation levels of taurine (0, 5, 10, and 15 g/kg) on the reproductive performance of Nile tilapia broodstock fed soybean meal-based diets. Spawning performances, including spawning frequencies, total number of spawning per tank, number of spawning per female, and absolute fecundity, were all significantly improved with increasing levels of dietary taurine up to 10 g/kg. Furthermore, the eggs produced from broodstock fed 10 g/kg taurine exhibited significantly higher hatchability and required a shorter time for hatching and for yolk sac absorption. Higher larval weight was also evident in response to this taurine dietary level [143].

The supplementation of vitamins, which cannot be synthesized by fish, may result in positive effects on gamete quality [144]. Vitamin C (ascorbic acid) and vitamin D have been the most used vitamins for improving broodstock diets, and positive effects on sperm motility have been reported in several fish species, including Nile tilapia [145], African catfish (*Clarias gariepinus*) [146], and Senegalese sole (*Solea senegalensis*) [86]. In addition, vitamin C has been shown to induce positive effects on sperm features in lebranche mullet (*Mugil liza*), in which dietary supplementation between 107 and 216 mg/kg optimizes the spermatic quality in this fish species, improving motility time and rate as well as spermatic density [147].

Finally, the dietary supplementation with probiotics, especially probiotic bacterial strains, led to a significant improvement in reproductive performance in farmed fish [148]. Particularly, the provision of 4×10^9 CFU/kg diet of probiotic Bio-Aqua® twice a day for 8 weeks before spawning season in rainbow trout resulted in an improvement in egg diameter, relative fecundities, fertilization and hatching rates, and eyed-egg and alevin survival rates, as well as in an anticipation of eyeing, hatching, and yolk sac absorbing stages [149]. Due to an increase in egg production and egg size in probiotic-fed treatments, this feed additive may represent a suitable ingredient to increase high-quality egg production in rainbow trout breeders [149].

3.5. Feed Restriction Practices

Given the importance of well-formulated diets for broodstock fish that ensure proper reproductive performance, it should be considered that the largest cost for modern aquaculture is represented by aquafeeds, especially those formulated for broodstock (representing between 60 and 80% of production cost), demonstrating the need for research focused on feed provision management [150]. In this regard, practices like feed restriction are of economic interest for fish farmers. However, restriction regimens superior or equal to half food ration can severely affect spawning success [151], as well as egg size and quality [152], and can lead to an arrest of reproductive development [153]. Interestingly, Cardona et al. [154] demonstrated that a moderate, yet significant, feeding restriction of 20% in rainbow trout, applied during the last five months before the second reproduction, did not affect egg production and egg quality, providing the first evidence of the acceptable level of feed restriction. This can represent a compromise between reducing the feed-associated costs and the maintenance of proper broodstock trout reproductive performance.

4. The Use of Model Species to Investigate Reproductive Challenges in Teleosts

Aquatic species, including teleosts, are especially susceptible to the toxic effects of pollutants due to their constant exposure to low concentrations of contaminants that continually enter the aquatic environment. In this regard, numerous studies have been conducted to investigate the reproductive toxicity of anthropogenic xenobiotics widely present in the environment, mainly using teleost model species, primarily *Danio rerio* for freshwater fish and, to a lesser extent, the Japanese medaka (*Oryzias melastigma* and *Oryzias latipes*) for saltwater fish. Model species, which are often chosen for their biological relevance, ease of maintenance in laboratory settings, and well-documented genetic and physiological profiles, provide valuable insights into the mechanisms by which pollutants impact fish health and ecosystems [155]. These species allow us to extrapolate findings to other, often more complex, wild and farmed fish species [95]. By examining the responses of model organisms to various contaminants, it is possible to identify potential risks, understand modes of action, and predict ecological consequences. This research is crucial for informing regulatory frameworks, improving environmental monitoring, and developing mitigation strategies to protect both wild fish populations and aquaculture operations from the adverse effects of pollution. Among these contaminants, Endocrine Disruptive Chemicals (EDCs) are those with the most detrimental effects on reproduction, due to their capacity to interfere with the hypothalamus–pituitary–gonad (HPG) axis by affecting endogenous hormone synthesis, secretion, transport, binding, or elimination [156]. The most prevalent EDCs in the environment are classified as plasticizers, primarily represented by bisphenols (i.e., BPA, BPS, BPF, BPC, BPB, and BPAF) and phthalates (i.e., DEHP, DiNP, DBP, DiDP, and DPHP). Per- and polyfluoroalkyl substances (PFASs) are another significant category of EDCs that are almost ubiquitous in the environment, with the main compounds being Perfluorooctanoic acid (PFOA) and Perfluorooctanesulfonic acid (PFOS). A vast body of the literature exists concerning the reproductive toxicity exerted by these pollutants on gametogenesis progression, gamete quality, and reproductive efficiency. Over the years, the findings have been summarized in numerous reviews [157–163].

In addition to plasticizers and PFAS, various other environmental contaminants can act as EDCs, leading to reproductive toxicity due to their widespread presence in the environment. Among these contaminants, heavy metals constitute a broad class of metallic elements characterized by high density and atomic weights, such as copper (Cu), mercury (Mg), cadmium (Cd), arsenic (Ar), zinc (Zn), iron (Fe), and chromium (Cr). These metals are extensively utilized across various industries due to their unique properties, finding applications in manufacturing, construction, electronics, and even healthcare [164]. In recent years, widespread industrialization, urbanization, and agricultural practices have resulted in increased heavy metal pollution across various environmental compartments, including soil, water, and air. Once released into the environment, heavy metals can endure for extended periods, causing chronic exposure and having adverse effects on various physiological processes, including reproduction.

Furthermore, the inadvertent release of pharmaceutical compounds into aquatic environments is also a source of concern due to their possible toxic effects on aquatic wildlife's reproductive activities. These chemicals include a wide range of medications, comprising antibiotics and antidepressants, and are able to enter the waterways through various routes, such as wastewater discharge, improper disposal, and excretion by humans and animals [165]. Fish can ultimately be exposed to pharmaceutical compounds directly from the water or through the food web, also leading to the bioaccumulation of these chemicals in their tissues [166,167]. Moreover, another cause of pollution in the aquatic ecosystem is represented nowadays by the intensive use of pesticides because of the widespread use of this class of compounds, primarily in agriculture, to control pests or unwanted organisms, such as invasive weeds, insects, and fungi. This results in their entry into waterways through runoff from fields, accidental spills, or improper disposal [168].

This section has focused on summarizing the information obtained regarding the reproductive toxicity resulting from exposure to heavy metals, commercial pharmaceuticals, and pesticides in two widely used teleost models, the zebrafish and the Japanese medaka. The reported effects of these compounds on reproductive success aim to illuminate their molecular mechanisms of action and potential implications for future generations.

4.1. Heavy Metals

Among heavy metals, copper, zinc, cadmium, mercury, iron, and chromium are the most used in different manufacturing activities, resulting in a wide presence in the environment. For this reason, many studies have evaluated their impact on fish reproduction.

Considering Cu, the findings highlight the endocrine disruptive ability of this xenobiotic by altering fish steroidogenesis and gonadal architecture in both females and males, despite not showing any effect on fertility and fecundity. In the form of Copper (II) sulfate ($CuSO_4$), this heavy metal finds applications in agriculture as a fungicide and fertilizer. Therefore, its runoff from agricultural fields could lead to its entrance in aquatic environments, posing a severe risk for teleost health. Recently, the effects of $CuSO_4$ on zebrafish reproduction were investigated, finding that 40 μg/L, a dose in the range of environmental concentrations, was able to alter the expression of master genes involved in reproduction along the HPG axis during chronic exposure (30 days) [169]. The alteration of genes involved in the HPG axis and the steroidogenesis finally resulted in a negative impact on the gonadal architecture, as shown by the decreased number of mature and secondary follicles in females and the number of spermatozoa in males. These results were also supported by the analysis of gonadosomatic index (GSI) which decreased in both sexes [169], suggesting the strong reproductive toxicity of $CuSO_4$. Similarly to $CuSO_4$, chronic exposure (20 days) to 50 and 100 μg/L Cu oxide nanoparticles, applied to industrial chemical catalysis [170], was observed to alter the mRNA levels of genes belonging to the HPG axis and encoding steroidogenic enzyme and hormone receptors in a sex-specific fashion, thus resulting in alterations in steroid hormone levels without exerting a sex-specific effect and, therefore, in both testicular and ovarian architecture alterations [171]. Despite the clear reproductive toxicity evidenced by both $CuSO_4$ and Cu oxide nanoparticles, their impact on the reproductive capacity of the fish, in terms of fertility and fecundity, was not explored. In this regard, to the best of our knowledge, only one study conducted in 2007 aimed to investigate if Cu food administration could affect zebrafish reproduction. Exposure to 100, 500, or 1000 μg Cu per g food in the form of $CuSO_4$ for 220 days did not change either the number of eggs spawned or fertilization of F0. Likewise, the F1 embryo's malformation up to 36 h post fertilization was not affected by Cu parental exposure [172].

Other than Cu, chemicals containing Zn are able to affect reproduction mainly by the induction of oxidative stress at the gonadal level, finally resulting in increased apoptosis levels. Considering this, zinc pyrithione (ZPT), with applications in the industry as algaecides and fungicides, has been used for this purpose in medical contexts, paint formulations, and textile manufacturing. Male zebrafish exposed to 0.15 and 0.30 μM concentrations of ZPT showed reduced activity of CAT, T-SOD, and GSH-PX and an increase in Malondialdehyde [173]. Concomitantly, TUNEL-positive cells and caspase 3 activity were increased. Collectively, these alterations led to a dose-dependent reduction in sperm [173]. This toxicity was also confirmed by the identification of 409 DEGs associated with testicular injury through RNA-seq and the analysis of steroid hormone levels also suggested the endocrine disruptive capacity of ZPT [173]. In the Japanese medaka, the exposure to 1, 5, and 10 μg/mL of zinc oxide nanoparticles (nZnOs) and 1, 5, 10, and 20 μg/mL $ZnSO_4$ for 1 week affected both reproductive behavior and steroidogenesis [174]. On the other hand, in males, nZnO was able to induce sperm ROS, causing sperm quality alterations, while in females the nZnO-induced ROS increase led to follicular growth arrest and atresia increase, finally leading to a subsequent fertility decrease caused by this contaminant in both sexes [174]. Finally, in order to evaluate the effects of Zn on sex differentiation and development, exposure of zebrafish to 650 μg/L of Zn during the first five days post fertilization (dpf) resulted in an increase in female to male ratio and hatching delay in the

F0 and, interestingly, also in the progeny F1 [175]. Considering this, Zn toxicity could last after recovery from the exposure and could also affect the next generations.

Cd is another heavy metal that is present in a wide range of environments. Exposure to Cd and Cd-containing molecules was demonstrated to affect fertility and embryo development. The toxicity of Cd on female reproduction was evaluated in *Danio rerio* by exposing the specimens to 10 μg/L Cd continuously, 20 μg/L Cd for 1 day every 2 days, or 30 μg/L Cd for 1 day every 3 days for a total of 48 days [176]. Exposure to the first two concentrations was able to decrease the percentage of spawned eggs, possibly as a consequence of the decreased plasma levels of VTG and E2 and the perturbation of mRNA levels of brain–liver–gonadal axis genes. A reduction in GSI and hatching rate, coupled with increased embryo mortality, was instead evident in all the Cd concentrations tested [176]. Chronic exposure (30 days) to 1 μmol/L of cadmium chloride ($CdCl_2$), an industry dye, also disrupted zebrafish female reproduction [177] by reducing the number of mature oocytes and causing the induction of developmental malformations in the offspring [177]. Moreover, deep-sequencing analysis shed light on the pathways underlying this reproductive toxicity, mainly affecting the estrogen metabolism, meiosis, and vitellogenin synthesis of oocytes [177]. In addition to the reproductive toxicity of Cd on female zebrafish, another study demonstrated that maternal exposure to 8.9, 17.8, and 35.6 μM of Cd^{2+} affected the pair-spawning success rate at all the concentrations tested, with the highest one also reducing the hatching percentage at 36 hpf [178]. Considering the effects of Cd on male fish reproduction, the exposure of zebrafish spermatozoa to Cd in vitro (0.5, 5, and 10 μg/L of Cd), demonstrated the direct effect of this pollutant on sperm motility as well as plasma membrane and DNA integrity [179]. Despite the interesting results found by these authors, no test was conducted on the capacity of the treated spermatozoa to fecundate the oocyte, letting us only hypothesize the possible implications of Cd exposure on the zebrafish fertility rate.

The presence of Hg in the aquatic environment represents another source of severe concern. For this reason, the impact of this heavy metal on female and male reproduction was evaluated in zebrafish, finding its ability to alter reproduction by increasing ROS levels and interfering with the HPG axis, also leading to detrimental effects on progeny. Chronic exposure (30 days) of zebrafish adults to 15 and 30 μg/L mercuric chloride ($HgCl_2$) increased the expression of genes encoding the antioxidant enzymes in both the ovaries and testes and higher MDA and GSH testicular levels. These results, together with changes in the HPG axis and male steroidogenesis, were correlated with the presence of atretic oocytes in females and a decrease in the number of spermatozoa in males [180]. The effects of zebrafish embryonic exposure to 0.6, 3, or 15 μg/L Hg^{2+} until 5 dpf caused a detrimental effect on the expression of genes related to hormone and neuropeptides synthesis, leading to reproductive behavior alteration and subsequent consequences on spawning and fertilization rate [181]. Interestingly, the aforementioned effects on reproduction and reproductive behavior were not evident in the untreated F1 generation [181], indicating that the toxic effects were not transmitted to the progeny. Hg^{2+} exposure decreased zebrafish fecundity and delayed gonadal development, resulting in the GSI decrease in F0 [182], a condition that persisted, to a lesser extent, in the F1 females and was absent in F1 males and in both sexes at F2 generation. Although these results may not represent a risk for future generations, the inter- and transgenerational effects of mercury could be influenced by the form in which this compound is applied and by the window of exposure. In this regard, 24 h early-life exposure to methylmercury (MeHg) at different concentrations (1, 3, 10, 30, and 100 nM) was observed to trigger sperm epimutations related to neurotoxicity at 30 nM in the F2 generation, even though the F0 was not affected [183].

Japanese medaka was used as model to determine the effects of two other heavy metals, Fe and Cr. Considering Fe, the effects of 21-day exposure to 5 and 20 mg/L of Fe^{2+} or iron oxide nanoparticles (nFe_3O_4) were assessed on *Oryzias melastigma* reproduction, finding that 20 mg/L of nFe_3O_4 decreased the fecundity and increased the occurrence of abnormal immature oocytes inside the ovaries, a mechanism that could be due to the contaminant inhibition of antioxidant SOD and GPX activities both in the ovaries and the brains of the specimens exposed [184]. Regarding Cr, the reproductive toxicity induced by 4 mg/L of Cr (VI) (administered as $K_2Cr_2O_7$) was investigated in Japanese medaka exposed for 3 months [185]. The results evidenced a decrease in male GSI, whereas in females, a reduction in spawned eggs and fertilization rate was observed [185], correlating well with an increased GSH/GSSG ratio in the ovaries and a reduction in both CAT and SOD levels in the testes [185].

4.2. Pharmaceuticals Compounds

A wide range of medications can enter the aquatic environment after human consumption and, among them, psychotropic molecules, such as antiepileptics, antidepressants, and antipsychotics are widely represented because of their chemical properties, which make their removal through wastewater treatment plants difficult. As a consequence, their presence is commonly found in different fish tissues and they were observed to be able to cause reproductive and developmental toxicity in zebrafish, as recently reviewed [186]. To the best of our knowledge, till date, only one study evaluated the reproductive alteration induced by psychoactive drugs on *Oryzias melastigma*. Fluoxetine, a selective serotonin reuptake inhibitor widely prescribed as an antidepressant, is highly environmentally present because of human consumption. Chronic exposure to environmentally relevant concentrations of Fluoxetine (0.1, 0.5, 1, and 5 μg/L) did not alter either egg production, fertilization, and the spawning rate or the hatchability of fertilized eggs, despite increased occurrence of developmental abnormalities [187]. Although parental GSI, liver vitellogenin content, and gonadal steroidogenesis were again unaffected, an effect on steroidogenesis was evidenced by the increase in E2 plasma levels by the lowest concentration, highlighting the disruptive chemical ability of this drug [187].

Analgesics represent another class of drugs abundantly found in water due to the anthropogenic action. Chronic exposure (28 days) to acetaminophen (ACE) at environmentally relevant concentrations (0.5, 5, and 50 μg/L) in zebrafish altered gonadal steroidogenesis in both sexes, increased vitellogenic follicles' dimension and number, and as follicular atresia in females, while in males, ACE triggered testis cellular apoptosis [188]. The reproductive toxicity of another anti-inflammatory drug, Ibuprofen, was evaluated by chronically exposing (36 days) zebrafish to 0.5, 2.4, 11.5, and 55.2 μg/L, finding no effects on F0 fertility and fecundity, as well as F1 generation hatching success [189]. On the other hand, Ibuprofen exposure (1, 10, and 100 μg/L) for 6 weeks was able to increase the number of eggs of *Oryzias melastigma* per reproductive event, despite decreasing the number of spawning events per week [190]. Finally, Diclofenac, another drug highly prescribed for its anti-inflammatory activity, led to a reduction in fertility and fecundity in medaka fish exposed to 37 and 78 μg/L for 14 days [191]. This effect was gradually rescued by maintaining the fish under standard conditions for other 14 days, suggesting that the effects exerted on reproduction could be reversible to a certain extent. Interestingly, another study showed that even though parental fertility was unaffected by chronic exposure (3 months) to 10 mg/L of Diclofenac, this contaminant still led to a complete impairment of F1 medaka embryo hatchability [192], indicating the transmission of this anti-inflammatory toxicity to the progeny.

Antibiotics and antibacterial compounds are widely, and sometimes inappropriately, prescribed and used nowadays to fight human and animal infections. Their extensive use over the years has led to the accumulation of this class of compounds in the environment, resulting in toxicity at different levels to non-target organisms. Over the years, evidence accumulates regarding their ability to impair fish reproduction by acting as EDCs and inducing oxidative stress. In this regard, long-term exposure (from 2 hpf to 120 dpf) to environmentally relevant concentrations (1 and 5 μg/L) of Sulfamethoxazole (SMZ), an antibiotic of clinical and animal husbandry, was observed to act on the G protein-coupled receptor 54, which in turn stimulates the synthesis and release of GnRH in the brain. This outcome led to an increase in FSH, LH, and E2 gonadal levels, thus resulting in an increase in adult mature oocyte proportions [193]. SMZ and norfloxacin (NOR), a fluoroquinolone, both singularly or in combination (2, 20, and 200 μg/L) were able to diminish zebrafish's daily egg production and increase embryo mortality and malformations at the highest dose [194]. Another antibiotic, Ciprofloxacin (CIP), was observed to act as an endocrine disruptor in zebrafish males exposed to 15.6 μg/mL for 21 days by affecting male steroidogenesis and altering the expression of testis gonadotropin receptors, genes related to the endocrine system, signal transduction, and membrane transport [195]. Moreover, zebrafish's exposure to 6.25, 12.5, and 25 mg/L to β-diketone antibiotics (DKAs) from 6 hpf to 144 hpf [196] showed a disruption in steroidogenesis in a dose-dependent fashion once these embryos reached adulthood and triggered gonadal ultrastructural damage in both sexes, indicated by TEM microscopy [196]. A reduction in egg production, increase in unfertilized eggs, and decrease in hatchability was also evident and, collectively, these alterations also led to transcriptome alteration in F1, highlighting that parental exposure could also result in detrimental effects for future generations [196]. The antibacterial and antifungal chemical Triclosan (TCS) is widely present in the environment due to its extensive use in personal care products. Its exposure at environmentally relevant concentrations (0.4, 4, or 40 μg/L) reduced the fertility and fecundity of F0 zebrafish, an effect that was no longer observed in the F1 generation [197]. Recently, in another study, zebrafish's exposure to TCS at 2, 20, and 200 μg/L for 150 days demonstrated that the highest dose reduced the male's fertilization capacity, while all the concentrations reduced embryo hatchability when mating treated males with untreated females [198]. Noteworthy, both female and male specimens also presented a sex-specific alteration of steroidogenesis and VTG whole-body levels, which could be link to the alteration of the testicular structure disruption and the reduction in mature spermatozoa in males and to the increased number of immature oocytes in female [198]. Finally, the long-term (120 days) effects of Cefadroxil (1, 7.8, 84.8, 718.9, and 8883.1 μg/L) and Cefradine (1, 7.1, 73.9, 724.6, and 7758.5 μg/L) on Japanese medaka reproduction were evaluated, showing that both contaminants decreased the number of eggs produced per reproductive pair, also evidencing their ability to affect steroidogenesis in a sex-specific fashion [199].

Nowadays, the link that interconnects the gut microbiota and reproduction is well known, highlighting that intestinal dysbiosis is able to negatively affect reproduction [200–203]. The results obtained regarding antibacterial and antibiotics exposure to zebrafish and medaka generally agree on the negative effects that these chemicals pose on reproduction. Despite this, since these chemicals could deeply affect gut microbiota composition, no evidence has been reported until now on the possible effects that these compounds could exert at the intestinal level and, consequently, at the reproductive level. Further investigations are needed to fill this gap of knowledge, building a comprehensive picture regarding this issue.

4.3. Pesticides

During the last decade, the effects of different classes of pesticides have been evaluated on the reproduction activity and early development of *Danio rerio*, demonstrating their negative effects on reproductive physiology at environmentally relevant concentrations. The results providing information about the pesticides' implications on reproduction and development in zebrafish, including those obtained through omics approaches, were recently reviewed [163], whereas those obtained using Japanese medaka as an experimental model will be summarized in this section.

Atrazine is a well-known herbicide widely present in the environment due to its massive applications in agriculture, with negative reproductive effects demonstrated on teleosts. In the Japanese medaka, atrazine (0.5, 5, and 50 μg/L) exposure for 38 days led to an alteration in steroidogenesis, altering the ovulation capacity and decreasing the egg production [204]. The same experimental design was also applied in a subsequent study to further investigate the mechanism underlying the reduced egg production, finding a dose-dependent decrease in the expression of genes involved in vitellogenesis and zona pellucida glycoprotein in the eggs, also affecting steroidogenesis and the HPG axis [205].

On the contrary, other studies, both using similar atrazine concentrations and exposure windows (0.6, 5.5, and 53 μg/L for 35 days) of the aforementioned study, or modifying the exposure regime (9.4, 48, 74, 97, and 244 μg/L for 28 days), found no reproductive effects of this herbicide on Japanese medaka reproduction [206,207]. Interestingly, even though the exposure to atrazine at 5 and 50 μg/L during the first 12 days of development (an important time window for epigenetic reprogramming of primordial germ cells and sex determination) did not affected F0 sperm quality, it reduced that of the F2 generation, thus leading to a reduction in the fertilization rate of the males exposed to the lowest concentration [208]. At the molecular level, the observed reproductive toxicity could be explained by the changes in DNA methylation and the expression of genes involved in steroidogenesis in the F2 generation [208], suggesting that even though atrazine reproductive effects are not evident in the parental generation, epigenetic modifications could be transmitted, leading to a reproductive toxicity of non-exposed generations.

The effects on medaka reproduction of another herbicide, Diuron (5, 50, 500, and 5000 ng/L) from 0 dpf to 180 dpf were evaluated, evidencing its detrimental effect on male reproduction by affecting HPG axis-related gene expression, thus altering gonadal architecture and reducing the fecundity and also leading to growth reduction in the F1 generation [209]. In this regard, another study from the same authors investigates the intergenerational effects of this contaminant (500, and 5000 ng/L), showing that Diuron affected F1 generation hatchability and delayed embryo development. Moreover, after reaching sex maturity, these individuals showed alterations in steroidogenesis, *VTG* transcription, and germ cells proportion [210]. This could be due to the increased mRNA levels of DNA methyltransferase along the HPG axis, suggesting the involvement of epigenetic modifications in the transmission of reprotoxic effects to the following generation [210].

Considering the reproductive toxicity of insecticides, Diazinon exposure at 2.9, 5.2, 10.3, 19.8, and 40.2 μg/L to female and male Japanese medaka (at 12-weeks post fertilization) from the F0 to F2 generation led to a dose-dependent reduction in the number of spawned, fertilized, and hatched eggs, also affecting the F1 growing parameters and secondary sexual characters in a dose-dependent manner [211]. These results agree with another study in which 20 μg/L of Diazinon was shown to increase the time to first spawn and reduced the fecundity, leaving the hatch rate of the embryos unchanged [212]. The exposure of female medaka fish to 1, 10, and 100 ng/L Diazinon showed that this contaminant affected the HPG axis in a similar way, with respect to males, and also affected female GSI and the percentage of mature oocytes, while the highest dose did not affect female medaka reproduction [213]. Similarly to males, the exposure of females to this toxicant affected F1 fecundity and viability [213], further confirming that parental exposure can act on the progeny, impairing embryo development.

Moving to anti-fungi chemicals, exposure to the azole fungicide Difenoconazole (1, 10, 100 and 1000 ng/L) for 180 days, starting right before fertilization, led to a negative impact on the transcription of genes involved in the HPG axis and a decreased GSI in males exposed to the highest concentration, while all concentrations led to an increase in spermatogonia and spermatocytes number and a concomitant decrease in the sperm cell number [214]. In the same study, the toxicity of this pollutant on F1 fertilization, hatchability, and sperm swim-up was also evaluated, finding that all these parameters were affected by the contaminant parental exposure [214]. In addition, the exposure to another azole fungicide, Triadimenol, at environmentally relevant concentrations (3–30 μg/L and 300 μg/L), from 0 dpf to 35 dpf (a critical window for gonad development), led to significant modifications of ovary development, such as an increase in previtellogenic oocytes and a decrease in fecundity, also affecting the F1 male to female sex ratio and hatchability [215], and confirming the capacity of azole fungicides to affect the subsequent generations.

5. Conclusions

As the Mediterranean Sea continues to undergo transformations, the study of teleost reproduction serves as a critical lens through which we can comprehend the broader implications of human activities on marine ecosystems and work towards their preservation. The challenges of teleost reproduction in the Mediterranean Sea are multifaceted, and addressing these challenges requires a comprehensive understanding of the complex interactions between environmental factors, reproductive biology, and anthropogenic impacts. Studies using model species often focus on specific stressors in isolation, whereas wild teleosts experience a combination of different stressors. Therefore, the cumulative impact of multiple stressors on reproduction may not be fully understood when studying only one or a few factors. In that regard, laboratory settings may not fully replicate the natural environmental conditions that teleosts experience in their habitats, thus providing limited information. However, the results obtained from studies carried out in model species, together with the use of mathematical models, could offer different predictions on energy assimilation and mobilization prior to and during reproduction, which will refine our comprehension of both wild and farmed teleosts. Recently, models that encompass entire life cycles and that quantify reproduction are becoming very useful for farming systems like hatcheries and nurseries, where the early stages of life are crucial.

The use of these new tools, together with an in-depth understanding of the variables that influence the reproduction of organisms, could enable better-informed conservation strategies, sustainable aquaculture practices, and management of wild populations in dynamic environments.

Author Contributions: Writing—original draft preparation, G.C., C.G., M.L. and M.Z; writing—review and editing, G.C., C.G., M.L. and M.Z. All authors have read and agreed to the published version of the manuscript.

Funding: This research received no funding.

Conflicts of Interest: The authors declare no conflicts of interest.

References

1. Coll, M.; Piroddi, C.; Steenbeek, J.; Kaschner, K.; Lasram, F.B.R.; Aguzzi, J.; Ballesteros, E.; Bianchi, C.N.; Corbera, J.; Dailianis, T.; et al. The biodiversity of the Mediterranean Sea: Estimates, patterns, and threats. *PLoS ONE* **2010**, *5*, e11842. [CrossRef] [PubMed]
2. Lionello, P.; Malanotte-Rizzoli, P.; Boscolo, R.; Alpert, P.; Artale, V.; Li, L.; Luterbacher, J.; May, W.; Trigo, R.; Tsimplis, M.; et al. The Mediterranean climate: An overview of the main characteristics and issues. *Dev. Earth Environ. Sci.* **2006**, *4*, 1–26. [CrossRef]
3. de la Hoz, C.F.; Ramos, E.; Puente, A.; Méndez, F.; Menéndez, M.; Juanes, J.A.; Losada, Í.J. Ecological typologies of large areas. An application in the Mediterranean Sea. *J. Environ. Manag.* **2018**, *205*, 59–72. [CrossRef] [PubMed]
4. Rodríguez-Rodríguez, D.; Rodríguez, J.; Abdul Malak, D.; Nastasi, A.; Hernández, P. Marine protected areas and fisheries restricted areas in the Mediterranean: Assessing "actual" marine biodiversity protection coverage at multiple scales. *Mar. Policy* **2016**, *64*, 24–30. [CrossRef]
5. Lasram, F.B.R.; Guilhaumon, F.; Mouillot, D. Fish diversity patterns in the Mediterranean Sea: Deviations from a mid-domain model. *Mar. Ecol. Prog. Ser.* **2009**, *376*, 253–267. [CrossRef]
6. Meynard, C.N.; Mouillot, D.; Mouquet, N.; Douzery, E.J.P. A phylogenetic perspective on the evolution of Mediterranean teleost fishes. *PLoS ONE* **2012**, *7*, e36443. [CrossRef] [PubMed]
7. Alix, M.; Kjesbu, O.S.; Anderson, K.C. From gametogenesis to spawning: How climate-driven warming affects teleost reproductive biology. *J. Fish Biol.* **2020**, *97*, 607–632. [CrossRef] [PubMed]
8. Planque, B.; Fromentin, J.M.; Cury, P.; Drinkwater, K.F.; Jennings, S.; Perry, R.I.; Kifani, S. How does fishing alter marine populations and ecosystems sensitivity to climate? *J. Mar. Syst.* **2010**, *79*, 403–417. [CrossRef]
9. Doney, S.C.; Ruckelshaus, M.; Emmett Duffy, J.; Barry, J.P.; Chan, F.; English, C.A.; Galindo, H.M.; Grebmeier, J.M.; Hollowed, A.B.; Knowlton, N.; et al. Climate change impacts on marine ecosystems. *Ann. Rev. Mar. Sci.* **2012**, *4*, 11–37. [CrossRef]
10. Prakash, S. Impact of climate change on aquatic ecosystem and its biodiversity: An overview. *Int. J. Biol. Innov.* **2021**, *3*, 312–317. [CrossRef]
11. Pankhurst, N.W.; Munday, P.L. Effects of climate change on fish reproduction and early life history stages. *Mar. Freshw. Res.* **2011**, *62*, 1015–1026. [CrossRef]
12. Mayer, I. The role of reproductive sciences in the preservation and breeding of commercial and threatened teleost fishes. *Adv. Exp. Med. Biol.* **2019**, *1200*, 187–224. [CrossRef] [PubMed]
13. Pörtner, H.O.; Farrell, A.P. Ecology. Physiology and climate change. *Science* **2008**, *322*, 690–692. [CrossRef] [PubMed]
14. Dahlke, F.T.; Wohlrab, S.; Butzin, M.; Pörtner, H.O. Thermal bottlenecks in the life cycle define climate vulnerability of fish. *Science* **2020**, *369*, 65–70. [CrossRef]

15. Kumar, P.; Babita, M.; Kailasam, M.; Muralidhar, M.; Hussain, T.; Behera, A.; Jithendran, K.P. Effect of changing environmental factors on reproductive cycle and endocrinology of fishes. In *Outlook of Climate Change and Fish Nutrition*; Springer Nature: Singapore, 2023; pp. 377–396, ISBN 9789811955006.
16. European Environment Agency. *Horizon 2020 Mediterranean Report—Toward Shared Environmenteal Information Systems*; EFA Technical Report No 6/2014; Publications Office of the European Union: Luxembourg, 2014; ISSN 1725-2237.
17. Hsieh, C.-H.; Yamauchi, A.; Nakazawa, T.; Wang, W.F. Fishing effects on age and spatial structures undermine population stability of fishes. *Aquat. Sci.* **2010**, *72*, 165–178. [CrossRef]
18. Geyer, H.J.; Rimkus, G.G.; Scheunert, I.; Kaune, A.; Schramm, K.-W.; Kettrup, A.; Zeeman, M.; Muir, D.C.G.; Hansen, L.G.; Mackay, D. Bioaccumulation and Occurrence of Endocrine-Disrupting Chemicals (EDCs), Persistent Organic Pollutants (POPs), and Other Organic Compounds in Fish and Other Organisms Including Humans. In *Bioaccumulation—New Aspects and Developments*; Springer: Berlin/Heidelberg, Germany, 2005; pp. 1–166, ISBN 978-3-540-68091-8.
19. Nyholm, J.R.; Norman, A.; Norrgren, L.; Haglund, P.; Andersson, P.L. Maternal transfer of brominated flame retardants in zebrafish (*Danio rerio*). *Chemosphere* **2008**, *73*, 203–208. [CrossRef] [PubMed]
20. Impellitteri, F.; Multisanti, C.R.; Rusanova, P.; Piccione, G.; Falco, F.; Faggio, C. Exploring the impact of contaminants of emerging concern on fish and invertebrates physiology in the Mediterranean Sea. *Biology* **2023**, *12*, 767. [CrossRef] [PubMed]
21. Miniero, R.; Abate, V.; Brambilla, G.; Davoli, E.; De Felip, E.; De Filippis, S.P.; Dellatte, E.; De Luca, S.; Fanelli, R.; Fattore, E.; et al. Persistent toxic substances in Mediterranean aquatic species. *Sci. Total Environ.* **2014**, *494–495*, 18–27. [CrossRef] [PubMed]
22. de Lucia, G.A.; Vianello, A.; Camedda, A.; Vani, D.; Tomassetti, P.; Coppa, S.; Palazzo, L.; Amici, M.; Romanelli, G.; Zampetti, G.; et al. Sea Water Contamination in the Vicinity of the Italian Minor Islands Caused by Microplastic Pollution. *Water* **2018**, *10*, 1108. [CrossRef]
23. Federation of European Aquaculture Producers (FEAP). DATA. Available online: https://feap.info/index.php/data/ (accessed on 20 November 2023).
24. Cascarano, M.C.; Stavrakidis-Zachou, O.; Mladineo, I.; Thompson, K.D.; Papandroulakis, N.; Katharios, P. Mediterranean aquaculture in a changing climate: Temperature effects on pathogens and diseases of three farmed fish species. *Pathogens* **2021**, *10*, 1205. [CrossRef]
25. Mylonas, C.; Zohar, Y. Controlling fish reproduction in a New Technologies in Aquaculture Improving Production Efficiency, Quality and Environmental Management quaculture. In *New Technologies in Aquaculture: Improving Production Efficiency, Quality and Environmental Management*; WoodHead Publishing: Oxford, UK, 2009; pp. 109–142. [CrossRef]
26. Callet, T.; Cardona, E.; Turonnet, N.; Maunas, P.; Larroquet, L.; Surget, A.; Corraze, G.; Panserat, S.; Marandel, L. Alteration of eggs biochemical composition and progeny survival by maternal high carbohydrate nutrition in a teleost fish. *Sci. Rep.* **2022**, *12*, 16726. [CrossRef]
27. Chen, Z.; Fei, S.; Duan, Y.; Liu, C.; Liu, H.; Han, D.; Jin, J.; Yang, Y.; Zhu, X.; Xie, S. Effects of dietary protein level on the growth, reproductive performance, and larval quality of female yellow catfish (*Pelteobagrus fulvidraco*) broodstock. *Aquac. Rep.* **2022**, *24*, 101102. [CrossRef]
28. Kause, A.; Nousiainen, A.; Koskinen, H. Improvement in feed efficiency and reduction in nutrient loading from rainbow trout farms: The role of selective breeding. *J. Anim. Sci.* **2022**, *100*, skac214. [CrossRef] [PubMed]
29. Furutani-Seiki, M.; Wittbrodt, J. Medaka and zebrafish, an evolutionary twin study. *Mech. Dev.* **2004**, *121*, 629–637. [CrossRef] [PubMed]
30. Ulloa, P.E.; Medrano, J.F.; Feijo, C.G. Zebrafish as animal model for aquaculture nutrition research. *Front. Genet.* **2014**, *5*, 313. [CrossRef]
31. Ulloa, P.E.; Iturra, P.; Neira, R.; Araneda, C. Zebrafish as a model organism for nutrition and growth: Towards comparative studies of nutritional genomics applied to aquacultured fishes. *Rev. Fish Biol. Fish.* **2011**, *21*, 649–666. [CrossRef]
32. Halpern, B.S.; Walbridge, S.; Selkoe, K.A.; Kappel, C.V.; Micheli, F.; D'Agrosa, C.; Bruno, J.F.; Casey, K.S.; Ebert, C.; Fox, H.E.; et al. A global map of human impact on marine ecosystems. *Science* **2008**, *319*, 948–952. [CrossRef] [PubMed]
33. Brander, K. Impacts of climate change on fisheries. *J. Mar. Syst.* **2010**, *79*, 389–402. [CrossRef]
34. Ma, J.; Hung, H.; Macdonald, R.W. The influence of global climate change on the environmental fate of persistent organic pollutants: A review with emphasis on the Northern Hemisphere and the Arctic as a receptor. *Glob. Planet. Chang.* **2016**, *146*, 89–108. [CrossRef]
35. Brown-Peterson, N.J.; Wyanski, D.M.; Saborido-Rey, F.; Macewicz, B.J.; Lowerre-Barbieri, S.K. A standardized terminology for describing reproductive development in fishes. *Mar. Coast. Fish.* **2011**, *3*, 52–70. [CrossRef]
36. Wernberg, T.; Smale, D.A.; Thomsen, M.S. A decade of climate change experiments on marine organisms: Procedures, patterns and problems. *Glob. Chang. Biol.* **2012**, *18*, 1491–1498. [CrossRef]
37. Servili, A.; Canario, A.V.M.; Mouchel, O.; Muñoz-Cueto, J.A. Climate change impacts on fish reproduction are mediated at multiple levels of the brain-pituitary-gonad axis. *Gen. Comp. Endocrinol.* **2020**, *291*, 113439. [CrossRef]
38. Miranda, L.A.; Chalde, T.; Elisio, M.; Strüssmann, C.A. Effects of global warming on fish reproductive endocrine axis, with special emphasis in pejerrey Odontesthes bonariensis. *Gen. Comp. Endocrinol.* **2013**, *192*, 45–54. [CrossRef] [PubMed]
39. Migaud, H.; Davie, A.; Taylor, J.F. Current knowledge on the photoneuroendocrine regulation of reproduction in temperate fish species. *J. Fish Biol.* **2010**, *76*, 27–68. [CrossRef]

40. Vergés, A.; McCosker, E.; Mayer-Pinto, M.; Coleman, M.A.; Wernberg, T.; Ainsworth, T.; Steinberg, P.D. Tropicalisation of temperate reefs: Implications for ecosystem functions and management actions. *Funct. Ecol.* **2019**, *33*, 1000–1013. [CrossRef]
41. Castro, J.S.; Braz-Mota, S.; Campos, D.F.; Souza, S.S.; Val, A.L. High temperature, pH, and hypoxia cause oxidative stress and impair the spermatic performance of the Amazon fish *Colossoma macropomum*. *Front. Physiol.* **2020**, *11*, 772. [CrossRef] [PubMed]
42. Cruz Vieira, A.B.; Weber, A.A.; Ribeiro, Y.M.; Luz, R.K.; Bazzoli, N.; Rizzo, E. Influence of salinity on spermatogenesis in adult Nile tilapia (*Oreochromis niloticus*) testis. *Theriogenology* **2019**, *131*, 1–8. [CrossRef]
43. Lehtonen, T.K.; Wong, B.B.M.; Kvarnemo, C. Effects of salinity on nest-building behaviour in a marine fish. *BMC Ecol.* **2016**, *16*, 7. [CrossRef] [PubMed]
44. Muir, D.C.G.; Howard, P.H. Are there other persistent organic pollutants? A challenge for environmental chemists. *Environ. Sci. Technol.* **2006**, *40*, 7157–7166. [CrossRef]
45. Mezzelani, M.; Gorbi, S.; Regoli, F. Pharmaceuticals in the aquatic environments: Evidence of emerged threat and future challenges for marine organisms. *Mar. Environ. Res.* **2018**, *140*, 41–60. [CrossRef]
46. Wuttke, W.; Jarry, H.; Seidlova-Wuttke, D. Definition, classification and mechanism of action of endocrine disrupting chemicals. *Hormones* **2010**, *9*, 9–15. [CrossRef]
47. Desbiolles, F.; Malleret, L.; Tiliacos, C.; Wong-Wah-Chung, P.; Laffont-Schwob, I. Occurrence and ecotoxicological assessment of pharmaceuticals: Is there a risk for the Mediterranean aquatic environment? *Sci. Total Environ.* **2018**, *639*, 1334–1348. [CrossRef] [PubMed]
48. Errico, S.; Nicolucci, C.; Migliaccio, M.; Micale, V.; Mita, D.G.; Diano, N. Analysis and occurrence of some phenol endocrine disruptors in two marine sites of the northern coast of Sicily (Italy). *Mar. Pollut. Bull.* **2017**, *120*, 68–74. [CrossRef]
49. Rios-Fuster, B.; Alomar, C.; Paniagua González, G.; Garcinuño Martínez, R.M.; Soliz Rojas, D.L.; Fernández Hernando, P.; Deudero, S. Assessing microplastic ingestion and occurrence of bisphenols and phthalates in bivalves, fish and holothurians from a Mediterranean marine protected area. *Environ. Res.* **2022**, *214*, 114034. [CrossRef]
50. Salgueiro-González, N.; Campillo, J.A.; Viñas, L.; Beiras, R.; López-Mahía, P.; Muniategui-Lorenzo, S. Occurrence of selected endocrine disrupting compounds in Iberian coastal areas and assessment of the environmental risk. *Environ. Pollut.* **2019**, *249*, 767–775. [CrossRef] [PubMed]
51. Miccoli, A.; Maradonna, F.; De Felice, A.; Caputo Barucchi, V.; Estonba, A.; Genangeli, M.; Vittori, S.; Leonori, I.; Carnevali, O. Detection of endocrine disrupting chemicals and evidence of their effects on the HPG axis of the European anchovy *Engraulis encrasicolus*. *Mar. Environ. Res.* **2017**, *127*, 137–147. [CrossRef] [PubMed]
52. Storelli, M.M.; Losada, S.; Marcotrigiano, G.O.; Roosens, L.; Barone, G.; Neels, H.; Covaci, A. Polychlorinated biphenyl and organochlorine pesticide contamination signatures in deep-sea fish from the Mediterranean Sea. *Environ. Res.* **2009**, *109*, 851–856. [CrossRef]
53. Storelli, M.M.; Storelli, A.; Barone, G.; Marcotrigiano, G.O. Polychlorinated biphenyl and organochlorine pesticide residues in *Lophius budegassa* from the Mediterranean Sea (Italy). *Mar. Pollut. Bull.* **2004**, *48*, 743–748. [CrossRef]
54. Brumovský, M.; Bečanová, J.; Kohoutek, J.; Borghini, M.; Nizzetto, L. Contaminants of emerging concern in the open sea waters of the Western Mediterranean. *Environ. Pollut.* **2017**, *229*, 976–983. [CrossRef]
55. Reboa, A.; Mandich, A.; Cutroneo, L.; Carbone, C.; Malatesta, A.; Capello, M. Baseline evaluation of metal contamination in teleost fishes of the Gulf of Tigullio (north-western Italy): Histopathology and chemical analysis. *Mar. Pollut. Bull.* **2019**, *141*, 16–23. [CrossRef]
56. Lounas, R.; Kasmi, H.; Chernai, S.; Amarni, N.; Ghebriout, L.; Hamdi, B. Heavy metal concentrations in wild and farmed gilthead sea bream from southern Mediterranean Sea—Human health risk assessment. *Environ. Sci. Pollut. Res.* **2021**, *28*, 30732–30742. [CrossRef] [PubMed]
57. Iamiceli, A.L.; Ubaldi, A.; Lucchetti, D.; Brambilla, G.; Abate, V.; De Felip, E.; De Filippis, S.P.; Dellatte, E.; De Luca, S.; Ferri, F.; et al. Metals in Mediterranean aquatic species. *Mar. Pollut. Bull.* **2015**, *94*, 278–283. [CrossRef] [PubMed]
58. Cardinale, M.; Scarcella, G. Mediterranean sea: A failure of the European fisheries management system. *Front. Mar. Sci.* **2017**, *4*, 72. [CrossRef]
59. Boudouresque, C.F. Marine biodiversity warming vs. biological invasions and overfishing in the Mediterranean Sea: Take care, 'One Train can hide another'. *MOJ Ecol. Environ. Sci.* **2017**, *2*, 172–183. [CrossRef]
60. van Overzee, H.M.J.; Rijnsdorp, A.D. Effects of fishing during the spawning period: Implications for sustainable management. *Rev. Fish Biol. Fish.* **2015**, *25*, 65–83. [CrossRef]
61. Fiorentino, F.; Badalamenti, F.; D'Anna, G.; Garofalo, G.; Gianguzza, P.; Gristina, M.; Pipitone, C.; Rizzo, P.; Fortibuoni, T. Changes in spawning-stock structure and recruitment pattern of red mullet, *Mullus barbatus*, after a trawl ban in the Gulf of Castellammare (central Mediterranean Sea). *ICES J. Mar. Sci.* **2008**, *65*, 1175–1183. [CrossRef]
62. Lloret, J.; Muñoz, M.; Casadevall, M. Threats posed by artisanal fisheries to the reproduction of coastal fish species in a Mediterranean marine protected area. *Estuar. Coast. Shelf Sci.* **2012**, *113*, 133–140. [CrossRef]
63. Colloca, F.; Scarcella, G.; Libralato, S. Recent trends and impacts of fisheries exploitation on Mediterranean stocks and ecosystems. *Front. Mar. Sci.* **2017**, *4*, 244. [CrossRef]
64. Albo-Puigserver, M.; Pennino, M.G.; Bellido, J.M.; Colmenero, A.I.; Giráldez, A.; Hidalgo, M.; Gabriel Ramírez, J.; Steenbeek, J.; Torres, P.; Cousido-Rocha, M.; et al. Changes in life history traits of small pelagic fish in the Western Mediterranean Sea. *Front. Mar. Sci.* **2021**, *8*, 570354. [CrossRef]

65. Hidalgo, M.; Rouyer, T.; Molinero, J.C.; Massutí, E.; Moranta, J.; Guijarro, B.; Stenseth, N.C. Synergistic effects of fishing-induced demographic changes and climate variation on fish population dynamics. *Mar. Ecol. Prog. Ser.* **2011**, *426*, 1–12. [CrossRef]
66. Coll, M.; Albo-Puigserver, M.; Navarro, J.; Palomera, I.; Dambacher, J.M. Who is to blame? Plausible pressures on small pelagic fish population changes in the northwestern Mediterranean Sea. *Mar. Ecol. Prog. Ser.* **2018**, *617*, 277–294. [CrossRef]
67. FAO. *The State of Mediterranean and Black Sea Fisheries*; General Fisheries Commission for the Mediterranean: Rome, Italy, 2023. [CrossRef]
68. Martín, P.; Sabatés, A.; Lloret, J.; Martin-Vide, J. Climate modulation of fish populations: The role of the Western Mediterranean Oscillation (WeMO) in sardine (*Sardina pilchardus*) and anchovy (*Engraulis encrasicolus*) production in the north-western Mediterranean. *Clim. Chang.* **2012**, *110*, 925–939. [CrossRef]
69. Maynou, F.; Sabatés, A.; Raya, V. Changes in the spawning habitat of two small pelagic fish in the Northwestern Mediterranean. *Fish. Oceanogr.* **2020**, *29*, 202–214. [CrossRef]
70. Basilone, G.; Ferreri, R.; Aronica, S.; Mazzola, S.; Bonanno, A.; Gargano, A.; Pulizzi, M.; Fontana, I.; Giacalone, G.; Calandrino, P.; et al. Reproduction and sexual maturity of European sardine (*Sardina pilchardus*) in the Central Mediterranean Sea. *Front. Mar. Sci.* **2021**, *8*, 715846. [CrossRef]
71. Chemello, G.; Cerrone, G.L.; Tavolazzi, V.; Donato, F.; Tiralongo, F.; Gioacchini, G. One year study on the reproductive biology, ovary characterization and age of the European sardine (*Sardina pilchardus*) in the middle-west Adriatic Sea. *Front. Mar. Sci.* **2023**, *10*, 1266894. [CrossRef]
72. Zorica, B.; Anđelić, I.; Keč, V.Č. Sardine (*Sardina pilchardus*) spawning in the light of fat content analysis. *Sci. Mar.* **2019**, *83*, 207–213. [CrossRef]
73. Hure, M.; Mustać, B. Feeding ecology of Sardina pilchardus considering co-occurring small pelagic fish in the eastern Adriatic Sea. *Mar. Biodivers.* **2020**, *50*, 40. [CrossRef]
74. Caballero-Huertas, M.; Frigola-Tepe, X.; Viñas, J.; Muñoz, M. Somatic condition and reproductive potential as a tandem in European sardine: An analysis with an environmental perspective in the Northern Adriatic (Gulf of Trieste). *Fishes* **2022**, *7*, 105. [CrossRef]
75. Mustać, B.; Sinovčić, G. Reproduction, length-weight relationship and condition of sardine, Sardina pilchardus (Walbaum, 1792), in the eastern Middle Adriatic Sea (Croatia). *Period. Biol.* **2010**, *112*, 133–138.
76. Fossi, M.C.; Casini, S.; Marsili, L.; Neri, G.; Mori, G.; Ancora, S.; Moscatelli, A.; Ausili, A.; Notarbartolo-Di-Sciara, G. Biomarkers for endocrine disruptors in three species of Mediterranean large pelagic fish. *Proc. Mar. Environ. Res.* **2002**, *54*, 667–671. [CrossRef]
77. De Metrio, G.; Corriero, A.; Desantis, S.; Zubani, D.; Cirillo, F.; Deflorio, M.; Bridges, C.R.; Eicker, J.; De La Serna, J.M.; Megalofonou, P.; et al. Evidence of a high percentage of intersex in the Mediterranean swordfish (*Xiphias gladius* L.). *Mar. Pollut. Bull.* **2003**, *46*, 358–361. [CrossRef] [PubMed]
78. Fossi, M.C.; Casini, S.; Ancora, S.; Moscatelli, A.; Ausili, A.; Notarbartolo-Di-Sciara, G. Do endocrine disrupting chemicals threaten Mediterranean swordfish? Preliminary results of vitellogenin and Zona radiata proteins in *Xiphias gladius*. *Mar. Environ. Res.* **2001**, *52*, 477–483. [CrossRef] [PubMed]
79. Desantis, S.; Corriero, A.; Cirillo, F.; Deflorio, M.; Brill, R.; Griffiths, M.; Lopata, A.L.; De La Serna, J.M.; Bridges, C.R.; Kime, D.E.; et al. Immunohistochemical localization of CYP1A, vitellogenin and Zona radiata proteins in the liver of swordfish (*Xiphias gladius* L.) taken from the Mediterranean Sea, South Atlantic, South Western Indian and Central North Pacific Oceans. *Aquat. Toxicol.* **2005**, *71*, 1–12. [CrossRef] [PubMed]
80. Bobe, J.; Labbé, C. Egg and sperm quality in fish. *Gen. Comp. Endocrinol.* **2010**, *165*, 535–548. [CrossRef]
81. Izquierdo, M.S.; Fernández-Palacios, H.; Tacon, A.G.J. Effect of broodstock nutrition on reproductive performance of fish. *Aquaculture* **2001**, *197*, 25–42. [CrossRef]
82. Kunz, Y.W. *Developmental Biology of Teleost Fishes*; Springer: Dordrecht, The Netherlands, 2004. [CrossRef]
83. Tocher, D.R. Omega-3 long-chain polyunsaturated fatty acids and aquaculture in perspective. *Aquaculture* **2015**, *449*, 94–107. [CrossRef]
84. Izquierdo, M.; Koven, W. Lipids. In *Larval Fish Nutrition*; Wiley-Blackwell: Oxford, UK, 2011; pp. 47–81.
85. Fei, S.; Chen, Z.; Duan, Y.; Haokun, L.; Jin, J.; Yang, Y.; Han, D.; Zhu, X.; Xie, S. Growth, reproduction, fatty acid profiles and offspring performance of broodstock yellow catfish *Pelteobagrus fulvidraco* fed diets with different lipid levels. *Aquaculture* **2024**, *580*, 740273. [CrossRef]
86. Beirão, J.; Soares, F.; Pousão-Ferreira, P.; Diogo, P.; Dias, J.; Dinis, M.T.; Herráez, M.P.; Cabrita, E. The effect of enriched diets on Solea senegalensis sperm quality. *Aquaculture* **2015**, *435*, 187–194. [CrossRef]
87. Samaee, S.M.; Manteghi, N.; Estévez, A. Zebrafish as a model to screen the potential of fatty acids in reproduction. *Zebrafish* **2019**, *16*, 47–64. [CrossRef]
88. Turchini, G.M.; Francis, D.S. Fatty acid metabolism (desaturation, elongation and β-oxidation) in rainbow trout fed fish oil- or linseed oil-based diets. *Br. J. Nutr.* **2009**, *102*, 69–81. [CrossRef]
89. Turchini, G.M.; Trushenski, J.T.; Glencross, B.D. Thoughts for the future of aquaculture nutrition: Realigning perspectives to reflect contemporary issues related to judicious use of marine resources in aquafeeds. *N. Am. J. Aquac.* **2019**, *81*, 13–39. [CrossRef]
90. Bandara, T. Alternative feed ingredients in aquaculture: Opportunities and challenges. *J. Entomol. Zool. Stud.* **2018**, *6*, 3087–3094.
91. Gatlin, D.M.; Barrows, F.T.; Brown, P.; Dabrowski, K.; Gaylord, T.G.; Hardy, R.W.; Herman, E.; Hu, G.; Krogdahl, Å.; Nelson, R.; et al. Expanding the utilization of sustainable plant products in aquafeeds: A review. *Aquac. Res.* **2007**, *38*, 551–579. [CrossRef]

92. Oliva-Teles, A.; Enes, P.; Peres, H. Replacing fishmeal and fish oil in industrial aquafeeds for carnivorous fish. In *Feed and Feeding Practices in Aquaculture*; Woodhead Publishing: Sawston, UK, 2015; pp. 203–233. [CrossRef]
93. Turchini, G.M.; Torstensen, B.E.; Ng, W.K. Fish oil replacement in finfish nutrition. *Rev. Aquac.* **2009**, *1*, 10–57. [CrossRef]
94. FAO. *The State of World Fisheries and Aquaculture 2020. Sustainability in Action*; FAO: Rome, Italy, 2020; ISBN 978-92-5-132692-3.
95. Ribas, L.; Piferrer, F. The zebrafish (*Danio rerio*) as a model organism, with emphasis on applications for finfish aquaculture research. *Rev. Aquac.* **2014**, *6*, 209–240. [CrossRef]
96. Yıldız, M.; Ofori-Mensah, S.; Arslan, M.; Yamaner, G.; Ekici, A.; Baltacı, M.A.; Korkmaz, F.; Tacer-Tanas, Ş. Effects of different dietary lipid resources on sperm quality and reproductive success in rainbow trout (*Oncorhynchus mykiss*). *Aquac. Res.* **2021**, *52*, 3804–3814. [CrossRef]
97. Yıldız, M.; Ofori-Mensah, S.; Arslan, M.; Ekici, A.; Yamaner, G.; Baltacı, M.A.; Tacer, Ş.; Korkmaz, F. Effects of different dietary oils on egg quality and reproductive performance in rainbow trout *Oncorhynchus mykiss*. *Anim. Reprod. Sci.* **2020**, *221*, 106545. [CrossRef] [PubMed]
98. Noori, F.; Agh, N.; Jafari, F.; Jalili, R.; Gisbert, E.; Torfi Mozanzadeh, M. Dietary fatty acid profiling in plant protein-rich diets affects the reproductive performance, egg fatty acid profile and haematological parameters in female rainbow trout (*Oncorhynchus mykiss*). *Aquac. Nutr.* **2019**, *25*, 1050–1062. [CrossRef]
99. Lazzarotto, V.; Corraze, G.; Leprevost, A.; Quillet, E.; Dupont-Nivet, M.; Médale, F. Three-year breeding cycle of rainbow trout (*Oncorhynchus mykiss*) fed a plant-based diet, totally free of marine resources: Consequences for reproduction, fatty acid composition and progeny survival. *PLoS ONE* **2015**, *10*, e0117609. [CrossRef]
100. Lee, K.-J.; Rinchard, J.; Dabrowski, K.; Babiak, I.; Ottobre, J.S.; Christensen, J.E. Long-term effects of dietary cottonseed meal on growth and reproductive performance of rainbow trout: Three-year study. *Anim. Feed Sci. Technol.* **2006**, *126*, 93–106. [CrossRef]
101. Vila-Donat, P.; Marín, S.; Sanchis, V.; Ramos, A.J. A review of the mycotoxin adsorbing agents, with an emphasis on their multi-binding capacity, for animal feed decontamination. *Food Chem. Toxicol.* **2018**, *114*, 246–259. [CrossRef] [PubMed]
102. Woźny, M.; Dobosz, S.; Hliwa, P.; Gomułka, P.; Król, J.; Obremski, K.; Blahova, J.; Svobodova, Z.; Michalik, O.; Ocalewicz, K.; et al. Feed-borne exposure to zearalenone impairs reproduction of rainbow trout. *Aquaculture* **2020**, *528*, 735522. [CrossRef]
103. Glencross, B.D. Exploring the nutritional demand for essential fatty acids by aquaculture species. *Rev. Aquac.* **2009**, *1*, 71–124. [CrossRef]
104. Tocher, D.R. Fatty acid requirements in ontogeny of marine and freshwater fish. *Aquac. Res.* **2010**, *41*, 717–732. [CrossRef]
105. Skjærven, K.H.; Adam, A.-C.; Saito, T.; Waagbø, R.; Espe, M. Epigenetics in fish nutritional programming. In *Epigenetics in Aquaculture*; John Wiley & Sons, Ltd.: Hoboken, NJ, USA, 2023; pp. 231–244, ISBN 9781119821946.
106. Clarkson, M.; Migaud, H.; Metochis, C.; Vera, L.M.; Leeming, D.; Tocher, D.R.; Taylor, J.F. Early nutritional intervention can improve utilisation of vegetable-based diets in diploid and triploid Atlantic salmon (*Salmo salar* L.). *Br. J. Nutr.* **2017**, *118*, 17–29. [CrossRef]
107. Turkmen, S.; Castro, P.L.; Caballero, M.J.; Hernández-Cruz, C.M.; Saleh, R.; Zamorano, M.J.; Regidor, J.; Izquierdo, M. Nutritional stimuli of gilthead seabream (*Sparus aurata*) larvae by dietary fatty acids: Effects on larval performance, gene expression and neurogenesis. *Aquac. Res.* **2017**, *48*, 202–213. [CrossRef]
108. Rocha, F.; Dias, J.; Geurden, I.; Dinis, M.T.; Panserat, S.; Engrola, S. Dietary glucose stimulus at larval stage modifies the carbohydrate metabolic pathway in gilthead seabream (*Sparus aurata*) juveniles: An in vivo approach using 14C-starch. *Comp. Biochem. Physiol. A Mol. Integr. Physiol.* **2016**, *201*, 189–199. [CrossRef]
109. Balasubramanian, M.N.; Panserat, S.; Dupont-Nivet, M.; Quillet, E.; Montfort, J.; Le Cam, A.; Medale, F.; Kaushik, S.J.; Geurden, I. Molecular pathways associated with the nutritional programming of plant-based diet acceptance in rainbow trout following an early feeding exposure. *BMC Genom.* **2016**, *17*, 449. [CrossRef]
110. Geurden, I.; Borchert, P.; Balasubramanian, M.N.; Schrama, J.W.; Dupont-Nivet, M.; Quillet, E.; Kaushik, S.J.; Panserat, S.; Médale, F. The positive impact of the early-feeding of a plant-based diet on its future acceptance and utilisation in rainbow trout. *PLoS ONE* **2013**, *8*, e83162. [CrossRef]
111. Fernández-Palacios, H.; Izquierdo, M.S.; Robaina, L.; Valencia, A.; Salhi, M.; Vergara, J. Effect of n − 3 HUFA level in broodstock diets on egg quality of gilthead sea bream (*Sparus aurata* L.). *Aquaculture* **1995**, *132*, 325–337. [CrossRef]
112. Izquierdo, M.S.; Turkmen, S.; Montero, D.; Zamorano, M.J.; Afonso, J.M.; Karalazos, V.; Fernández-Palacios, H. Nutritional programming through broodstock diets to improve utilization of very low fishmeal and fish oil diets in gilthead sea bream. *Aquaculture* **2015**, *449*, 18–26. [CrossRef]
113. Turkmen, S.; Zamorano, M.J.; Fernández-Palacios, H.; Hernández-Cruz, C.M.; Montero, D.; Robaina, L.; Izquierdo, M. Parental nutritional programming and a reminder during juvenile stage affect growth, lipid metabolism and utilisation in later developmental stages of a marine teleost, the gilthead sea bream (*Sparus aurata*). *Br. J. Nutr.* **2017**, *118*, 500–512. [CrossRef] [PubMed]
114. Fernández-Palacios, H.; Norberg, B.; Izquierdo, M.; Hamre, K. Effects of broodstock diet on eggs and larvae. In *Larval Fish Nutrition*; Holt, G.J., Ed.; Wiley-Blackwell: Chichester, UK, 2011; ISBN 9780813817927.
115. Xu, H.; Turkmen, S.; Rimoldi, S.; Terova, G.; Zamorano, M.J.; Afonso, J.M.; Sarih, S.; Fernández-Palacios, H.; Izquierdo, M. Nutritional intervention through dietary vegetable proteins and lipids to gilthead sea bream (*Sparus aurata*) broodstock affects the offspring utilization of a low fishmeal/fish oil diet. *Aquaculture* **2019**, *513*, 734402. [CrossRef]

116. Ferosekhan, S.; Xu, H.; Turkmen, S.; Gómez, A.; Afonso, J.M.; Fontanillas, R.; Rosenlund, G.; Kaushik, S.; Izquierdo, M. Reproductive performance of gilthead seabream (*Sparus aurata*) broodstock showing different expression of fatty acyl desaturase 2 and fed two dietary fatty acid profiles. *Sci. Rep.* **2020**, *10*, 15547. [CrossRef] [PubMed]
117. Zarantoniello, M.; Pulido Rodriguez, L.F.; Randazzo, B.; Cardinaletti, G.; Giorgini, E.; Belloni, A.; Secci, G.; Faccenda, F.; Pulcini, D.; Parisi, G.; et al. Conventional feed additives or red claw crayfish meal and dried microbial biomass as feed supplement in fish meal-free diets for rainbow trout (*Oncorhynchus mykiss*): Possible ameliorative effects on growth and gut health status. *Aquaculture* **2022**, *554*, 738137. [CrossRef]
118. El-Gamal, M.M.; Othman, S.I.; Abdel-Rahim, M.M.; Mansour, A.T.; Alsaqufi, A.S.; El Atafy, M.M.; Mona, M.H.; Allam, A.A. Palaemon and artemia supplemented diet enhances sea bass, *Dicentrarchus labrax*, broodstock reproductive performance and egg quality. *Aquac. Rep.* **2020**, *16*, 100290. [CrossRef]
119. Sahin, T.; Yılmaz, S.; Gürkan, M.; Ergün, S. Effects of Rapana venosa meal-supplemented diets on reproduction, histopathology and some blood parameters of rainbow trout (*Oncorhynchus mykiss*) broodstock. *Aquac. Res.* **2021**, *52*, 4897–4910. [CrossRef]
120. Zarantoniello, M.; de Oliveira, A.A.; Sahin, T.; Freddi, L.; Torregiani, M.; Tucciarone, I.; Chemello, G.; Cardinaletti, G.; Gatto, E.; Parisi, G.; et al. Enhancing rearing of European seabass (*Dicentrarchus labrax*) in aquaponic systems: Investigating the effects of enriched black soldier fly (*Hermetia illucens*) prepupae meal on fish welfare and quality traits. *Animals* **2023**, *13*, 1921. [CrossRef]
121. Ratti, S.; Zarantoniello, M.; Chemello, G.; Giammarino, M.; Palermo, F.A.; Cocci, P.; Mosconi, G.; Tignani, M.V.; Pascon, G.; Cardinaletti, G.; et al. Spirulina-enriched substrate to rear black soldier fly (*Hermetia illucens*) prepupae as alternative aquafeed ingredient for rainbow trout (*Oncorhynchus mykiss*) diets: Possible effects on zootechnical performances, gut and liver health status, and fillet quality. *Animals* **2023**, *13*, 173. [CrossRef]
122. Shah, M.R.; Lutzu, G.A.; Alam, A.; Sarker, P.; Kabir Chowdhury, M.A.; Parsaeimehr, A.; Liang, Y.; Daroch, M. Microalgae in aquafeeds for a sustainable aquaculture industry. *J. Appl. Phycol.* **2017**, *30*, 197–213. [CrossRef]
123. Alagawany, M.; Taha, A.E.; Noreldin, A.; El-Tarabily, K.A.; Abd El-Hack, M.E. Nutritional applications of species of Spirulina and Chlorella in farmed fish: A review. *Aquaculture* **2021**, *542*, 736841. [CrossRef]
124. Carneiro, W.F.; Castro, T.F.D.; Orlando, T.M.; Meurer, F.; de Jesus Paula, D.A.; do Carmo Rodrigues Virote, B.; da Cunha Barreto Vianna, A.R.; Murgas, L.D.S. Replacing fish meal by *Chlorella* sp. meal: Effects on zebrafish growth, reproductive performance, biochemical parameters and digestive enzymes. *Aquaculture* **2020**, *528*, 735612. [CrossRef]
125. Teimouri, M.; Yeganeh, S.; Mianji, G.R.; Najafi, M.; Mahjoub, S. The effect of Spirulina platensis meal on antioxidant gene expression, total antioxidant capacity, and lipid peroxidation of rainbow trout (*Oncorhynchus mykiss*). *Fish Physiol. Biochem.* **2019**, *45*, 977–986. [CrossRef] [PubMed]
126. Chong, A.S.C.; Ishak, S.D.; Osman, Z.; Hashim, R. Effect of dietary protein level on the reproductive performance of female swordtails *Xiphophorus helleri* (Poeciliidae). *Aquaculture* **2004**, *234*, 381–392. [CrossRef]
127. Pati, D.; Habibi, H.R. Involvement of protein kinase C and arachidonic acid pathways in the gonadotropin-releasing hormone regulation of oocyte meiosis and follicular steroidogenesis in the goldfish ovary. *Biol. Reprod.* **2002**, *66*, 813–822. [CrossRef] [PubMed]
128. Khanzadeh, M.; Esmaeili Fereidouni, A.; Seifi Berenjestanaki, S. Effects of partial replacement of fish meal with Spirulina platensis meal in practical diets on growth, survival, body composition, and reproductive performance of three-spot gourami (*Trichopodus trichopterus*) (Pallas, 1770). *Aquac. Int.* **2016**, *24*, 69–84. [CrossRef]
129. Güroy, B.; Şahin, İ.; Mantoğlu, S.; Kayalı, S. Spirulina as a natural carotenoid source on growth, pigmentation and reproductive performance of yellow tail cichlid *Pseudotropheus acei*. *Aquac. Int.* **2012**, *20*, 869–878. [CrossRef]
130. Takeuchi, T.; Lu, J.; Yoshizaki, G.; Satoh, S. Effect on the growth and body composition of juvenile tilapia *Oreochromis niloticus* fed raw Spirulina. *Fish. Sci.* **2002**, *68*, 34–40. [CrossRef]
131. Jovanovic, S.; Dietrich, D.; Becker, J.; Kohlstedt, M.; Wittmann, C. Microbial production of polyunsaturated fatty acids—High-value ingredients for aquafeed, superfoods, and pharmaceuticals. *Curr. Opin. Biotechnol.* **2021**, *69*, 199–211. [CrossRef]
132. Patel, A.; Karageorgou, D.; Katapodis, P.; Sharma, A.; Rova, U.; Christakopoulos, P.; Matsakas, L. Bioprospecting of thraustochytrids for omega-3 fatty acids: A sustainable approach to reduce dependency on animal sources. *Trends Food Sci. Technol.* **2021**, *115*, 433–444. [CrossRef]
133. Katerina, K.; Berge, G.M.; Turid, M.; Aleksei, K.; Grete, B.; Trine, Y.; Mats, C.; John, S.; Bente, R. Microalgal *Schizochytrium limacinum* biomass improves growth and filet quality when used long-term as a replacement for fish oil, in modern salmon diets. *Front. Mar. Sci.* **2020**, *7*, 57. [CrossRef]
134. Cardona, E.; Segret, E.; Cachelou, Y.; Vanderesse, T.; Larroquet, L.; Hermann, A.; Surget, A.; Corraze, G.; Cachelou, F.; Bobe, J.; et al. Effect of micro-algae *Schizochytrium* sp. supplementation in plant diet on reproduction of female rainbow trout (*Oncorhynchus mykiss*): Maternal programming impact of progeny. *J. Anim. Sci. Biotechnol.* **2022**, *13*, 33. [CrossRef] [PubMed]
135. Zarantoniello, M.; Randazzo, B.; Cardinaletti, G.; Truzzi, C.; Chemello, G.; Riolo, P.; Olivotto, I. Possible dietary effects of insect-based diets across zebrafish (*Danio rerio*) generations: A multidisciplinary study on the larval phase. *Animals* **2021**, *11*, 751. [CrossRef] [PubMed]
136. Chemello, G.; Zarantoniello, M.; Randazzo, B.; Gioacchini, G.; Truzzi, C.; Cardinaletti, G.; Riolo, P.; Olivotto, I. Effects of black soldier fly (*Hermetia illucens*) enriched with *Schizochytrium* sp. on zebrafish (*Danio rerio*) reproductive performances. *Aquaculture* **2022**, *550*, 737853. [CrossRef]

137. Ewald, N.; Vidakovic, A.; Langeland, M.; Kiessling, A.; Sampels, S.; Lalander, C. Fatty acid composition of black soldier fly larvae (*Hermetia illucens*)—Possibilities and limitations for modification through diet. *Waste Manag.* **2020**, *102*, 40–47. [CrossRef] [PubMed]
138. Randazzo, B.; Zarantoniello, M.; Gioacchini, G.; Giorgini, E.; Truzzi, C.; Notarstefano, V.; Cardinaletti, G.; Huyen, K.T.; Carnevali, O.; Olivotto, I. Can insect-based diets affect Zebrafish (*Danio rerio*) reproduction? A multidisciplinary study. *Zebrafish* **2020**, *17*, 287–304. [CrossRef]
139. Becker, E.W. Micro-algae as a source of protein. *Biotechnol. Adv.* **2007**, *25*, 207–210. [CrossRef]
140. Sarker, P.K.; Kapuscinski, A.R.; Lanois, A.J.; Livesey, E.D.; Bernhard, K.P.; Coley, M.L. Towards sustainable aquafeeds: Complete substitution of fish oil with marine microalga *Schizochytrium* sp. improves growth and fatty acid deposition in juvenile Nile tilapia (*Oreochromis niloticus*). *PLoS ONE* **2016**, *11*, e0156684. [CrossRef]
141. Krogdahl, Å.; Hemre, G.I.; Mommsen, T.P. Carbohydrates in fish nutrition: Digestion and absorption in postlarval stages. *Aquac. Nutr.* **2005**, *11*, 103–122. [CrossRef]
142. Cardinaletti, G.; Messina, M.; Bruno, M.; Tulli, F.; Poli, B.M.; Giorgi, G.; Chini-Zittelli, G.; Tredici, M.; Tibaldi, E. Effects of graded levels of a blend of *Tisochrysis lutea* and *Tetraselmis suecica* dried biomass on growth and muscle tissue composition of European sea bass (*Dicentrarchus labrax*) fed diets low in fish meal and oil. *Aquaculture* **2018**, *485*, 173–182. [CrossRef]
143. Al-Feky, S.S.A.; El-Sayed, A.-F.M.; Ezzat, A.A. Dietary taurine improves reproductive performance of Nile tilapia (*Oreochromis niloticus*) broodstock. *Aquac. Nutr.* **2016**, *22*, 392–399. [CrossRef]
144. Gallego, V.; Asturiano, J.F. Fish sperm motility assessment as a tool for aquaculture research: A historical approach. *Rev. Aquac.* **2019**, *11*, 697–724. [CrossRef]
145. Sarmento, N.; Martins, E.F.F.; Costa, D.C.; Silva, W.S.; Mattioli, C.C.; Luz, M.R.; Luz, R.K. Effects of supplemental dietary vitamin C on quality of semen from Nile tilapia (*Oreochromis niloticus*) breeders. *Reprod. Domest. Anim.* **2017**, *52*, 144–152. [CrossRef]
146. Dada, A.A. Effect of ascorbic acid supplementation in broodstock feed on sperm quality of african sharptooth catfish (*Clarias gariepinus*). *Indian J. Anim. Res.* **2012**, *46*, 213–218.
147. Castro, J.; Magnotti, C.; Angelo, M.; Sterzelecki, F.; Pedrotti, F.; Oliveira, M.F.; Soligo, T.; Fracalossi, D.; Cerqueira, V.R. Effect of ascorbic acid supplementation on zootechnical performance, haematological parameters and sperm quality of lebranche mullet *Mugil liza*. *Aquac. Res.* **2019**, *50*, 3267–3274. [CrossRef]
148. Rohani, M.F.; Islam, S.M.M.; Hossain, M.K.; Ferdous, Z.; Siddik, M.A.B.; Nuruzzaman, M.; Padeniya, U.; Brown, C.; Shahjahan, M. Probiotics, prebiotics and synbiotics improved the functionality of aquafeed: Upgrading growth, reproduction, immunity and disease resistance in fish. *Fish Shellfish Immunol.* **2022**, *120*, 569–589. [CrossRef] [PubMed]
149. Akbari Nargesi, E.; Falahatkar, B.; Sajjadi, M.M. Dietary supplementation of probiotics and influence on feed efficiency, growth parameters and reproductive performance in female rainbow trout (*Oncorhynchus mykiss*) broodstock. *Aquac. Nutr.* **2020**, *26*, 98–108. [CrossRef]
150. Py, C.; Elizondo-González, R.; Peña-Rodríguez, A. Compensatory growth: Fitness cost in farmed fish and crustaceans. *Rev. Aquac.* **2022**, *14*, 1389–1417. [CrossRef]
151. Imsland, A.K.; Gunnarsson, S. Growth and maturation in Arctic charr (*Salvelinus alpinus*) in response to different feed rations. *Aquaculture* **2011**, *318*, 407–411. [CrossRef]
152. Cleveland, B.M.; Kenney, P.B.; Manor, M.L.; Weber, G.M. Effects of feeding level and sexual maturation on carcass and fillet characteristics and indices of protein degradation in rainbow trout (*Oncorhynchus mykiss*). *Aquaculture* **2012**, *338–341*, 228–236. [CrossRef]
153. Caldwell, L.K.; Pierce, A.L.; Riley, L.G.; Duncan, C.A.; Nagler, J.J. Plasma Nesfatin-1 is not affected by long-term food restriction and does not predict rematuration among iteroparous female rainbow trout (*Oncorhynchus mykiss*). *PLoS ONE* **2014**, *9*, e85700. [CrossRef]
154. Cardona, E.; Bugeon, J.; Guivarc'h, F.; Goardon, L.; Panserat, S.; Labbé, L.; Corraze, G.; Skiba-Cassy, S.; Bobe, J. Positive impact of moderate food restriction on reproductive success of the rainbow trout *Oncorhynchus mykiss*. *Aquaculture* **2019**, *502*, 280–288. [CrossRef]
155. Kar, S.; Sangem, P.; Anusha, N.; Senthilkumaran, B. Endocrine disruptors in teleosts: Evaluating environmental risks and biomarkers. *Aquac. Fish.* **2021**, *6*, 1–26. [CrossRef]
156. Kavlock, R.J.; Daston, G.P.; DeRosa, C.; Fenner-Crisp, P.; Gray, L.E.; Kaattari, S.; Lucier, G.; Luster, M.; Mac, M.J.; Maczka, C.; et al. Research needs for the risk assessment of health and environmental effects of endocrine disruptors: A report of the U.S. EPA-sponsored workshop. *Environ. Health Perspect.* **1996**, *104*, 715–740. [CrossRef]
157. Carnevali, O.; Santangeli, S.; Forner-Piquer, I.; Basili, D.; Maradonna, F. Endocrine-disrupting chemicals in aquatic environment: What are the risks for fish gametes? *Fish Physiol. Biochem.* **2018**, *44*, 1561–1576. [CrossRef] [PubMed]
158. Santoro, A.; Chianese, R.; Troisi, J.; Richards, S.; Nori, S.L.; Fasano, S.; Guida, M.; Plunk, E.; Viggiano, A.; Pierantoni, R.; et al. Neuro-toxic and reproductive effects of BPA. *Curr. Neuropharmacol.* **2019**, *17*, 1109–1132. [CrossRef]
159. Zhang, T.; Shen, W.; De Felici, M.; Zhang, X.F. Di(2-ethylhexyl)phthalate: Adverse effects on folliculogenesis that cannot be neglected. *Environ. Mol. Mutagen.* **2016**, *57*, 579–588. [CrossRef] [PubMed]
160. Zhang, W.; Sheng, N.; Wang, M.; Zhang, H.; Dai, J. Zebrafish reproductive toxicity induced by chronic perfluorononanoate exposure. *Aquat. Toxicol.* **2016**, *175*, 269–276. [CrossRef]

161. Maradonna, F.; Meccariello, R. EDCs: Focus on reproductive alterations in mammalian and nonmammalian models. In *Environmental Contaminants and Endocrine Health*; Academic Press: New York, NY, USA, 2023; pp. 89–108. [CrossRef]
162. Herráez, M.P.; Lombó, M.; González-Rojo, S. The EDCs as epigenetic disruptors: Implications for development and health. In *Environmental Contaminants and Endocrine Health*; Academic Press: New York, NY, USA, 2023; pp. 109–124. [CrossRef]
163. Cui, J.; Tian, S.; Gu, Y.; Wu, X.; Wang, L.; Wang, J.; Chen, X.; Meng, Z. Toxicity effects of pesticides based on zebrafish (*Danio rerio*) models: Advances and perspectives. *Chemosphere* **2023**, *340*, 139825. [CrossRef]
164. Zamora-Ledezma, C.; Negrete-Bolagay, D.; Figueroa, F.; Zamora-Ledezma, E.; Ni, M.; Alexis, F.; Guerrero, V.H. Heavy metal water pollution: A fresh look about hazards, novel and conventional remediation methods. *Environ. Technol. Innov.* **2021**, *22*, 101504. [CrossRef]
165. Giunchi, V.; Fusaroli, M.; Linder, E.; Villén, J.; Wettermark, B.; Nekoro, M.; Raschi, E.; Lunghi, C.; Poluzzi, E. The environmental impact of pharmaceuticals in Italy: Integrating healthcare and eco-toxicological data to assess and potentially mitigate their diffusion to water supplies. *Br. J. Clin. Pharmacol.* **2023**, *89*, 2020–2027. [CrossRef] [PubMed]
166. Grabicova, K.; Grabic, R.; Fedorova, G.; Fick, J.; Cerveny, D.; Kolarova, J.; Turek, J.; Zlabek, V.; Randak, T. Bioaccumulation of psychoactive pharmaceuticals in fish in an effluent dominated stream. *Water Res.* **2017**, *124*, 654–662. [CrossRef] [PubMed]
167. Manjarrés-López, D.P.; Peña-Herrera, J.M.; Benejam, L.; Montemurro, N.; Pérez, S. Assessment of wastewater-borne pharmaceuticals in tissues and body fluids from riverine fish. *Environ. Pollut.* **2023**, *324*, 121374. [CrossRef]
168. Tang, F.H.M.; Lenzen, M.; McBratney, A.; Maggi, F. Risk of pesticide pollution at the global scale. *Nat. Geosci.* **2021**, *14*, 206–210. [CrossRef]
169. Cao, J.; Wang, G.; Wang, T.; Chen, J.; Wenjing, G.; Wu, P.; He, X.; Xie, L. Copper caused reproductive endocrine disruption in zebrafish (*Danio rerio*). *Aquat. Toxicol.* **2019**, *211*, 124–136. [CrossRef] [PubMed]
170. Gawande, M.B.; Goswami, A.; Felpin, F.X.; Asefa, T.; Huang, X.; Silva, R.; Zou, X.; Zboril, R.; Varma, R.S. Cu and Cu-Based nanoparticles: Synthesis and applications in catalysis. *Chem. Rev.* **2016**, *116*, 3722–3811. [CrossRef] [PubMed]
171. Wu, G.; Gao, L.; Zhang, S.; Du, D.; Xue, Y. Effects of copper oxide nanoparticles on reproductive system of zebrafish. *Ecotoxicol. Environ. Saf.* **2023**, *263*, 115252. [CrossRef]
172. Alsop, D.; Brown, S.; Van Der Kraak, G. The effects of copper and benzo[a]pyrene on retinoids and reproduction in zebrafish. *Aquat. Toxicol.* **2007**, *82*, 281–295. [CrossRef] [PubMed]
173. Hu, J.; Luo, X.; Panga, M.J.; Appiah, C.; Retyunskiy, V.; Zhu, L.; Zhao, Y. Toxic effects and potential mechanisms of zinc pyrithione (ZPT) exposure on sperm and testicular injury in zebrafish. *J. Hazard. Mater.* **2024**, *461*, 132575. [CrossRef]
174. Paul, V.; Krishnakumar, S.; Gowd, G.S.; Nair, S.V.; Koyakutty, M.; Paul-Prasanth, B. Sex-dependent bioaccumulation of nano zinc oxide and its adverse effects on sexual behavior and reproduction in Japanese medaka. *ACS Appl. Bio Mater.* **2021**, *4*, 7408–7421. [CrossRef]
175. Zheng, J.L.; Zhu, Q.L.; Hu, X.C.; Parsons, D.; Lawson, R.; Hogstrand, C. Transgenerational effects of zinc in zebrafish following early life stage exposure. *Sci. Total Environ.* **2022**, *828*, 154443. [CrossRef] [PubMed]
176. Zheng, J.L.; Peng, L.B.; Xia, L.P.; Li, J.; Zhu, Q.L. Effects of continuous and intermittent cadmium exposure on HPGL axis, GH/IGF axis and circadian rhythm signaling and their consequences on reproduction in female zebrafish: Biomarkers independent of exposure regimes. *Chemosphere* **2021**, *282*, 130879. [CrossRef] [PubMed]
177. Liu, P.; Zhao, Y.; Wang, S.; Xing, H.; Dong, W.F. Effect of combined exposure to silica nanoparticles and cadmium chloride on female zebrafish ovaries. *Environ. Toxicol. Pharmacol.* **2021**, *87*, 103720. [CrossRef] [PubMed]
178. Wu, S.M.; Tsai, P.J.; Chou, M.Y.; Wang, W.-D. Effects of maternal cadmium exposure on female reproductive functions, gamete quality, and offspring development in Zebrafish (*Danio rerio*). *Arch. Environ. Contam. Toxicol.* **2013**, *65*, 521–536. [CrossRef] [PubMed]
179. Acosta, I.B.; Junior, A.S.V.; e Silva, E.F.; Cardoso, T.F.; Caldas, J.S.; Jardim, R.D.; Corcini, C.D. Effects of exposure to cadmium in sperm cells of zebrafish, *Danio rerio*. *Toxicol. Rep.* **2016**, *3*, 696–700. [CrossRef] [PubMed]
180. Zhang, Q.F.; Li, Y.W.; Liu, Z.H.; Chen, Q.L. Reproductive toxicity of inorganic mercury exposure in adult zebrafish: Histological damage, oxidative stress, and alterations of sex hormone and gene expression in the hypothalamic-pituitary-gonadal axis. *Aquat. Toxicol.* **2016**, *177*, 417–424. [CrossRef] [PubMed]
181. Chen, Q.; An, J.; Xie, D.; Gong, S.; Lian, X.; Liu, Z.; Shen, Y.; Li, Y. Suppression and recovery of reproductive behavior induced by early life exposure to mercury in zebrafish. *Comp. Biochem. Physiol. C Toxicol. Pharmacol.* **2021**, *239*, 108876. [CrossRef] [PubMed]
182. Xie, D.; Chen, Q.; Gong, S.; An, J.; Li, Y.; Lian, X.; Liu, Z.; Shen, Y.; Giesy, J.P. Exposure of zebrafish to environmentally relevant concentrations of mercury during early life stages impairs subsequent reproduction in adults but can be recovered in offspring. *Aquat. Toxicol.* **2020**, *229*, 105655. [CrossRef] [PubMed]
183. Carvan, M.J.; Kalluvila, T.A.; Klingler, R.H.; Larson, J.K.; Pickens, M.; Mora-Zamorano, F.X.; Connaughton, V.P.; Sadler-Riggleman, I.; Beck, D.; Skinner, M.K. Mercury-induced epigenetic transgenerational inheritance of abnormal neurobehavior is correlated with sperm epimutations in zebrafish. *PLoS ONE* **2017**, *12*, e0176155. [CrossRef]
184. Yang, C.H.; Kung, T.A.; Chen, P.J. Differential alteration in reproductive toxicity of medaka fish on exposure to nanoscale zerovalent iron and its oxidation products. *Environ. Pollut.* **2019**, *252*, 1920–1932. [CrossRef]
185. Chen, H.; Cao, J.; Li, L.; Wu, X.; Bi, R.; Klerks, P.L.; Xie, L. Maternal transfer and reproductive effects of Cr(VI) in Japanese medaka (*Oryzias latipes*) under acute and chronic exposures. *Aquat. Toxicol.* **2016**, *171*, 59–68. [CrossRef] [PubMed]

186. Yan, W.; Hamid, N.; Deng, S.; Jia, P.P.; Pei, D.S. Individual and combined toxicogenetic effects of microplastics and heavy metals (Cd, Pb, and Zn) perturb gut microbiota homeostasis and gonadal development in marine medaka (*Oryzias melastigma*). *J. Hazard. Mater.* **2020**, *397*, 122795. [CrossRef] [PubMed]
187. Zhu, X.; Luo, T.; Wang, D.; Zhao, Y.; Jin, Y.; Yang, G. The occurrence of typical psychotropic drugs in the aquatic environments and their potential toxicity to aquatic organisms—A review. *Sci. Total Environ.* **2023**, *900*, 165732. [CrossRef] [PubMed]
188. Foran, C.M.; Weston, J.; Slattery, M.; Brooks, B.W.; Huggett, D.B. Reproductive assessment of Japanese medaka (*Oryzias latipes*) following a four-week fluoxetine (SSRI) exposure. *Arch. Environ. Contam. Toxicol.* **2004**, *46*, 511–517. [CrossRef] [PubMed]
189. Moreira, D.P.; Ribeiro, Y.M.; Ferreira, C.S.; dos Santos Nassif Lacerda, S.M.; Rizzo, E. Exposure to acetaminophen impairs gametogenesis and fertility in zebrafish (*Danio rerio*). *Arch. Toxicol.* **2023**, *97*, 263–278. [CrossRef] [PubMed]
190. Morthorst, J.E.; Lund, B.F.; Holbech, H.; Bjerregaard, P. Two common mild analgesics have no effect on general endocrine mediated endpoints in zebrafish (*Danio rerio*). *Comp. Biochem. Physiol. Part C Toxicol. Pharmacol.* **2018**, *204*, 63–70. [CrossRef] [PubMed]
191. Wang, H.; Tan, J.T.T.; Emelyanov, A.; Korzh, V.; Gong, Z. Hepatic and extrahepatic expression of vitellogenin genes in the zebrafish, *Danio rerio*. *Gene* **2005**, *356*, 91–100. [CrossRef] [PubMed]
192. Constantine, L.A.; Green, J.W.; Schneider, S.Z. Ibuprofen: Fish short-term reproduction assay with zebrafish (*Danio rerio*) based on an extended OECD 229 protocol. *Environ. Toxicol. Chem.* **2020**, *39*, 1534–1545. [CrossRef] [PubMed]
193. Flippin, J.L.; Huggett, D.; Foran, C.M. Changes in the timing of reproduction following chronic exposure to ibuprofen in Japanese medaka, *Oryzias latipes*. *Aquat. Toxicol.* **2007**, *81*, 73–78. [CrossRef]
194. Yokota, H.; Higashi, K.; Hanada, E.; Matsuzaki, E.; Tsuruda, Y.; Suzuki, T.; Nakano, E.; Eguchi, S. Recovery from reproductive and morphological abnormalities in medaka (*Oryzias latipes*) following a 14-day exposure to diclofenac. *Environ. Toxicol. Chem.* **2017**, *36*, 3277–3283. [CrossRef]
195. Yu, K.; Qiu, Y.; Shi, Y.; Yu, X.; Dong, T.; Wu, Y.; Li, H.; Huang, L. Association of long-term effects of low-level sulfamethoxazole with ovarian lipid and amino acid metabolism, sex hormone levels, and oocyte maturity in zebrafish. *Ecotoxicol. Environ. Saf.* **2022**, *247*, 114234. [CrossRef]
196. Yan, Z.; Lu, G.; Ye, Q.; Liu, J. Long-term effects of antibiotics, norfloxacin, and sulfamethoxazole, in a partial life-cycle study with zebrafish (*Danio rerio*): Effects on growth, development, and reproduction. *Environ. Sci. Pollut. Res.* **2016**, *23*, 18222–18228. [CrossRef]
197. Stenzel, A.; Wirt, H.; Patten, A.; Theodore, B.; King-Heiden, T. Larval exposure to environmentally relevant concentrations of triclosan impairs metamorphosis and reproductive fitness in zebrafish. *Reprod. Toxicol.* **2019**, *87*, 79–86. [CrossRef] [PubMed]
198. Qiao, Y.; He, J.; Han, P.; Qu, J.; Wang, X.; Wang, J. Long-term exposure to environmental relevant triclosan induces reproductive toxicity on adult zebrafish and its potential mechanism. *Sci. Total Environ.* **2022**, *826*, 154026. [CrossRef]
199. Kim, B.; Ji, K.; Kho, Y.; Kim, P.G.; Park, K.; Kim, K.; Kim, Y.; Kim, K.T.; Choi, K. Effects of chronic exposure to cefadroxil and cefradine on *Daphnia magna* and *Oryzias latipes*. *Chemosphere* **2017**, *185*, 844–851. [CrossRef]
200. Velmurugan, G.; Ramprasath, T.; Gilles, M.; Swaminathan, K.; Ramasamy, S. Gut microbiota, endocrine-disrupting chemicals, and the diabetes epidemic. *Trends Endocrinol. Metab.* **2017**, *28*, 612–625. [CrossRef]
201. He, S.; Li, H.; Yu, Z.; Zhang, F.; Liang, S.; Liu, H.; Chen, H.; Lü, M.H. The gut microbiome and sex hormone-related diseases. *Front. Microbiol.* **2021**, *12*, 711137. [CrossRef] [PubMed]
202. Baker, J.M.; Al-Nakkash, L.; Herbst-Kralovetz, M.M. Estrogen–gut microbiome axis: Physiological and clinical implications. *Maturitas* **2017**, *103*, 45–53. [CrossRef]
203. Ervin, S.M.; Li, H.; Lim, L.; Roberts, L.R.; Liang, X.; Mani, S.; Redinbo, M.R. Gut microbial β-glucuronidases reactivate estrogens as components of the estrobolome that reactivate estrogens. *J. Biol. Chem.* **2019**, *294*, 18586–18599. [CrossRef]
204. Papoulias, D.M.; Tillitt, D.E.; Talykina, M.G.; Whyte, J.J.; Richter, C.A. Atrazine reduces reproduction in Japanese medaka (*Oryzias latipes*). *Aquat. Toxicol.* **2014**, *154*, 230–239. [CrossRef]
205. Richter, C.A.; Papoulias, D.M.; Whyte, J.J.; Tillitt, D.E. Evaluation of potential mechanisms of atrazine-induced reproductive impairment in fathead minnow (*Pimephales promelas*) and Japanese medaka (*Oryzias latipes*). *Environ. Toxicol. Chem.* **2016**, *35*, 2230–2238. [CrossRef] [PubMed]
206. Brain, R.A.; Schneider, S.Z.; Anderson, J.C.; Knopper, L.D.; Wolf, J.C.; Hanson, M.L. Extended fish short term reproduction assays with the fathead minnow and Japanese medaka: No evidence of impaired fecundity from exposure to atrazine. *Chemosphere* **2018**, *205*, 126–136. [CrossRef] [PubMed]
207. Hosmer, A.J.; Schneider, S.Z.; Anderson, J.C.; Knopper, L.D.; Brain, R.A. Fish short-term reproduction assay with atrazine and the Japanese medaka (*Oryzias latipes*). *Environ. Toxicol. Chem.* **2017**, *36*, 2327–2334. [CrossRef] [PubMed]
208. Cleary, J.A.; Tillitt, D.E.; vom Saal, F.S.; Nicks, D.K.; Claunch, R.A.; Bhandari, R.K. Atrazine induced transgenerational reproductive effects in medaka (*Oryzias latipes*). *Environ. Pollut.* **2019**, *251*, 639–650. [CrossRef] [PubMed]
209. Zhou, Y.; Zhu, K.; Wang, Q.; Chen, M.; He, C.; Yang, C.; Zuo, Z. Aryl hydrocarbon receptor agonist diuron and its metabolites cause reproductive disorders in male marine medaka (*Oryzias melastigma*). *Chemosphere* **2022**, *305*, 135388. [CrossRef] [PubMed]
210. Bao, Y.; Zhou, Y.; Tang, R.; Yao, Y.; Zuo, Z.; Yang, C. Parental diuron exposure causes lower hatchability and abnormal ovarian development in offspring of medaka (*Oryzias melastigma*). *Aquat. Toxicol.* **2022**, *244*, 106106. [CrossRef] [PubMed]

211. Flynn, K.; Lothenbach, D.; Whiteman, F.; Hammermeister, D.; Swintek, J.; Etterson, M.; Johnson, R. The effects of continuous diazinon exposure on growth and reproduction in Japanese medaka using a modified Medaka Extended One Generation Reproduction Test (MEOGRT). *Ecotoxicol. Environ. Saf.* **2018**, *162*, 438–445. [CrossRef] [PubMed]
212. Flynn, K.; Johnson, R.; Lothenbach, D.; Swintek, J.; Whiteman, F.; Etterson, M. The effects of combinations of limited ration and diazinon exposure on acetylcholinesterase activity, growth and reproduction in Oryzias latipes, the Japanese medaka. *J. Appl. Toxicol.* **2020**, *40*, 535–547. [CrossRef] [PubMed]
213. Dong, X.; Zhang, L.; Chen, M.; Yang, Z.; Zuo, Z.; Wang, C. Exposure to difenoconazole inhibits reproductive ability in male marine medaka (*Oryzias melastigma*). *J. Environ. Sci.* **2018**, *63*, 126–132. [CrossRef]
214. Dong, X.; Zuo, Z.; Guo, J.; Li, H.; Zhang, L.; Chen, M.; Yang, Z.; Wang, C. Reproductive effects of life-cycle exposure to difenoconazole on female marine medaka (*Oryzias melastigma*). *Ecotoxicology* **2017**, *26*, 772–781. [CrossRef]
215. Chu, S.H.; Liao, P.H.; Chen, P.J. Developmental exposures to an azole fungicide triadimenol at environmentally relevant concentrations cause reproductive dysfunction in females of medaka fish. *Chemosphere* **2016**, *152*, 181–189. [CrossRef] [PubMed]

animals

Review

Histomorphological Changes in Fish Gut in Response to Prebiotics and Probiotics Treatment to Improve Their Health Status: A Review

Giuseppe De Marco, Tiziana Cappello * and Maria Maisano

Department of Chemical, Biological, Pharmaceutical and Environmental Sciences, University of Messina, 98166 Messina, Italy; gdemarco@unime.it (G.D.M.); mmaisano@unime.it (M.M.)
* Correspondence: tcappello@unime.it; Tel.: +39-090-6765741

Simple Summary: Activities such as the digestion and absorption of feeds occur into the gastrointestinal tract (GIT), which also serves to excrete waste products of digestion. These processes occur thanks to the different species of microorganisms inhabiting the GIT, the microbiota, which contribute to the health status of fish by providing metabolic benefits and counteracting pathogen infection. The microbiota is affected by environmental conditions and by the dietary habits of fish species, and it may be modulated by the administration of feed additives based on prebiotics and probiotics. These represent a very useful tool to improve the health status of fish since they are able to enhance gut efficiency, nutrient uptake, defense against pathogens, and growth performance, as may also be assessed by histological endpoints. Overall, a better understanding of the appropriate administration of feed supplements for individual fish species is a promising strategy for improving of the quality and sustainability of the aquaculture sector.

Abstract: The gastrointestinal tract (GIT) promotes the digestion and absorption of feeds, in addition to the excretion of waste products of digestion. In fish, the GIT is divided into four regions, the headgut, foregut, midgut, and hindgut, to which glands and lymphoid tissues are associated to release digestive enzymes and molecules involved in the immune response and control of host-pathogens. The GIT is inhabited by different species of resident microorganisms, the microbiota, which have co-evolved with the host in a symbiotic relationship and are responsible for metabolic benefits and counteracting pathogen infection. There is a strict connection between a fish's gut microbiota and its health status. This review focuses on the modulation of fish microbiota by feed additives based on prebiotics and probiotics as a feasible strategy to improve fish health status and gut efficiency, mitigate emerging diseases, and maximize rearing and growth performance. Furthermore, the use of histological assays as a valid tool for fish welfare assessment is also discussed, and insights on nutrient absorptive capacity and responsiveness to pathogens in fish by gut morphological endpoints are provided. Overall, the literature reviewed emphasizes the complex interactions between microorganisms and host fish, shedding light on the beneficial use of prebiotics and probiotics in the aquaculture sector, with the potential to provide directions for future research.

Keywords: fish gastrointestinal tract (GIT); feed additives; microbial communities; fish welfare assessment; histological assays; aquaculture

1. Introduction

In teleosts, the gastrointestinal tract (GIT) consists of a continuous hollow tube that opens to the outside at both ends, through the mouth at one end and through the anus at the other. Along this structure, it is possible to recognize areas that differ in histology and function since several activities occur, including digestion, absorption, and excretion of waste materials. As in mammals, a number of endogenous enzymes and key molecules produced by associated glands, such as the liver, pancreas, and cells of the intestinal wall, are released into the GIT to contribute to all of these actions [1]. The role of these enzymes is coupled with a variety of fermentation processes carried out by microorganisms (bacteria, fungi, and yeasts) colonizing the GIT that constitute the so-called "microbiota", which has co-evolved with the host in a symbiotic relationship and is responsible for providing metabolic benefits and counteracting pathogen infection.

In fish, the number of microorganisms inhabiting the GIT has been estimated in the range of 10^7–10^8 per gram [2]. However, the GIT microbiota is highly variable and its normal variation in abundance and diversity of microorganisms is affected by several factors [3] including age, diet, host genetics, and the environment (freshwater, seawater). Feeding habits are surely one of the most relevant factors affecting the composition of the GIT microbiota among fish species. Indeed, it is well known how different dietary habits (herbivores, carnivores, omnivores, detritivores) come together with variations in both fish GIT morphology and microbial populations [3–5]. The strict correlation between the structures of the digestive apparatus and the feeding habit of fishes [6], as revealed in some morphometric parameters such as the intestinal length and area, highly influences the microbial populations detected along the alimentary canal [7]. On the other hand, microorganisms may also affect the digestive morphological structures and functions, both negatively [8], and positively, as in the case of probiotics [9].

In accordance with the currently adopted definition by the Food and Agricultural Organization of the United Nations (FAO) and the World Health Organization (WHO), probiotics are live microorganisms that, when administered in adequate amounts, confer health benefits to the host [10]. In fact, after administration, probiotics are able to colonize and multiply in the gut of the host and enact numerous beneficial effects by modulating various biological systems in the host [11,12], including immune status, growth performance, disease resistance, and feed conversion, with an overall improvement of the fish health status [13,14]. Probiotics, together with other additives such as prebiotics, which are referred to as the food or energy provider for good microorganisms [15], aim to optimize the host–microbiota ratio, which plays a key role in ensuring the proper functioning of the fish GIT [3,16]. Therefore, for all the above-mentioned reasons, the use of these feed additives has gained a key strategic role in the aquaculture sector [14,16].

Overall, this review focuses on the current knowledge regarding the use of a variety of probiotics and prebiotics in different fish species for the improvement of their health and gut efficiency. Furthermore, particular attention was paid to the mutual influences between the microbial populations and morphological features of the fish GIT areas, especially of the gut tract, which is usually investigated by histological assays to evaluate the performance of various feed additives in aquaculture.

2. The Gastrointestinal Tract (GIT) in Fish

As mentioned above, along the "tube"-termed GIT, it is possible to recognize different areas distinguishable from a functional and histological point of view. The main areas of the fish GIT can be summarized as follows:

- Headgut: the area where it is possible to distinguish between a buccal and a pharyngeal cavity;
- Foregut: the area that begins at the posterior edge of the gills and includes the oesophagus, the stomach, and the pylorus;
- Midgut: the anterior intestine, which includes a variable number of pyloric caeca or appendages which are useful for increasing the surface area and optimizing the absorption of nutrients;
- Hindgut: the area including the distal intestine and the anus.

The headgut, mainly characterized by the mouth cavity, plays the key role of ensuring feed acquisition; therefore, it is not unusual to observe differential features of this area among fish species in relation to their different feed habits. For instance, in lepidophagous fishes, the occurrence of a large sub-terminal mouth coupled to a unique arrangement of teeth on the jaws enables this species to perform its specialized feeding behaviour, consisting of scraping and eating the scales of other fish [7–18]. Additionally in the headgut, it is possible to find other distinctive anatomical structures such as the pharyngeal jaw apparatus that, combined with the oral jaws, allows some fish species to crush shells, mince feed, and separate edible material from unwanted debris, thus expanding their feeding capabilities [19].

The first part of the foregut is characterised by the oesophagus, with a squamous epithelium in the anterior part and a columnar one in the posterior part [20]. Noteworthy, the epithelium of the foregut has a different origin from that of the midgut, namely ectodermal for the foregut and endodermal for the midgut [21]. In addition, the foregut is characterized by the presence of goblet cells that, by producing mucous, facilitate the acquisition of food, preventing abrasion as well as invasion of pathogens [22]. These cells contribute to form the mucous layer, beneath which, a submucosal layer composed of a thick mass of loose connective tissue combined with a muscular layer is immediately arranged, which is crucial for the integrity of the oesophagus wall [23]. The occurrence of a muscular layer, which is also coupled with long branched folds in some species, suggests a certain elasticity and stretching capacity of the luminal surface area of the oesophagus, particularly in predatory fish [20,24]. Furthermore, in some fish species, the oesophagus is characterized by the presence, especially in the cranial portion, of several taste buds that are helpful for food selection [25].

Following the oesophagus, as in other vertebrates, it is quite common to recognise the stomach. Nevertheless, it is worthy to highlight that the stomach is lacking in 20% of fish species (i.e., Gobiidae, Blennidae, Cyprinadae), where it is partly replaced by other adaptations such as well-developed pharyngeal teeth, pharyngeal pockets, secretory glands present in the oesophagus, or a muscular gizzard [26,27]. In fish species where the stomach occurs, it may assume different conformations (straight, U-shaped, or Y-shaped), normally linked to the feeding habits of the specimens. For instance, the U-shaped stomach is typical of herbivores/omnivores, whereas the Y-shaped stomach is present in some predators since it allows the storage of large amounts of food [7,28]. Despite the different morphological structures, it is possible to recognize three distinct regions in fish stomach, defined as cardiac, fundic, and pyloric [17]. The primary roles of the fish stomach are the storage of food and the production of hydrochloric acid (HCl), which is necessary to promote the digestion process. The HCl secreted by the parietal cells of gastric glands generates the strongly acidic environment (pH < 2) of the gastric lumen, which is crucial for the correct activity of several digestive enzymes [21,22]. The gastric gland allocation appears to be greater in the anterior than in the posterior part of the stomach, the latter represented by the pyloric region that performs a key role in food storage rather than digestion. The occurrence of neutral mucins throughout the stomach, produced by the columnar epithelial cells, ensures a dual protective function against the presence of microorganisms and the high acidity of the gastric lumen [17,24,28]. Even though digestion into the stomach is primarily carried out by digestive enzymes, in some fish species (i.e., Clupeoidei, Channidae, Mugilidae, Acipenseridae, Coregoninae, and Chanidae), it is possible to notice a pyloric stomach able to grind food, acting as a sort of "gizzard" [5,21].

In the midgut, the forward part of the gut, the assimilation of digested food into the stomach begins. In order to fulfil this task, the pyloric caeca are located in this portion. The pyloric caeca are finger-like outgrowths whose shape permits an increase of the absorption surface area without promoting fermentation and storage, as demonstrated by [29]. Similar to the other portions of the GIT, differences in regard to size, state of branching, and connection to the gut are detectable in the various fish species. For instance, in the Mugilidae, the number of pyloric caeca is a crucial element for the identification of the species [30]. However, no clear correlation seems to exist between the pyloric caeca shape and feed habits of fish [31,32]. Interestingly, the midgut is the longest portion of the gut and its length is extremely variable and generally correlated with the feeding habits of the fish species. For instance, carnivorous species frequently possess intestines that are shorter than herbivorous fish, with large coils occurring in the former [33]. From a histological perspective, three layers, namely the mucosa, muscularis externa, and serosa, may be observed in the gut [26]. In the mucosa layer, the columnar epithelium plays the absorptive role, as testified by the presence of a distinctive brush border in the apical part. In the same layer, it is possible to observe other types of cells such as goblet-type mucous cells, lymphocytes, enteroendocrine cells, and rodlet cells [34], each one performing a different role. Indeed, the goblet cells are mucous cells [21], the enteroendocrine cells produce neuroendocrine substances [35], and the lymphocytes and rodlet cells have an immune function [36,37]. In contrast to mammals, in the fish gut, it is not possible to identify the villi organisation, but rather the existence of irregular random folds without crypts. Lieberkuhn crypt-like structures are encountered only in the Gadidae, such as cods [21,26].

The terminal part of the GIT is the hindgut. The hindgut of most fish is short and very difficult to distinguish from the midgut in terms of histomorphology, in particular with respect to changes in diameter or epithelial morphology. Additionally, in the hindgut, unlike the midgut, it is possible to observe an increased level of mucus production [5,32] and variations with regards to the mucosal fold and muscle thickness [21]. In the posterior intestine, unlike the midgut, a higher amount of goblet cells may be observed, resulting in greater mucous production. The rise in the presence of goblet cells from the anterior to posterior intestine, common in various fish species, provides epithelium protection and lubrification in order to facilitate food flow and defecation [17,22,38,39].

3. The Interaction between Fish Immune System and Microbiota

As previously described, a mucous layer is present, lining the entire lumen of the GIT. This mucous system allows for the development of an immune system, able to enhance innate or adaptive immune responses [40]. The immune system avoids the invasion of opportunistic and pathogenic microorganisms, whilst simultaneously preserving and enhancing the proliferation of commensal bacteria [41–43]. The innate immunity represents the main defence line for fish, despite the occurrence of an adaptive immune system, which has a restricted antibody repertoire and slower lymphocyte maturation compared to mammals [44]. The innate immunity is composed of physical barriers, humoral, and cellular components [40]. The physical barriers are characterized by a series of mucosal structures located in the skin, gills, and gut, which are able to prevent pathogen invasion. In the mucosal secretions, it is possible to observe several molecules such as:

- Antimicrobial peptides (AMPs), which are small peptides (up to 50 amino acids residues). In fish species, it is possible to find a variety of major groups of AMPs such as the piscidins, cathelicidins, defensins, hepcidins, and high-density lipoproteins [44]. AMPs can guarantee general microbiota homeostasis thanks to their antimicrobial activity of immunity modulation [45].
- Lysozymes, which are other useful components against pathogenic microorganisms. Indeed, these enzymes are able to cleave the glycosidic bond between the N-acetylmura mic acid and N-acetylglucosamine of bacterial cell wall peptidoglycans [46]. In fish, the role of lysozymes against pathogens has been demonstrated in several studies [47–50], which reported the modulation in their expression and activity versus several fish pathogen bacteria, such as *Streptococcus iniae* or *Vibrio alginolyticus*.
- Complement system, which represents the element able to connect the innate and adaptive immunity. Approximately 30 inactive circulating proteins and membrane-bound receptors belong to this system [51]. The activation of these proteins is related to three different pathways: the classical pathway, where antigen-antibody complexes act as activators; the alternative pathway, where the activation is caused by the presence of several molecules on the surface of microorganisms without the antibodies attendance; and the lectin pathway, where the activation occurs after the mannose-binding lectin binds to the cell-surface carbohydrates of microorganisms [52,53]. Regardless of the three complement pathways, the C3 protein always plays the main role in the function

of the complement system. Indeed, the C3 cleavage of C3 generates two protein subunits, namely C3a and C3b. C3b is fundamental because it activates the lytic pathway by attaching itself to the pathogen surface. In addition, C3b permits the sequential connection of C5b, C6, C7, C8, and C9 proteins to form the membrane-attack complex (MAC), able to provoke cell lysis [53].
- Other useful compounds, such as transferrin, pentraxins, and lectins. These compounds represent further useful weapons to avoid the growth of dangerous bacteria and allow for beneficial microbiota homeostasis [43,44,51].

Among the physical barriers and antimicrobial compounds, several kinds of cells, such as the monocytes/macrophages, neutrophils, dendritic cells, and natural killer cells, represent the cellular component of the innate immune system in fish species. These cells are present in the Gut-Associated Lymphoid Tissue (GALT), where it is also possible to observe the lymphocytes, fundamental in the adaptive immune system. In contrast to mammals, in the fish gut, these cells are located between the epithelium and lamina propria, without a tissue organization [43]. Additionally, the epithelium is able to corroborate to the immune responses, producing specific receptors called Toll-like receptors (TLRs), which are important mediators of inflammatory pathways in the gut, playing a key role to recognize specific microbial patterns [54]. When this pattern is linked to the receptor, the cells start to produce several signal molecules (i.e., tumor necrosis factors (TNF), interleukins, and chemokines) able to guarantee the recruitment of innate immune cells. The innate immune cells (i.e., monocytes/macrophages, neutrophils, dendritic cells, and natural killer cells) use differential mechanisms of action to inhibit microbial proliferation. Indeed, in order to destroy pathogenic organisms, the innate immunity involves the activation of phagocytose (mediated by monocytes/macrophages) and the release of cytotoxic molecules (mediated by natural killer cells) or cytokines for the recruitment of adaptive immune cells [51]. For several years, researchers had supposed that adaptive immune cells did not have a specific role in the microbiota balance; however, more recent studies [55,56] suggested a possible implication in this context, mediated mainly by lymphocytes [57]. Although the role of the B and T lymphocytes seems to be not so different, some studies indicated that T and Z immunoglobulins (Ig) produced by the B lymphocyte are able to guarantee correct microbial colonization [43,58]. It has been reported that IgT is the primary immunoglobulin produced in response to the pathogenic microbiome in the fish gut and prevents any damaging effects of pathogens on the fish [3]. In general, the presence of specific bacteria is required for normal gut development in terms of the proliferation of epithelial cells and maturation of the immune system [43]. As a matter of fact, studies of the gnotobiotic zebrafish have pointed out the relevance of the microbiota in the proliferation of the intestinal epithelial cells by involving MyD88 [59] and in promoting the movement of immune cells to the intestine, such as the migration of neutrophils induced by commensal microorganisms activating serum amyloid A [60]. Additionally, the microbiota is crucial for regulating the neutrophil activity by enhancing the expression of several pro-inflammatory genes [61,62], thus improving infection resistance [43].

4. The Interaction between Fish GIT and Microbiota

The microorganisms that belong to fish microbiota can colonize several parts of the fish body such as the skin, gills, and obviously the gastrointestinal tract. Indeed, although the gut, with its morphology, helps the microbial colonization, other portions of the fish GIT can host different microorganisms, such as bacteria, fungi, and viruses [7,63]. Bacteria represent the most abundant component throughout the fish GIT, with an amount ranging from 10^7 to 10^{11} microorganisms per gram of intestinal content [2,12,64]. As a result, the majority of the research studies are primarily concerned with them. As in other animals, the microbiota colonization withstands the influence of different factors. Agents such as the environment (freshwater or saltwater), the feed habits (herbivores, carnivores, omnivores, or detritivores), and GIT shape are among the main ones [3]. The influence of the environment is confirmed by various studies [2,63] that highlighted how the genera

Alteromonas, *Flavobacterium*, *Pseudomonas*, and *Vibrio* are more abundant in saltwater fish species, while in freshwater ones, the main microbial genera are represented by *Aeromonas*, *Lactococcus*, *Pseudomonas*, and *Clostridium*. A comparison of the main microorganisms found in the GIT of freshwater and saltwater fishes is reported in Table 1, together with data about the dominant bacteria stains typically detected in mammals.

Table 1. Comparison of the main microorganisms found in the gastrointestinal tract (GIT) of mammals, freshwater, and saltwater fishes.

	Amount of Bacteria in GIT	Dominant Microbial Phyla	Less Abundant Microbial Phyla	References
Mammals	10^8 microbial cells for gram of intestinal content	Bacteroidetes, Firmicutes	Proteobacteria, Actinobacteria, Fusobacteria, Cyanobacteria	[65–67]
Freshwater fishes	10^7 to 10^{11} microbial cells for gram of intestinal content	Proteobacteria, Fusobacteria, Firmicutes, Bacteroidetes, Actinobacteria *Aeromonas*, *Pseudomonas*, *Bacteroides* type A	Verrucomicrobia, Enterobacteriaceae, *Micrococcus*, *Acinetobacter*, *Clostridium*, *Bacteroides* type B, *Fusarium*, *Plesiomonas*	[11,12,68–71]
Saltwater fishes	10^7 to 10^{11} microbial cells for gram of intestinal content	Proteobacteria, Fusobacteria, Firmicutes, Bacteroidetes, Actinobacteria *Vibrio*, *Pseudomonas*, *Acinetobacter*, *Corynebacterium*, *Alteromonas*, *Flavobacterium*, *Micrococcus*	Verrucomicrobia	[11,12,68–70]

As with other species, even among fish, it is possible to distinguish among different feed habits. The diet of a species is the result of specific adaptations that are also reflected in the digestive system, including the occurrence of digestive enzymes. Since a significant part of the digestive enzymes are synthesized by bacteria, the connection between the trophic level of a species and microbiota is not unusual [72]. Indeed, different studies [4,7,73] have documented the prevalence of some bacteria in fish species with similar feed habits [63], as also highlighted by [72], which reported the prevalence of cellulose-degrading bacteria such as *Clostridium*, *Citrobacter*, and *Leptotrichia* in the herbivores, while protease-producing bacteria, such as *Cetobacterium* and *Halomonas*, were reported in the carnivorous fish species. A more detailed list with the dominant microbial phyla found in the GIT of fish species of commercial relevance, grouped according to their feed habits, is reported in Table 2.

Table 2. Comparison of the main microorganisms found in the gastrointestinal tract of fish species according to their feed habits. (SW—seawater; FW—freshwater; AE—anadromous euryhaline).

Feed Habits	Fish Species	Dominant Microbial Phyla	References
Herbivores	*Acanthurus nigricans* SW	Bacteroidetes, Proteobacteria, Firmicutes	[74]
	Acanthurus sp. SW	*Epulopiscium*	[75]
	Aplodactylus arctidens SW	*Clostridium, Eubacterium desmolans, Papillibacter cinnaminovorans*	[76]
	Chlorurus sordidus SW	*Vibrio, Photobacterium*	[74]
	Hermosilla azurea SW	*Enterovibrio, Bacteroides, Faecalibacterium, Desulfovibrio*	[77]
	Kyphosus spp. SW	Gamma-proteobacteria, Clostridia, Bacteroidia, Erysipelotrichia (*Vibrionaceae, Pasteurellaceae, Erysipelotrichaceae*)	[78]
	Odax pullus SW	*Clostridium*	[76]
	Ctenopharyngodon idellus FW	*Clostridium, Citrobacter, Leptotrichia*	[72]
	Megalobrama amblycephala FW	*Brevundimonas, Massilia, Curvibacter, Acinetobacter, Clostridium, Citrobacter, Leptotrichia*	[72,79]
Carnivores	*Acipenser baerii* SW	*Cetobacterium*	[80]
	Chaenocephalus aceratus SW	*Photobacterium*	[81]
	Cynoscion nebulosus SW	*Escherichia*	[82]
	Epinephelus coioides SW	*Bacillus, Vibrio, Delftia, Psychrobacter, Acinetobacter, Pseudomonas*	[83]
	Fugu niphobles SW	*Vibrio, Pseudomonas, Flavobacterium*	[84]
	Gadus morhua SW	*Clostridium, Vibrio*	[85]
	Hippoglossus hippoglossus SW	*Vibrionaceae* (in larval and juvenile stages), *Photobacterium phosphoreum* (in adults)	[71]
	Lutjanusn bohar SW	*Vibrio, Photobacterium*	[74]
	Morone saxatilis SW	*Aeromonas, Pseudomonas, Vibrio*	[86]
	Notothenia coriiceps SW	*Photobacterium, Vibrio*	[81]
	Paralichthys lethostigma SW	*Clostridium, Photobacterium,*	[82]
	Pomatomus saltatrix SW	*Vibrio, Pseudomonas, Enterobacteraceae*	[87]
	Sciaenops ocellatus SW	*Mycoplasmataceae, Photobacterium, Cetobacterium, Clostridiaceae, Vibrio*	[82]
	Sparus aurata SW	*Pseudomonas*	[88]
	Salmo trutta trutta SW	*Aeromonas, Pseudomonas*	[89]
	Salmo salar AE	*Acinetobacter junii, Mycoplasma, Lactobacillus, Photobacterium phosphoreum, Lactococcus, Bacillus*	[90]
	Culter alburnus FW	*Cetobacterium, Halomonas*	[72]

Table 2. *Cont.*

Feed Habits	Fish Species	Dominant Microbial Phyla	References
	Pelteobagrus fulvidraco FW	*Clostridium, Yersinia, Aeromonas, Clostridiales*	[79]
	Siniperca chuatsi FW	*Cetobacterium, Halomonas*	[72]
Onnivores	*Gillichthys mirabilis* SW	*Mycoplasma*	[91]
	Lagodon rhomboides SW	*Clostridium, Mycoplasma, Photobacterium, Propionibacterium, Staphylococcus, Pseudomonas, Corynebacterium*	[82]
	Cariassus auratus FW	*Brevundimonas, Massilia, Curvibacter, Delftia Clostridium, Cetobacterium Halomonas*	[72,79]
	Cyprinus carpio FW	*Brevundimonas, Massilia, Curvibacter, Sphingobacteriales, Cetobacterium, Aeromonas, Chitinophaga, Halomonas, Clostridium*	[72,79]
Zooplanktivores	*Apogonidae* SW	*Vibrionaceae, Pasteurellaceae, Vibrio harveyi, Shewanella* sp., *Endozoicomonas* sp.	[73]
	Clupea harengis SW	*Pseudomonas, Alteromonas, Psychrobacter*	[92]
	Pomacentridae SW	*Vibrionaceae, Pasteurellaceae, Vibrio harveyi, Shewanella* sp., *Endozoicomonas* sp.	[73]
	Sardinella longiceps SW	*Achromobacter, Vibrio, Pseudomonas*	[93]
	Scomber scombrus SW	*Psychrobacter, Vibrio, Shewanella*	[94]
	Syngnathus scovelli SW	*Proteobacteria*	[82]

As described previously, in the fish GIT, is possible to recognize different regions, and each one is characterized by a different microbial density. Indeed, a certain amount of bacteria is detectable along the entire gastrointestinal tract, but some parts of the GIT are more colonized than others. In general, in fish, the bacterial colonization seems to follow an increasing trend from the stomach to the hindgut [63]. Nevertheless, this tendency is also associated with other different situations of constant microbial density along the fish GIT [95], or a reduction in the amount of microbiota from the mouth to the anus [96]. In addition to the previously reported quantitative differences, a wide range of microbial species can be identified in the various parts of the digestive tract in different fish species, as reported in Table 3.

Table 3. Distribution of microorganisms along the gastrointestinal tract (GIT) of different fish species. (SW—seawater; FW—freshwater; AE—anadromous euryhaline).

GIT	Fish Species	Dominant Phyla	References
Stomach	*Sparus aurata* SW	Firmicutes, Proteobacteria, Bacteroidetes (*Bacillales, Flavobacteriaceae*)	[97]
	Sparus aurata SW	Firmicutes, Proteobacteria, Actinobacteria	[98]
	Epinephelus awoara SW	Firmicutes, Proteobacteria, Actinobacteria, *Deinococcus-Thermus, Planctomycete*	[99]
	Scophthalmus maximus SW	Proteobacteria, Firmicutes, Bacteroidetes, Tenericutes, Actinobacteria, Cyanobacteria	[100]
	Oncorhynchus mykiss FW	Proteobacteria, Firmicutes, Fusobacteria (*Cetobacterium dominanti*)	[101]

Table 3. *Cont.*

GIT	Fish Species	Dominant Phyla	References
Foregut (oesophagus, stomach, pylorus)	*Salmo salar* AE *Gadus morhua* SW	Proteobacteria (*Janthinobacterium, Pseudomonas, Acinetobacter, Vibrio)*	[70,90]
	Scophthalmus maximus SW	Proteobacteria, Firmicutes, Tenericutes, Actinobacteria, Cyanobacteria	[100]
Midgut	*Salmo salar* AE *Gadus morhua* SW	Proteobacteria (*Photobacterium phosphoreum, Pseudomonas)*	[70,90]
	Pelteobagrus fulvidraco FW	Firmicutes, Proteobacteria, Bacteriodetes, Fusobacteria	[102]
	Oncorhynchus mykiss FW	Proteobacteria, Firmicutes, Fusobacteria, *Bacteroidetes* (*Bacillus*)	[101]
	Ctenopharyngodon idella FW	Proteobacteria, Firmicutes, Actinobacteria (*Anoxybacillus, Leuconostoc, Clostridium, Actinomyces, Citrobacter*)	[103]
Hindgut	*Salmo salar* AE *Gadus morhua* SW	*Vibrio, P. phosphoreum*	[70,90]
	Scophthalmus maximus SW	Proteobacteria, Firmicutes, Tenericutes	[100]
	Acipenser baerii AE	Fusobacteria/Firmicutes, Chlamydiae, Bacteriodetes, Actinobacteria	[80]
	Danio rerio FW	Firmicutes, Proteobalceria, Bacteriodetes	[104]

Data reported in the previous published articles allow us to create a sort of "map" detailing one microbial species is more commonly present than another. The foregut, in particular the stomach, is not often subjected to microbiota analysis in comparison to the other regions of the fish GIT because it is thought that the data obtained can be easily affected by the fast movement of the food. Nevertheless, some research studies [7,97–99] have focused on the microbiota in the stomach, demonstrating the dominance of the phyla Firmicutes, Proteobacteria, and Actinobacteria in fish species such as *Sparus aurata* and *Epinephelus awoara*. These findings may suggest how these microbial phyla could likely adapt to both the low gastric pH and the activity of proteolytic enzymes [99].

In the other regions of the fish GIT, namely the midgut and hindgut, the physico-chemical conditions are more suitable for microbial proliferation than those observed in the foregut, particularly in stomach, and therefore, the bacteria present in the gut differ from those found in this area [105]. In certain cases, even the microbial population detected in the midgut and hindgut might differ, as different bacteria genera, such as *Pseudomonas* for the midgut and *Vibrio* for the hindgut, were, for instance, detected in *Salmo salar* and *Gadus morhua* [70,90]. Given the natural connection established between the host's microbiota and its health status [43], the knowledge gained so far regarding the microbial communities offers useful information to optimize strategies of gut microbiota handling with the aim of enhancing the health status of fish, especially in the aquaculture sector [7].

5. Use of Feed Additives for Fish Health Improvement and Gut Efficiency

Over the last decades, the increased development of aquaculture activity has led to a growing attention on the wellness of fish species of commercial relevance. Nevertheless, the productivity growth of the aquaculture sector appears to be challenged by the frequent infections impacting the commercial fish species triggered by pathogens. Among the numerous forms of stressors, bacterial infections are one of the most widespread causes of harm in fish farming, with considerable economic implications [106]. To counteract this serious issue, as in other kinds of rearing, antibiotics are routinely used in the aquaculture sector. However, the extensive use of antibiotics to avoid the onset of diseases in aquaculture facilities may lead to the leakage of these drugs into wastewater, and consequently into rivers and seas [107,108]. Antibiotics, particularly those present in rivers, have the ability to permeate soil and enhance the selection and growth of antibiotic-resistant bacteria [109]. The variety of environmental implications notwithstanding, antibiotic usage can lead to a substantial loss of gut microbiota variability [110]. Generally, the conditions which fish are subjected to in fishery farms might be a cause of stress. Indeed, the continuous maintenance of different environmental factors such as temperature, photoperiod, and salinity in aquaculture plants is not always straightforward, and significant fluctuations in these parameters are a very common occurrence. As a result, the stress exposure in farmed fish may have a negative impact on a variety of physiological mechanisms, including the immune system [111]. Therefore, due to the huge impact that antibiotics have on the environment and health of farmed fish, it is crucial to establish alternative solutions capable of preserving the well being of farmed species and at the same time, reducing the ecological harm [112].

According to numerous recent studies [13,111–114], feed additives such as prebiotics and probiotics provide a tool capable of boosting aquaculture plant productivity by acting on a variety of ecological and biological aspects while minimizing the antibiotic consumption. Besides being identified as a very good alternative to the usage of antibiotics, the use of these feed additives in aquaculture represents a useful strategy for improving the overall performance of the aquaculture sector since it affects a variety of factors including rearing water quality [115,116], food absorption, and digestion, and has shown a positive effect on the growth performance of fish [117–119]. From an immunological perspective, the activity of these feed additives does not merely promote the proliferation of commensal bacteria at the detriment of pathogens. In actuality, it has been demonstrated that prebiotics and probiotics may enhance and improve the immune response of fish species by either increasing the leukocyte amount or the activity of key proteins such as lysozymes, as well as the complement system [120–123]. More in detail, data from the literature on the administration of specific prebiotics (Table 4) and probiotics (Table 5) acting as effective immunostimulants on aquaculture fish species from seawater and freshwater are listed in the following tables.

Table 4. Administration of prebiotics acting as effective immunostimulants in different aquaculture fish species. (SW—seawater; FW—freshwater; AE—anadromous euryhaline; MOS—mannan-oligosaccharides; FOS—fructo-oligosaccharides; COS—chito-oligosaccharides; GOS—galacto-oligosaccharides).

Prebiotics	Fish Species	Dosage and Timing of Administration	Initial Weight (g)	Effects	References
Arabinoxylan-oligosaccharides	*Acipenser baerii* AE	2% for 28 days and 12 weeks	20.00	Increase of phagocytic activity of leukocytes	[124]
β-glucan	*Epinephelus coioides* SW	1, 2 g/kg for 30 days	6.40 ± 0.65	Enhancing of innate immune responses	[125]
β-glucan + MOS	*Cyprinus carpio* FW	Immunogen® at 0.5, 1, 1.5, 2.5 g/kg for 8 weeks	11.12 ± 0.55	Increase of leukocyte count and resistance to *Aereomonas hydrophila*	[126]
COS	*Cyprinus carpio* FW	0.2% for 8 weeks	24.90 ± 0.52	Increase of lysozyme level	[127]

Table 4. *Cont.*

Prebiotics	Fish Species	Dosage and Timing of Administration	Initial Weight (g)	Effects	References
COS	*Scophthalmus maximus* SW	75, 150, 300, 600, 1200 mg/kg for 56 days	2.10 ± 0.10	Increase of phagocytic activity at doses > 300 mg/kg	[128]
COS	*Scophthalmus maximus* SW	2 g/kg for 56 days	11.00	Reduction of gene expression of pro-inflammatory cytokines	[129]
FOS	*Atractosteus tropicus* SW (Larvae)	7.5 g/kg for 15 days	0.03 ± 0.006	No significant increase in gene expression of pro-inflammatory cytokines	[130]
FOS	*Megalobrama terminalis* FW	0.3, 0.6% for 8 weeks	30.50 ± 0.50	Increase of various factors of immune system (IgM and complement proteins)	[131]
FOS	*Oreochromis niloticus* FW	0.5, 1, 2, 4 g/kg for 8 weeks	5.00 ± 0.02 g	Enhancing of immune and antioxidant responses, and resistance to *A. hydrophila* (optimum at 1 g/kg)	[132]
FOS	*Salmo salar* AE	1% for 4 months	200.20 ± 0.60	No changes on immune system parameters	[133]
FOS	*Rutilus rutilus* FW	1%, 2%, 3% for 7 weeks	0.67 ± 0.03	Increase of immunoglobulins and lysozyme activity	[118]
FOS	*Paralichthys olivaceus* SW	0.5% for 56 days	21.00	Increase of lysozyme activity	[134]
FOS	*Acipenser stellatus* AE	1%, 2% for 11 weeks	30.16 ± 0.14	Increase of lysozyme activity at 1% dose	[135]
FOS	*Trachinotus ovatus* SW	2, 4 g/kg for 56 days	10.32	Increase of lysozyme activity and immunoglobulins	[136]
GOS	*Danio rerio* FW	0.5%, 1%, 2% for 56 days	0.045 ± 0.001	Enhancement of immune system parameters	[137]
Inulin	*Cyprinus carpio* FW	1, 2 g/kg for 60 days	25.37 ± 0.22	Enhancement of immunoglobulins of epithelial mucosa (optimum at 2 g/kg)	[138]
Inulin	*Huso huso* AE (juveniles)	1%, 2%, 3% for 8 weeks	16.14 ± 0.38	Increase of leukocyte count	[139]
Inulin	*Sparus orata* SW	10 g/kg for 2 and 4 weeks	50.00	Increase of immunglobulins IgM and immune responses, resistance to *P. damselae*	[140]
Inulin	*Oreochromis niloticus* FW	5 g/kg for 1 and 2 months	11.00	Increase of di haematocrit level, lysozyme activity and resistance to *A. hydrophila*	[120]

Table 4. *Cont.*

Prebiotics	Fish Species	Dosage and Timing of Administration	Initial Weight (g)	Effects	References
Inulin	*Lates calcarifer* FW	15, 20 g/kg for 60 days	7.14 ± 0.05	Enhancement of immune and blood parameters	[141]
Levan	*Cyprinus carpio* FW	0.1%, 0.2%, 0.5% for 75 days	9.00 ± 0.50	Increase of lysozyme level (optimum at 0.5% dose)	[142]
Levan	*Cyprinus carpio* FW	0.75% for 60 days	3.31 ± 0.52	Increase of myeloperoxidases and immunoglobulins	[143]
Levan	*Labeo rohita rohu* FW	0.1%, 0.25%, 0.50%, 0.75%, 1%, 1.25% for 60 days	4.50 ± 0.14	Increase of lysozyme level (optimum at 1.25% dose)	[144]
Levan	*Epinephelus coioides* SW	5, 10, 25, 50 g/kg for 12 weeks	6.00	Increase of lysozyme level (optimum at 25 g/kg)	[145]
MOS	*Salmo salar* AE	10 g/kg for 4 months	200.20 ± 0.60	Reduction of lysozyme activity	[133]
MOS	*Dicentrarchus labrax* SW	2%, 4% for 8 weeks	116.00	Increase of leukocyte activity and resistance to *Vibrio anguillarum* at 4% dose	[123]
MOS	*Oncorhynchus mykiss* FW	2 g/kg for 42 days	30.00	Increase of antibody titers and lysozyme	[146]
MOS	*Oncorhynchus mykiss* FW	0.4% for 12 weeks	13.20	Increase of phagocytic activity, haematocrit and resistance to *Vibrio anguillarum*	[147]
MOS	*Oncorhynchus mykiss* FW	0.25%, 0.5% for 12 weeks	36.27 ± 0.42	Increase of phagocytic activity, haematocrit and resistance to *Aereomonas salmonicida* at 0.5% dose	[148]
MOS	*Clarias gariepinus* FW	10 g/kg for 45 days	35.00	Increase of lysozyme activity	[149]
MOS	*Channa striata* FW	2 g/Kg for 12 weeks	10.00	Increase of lysozyme activity	[150]
MOS	*Pangasianodon hypophthalmus* FW	0.2%, 0.4%, 0.6%, 0.8% for 12 weeks	20.41 ± 1.64	Enhancement of immune parameters (immunoglobulins, lysozyme, leukocyte count) and resistance to *A. hydrophila*	[151]
MOS	*Sciaenops ocellatus* FW	10 g/kg for 6–8 weeks	10.00	Increase of lysozyme activity	[152]
MOS	*Gadus morhua* SW	1 g/kg for 5 weeks	90.00	Increase of expression of interleukins and resistance to *V. anguillarum*	[153]

Table 4. *Cont.*

Prebiotics	Fish Species	Dosage and Timing of Administration	Initial Weight (g)	Effects	References
Nucleotides (NucleoforceFish™)	*Sparus aurata* SW	250, 500 mg/kg for 150 days	0.36 ± 0.002	Increase of gene expression of interleukins, hepcidin and B receptors of T cells	[154]
Sodium alginate	*Epinephelus coioides* SW	10 g/kg for 12 days	19.50 ± 0.50	Increase of immune responses (complement system, phagocytic activity, lysozyme)	[155]
Stachyose	*Scophthalmus maximus* SW	1.25%, 5% for 12 weeks	4.63 ± 0.01	Inhibition of microbial pathogen growth and increase of beneficial ones (optimum at 5% dose)	[156]
Thymol and carvacrol	*Sparus aurata* SW	0.01% for 9 weeks	26.00 ± 1.00	Increase of interleukins, cytokines and inflammatory responses	[157]
Thymol, carvacrol and essential oils from *Allium* spp.	*Sparus aurata* SW	0.5%	40.30 ± 0.1	Increase of immune innate responses at epithelial level and cortisol-mediated response, and lower microbial growth	[158]

Table 5. Administration of probiotics acting as effective immunostimulants in different aquaculture fish species. (SW—seawater; FW—freshwater; AE—anadromous euryhaline; CFU—colony-forming unit).

Probiotics	Fish Species	Dosage and Timing of Administration	Initial Weight (g)	Effects	References
Aspergillus oryzae	*Oreochromis niloticus* FW	1×10^{10} CFU/g	19.50 ± 0.50	Increase of non-specific immune and antioxidant responses	[159]
Bacillus aerophilus	*Labeo rohita* FW	1×10^6, 1×10^7, 1×10^8, 1×10^9 CFU/g for 6 weeks	35.00–40.00	Increase of lysozyme and phagocytic activity, and resistance to *A. hydrophila* (at 1×10^8 CFU/g)	[160]
Bacillus amyloliquefaciens	*Labeo rohita* FW	1×10^5, 1×10^7, 1×10^9 CFU/g for 70 days	20.23	Increase of antibody titers	[161]
Bacillus amyloliquefacien, *Bacillus pumilus*	*Pangasianodon Hypophthalmus* FW	1×10^8, 3×10^8, 5×10^8 CFU/g for 90 days	15.30 ± 1.20	Increase of lysozyme and phagocytic activity, and resistance to *E. ictaluri* with reduced death rate (optimum at 5×10^8 CFU/g)	[162]

Table 5. *Cont.*

Probiotics	Fish Species	Dosage and Timing of Administration	Initial Weight (g)	Effects	References
Bacillus licheniformis	*Cyprinus carpio* FW	1×10^6, 1×10^7, 1×10^8 CFU/g for 60 days	38.58 ± 0.42	Increase of pro-and anti-inflammatory citokines and resistance to *A. hydrophila* (optimum at 1×10^8 CFU/g)	[163]
Bacillus licheniformis	*Oreochromis mossambicus* FW	1×10^5, 1×10^7 CFU/g for 4 weeks	24.00 ± 2.50	Increase of lysozyme activity and resistance to a *A. hydrophila*	[164]
Bacillus licheniformis, *Bacillus amyloliquefaciens*	*Centropomus undecimalis* SW (Larvae)	1×10^{13} CFU/g for 28 days		Reduction of lysozyme activity and not significant increase of SOD	[165]
Bacillus licheniformis, *Bacillus pumilus*	*Labeo rohita* FW	1×10^8 CFU/g for 14 days	50.00–60.00	Increase of immune response and resistance to *A. hydrophila*	[122]
Bacillus licheniformi, *Bacillus subtilis*	*Ctenopharyngodon idella* FW	1×10^8 CFU/g for 4 weeks	45.00	General enhancement of immune parameters	[166]
Bacillus licheniformis Dahb1	*Pangasius hypophthalmus* FW	1×10^5, 1×10^7 CFU/mL for 24 days	15.00 ± 2.50	Increase of lysozyme activity and myeloperoxidase (optimum at 1×10^5 CFU/mL)	[167]
Bacillus methylotrophicus, *Bacillus amyloliquefaciens*, *Bacillus licheniformis*	*Labeo rohita* FW	1×10^7 cells/g for 60 days	~50.00	Increase of immune and blood parameters, and resistance to *A. hydrophila*	[168]
Bacillus pumilus	*Oreochromis niloticus* FW	1×10^6, 1×10^7, 1×10^8, 1×10^9 CFU/kg for 30 days and 4 months	50.00	Increase of phagocytic activity and resistance to *Streptococcus agalactiae* (at 1×10^8, 1×10^9 CFU/kg)	[169]
Bacillus pumilus A97	*Trachinotus ovatus* SW	1×10^8 CFU/g for 56 days	5.95 ± 1.69	Increase of expression of genes of non-specific immune responses (Toll-like receptor) and resistance to *Vibrio ponticus*	[170]
Bacillus pumilus, *Bacillus clausii*	*Epinephelus coioides* SW (Larvae)	1×10^6 bacteria/mL for 28 days		Increased resistance to *Aliivibrio fischeri*, *Vibrio scophthalmi* and *Vibrio* sp.	[171]
Bacillus spp., *Enterococcus* spp., *Lactobacillus* spp. (AquaStar®)	*Oreochromis niloticus* FW	3 g/kg for 6 weeks	29.02 ± 0.33	Increase of expression of genes of immune responses (interleukins and cytokines), intraepithelial leukocytes and intestinal mucous cells	[172]

Table 5. *Cont.*

Probiotics	Fish Species	Dosage and Timing of Administration	Initial Weight (g)	Effects	References
Bacillus subtilis	*Oncorhynchus mykiss* FW	1×10^4, 1×10^5, 1×10^6, 1×10^7, 1×10^8, 1×10^9 CFU/g for 14 days	30.00	Increase of leukocytes, lysozyme activity and resistance to *Aeromonas* sp. (optimum at 1×10^7 CFU/g)	[121]
Bacillus subtilis	*Anguilla japonica* FW	1×10^6, 1×10^7, 1×10^8 CFU/g for 8 weeks	8.29 ± 0.06	Increase of lysozyme activity and IgM (optimum at 1×10^7 CFU/g)	[173]
Bacillus thuringiensis, *Bacillus cereus*	*Lates calcarifer* FW	1×10^{12} CFU/kg for 35 days	75.00 ± 0.60	Enhancement of antioxidant response resistance to pathogens (*V. harvey*)	[174]
Bacillus velezensis TPS3N, *Bacillus subtilis TPS*, *Bacillus amyloliquefaciens TPS1*	*Oreochromis niloticus* FW	1×10^8 CFU/mL for 4 weeks	46.24 ± 0.48	Increase of lysozyme activity, IgM at epithelial mucosa and intestine, survival rate to *A. hydrophila*	[175]
Bacillus amyloliquefaciens	*Oreochromis niloticus* FW	2×10^6 CFU/g for 60 days	1.49 ± 0.15	Enhancement of immune parameters and resistance to *A. hydrophila*	[176]
Bacillus cereus, *Bacillus subtilis*	*Oreochromis niloticus* FW	0.5×10^8, 1×10^8 CFU/g for 90 days	0.20 ± 0.05	Enhancement of immune status and resistance to diseases (optimum at 1×10^8 CFU/g)	[177]
Bacillus circulans	*Catla catla* SW	2×10^4, 2×10^5, 2×10^6 CFU/100 g for 60 days	6.48 ± 0.04	Enhancement of non specific immune parameters (optimum at 2×10^5 CFU/100 g)	[178]
Bacillus pumilus, *Bacillus clausii*	*Epinephelus coioides* SW (Larvae)	2×10^6 bacetria/mL for 28 days		Increased resistance to *Aliivibrio fischeri*, *Vibrio scophthalmi*, and *Vibrio* sp.	[171]
Bacillus spp. + *Lactobacillus* spp. + *S. cerevisiae*	*Paralichthys olivaceus* SW	1×10^8, 1×10^9 CFU/kg for 12 weeks	13.50 ± 0.01	Enhancement of immune parameters	[179]
Bacillus subtilis	*Oreochromis niloticus* FW	0, 1, 2, 3, 4 g (1.19 $\times 10^8$ CFU/g)/kg for 50 days	14.82 ± 0.42	Enhancement of immune responses (optimum at 3 g/kg)	[180]
Bacillus subtilis	*Oreochromis niloticus* FW	1.1×10^5 CFU/g for 84 days	5.26 ± 0.06	Enhancement of blood parameters (haematocrit and leukocytes)	[181]
Bacillus aerius	*Pangasius bocourti* FW	1×10^7 CFU/g for 60 days	69.00	Increase of phagocytic activity, lysozyme, complement, and resistance to *A. hydrophila*	[182]

Table 5. *Cont.*

Probiotics	Fish Species	Dosage and Timing of Administration	Initial Weight (g)	Effects	References
Debaryomyces hansenii BCS004	*Sparus aurata* SW	10^6 CFUg^{-1} for 4 weeks	80.00 ± 5.00	Enhancement of immune parameters (phagocytic activity, increase of IgM)	[183]
Enterococcus casseliflavus EC-001	*Cyprinus carpio* FW	1×10^{10}, 1×10^{11}, 1×10^{12} CFU/kg for 56 days	12.00 ± 0.50	Increase of immune response and resistance to infections (optimum at 1×10^{12} CFU/kg)	[184]
Enterococcus faecalis	*Oncorhynchus mykiss* FW	5×10^8 CFU/g for 30 days	50.00	Increase of resistance to *L. garvieae CECT 527*	[185]
Enterococcus faecium	*Rutilus rutilus caspicus* AE	1×10^7, 1×10^8 CFU/g	12.00	Increase of immunoglobulins and lysozyme activity	[186]
Enterococcus faecium	*Oreochromis niloticus* FW	1×10^7 CFU/mL for 40 days	6.83 ± 0.18	Enhancement of immune parameters	[116]
L. delbrueckii sp. *delbrueckii AS13B*	*Dicentrarchus labrax* SW (Larvae)	1×10^5 bacteria/cm^3 for 74 days		Increase of T cells and granulocytes of intestinal mucosa	[187]
Lactobacillus plantarum (alone); *Bacillus subtilis*, *L. plantarum*, *L. rhamnosus*, *L. acidophilus*, *L. delbrueckii* (in mixture)	*Oreochromis niloticus* FW	1×10^8 CFU/g for 112 days (by biofloc system)	8.63 ± 3.35	Increase of immune and antioxidant response, and resistance to pathogens (*A. hydrophila)*	[188]
Lactobacillus rhamnosus ATCC 7469	*Oncorhynchus mykiss* FW	1×10^9 CFU/kg for 60 days	18.41 ± 0.32	Enhancement of non-specific immune parameters, gene expression of IL-1 and TNF, antioxidant activity and resistance to pathogens (*Yersinia ruckeri)*	[189]
Lactobacillus acidophilus	*Xiphophorus helleri* FW	1.5×10^8, 3×10^8, 6×10^8 CFU/g for 70 days	0.03 ± 0.001	Increase of gene expression of immune responses (optimum at 6×10^8 CFU/g)	[190]
Lactobacillus acidophilus	*Clarias gariepinus* FW	3×10^7 CFU/g for 21 days	5.31 ± 0.10	Enhancement of blood parameters	[191]
Lactobacillus acidophilus	*Cyprinus carpio* FW	1×10^2, 1×10^4, 1×10^6 CFU/kg for 56 days	21.34 ± 1.85	Increase of innate immune and antioxidant responses (optimum at 1×10^6 CFU/kg)	[192]
Lactobacillus delbrueckii	*Cyprinus Carpio* FW	1×10^6, 1×10^7 CFU/g for 8 weeks	1.05 ± 0.03	Enhancement of intestinal immune parameters and resistance to *A. hydrophila*	[193]

Table 5. *Cont.*

Probiotics	Fish Species	Dosage and Timing of Administration	Initial Weight (g)	Effects	References
Lactobacillus fermentum	*Cyprinus carpio* FW	1×10^8 CFU/g for 56 days	3.90 ± 0.20	Enhancement of immune responses and resistance to pathologies	[194]
Lactobacillus plantarum	*Oreochromis niloticus* FW	3.4×10^8, 6.8×10^8, 1.3×10^9 CFU/g for 40 days	24.50	Increase of gene expression of interleukins and cytokines (optimum at 6.8×10^8 CFU/g)	[195]
Lactobacillus plantarum	*Oreochromis niloticus* FW	1×10^8 CFU/g for 12 weeks	4.90 ± 0.04	Increase of lysozyme level in serum and epithelial mucosa, phagocytic activity and resistance to *Streptococcus agalctiae*	[112]
Lactobacillus plantarum	*Oreochromis niloticus* FW	1×10^8 CFU/g for 84 days	5.92 ± 0.08	Increase of immune response in serum and epithelial mucosa	[196]
Lactobacillus plantarum, Bacillus velezensis H3.1	*Oreochromis niloticus* FW	1×10^7 CFU/g *Bacillus velezensis*; 1×10^8 CFU/g *Lactobacillus plantarum* for 30 days	21.80 ± 0.03	Increase of lysozyme and phagocytic activities, complement proteins and resistance to *Streptococcus agalactiae*	[197]
Lactobacillus rhamnosus JCM1136, Lactococcus lactis subsp. *lactis JCM5805*	*Oreochromis niloticus* FW (Juveniles)	1×10^8 CFU/g for 6 weeks	0.20 ± 0.05	Increase of gene expression of immune response (interleukins, interferons) and survival rate	[198]
Lactobacillus spp., *Bacillus subtilis, Bifidobacterium bifidum*	*Acipenser buerii* AE (Fingerlings)	1×10^6, 2×10^6, 3×10^6 CFU/g for 56 days	10.50 ± 0.14	Increase of lysozyme and IgM levels (optimum at 3×10^6 CFU/g)	[199]
Lactococcus garvieae	*Oreochromis niloticus* FW	1×10^7 CFU/g for 2 weeks	50.00 ± 5.00	Increase of IgM and lysozyme activity	[200]
Lactococcus lactis	*Paralichthys olivaceus* SW	1×10^9 CFU/g for 8 weeks	80.84 ± 9.37	Enhancement of innate immune response and resistance to *Streptococcus*	[201]
Lactococcus lactis	*Cyprinus carpio* FW	5×10^8 CFU/g for 8 weeks	33.07 ± 0.55	Increase of pro-inflammatory cytokines	[202]
Lactococcus lactis	*Sparus aurata* SW	2×10^9, 5×10^9 CFU/kg for 12 weeks	70.00–90.00	Increase of gene expression of immune responses	[9]
Lactobacillus rhamnosus GG	*Oreochromis niloticus* FW	1×10^8 CFU/g for 2 weeks	14.05 ± 0.42	Increase of lysozyme activity, mucous cells at intestine and resistance to *Aeromonas veroni*	[203]

Table 5. *Cont.*

Probiotics	Fish Species	Dosage and Timing of Administration	Initial Weight (g)	Effects	References
Lactobacillus sakei PO11, Lb. plantarum PO23	*Paralichthys olivaceus* SW	1×10^{11} CFU/g for 27 days	35.00 ± 5.00	Increase of gene expression of immune responses (interleukins and immunoglobulins)	[204]
Bacillus subtilis and *Saccharomyces cerevisiae*	*Oreochromis niloticus* FW	6×10^{7} CFU/g for 56 days	24.01 ± 0.02	Enhancement of innate immune responses and resistance to pathologies	[205]
Paenibacillus ehimensis	*Oreochromis niloticus* FW	1×10^{6}, 1×10^{7} CFU/g for 60 days	5.53 ± 0.45	Enhancement of immune response and resistance to pathologies	[206]
Paenibacillus polymyxa	*Cyprinus carpio* FW	1×10^{3}, 1×10^{4}, 1×10^{5} CFU/mL for 8 weeks	23.17	Increase of lysozyme activity, myeloperoxidase, and resistance to *A. hydrophila* (optimum at 1×10^{3} CFU/mL)	[207]
Pediococcus acidilactici	*Oreochromis niloticus* FW	1×10^{10} CFU/kg for 32 days	175.00	Increase of lysozyme activity and leukocytes	[208]
Pediococcus acidilactici	*Oreochromis niloticus* FW	2.81×10^{6} CFU/g for 6 weeks	9.19 ± 0.04	Increase of leukocytes and intestinal mucous cells	[209]
Pediococcus acidilactici	*Cyprinus carpio* FW	6×10^{8} CFU/g for 60 days	10.00 ± 2.50	Increase of Ig and proteases of epithelial mucosa, gene expression of lysozyme and TNF-α	[210]
Pediococcus pentosaceus	*Ctenopharyngodon idella* FW	1×10^{9} CFU/g for 30 days	32.10 ± 9.00	Increase of gene expression of immune response and resistance to *A. hydrophila*	[211]
Pseudomonas fluorescens biovar I, II and III	*Oreochromis niloticus* FW	1×10^{11} CFU/kg for 45 days	2.93 ± 0.22	Enhancement of non-specific immune and blood parameters	[212]
Psychrobacter namhaensis SO89	*Oreochromis niloticus* FW	2.8×10^{7}, 5.6×10^{7} for 50 days	4.58 ± 0.14	Increase of haematocrit, leukocytes and other immune parameters	[117]
Saccharomyces cerevisiae	*Pangasianodon hypophthalmus* FW	1×10^{9}, 1×10^{11} CFU/kg for 120 days	55.00–65.00	Enhancement of immune parameters (immunglobulines, lysozyme activity)	[213]
Saccharomyces cerevisiae	*Oreochromis niloticus* FW	0.25, 0.50, 1, 2, 5 g/kg for 12 weeks	0.33	Enhancement of immune parameters and resistance to pathologies (optimum at 1 g/kg)	[214]
Shewanella putrefaciens Pdp11	*Sparus orata* SW	1×10^{8} CFU/g for 15–30 days	104.20 ± 7.40	Increase of lysozyme activity and gene expression of immune responses	[215]

Table 5. *Cont.*

Probiotics	Fish Species	Dosage and Timing of Administration	Initial Weight (g)	Effects	References
Shewanella xiamenensis, *Aeromonas veronii*	*Ctenopharyngodon idella* FW	1×10^8 cell/g for 28 days	35.00 ± 5.00	Enhancement of immune parameters, and resistance to *A. hydrophila*	[216]
Vibrio lentus	*Dicentrarchus labrax* SW (Larvae)	1×10^6 CFU/mL for 10 days post-hatching		Increase of transcription of genes of immune responses	[217]
Weissella confusa	*Oncorhynchus mykiss* FW	1.5×10^{10}, 3×10^{10}, 4.5×10^{10} CFU/kg for 60 days	115.00 ± 2.60	Increase of lysozyme activity and expression of INF-γ, TNF-α and IL-8 at 3×10^{10} CFU/kg	[218]

Ensuring the health of aquaculture species has an impact on parameters such as nutrient uptake and the growth performance of fish. Growth performance improvement is another relevant target for dietary supplements, since it can lead to a higher output, and hence to higher earnings for the fishery plant. The purpose of new feed formulations is, therefore, to assure an optimum feed absorption and performance of fish growth whilst reducing the rearing costs. In many cases, it seems fundamental to increase the digestive enzymes in order to maximize nutrient absorption. Indeed, numerous studies [117,156,219,220] have shown that the addition of prebiotics and probiotics enhances the nutrient absorption from feed, since they act on a variety of enzymes such as amylases and proteases [119]. However, in other cases, it was also observed that the new diet formulations could have a deleterious impact on the morphological structure of the fish gut, triggering some inflammatory responses [221]. This condition, especially if maintained for a prolonged period, might disrupt the architecture of the intestine, and therefore lead to the upset of the general health of the microbiota and the organism itself. To overcome this situation, the use of prebiotics and probiotics might be helpful. As a matter of fact, several studies [9,222,223] have demonstrated how the intake of prebiotics and probiotics may preserve and, in some cases, improve the morphology and the function of gut, often affected by the rearing conditions. More in detail, data from the literature on the administration of specific prebiotics (Table 6) and probiotics (Table 7) enhancing the gut efficiency and growth performance of aquaculture fish species from seawater and freshwater are listed in the following tables.

Table 6. Administration of prebiotics enhancing the gut efficiency and growth performance in different aquaculture fish species. (SW—seawater; FW—freshwater; AE—anadromous euryhaline; MOS—mannan-oligosaccharides; GOS—galacto-oligosaccharides; COS—chito-oligosaccarides; FOS—fructo-oligosaccharides).

Prebiotics	Fish Species	Dosage and Timing of Administration	Initial Weight (g)	Effects	References
β-glucan + MOS	*Oreochromis niloticus* FW (Fingerlings)	1.5, 3 g/kg for 60 days	8.70 ± 0.40	Enhancement of growth performance and histological parameters of the intestine	[224]
β-glucan, GOS, MOS	*Channa striata* FW	2 g/kg (β-glucan), 5 g/kg (GOS), 5 g/kg (MOS) for 16 weeks	22.40	Enhancement of growth performance and protein digestibility	[225]

Table 6. *Cont.*

Prebiotics	Fish Species	Dosage and Timing of Administration	Initial Weight (g)	Effects	References
β-glucan + MOS	*Tor grypus* FW	1.5% (MOS), 1.5% (β-glucan) for 90 days	35.00 ± 1.20	Enhancement of growth performance and increase of protein content in fish	[226]
COS	*Scophthalmus maximus* SW	2 g/kg for 56 days	11.00	Enhancement of growth performance and positive effects on intestine structure	[129]
FOS	*Atractosteus tropicus* SW (Larvae)	7.5 g/kg for 15 days	0.03	Enhancement of growth performance and digestive capacity	[130]
FOS	*Sciaenops ocellatus* FW	1% for 8 weeks	7.00	Increase in height of intestinal microvilli	[152]
FOS	*Oreochromis niloticus* FW	0.5, 1, 2, 4 g/kg for 8 weeks	5.00 ± 0.02	Increase of amylase enzyme and height of anterior intestinal microvilli with enhancement of growth parameters (optimum at 1 g/kg)	[132]
GOS	*Sciaenops ocellatus* FW	1% for 8 weeks	7.00	Increase in height of intestinal microvilli	[152]
GOS	*Pangasius hypophthalmus* FW	10 g/kg for 12 weeks	16.45 ± 0.07	Increase of digestive enzymes activity and length of posterior intestinal microvilli	[227]
Inosine, monophosphate inosine	*Pagrus major* SW	0.2%, 0.4%, 0.6%, 0.8% for 10 weeks	6.60	Increase in height of enterocytes, intestinal folds and microvilli (optimum at 0.4%)	[228]
Inulin	*Sciaenops ocellatus* FW	1% for 8 weeks	7.00	Increase in height of intestinal microvilli	[152]
Inulin	*Lates calcarifer* FW	15, 20 g/kg for 60 days	7.14 ± 0.05	Increase of intestinal adsorbent surface	[141]
MOS	*Sciaenops ocellatus* FW	1% for 8 weeks	7.00	Increase in height of intestinal microvilli	[152]
MOS	*Ctenopharyngodon idella* FW	200, 400, 600, 800 and 1000 mg/kg for 60 days	215.85 ± 0.30	Enhancement of the intestinal health status in fish with enteritis by *Aeromonas hydrophila* (optimum at 400 mg/kg)	[229]
Nucleotides	*Sciaenops ocellatus* FW	1% for 6 weeks	7.10	Increase in height of intestinal folds and microvilli	[230]
Nucleotides (NucleoforceFish™)	*Sparus aurata* SW	250, 500 mg/kg for 150 days	0.36 ± 0.002	Enhancement of growth parameters	[154]

Table 6. *Cont.*

Prebiotics	Fish Species	Dosage and Timing of Administration	Initial Weight (g)	Effects	References
Essential oils of *Ocimum basilicum* (EOOB)	*Oreochromis niloticus* FW	0.25, 0.5, 1, 2 mL of EOOB/kg for 45 days	12.13 ± 0.11	Enhancement of digestive enzymes activity (amylases) (optimum at 1 mL/kg)	[219]
Stachyose	*Scophthalmus maximus* L. SW	1.25%, 5% for 12 weeks	4.63 ± 0.01	Increase of gene expression of occludins and ZO-1, and intestinal health (optimum at 5%)	[156]
Thymol and carvacrol	*Sparus aurata* SW	From 0.005 to 0.03% for 9 weeks	26.00–27.00	Increase in number of enterocytes e mucous cells in intestinal epithelium, and enhancement of feed gain ratio at 0.01%	[157]

Table 7. Administration of probiotics enhancing the gut efficiency and growth performance in different aquaculture fish species. (SW—seawater; FW—freshwater; AE—anadromous euryhaline).

Probiotics	Fish Species	Dosage and Timing of Administration	Initial Weight (g)	Effects	References
Acinetobacter KU011TH	*Clarias macrocephalus* FW	1×10^5, 1×10^7, 1×10^9 CFU/kg and 1×10^3, 1×10^4, 1×10^5 CFU/mL	150.00	Enhancement of growth performance at 1×10^9 CFU/kg and 1×10^3 CFU/mL	[231]
Acinetobacter, Vibrio, Bacillus, Alcaligens, Photobacterium, Flavobacterium	*Oncorhynchus mykiss* FW	3.9×10^9 CFU/g for 56 days	113.00 ± 10.40	Enhancement of growth performance	[232]
Bacillus amyloliquefaciens	*Oreochromis niloticus* FW	60 mg/kg for 42 days	39.00	Enhancement of growth performance	[233]
Bacillus coagulans, Rhodopseudomonas palustris, Lactobacillus acidophilus	*Ctenopharyngodon idella* FW	1×10^6 CFU/g for 60 days	2.10 ± 0.09	Increase in weight and digestive enzymes activity	[234]
Bacillus licheniformis	*Cyprinus carpio* FW	1×10^6, 1×10^7, 1×10^8 CFU/g for 60 days	38.58 ± 0.42	Enhancement of growth performance and increase in height of intestinal microvilli (optimum at 1×10^8 CFU/g)	[163]

Table 7. *Cont.*

Probiotics	Fish Species	Dosage and Timing of Administration	Initial Weight (g)	Effects	References
Bacillus thuringiensis, *Bacillus cereus*	*Lates calcarifer* FW	1×10^{12} CFU/kg for 42 days	50.00 ± 0.50	Enhancement of growth performance	[174]
Bacillus velezensis TPS3N, *Bacillus subtilis TPS4*, *Bacillus amyloliquefaciens*	*Oreochromis niloticus* FW	1×10^{8} CFU/mL for 4 weeks	46.24 ± 0.48	Enhancement of digestive enzymes activity and increase in thickness and length of intestinal microvilli, and number of mucous cells	[175]
Bacillus sp.	*Cyprinus carpio* FW	~1×10^{11} CFU/g for 60 days	5.90–7.10	Enhancement of digestive enzymes activity and food conversion rate	[235]
Bacillus sp., *Alcaligenes* sp.	*Tor tambroides* FW	1×10^{8} CFU/g for 90 days	1.39 ± 0.06	Enhancement of growth rate and digestive enzymes activity, and positive effects on intestinal morphology	[236]
Bacillus sp. *SJ-10*, *Lactobacillus plantarum*	*Paralichthys olivaceus* SW	1×10^{8} CFU/g for 8 weeks	14.92 ± 0.21	Not significant enhancement of some histo-morphological intestinal parameters and increase in some digestive enzymes activity	[237]
Bacillus subtilis	*Oreochromis niloticus* FW	1.1×10^{5} CFU/g for 84 days	5.26 ± 0.06	Enhancement of growth performance	[181]
Bacillus subtilis and *Lactobacillus rhamnosus*	*Labeo rohita* FW (Fingerlings)	1×10^{7} CFU/g for 60 days	0.38 ± 0.015	Enhancement of growth performance and nutrients usage	[238]
Bacillus subtilis, *Bacillus licheniformis*, *Bacillus pumilus*	*Oreochromis niloticus* FW	20 mg/kg for 49 days	34.56 ± 0.05	Increase in number of microvilli and lysozyme activity, no changes in growth performance and blood and immune parameters	[239]
Bacillus subtilis, *Bacillus licheniformis*, *Bacillus pumilus*	*Panganodon hypophthalmus* FW	1×10^{10} CFU/g for 110 days	0.03	Enhancement of growth performance and survival rate	[240]
Bacillus, *Bifidobacterium*, *Enterococcus*, *Lactobacillus*, *Pediococcus* sp.	*Oreochromis niloticus* FW (Fingerlings)	1, 2 g/kg for 90 days	5.00	Increase in final weight and muscle mass	[241]
Enterococcus casseliflavus (EC-001)	*Cyprinus carpio* FW	1×10^{10}, 1×10^{11}, 1×10^{12} CFU/kg for 56 days	12.00 ± 0.50	Increase in weight at 1×10^{11} and 1×10^{12} CFU/kg	[184]

Table 7. *Cont.*

Probiotics	Fish Species	Dosage and Timing of Administration	Initial Weight (g)	Effects	References
Enterococcus faecium	*Oreochromis niloticus* FW	1×10^7 CFU/mL for 40 days	6.83 ± 0.18	Increase in final weight	[116]
Lactobacillus. acidophilus	*Cyprinus carpio* FW	1×10^2, 1×10^4, 1×10^6 CFU/kg for 56 days	21.34 ± 1.85	Enhancement of length and width of microvilli of growth performance (optimum at 1×10^6 CFU/kg)	[192]
Lactobacillus rhamnosus ATCC 7469	*Oncorhynchus mykiss* FW	1×10^9 CFU/kg for 60 days	18.41 ± 0.32	Enhancement of growth parameters	[189]
Lactobacillus acidophilus	*Xiphophorus helleri* FW	1.5×10^8, 3×10^8, 6×10^8 CFU/g for 70 days	3.90 ± 0.20	Enhancement of growth performance (optimum at 6×10^8 CFU/g)	[190]
Lactobacillus casei	*Cyprinus carpio* FW	1×10^8 CFU/g for 30 days	68.40 ± 5.90	Enhancement of intestinal enzyme parameters	[242]
Lactobacillus casei (Yakult®)	*Tor tambra* FW (Larvae)	0, 5, 10, 15 mL/kg for 80 days	0.03	Enhancement of growth performance and nutrients usage (optimum at 10 mL/kg)	[243]
Lactobacillus delbrueckii	*Poecilia sphenops* FW	0.8 g/kg for 30 days	0.90 ± 0.02	Enhancement of growth performance and survival rate	[244]
Lactobacillus plantarum	*Oreochromis niloticus* FW	1×10^8 CFU/g for 84 days	5.92 ± 0.08	Enhancement of growth performance	[196]
Lactobacillus. rhamnosus GG	*Oreochromis niloticus* FW	1×10^8 CFU/g for 2 weeks	14.05 ± 0.42	Enhancement of histo-morphological intestinal parameters	[203]
Lactobacillus rhamnosus JCM1136, Lactobacillus lactis subsp. *lactis JCM1136*	*Oreochromis niloticus* FW	1×10^8 CFU/g for 6 weeks	0.20 ± 0.05	Increase in length and density of intestinal microvilli	[198]
Micrococcus luteus	*Oreochromis niloticus* FW	1×10^7 cells/g for 90 days	2.35 ± 0.10	Enhancement of growth performance	[245]
Micrococcus MCCB 104, Bacillus MCCB 101	*Oreochromis mossambicus* FW	1×10^3 CFU per individual for 28 days		Increase of intestinal and hepato-pancreatic enzymes activity	[246]
Pediococcus acidilactici	*Oreochromis niloticus* FW	1×10^{10} CFU/kg for 32 days	175.00	No significant change in growth parameters	[208]

Table 7. *Cont.*

Probiotics	Fish Species	Dosage and Timing of Administration	Initial Weight (g)	Effects	References
Pediococcus pentosaceus SL001	*Ctenopharyngodon idella* FW	1×10^9 CFU/g for 30 days	32.10 ± 9.00	Increase in number of goblet cells and length of intestinal villi	[211]
Pseudomonas fluorescens biovar I, II and *III*	*Oreochromis niloticus* FW	1×10^{11} CFU/kg for 45 days	2.93 ± 0.22	Enhancement of growth performance and blood and non-specific immune parameters	[212]
Saccharomyces cerevisiae	*Pangasianodon hypophthalmus* FW	1×10^9, 1×10^{11} CFU/kg for 120 days	55.00–65.00	Enhancement of growth rate, food conversion rate and immune parameters	[213]
Saccharomyces cerevisiae (Angel Yeast Co. Ltd., China)	*Oreochromis niloticus* FW (Fingerlings)	0, 1, 2, 4 g/kg for 60 days	7.55 ± 1.25	Increase of growth performance and change of histo-morphological intestinal parameters (optimum at 4 g/kg)	[247]
Saccharomyces cerevisiae	*Labeo rohita* FW (Fingerlings)	0, 1, 2, 4 g/kg for 90 days	5.69 ± 0.02	Enhancement of growth rate, histological parameters and nutrients usage (optimum at 4 g/Kg)	[248]
Saccharomyces cerevisiae	*Mystus cavasius* FW	0, 0.5, 1, 1.5 g/kg for 75 days	0.50 ± 0.20	Enhancement of growth performance (optimum at 1 g/kg)	[249]
Streptomyces spp.	*Xiphophorus helleri* FW	10 g/kg for 50 days	0.60	Enhancement of growth performance and nutrients usage	[250]

6. Use of Histomorphological Assays to Estimate the Quality of Feed Additives

As previously stated, the selection of new feed formulations plays an essential role in the well being of the specimens, and hence, the yield of the aquaculture activity [106]. Nevertheless, the formulation of optimal feeding diets for the wide variety of aquaculture fish species requires the implementation of different types of analyses to verify their impact on the health status of the specimens [251]. Among the different approaches employed [194,252], histomorphological assays represent a good biomarker for the assessment of the welfare of aquatic organisms [253–257] since it is able to promptly provide insights into the overall health status of individuals under examination. In particular, in the histological assays of fish guts, endpoints such as mucosal fold length (villi), muscle thickness (MT), and crypt depth (CD) might provide useful information regarding the efficiency of the gut in terms of nutrient absorption [223], while evaluation of the number of goblet cells and leukocytes may be useful for estimating the state of the immune system response, which is crucial for gut microbiota balance [222]. The evaluation of these endpoints may be performed by using both optical and electron microscopy, which offer useful information for an accurate assessment of the actions of prebiotics and probiotics at the fish-GIT-tissue level, as summarized in Table 8.

Table 8. The impact of some prebiotics/probiotics on the gut of aquaculture fish species assessed by histological endpoints. (↑: increase; ↓: decrease; =: no change).

Prebiotics/Probiotics	Fish Species	Histological Endpoints	References
Lactic acid bacteria Saccharomyces	*Oreochromis niloticus*	↑ intestinal perimeter ratio, mucosal fold length, goblet cells and leukocytes	[222]
P. acidilactici MA18/5M + short-chain fructo-oligosaccharides	*Salmo salar*	↑ mucosal fold height, intraepithelial leukocytes, granulocytes, and goblet cells	[258]
Bacillus coagulans, B. licheniformis and *Paenibacillus polymyxa*	*Sillago sihama*	↑ villus height, villus width, muscle thickness, and crypt depth	[223]
Bacillus subtilis + inulin	*Sparus aurata L.*	↑ enterocyte vacuolisation, microvilli disruption	[221]
Lactococcus lactis	*Sparus aurata*	= mucosal folds, connective tissue, lamina propria of simple folds, supranuclear vacuoles	[9]
Nucleotides	*Sciaenops ocellatus*	↑ microvillus height, enterocyte height, = fold height	[230]
Lactobacillus brevis and *L. buchneri*	*Seriola dumerili*	↓ thickness of submucosa layer	[259]
Bacillus licheniformis, Bacillus subtilis + Saccharomyces cerevisiae	*Acipenser persicus*	↑ goblet and mucosal cell number ↑ villus length and width	[260]
Aquablend® + *GroBiotic®*	*Totoaba macdonaldi*	↑ fold length, = fold width, = enterocyte height, = microvilli height	[261]
Short-chain fructo-oligosaccharides and xylo-oligosaccharides	*Dicentrarchus labrax*	↓ mucosal folds, = enterocytes nucleous, ↑ supranuclear absorptive vacuolisation, = hyperplasia in the lamina propria and submucosa, ↑ intraepithelial leucocytes	[262]
Bacillus velezensis TPS3N, Bacillus subtilisTPS4, Bacillus amyloliquefaciens TPS17 (as singular and combined suspension)	*Oreochromis niloticus*	↑ villus length and width, goblet cells count, intestinal epithelial muscle thickness	[175]
AquaStar® Growout *(Lactobacillus reuteri, Bacillus subtilis, Enterococcus faecium* and *Pediococcus acidilactici)*	*Oreochromis niloticus*	↑ intraepithelial leucocytes, microvilli density	[172,209,263]

For instance, a number of histomorphological parameters were used to assess the impact of lactic acid bacteria and yeast on *Oreochromis niloticus* [222], including the haematoxylin and eosin (H/E) and periodic acid-Sciff (PAS) staining, which allowed for an observation of a relevant increase in the intestinal perimeter ratio and mucosal fold length (villi) after the probiotic treatment. These kinds of alterations in the fish gut structure may result in a facilitated and increased feed absorption and growth performance [264]. In parallel, the number of goblet cells and leukocytes in treated samples were found to be higher than those in control fish, highlighting the positive effect of the administrated microorganism on the immune system of fish. Similar results were obtained in *O. niloticus* challenged with the combined treatment of the probiotic *P. acidilactici MA18/5M* (Bactocell, Lallemand SAS, France) and the prebiotic short-chain fructo-oligosaccharides (FOS), usually termed as symbiotic [265], which enhanced the intestinal structure with increases in the mucosal fold (villi) height and the immune system with elevation of intraepithelial leukocytes, granulocytes, and goblet cells [258]. Notably, the combined administration of three probiotic strains (*Bacillus coagulans, B. licheniformis* and *P. polymyxa*) on the northern whiting *Sillago sihama Forsskål* induced positive effects in terms of gut wellness, resulting in higher uptake of intestinal nutrients and better growth performance, as demonstrated by an evaluation of some histological parameters (i.e., villus height and width, muscle thickness, and crypt depth) and growth and immune response endpoints [223].

The primary aim of these current research studies is to determine the optimal feed additives for each aquaculture species while also considering the dosage [266]. Indeed, the inappropriate usage of certain feed additives may have negative effects in certain circumstances, as reported in gilthead sea bream *Sparus aurata*, in which the combined intake of inulin and *B. subtilis* caused oedema and inflammations in the fish gut, as observed by transmission electron microscopy (TEM) [221]. Contrarily, Moroni et al. [9] documented in the same fish species that the probiotic *Lactococcus lactis* does not provoke any impairments in gut morphology nor inflammation. Therefore, these findings suggest the need to employ the most appropriate feed additive for each individual fish species. Interestingly, the supplementation of nucleotides in the diet of red drum *Sciaenops ocellatus* was capable of improving several histomorphological parameters in different gut tracts (distal intestine, mid-intestine, proximal intestine, pyloric caeca), enhancing the general health status of the fish and its performance in terms of nutrient absorption [230].

Feed additives can be therefore considered as a useful tool in the design of new food formulae for a more sustainable aquaculture. Indeed, the inclusion of the probiotics *Lactobacillus brevis* and *L. buchneri* in the diets of *Seriola dumerili*, in which fish oil was replaced by vegetable oils, reduced the thickness of submucosa layer of the posterior intestine, thus lowering inflammatory condition [259]. *Lactobacillus* spp., along with *Bacillus* spp., are probably among the most widely used probiotics in aquaculture [170,175,267,268]. Moreover, ref. [264] elucidated the ability of the probiotics *B. licheniformis* and *B. subtilis* coupled to the yeast *S. cerevisiae* to enhance several histological parameters such as the number of goblet and mucosal cells, and villus length and width. Interestingly, ref. [261] documented that although the individual use of the commercial probiotic Aquablend® had a significant effect on fold length in the proximal section of the GIT in *Totoaba macdonaldi*, its use in combination with the prebiotic GroBiotic®-A showed an increase in the values of these parameters, though not significantly, as a signal that the formulation used needs to be revised.

To accurately integrate and interpret the histological results, the use of a continuous scale scoring system may be helpful, as applied and reported in [262] for the evaluation of a number of histological parameters (i.e., intestinal folds, changes in enterocytes nucleous, supranuclear absorptive vacuolisation, connective tissue hyperplasia in the lamina propria and submucosa, infiltration of inflammatory intraepithelial leukocytes), that allowed for an understanding of the inefficiency of prebiotics (FOS and xylo-oligosaccharides) to counteract the negative impact provoked by diets including plant feedstuffs in the distal intestine of juveniles *D. labrax*.

Besides enhancing the general wellness of aquaculture fish species, feed additives (prebiotics and probiotics) play a pivotal role in challenging infections, providing a valid alternative to the use of antibiotics that may lead to antibiotic resistance after a prolonged period of time [175,256,257,269]. In [175], the efficiency of three *Bacillus* species in *O. niloticus* against *A. hydrophila* infection was assessed, resulting in improvements in some histological parameters of the fish gut (i.e., villus length and width, goblet cells count, intestinal epithelial muscle thickness). Noteworthy, the rise in the number of goblet cells is relevant to challenging infection, as these cells secrete mucus-containing bactericidal compounds which are useful against pathogens [115]. Therefore, this histological parameter, combined with other endpoints such as mortality rate, could provide relevant information on the immunostimulant and pathogen resistance effects of the tested feed additives. A further relevant histological parameter to estimate the pathogen resistance effects of feed additives is the level of intraepithelial leukocytes, as reported by [172,263] after using the commercial probiotic AquaStar® Growout (*L. reuteri*, *B. subtilis*, *E. faecium* and *P. acidilactici*) in tilapia *O. niloticus*. Finally, the evaluation of microvilli density by TEM, besides providing information on the absorptive surface area index (combined with numerical data of microvilli length and perimeter ratio), may also reveal tight junctions between enterocytes, and thus, the occurrence of a barrier against pathogens [263]. Therefore, these results point out the potential of the histological approach for obtaining a variety of precious information to delineate the gut health status of fish species treated with feed additives.

7. Conclusions

In this review, the close relationship between the microbiota and the GIT of fish is highlighted. The occurrence of different environmental conditions (freshwater and seawater) and dietary habits (herbivores, carnivores, omnivores, detritivores) are able to influence this ratio, resulting in changes both in terms of microbial colonization and histomorphological patterns, with repercussions in fish GIT efficiency. Therefore, the use of histological endpoints to assess the general health status of aquaculture fish species treated with feed additives (prebiotics and probiotics) represents a very useful strategy for understanding the capacity of these treatments to promote the interface between the microbiota and the gut, which is crucial for its proper performance in terms of nutrient uptake and defense against pathogens. Overall, an in-depth knowledge of the appropriate administration of feed supplements for individual fish species is a promising strategy to greatly enhance the aquaculture sector, both in terms of quality and sustainable production. Therefore, this review has the potential to provide directions for future research in this field.

Author Contributions: Conceptualization, T.C. and M.M.; data curation, G.D.M., T.C. and M.M.; writing—original draft preparation, G.D.M.; writing—review and editing, T.C. and M.M.; visualization, T.C. and G.D.M.; supervision, M.M. All authors have read and agreed to the published version of the manuscript.

Funding: G.D.M. received a scholarship within the project "FISH PATH NET—Enhancement of Sicilian ichthyopathological centres", funded by PO FEAMP 2014/2020.

Institutional Review Board Statement: Not applicable.

Data Availability Statement: The data presented in this study are available on request from the corresponding author.

Conflicts of Interest: The authors declare no conflict of interest.

References

1. Ringø, E.; Zhou, Z.; Vecino, J.L.G.; Wadsworth, S.; Romero, J.; Krogdahl, Å.; Olsen, R.E.; Dimitroglou, A.; Foey, A.; Davies, S.; et al. Effect of dietary components on the gut microbiota of aquatic animals. A never-ending story? *Aquac. Nutr.* **2016**, *22*, 219–282. [CrossRef]
2. Pérez, T.; Balcázar, J.L.; Ruiz-Zarzuela, I.; Halaihel, N.; Vendrell, D.; de Blas, I.; Múzquiz, J.L. Host–microbiota interactions within the fish intestinal ecosystem. *Mucosal Immunol.* **2010**, *3*, 355–360. [CrossRef] [PubMed]
3. Diwan, A.D.; Harke, S.N.; Panche, A.N. Host-microbiome interaction in fish and shellfish: An overview. *Fish Shellfish Immunol. Rep.* **2023**, *4*, 100091. [CrossRef] [PubMed]
4. Clements, K.D.; Angert, E.R.; Montgomery, W.L.; Choat, J.H. Intestinal microbiota in fishes: What's known and what's not. *Mol. Ecol.* **2014**, *23*, 1891–1898. [CrossRef]
5. Ray, A.K.; Ringø, E. The Gastrointestinal Tract of Fish. In *Aquaculture Nutrition: Gut Health, Probiotcs and Prebiotic*; Merrifield, D., Ringø, E., Eds.; Wiley & Sons, Ltd.: Hoboken, NJ, USA, 2014.
6. Becker, A.G.; Gonçalves, J.F.; Oliveira Garcia, L.; Behr, E.R.; Graça, D.L.; Kurtz Filho, M.; Martins, T.; Baldisserotto, B. Morphometric parameters comparisons of the digestive tract of four teleosts with different feeding habits. *Biol. Cienc. Rural* **2010**, *40*, 862–866. [CrossRef]
7. Egerton, S.; Culloty, S.; Whooley, J.; Stanton, C.; Ross, R.P. The gut microbiota of marine fish. *Front. Microbiol.* **2018**, *9*, 873. [CrossRef]
8. Deng, M.; Yu, Z.; Geng, Y.; Wang, K.; Chen, D.; Huang, X.; Ou, Y.; Chen, Z.; Zhong, Z.; Lai, W. Outbreaks of Streptococcosis associated with *Streptococcus iniae* in Siberian sturgeon (*Acipenser baerii*) in China. *Aquac. Res.* **2017**, *48*, 909–919. [CrossRef]

9. Moroni, F.; Naya-Català, F.; Piazzon, M.C.; Rimoldi, S.; Calduch-Giner, J.; Giardini, A.; Martínez, I.; Brambilla, F.; Pérez-Sánchez, J.; Terova, G. The effects of nisin-producing *Lactococcus lactis* strain used as probiotic on gilthead sea bream (*Sparus aurata*) growth, gut microbiota, and transcriptional response. *Front. Mar. Sci.* **2021**, *8*, 659519. [CrossRef]
10. Food and Agricultural Organization of the United Nations; World Health Organization. Health and Nutrition Properties of Probiotics in Food including Powder Milk with Live Lactic Acid Bacteria. In Proceedings of the Report of a Joint FAO/WHO Expert Consultation on Evaluation of Health and Nutritional Properties of Probiotics in Food Including Powder Milk with Live Lactic Acid Bacteria, Amerian Córdoba Park Hotel, Cordoba, Argentina, 1–4 October 2001.
11. Nayak, S.K. Probiotics and immunity: A fish perspective. *Fish Shellfish Immunol.* **2010**, *29*, 2–14. [CrossRef]
12. Nayak, S.K. Role of gastrointestinal microbiota in fish. *Aquac. Res.* **2010**, *41*, 1553–1573. [CrossRef]
13. Akhter, N.; Wu, B.; Memon, A.M.; Mohsin, M. Probiotics and prebiotics associated with aquaculture: A review. *Fish Shellfish Immunol.* **2015**, *45*, 733–741. [CrossRef]
14. Wee, W.; Hamid, N.K.A.; Mat, K.; Khalif, R.I.A.; Rusli, N.D.; Rahman, M.M.; Kabir, M.A.; Wei, L.S. The effects of mixed prebiotics in aquaculture: A review. *Aquac. Fish.* 2022, *in press*. [CrossRef]
15. Tran, N.T.; Li, S. Potential role of prebiotics and probiotics in conferring health benefits in economically important crabs. *Fish Shellfish Immunol. Rep.* **2022**, 3, 100041. [CrossRef] [PubMed]
16. Hancz, C. Application of probiotics for environmentally friendly and sustainable aquaculture: A Review. *Sustainability* **2022**, *14*, 15479. [CrossRef]
17. Gosavi, S.M.; Verma, C.R.; Kharat, S.S.; Pise, M.; Kumkar, P. Structural adequacy of the digestive tract supports dual feeding habit in catfish *Pachypterus khavalchor* (Siluriformes: Horabagridae). *Acta Histochem.* **2019**, *121*, 437–449. [CrossRef]
18. Takahashi, T.; Koblmüller, S. The adaptive radiation of cichlid fish in Lake Tanganyika: A morphological perspective. *Int. J. Evol. Biol.* **2011**, *2011*, 620754. [CrossRef] [PubMed]
19. Wainwright, P.C. Functional morphology of the pharyngeal jaw apparatus. *Fish Physiol.* **2005**, *23*, 77–101.
20. Kalhoro, H.; Tong, S.; Wang, L.; Hua, Y.; Volatiana, J.A.; Shao, Q. Morphological study of the gastrointestinal tract of *Larimichthys crocea* (Acanthopterygii: Perciformes). *Zoologia* **2018**, *35*, e25171. [CrossRef]
21. Wilson, J.M.; Castro, L.F.C. Morphological Diversity of The Gastrointestinal Tract in Fishes. In *The Multifunctional Gut of Fish;* Grosell, M., Farrell, A.P., Brauner, C.J.B.T.-F.P., Eds.; Academic Press: Cambridge, MA, USA, 2010; pp. 1–55.
22. Okuthe, G.E.; Bhomela, B. Morphology, histology and histochemistry of the digestive tract of the Banded tilapia, *Tilapia sparrmanii* (Perciformes: Cichlidae). *Zoologia* **2021**, *37*, e51043. [CrossRef]
23. Abdulhadi, H.A. Some comparative histological studies on alimentary tract of Tilapia fish (*Tilapia spilurus*) and sea bream (*Mylio cuvieri*). *Egypt. J. Aquat. Res.* **2015**, *31*, 387–398.
24. Alves, A.P.C.; Pereira, R.T.; Rosa, P.V. Morphology of the digestive system in carnivorous freshwater dourado *Salminus brasiliensis*. *J. Fish Biol.* **2021**, *99*, 1222–1235. [CrossRef]
25. Mello, G.C.G.; Santos, M.L.; Arantes, F.P.; Pessali, T.C.; Brito, M.F.G.; Santos, J.E. Morphological characterisation of the digestive tract of the catfish *Lophiosilurus alexandri* Steindachner, 1876 (Siluriformes, Pseudopimelodidae). *Acta Zool.* **2019**, *100*, 14–23. [CrossRef]
26. Flores, E.M.; Nguyen, A.T.; Odem, M.A.; Eisenhoffer, G.T.; Krachler, A.M. The zebrafish as a model for gastrointestinal tract–microbe interactions. *Cell. Microbiol.* **2020**, *22*, e13152. [CrossRef] [PubMed]
27. Stevens, C.E.; Hume, I.D. *Comparative Physiology of the Vertebrate Digestive System;* Cambridge University Press: Cambridge, UK, 2004.
28. Johnson, K.S.; Clements, K.D. Histology and ultrastructure of the gastrointestinal tract in four temperate marine herbivorous fishes. *J. Morphol.* **2022**, *283*, 16–34. [CrossRef] [PubMed]
29. Buddington, R.K.; Diamond, J.M. Pyloric ceca of fish: A "new" absorptive organ. *Am. J. Physiol. Liver Physiol.* **1987**, *252*, G65–G76. [CrossRef]
30. Farrag, M.G.; Azab, D.M.; Alabssawy, A.N. Comparative study on the histochemical structures of stomach, pyloric caeca and anterior intestine in the grey mullet, *Mugil cephalus* (Linnaeus, 1758). *Egypt. J. Aquat. Biol. Fish.* **2020**, *24*, 1055–1071. [CrossRef]
31. Olsen, R.; Ringø, E. Lipid digestibility in fish: A review. *Recent Res. Dev. Lipid Res.* **1997**, *1*, 199–265.
32. Ringø, E.; Olsen, R.E.; Mayhew, T.M.; Myklebust, R. Electron microscopy of the intestinal microflora of fish. *Aquaculture* **2003**, *227*, 395–415. [CrossRef]
33. Banan Khojasteh, S.M. The morphology of the post-gastric alimentary canal in teleost fishes: A brief review. *Int. J. Aquat. Sci.* **2012**, *3*, 71–88.
34. Kapoor, B.G.; Smit, H.; Verighina, I.A. *The Alimentary Canal and Digestion in Teleosts;* Russell, F.S., Yonge, M.B.T.-A., Eds.; Academic Press: Cambridge, MA, USA, 1976; pp. 109–239.
35. Holmgren, S.; Olsson, C.B.T.-F.P. The neuronal and endocrine regulation of gut function. *Fish Physiol.* **2009**, *28*, 467–512.
36. Manera, M.; Dezfuli, B.S. Rodlet cells in teleosts: A new insight into their nature and functions. *J. Fish Biol.* **2004**, *65*, 597–619. [CrossRef]
37. Reite, O.B.; Evensen, O. Inflammatory cells of teleostean fish: A review focusing on mast cells/eosinophilic granule cells and rodlet cells. *Fish Shellfish Immunol.* **2006**, *20*, 192–208. [CrossRef] [PubMed]
38. Cho, J.-H.; Park, J.W.; Ryu, Y.-W.; Kim, K.-W.; Hur, S.-W. Morphology, histology, and histochemistry of the digestive tract of the marbled flounder *Pseudopleuronectes yokohamae*. *Animals* **2023**, *13*, 936. [CrossRef] [PubMed]

39. Alabssawy, A.N.; Khalaf-Allah, H.M.M.; Gafar, A.A. Anatomical and histological adaptations of digestive tract in relation to food and feeding habits of lizardfish, *Synodus variegatus* (Lacepède, 1803). *Egypt. J. Aquat. Res.* **2019**, *45*, 159–165. [CrossRef]
40. Rauta, P.R.; Nayak, B.; Das, S. Immune system and immune responses in fish and their role in comparative immunity study: A model for higher organisms. *Immunol. Lett.* **2012**, *148*, 23–33. [CrossRef]
41. Gomez, D.; Sunyer, J.O.; Salinas, I. The mucosal immune system of fish: The evolution of tolerating commensals while fighting pathogens. *Fish Shellfish Immunol.* **2013**, *35*, 1729–1739. [CrossRef]
42. Kelly, C.; Salinas, I. Under pressure: Interactions between commensal microbiota and the teleost immune system. *Front. Immunol.* **2017**, *8*, 559. [CrossRef]
43. Yu, Y.-Y.; Ding, L.-G.; Huang, Z.-Y.; Xu, H.-Y.; Xu, Z. Commensal bacteria-immunity crosstalk shapes mucosal homeostasis in teleost fish. *Rev. Aquac.* **2021**, *13*, 2322–2343. [CrossRef]
44. Valero, Y.; Saraiva-Fraga, M.; Costas, B.; Guardiola, F.A. Antimicrobial peptides from fish: Beyond the fight against pathogens. *Rev. Aquac.* **2020**, *12*, 224–253. [CrossRef]
45. Valero, Y.; Chaves-Pozo, E.; Meseguer, J.; Esteban, M.; Cuesta, A. Biologial Role of Fish Antimicrobial Peptides. In *Antimicrobial Peptides: Properties, Functions and Role in Immune Response*; Seong, M.D., Hak, Y.I., Eds.; Nova Science Publisher: Hauppauge, NY, USA, 2013; pp. 31–60.
46. Vermassen, A.; Leroy, S.; Talon, R.; Provot, C.; Popowska, M.; Desvaux, M. Cell wall hydrolases in bacteria: Insight on the diversity of cell wall amidases, glycosidases and peptidases toward peptidoglycan. *Front. Microbiol.* **2019**, *10*, 331. [CrossRef]
47. Gao, C.; Fu, Q.; Zhou, S.; Song, L.; Ren, Y.; Dong, X.; Su, B.; Li, C. The mucosal expression signatures of g-type lysozyme in turbot (*Scophthalmus maximus*) following bacterial challenge. *Fish Shellfish Immunol.* **2016**, *54*, 612–619. [CrossRef]
48. Li, S.; Wang, D.; Liu, H.; Yin, J.; Lu, T. Expression and antimicrobial activity of c-type lysozyme in taimen (*Hucho taimen*, Pallas). *Dev. Comp. Immunol.* **2016**, *63*, 156–162. [CrossRef]
49. Wei, S.; Huang, Y.; Huang, X.; Cai, J.; Wei, J.; Li, P.; Ouyang, Z.; Qin, Q. Molecular cloning and characterization of a new G-type lysozyme gene (Ec-lysG) in orange-spotted grouper, *Epinephelus coioides*. *Dev. Comp. Immunol.* **2014**, *46*, 401–412. [CrossRef]
50. Yu, L.; Sun, B.; Li, J.; Sun, L. Characterization of a c-type lysozyme of Scophthalmus maximus: Expression, activity, and antibacterial effect. *Fish Shellfish Immunol.* **2013**, *34*, 46–54. [CrossRef] [PubMed]
51. Smith, N.C.; Rise, M.L.; Christian, S.L. A comparison of the innate and adaptive immune systems in cartilaginous fish, ray-finned fish, and lobe-finned fish. *Front. Immunol.* **2019**, *10*, 2292. [CrossRef]
52. Abram, Q.H.; Dixon, B.; Katzenback, B.A. Impacts of low temperature on the teleost immune system. *Biology* **2017**, *6*, 39. [CrossRef] [PubMed]
53. Natnan, M.E.; Low, C.-F.; Chong, C.-M.; Bunawan, H.; Baharum, S.N. Integration of omics tools for understanding the fish immune response due to microbial challenge. *Front. Mar. Sci.* **2021**, *8*, 668771. [CrossRef]
54. Sahoo, B.R. Structure of fish Toll-like receptors (TLR) and NOD-like receptors (NLR). *Int. J. Biol. Macromol.* **2020**, *161*, 1602–1617. [CrossRef] [PubMed]
55. Nutsch, K.M.; Hsieh, C.-S. T cell tolerance and immunity to commensal bacteria. *Curr. Opin. Immunol.* **2012**, *24*, 385–391. [CrossRef] [PubMed]
56. Yu, Y.-Y.; Kong, W.-G.; Xu, H.-Y.; Huang, Z.-Y.; Zhang, X.-T.; Ding, L.-G.; Dong, S.; Yin, G.-M.; Dong, F.; Yu, W.; et al. Convergent evolution of mucosal immune responses at the buccal cavity of teleost fish. *iScience* **2019**, *19*, 821–835. [CrossRef]
57. Dawood, M.A.O. Nutritional immunity of fish intestines: Important insights for sustainable aquaculture. *Rev. Aquac.* **2021**, *13*, 642–663. [CrossRef]
58. Bilal, S.; Etayo, A.; Hordvik, I. Immunoglobulins in teleosts. *Immunogenetics* **2021**, *73*, 65–77. [CrossRef] [PubMed]
59. Cheesman, S.E.; Neal, J.T.; Mittge, E.; Seredick, B.M.; Guillemin, K. Epithelial cell proliferation in the developing zebrafish intestine is regulated by the Wnt pathway and microbial signaling via Myd88. *Proc. Natl. Acad. Sci. USA* **2011**, *108*, 4570–4577. [CrossRef] [PubMed]
60. Kanther, M.; Tomkovich, S.; Xiaolun, S.; Grosser, M.R.; Koo, J.; Flynn, E.J., III; Jobin, C.; Rawls, J.F. Commensal microbiota stimulate systemic neutrophil migration through induction of serum amyloid A. *Cell. Microbiol.* **2014**, *16*, 1053–1067. [CrossRef] [PubMed]
61. Galindo-Villegas, J.; García-Moreno, D.; De Oliveira, S.; Meseguer, J.; Mulero, V. Regulation of immunity and disease resistance by commensal microbes and chromatin modifications during zebrafish development. *Proc. Natl. Acad. Sci. USA* **2012**, *109*, E2605–E2614. [CrossRef] [PubMed]
62. Rolig, A.S.; Parthasarathy, R.; Burns, A.R.; Bohannan, B.J.M.; Guillemin, K. Individual members of the microbiota disproportionately modulate host innate immune responses. *Cell Host Microb.* **2015**, *18*, 613–620. [CrossRef]
63. Wang, A.R.; Ran, C.; Ringø, E.; Zhou, Z.G. Progress in fish gastrointestinal microbiota research. *Rev. Aquac.* **2018**, *10*, 626–640. [CrossRef]
64. Rombout, J.H.W.M.; Abelli, L.; Picchietti, S.; Scapigliati, G.; Kiron, V. Teleost intestinal immunology. *Fish Shellfish Immunol.* **2011**, *31*, 616–626. [CrossRef]
65. Fujimura, K.E.; Slusher, N.A.; Cabana, M.D.; Lynch, S.V. Role of the gut microbiota in defining human health. *Expert Rev. Anti. Infect. Ther.* **2010**, *8*, 435–454. [CrossRef]
66. Hsiao, W.W.; Metz, C.; Singh, D.P.; Roth, J. The microbes of the intestine: An introduction to their metabolic and signaling capabilities. *Endocrinol. Metab. Clin. N. Am.* **2008**, *37*, 857–871. [CrossRef]

67. Rajilic-Stojanovic, M.; Smidt, H.; de Vos, W.M. Diversity of the human gastrointestinal tract microbiota revisited. *Environ. Microbiol.* **2007**, *9*, 2125–2136. [CrossRef] [PubMed]
68. Carda-Dieguez, M.; Mira, A.; Fouz, B. Pyrosequencing survey of intestinal microbiota diversity in cultured sea bass (*Dicentrarchus labrax*) fed functional diets. *FEMS Microbiol. Ecol.* **2014**, *87*, 451–459. [CrossRef] [PubMed]
69. Desai, A.R.; Links, M.G.; Collins, S.A.; Mansfield, G.S.; Drew, M.D.; Van Kessel, A.G.; Hill, J.E. Effects of plant-based diets on the distal gut microbiome of rainbow trout (*Oncorhynchus mykiss*). *Aquaculture* **2012**, *350*, 134–142. [CrossRef]
70. Ringø, E.; Sperstad, S.; Myklebust, R.; Refstie, S.; Krogdahl, Å. Characterisation of the microbiota associated with intestine of Atlantic cod (*Gadus morhua* L.): The effect of fish meal, standard soybean meal and a bioprocessed soybean meal. *Aquaculture* **2006**, *261*, 829–841. [CrossRef]
71. Verner-Jeffreys, D.W.; Shields, R.J.; Bricknell, I.R.; Birkbeck, T.H. Changes in the gut-associated microflora during the development of Atlantic halibut (*Hippoglossus hippoglossus* L.) larvae in three British hatcheries. *Aquaculture* **2003**, *219*, 21–42. [CrossRef]
72. Liu, H.; Guo, X.; Gooneratne, R.; Lai, R.; Zeng, C.; Zhan, F.; Wang, W. The gut microbiome and degradation enzyme activity of wild freshwater fishes influenced by their trophic levels. *Sci. Rep.* **2016**, *6*, 24340. [CrossRef] [PubMed]
73. Parris, D.J.; Brooker, R.M.; Morgan, M.A.; Dixson, D.L.; Stewart, F.J. Whole gut microbiome composition of damselfish and cardinalfish before and after reef settlement. *PeerJ* **2016**, *4*, e2412. [CrossRef]
74. Smriga, S.; Sandin, S.A.; Azam, F. Abundance, diversity, and activity of microbial assemblages associated with coral reef fish guts and feces. *FEMS Microbiol. Ecol.* **2010**, *73*, 31–42. [CrossRef]
75. Miyake, S.; Ngugi, D.K.; Stingl, U. Diet strongly influences the gut microbiota of surgeonfishes. *Molecul. Ecol.* **2015**, *24*, 656–672. [CrossRef]
76. Clements, K.D.; Pasch, I.B.; Moran, D.; Turner, S.J. Clostridia dominate 16S rRNA gene libraries prepared from the hindgut of temperate marine herbivorous fishes. *Mar. Biol.* **2007**, *150*, 1431–1440. [CrossRef]
77. Fidopiastis, P.M.; Bezdek, D.J.; Horn, M.H.; Kandel, J.S. Characterizing the resident, fermentative microbial consortium in the hindgut of the temperate-zone herbivorous fish, *Hermosilla azurea* (Teleostei: Kyphosidae). *Mar. Biol.* **2006**, *148*, 631–642. [CrossRef]
78. Sparagon, W.J.; Gentry, E.C.; Minich, J.J.; Vollbrecht, L.; Laurens, L.M.L.; Allen, E.E.; Sims, N.A.; Dorrestein, P.C.; Wegley, K.L.; Nelson, C.E. Fine scale transitions of the microbiota and metabolome along the gastrointestinal tract of herbivorous fishes. *Anim. Microbiome* **2022**, *4*, 33. [CrossRef] [PubMed]
79. Li, J.; Ni, J.; Li, J.; Wang, C.; Li, X.; Wu, S.; Zhang, T.; Yu, Y.; Yan, Q. Comparative study on gastrointestinal microbiota of eight fish species with different feeding habits. *J. Appl. Microbiol.* **2014**, *117*, 1750–1760. [CrossRef]
80. Geraylou, Z.; Souffreau, C.; Rurangwa, E.; Maes, G.E.; Spanier, K.I.; Courtin, C.M.; Delcour, J.A.; Buyse, J.; Ollevier, F. Prebiotic effects of arabinoxylan oligosaccharides on juvenile Siberian sturgeon (*Acipenser baerii*) with emphasis on the modulation of the gut microbiota using 454 pyrosequencing. *FEMS Microbiol. Ecol.* **2013**, *86*, 357–371. [CrossRef]
81. Ward, N.L.; Steven, B.; Penn, K.; Methe, B.A.; Detrich, W.H. Characterization of the intestinal microbiota of two Antarctic notothenioid fish species. *Extremophiles* **2009**, *13*, 679–685. [CrossRef] [PubMed]
82. Ransom, B.L. *Intestinal Microbial Community Composition of Six Actinopterygii Fish Species in The Southeastern United States*; University of Georgia: Athens, GA, USA, 2008.
83. Sun, Y.; Yang, H.; Ling, Z.; Chang, J.; Ye, J. Gut microbiota of fast and slow growing grouper *Epinephelus coioides*. *Afric. J. Microbiol. Res.* **2009**, *3*, 637–640.
84. Sugita, H.; Iwata, J.; Miyajima, C.; Kubo, T.; Noguchi, T.; Hashimoto, K.; Deguchi, Y. Changes in microflora of a puffer fish *Fugu niphobles*, with different water temperatures. *Mar. Biol.* **1989**, *101*, 299–304. [CrossRef]
85. Star, B.; Haverkamp, T.H.; Jentoft, S.; Jakobsen, K.S. Next generation sequencing shows high variation of the intestinal microbial species composition in Atlantic cod caught at a single location. *BMC Microbiol.* **2013**, *13*, 248. [CrossRef]
86. MacFariane, R.D.; McLaughlin, J.J.; Bullock, G.L. Quantitative and qualitative studies of gut flora in striped bass from estuarine and coastal marine environments. *J. Wildl. Dis.* **1986**, *22*, 344–348. [CrossRef]
87. Newman Jr, J.T.; Cosenza, B.J.; Buck, J.D. Aerobic microflora of the bluefish (*Pomatomus saltatrix*) intestine. *J. Fish. Res. Board Can.* **1972**, *29*, 333–336. [CrossRef]
88. Floris, R.; Manca, S.; Fois, N. Microbial ecology of intestinal tract of gilthead sea bream (*Sparus aurata* Linnaeus, 1758) from two coastal lagoons of Sardinia (Italy). *Transitional Waters Bull.* **2013**, *7*, 4–12.
89. Skrodenytė-Arbačiauskienė, V.; Sruoga, A.; Butkauskas, D.; Skrupskelis, K. Phylogenetic analysis of intestinal bacteria of freshwater salmon *Salmo salar* and sea trout *Salmo trutta trutta* and diet. *Fish. Sci.* **2008**, *74*, 1307–1314. [CrossRef]
90. Hovda, M.B.; Lunestad, B.T.; Fontanillas, R.; Rosnes, J.T. Molecular characterization of the intestinal microbiota of farmed Atlantic salmon (*Salmo salar* L.). *Aquaculture* **2007**, *272*, 581–588. [CrossRef]
91. Bano, N.; DeRae Smith, A.; Bennett, W.; Vasquez, L.; Hollibaugh, J.T. Dominance of mycoplasma in the guts of the long-jawed mudsucker, *Gillichthys mirabilis*, from five California salt marshes. *Environ. Microbiol.* **2007**, *9*, 2636–2641. [CrossRef] [PubMed]
92. Curson, A.R.; Sullivan, M.J.; Todd, J.D.M.; Johnston, A.W. Identification of genes for dimethyl sulfide production in bacteria in the gut of Atlantic Herring (*Clupea harengus*). *ISME J.* **2010**, *4*, 144–146. [CrossRef] [PubMed]
93. Karthiayani, T.; Mahadeva Iyer, K. Quantitative and qualitative studies on the bacterial flora of fresh sardines. *Fish. Technol.* **1967**, *4*, 89–97.

94. Svanevik, C.S.; Lunestad, B.T. Characterization of the microbiota of Atlantic mackerel (*Scomber scombrus*). *Int. J. Food Microbiol.* **2011**, *151*, 164–170. [CrossRef]
95. Ringø, E.; Strøm, E. Microflora of Arctic charr, *Salvelinus alpinus* (L.): Gastrointestinal microflora of free-living fish and effect of diet and salinity on intestinal microflora. *Aquac. Res.* **1994**, *25*, 623–629. [CrossRef]
96. Austin, B.; Al-Zahrani, A.M.J. The effect of antimicrobial compounds on the gastrointestinal microflora of rainbow trout, *Salmo gairdneri* Richardson. *J. Fish Biol.* **1988**, *33*, tb05444. [CrossRef]
97. Silva, F.C.P.; Nicoli, J.R.; Zambonino-Infante, J.L.; Kaushik, S.; Gatesoupe, F.-J. Influence of the diet on the microbial diversity of faecal and gastrointestinal contents in gilthead sea bream (*Sparus aurata*) and intestinal contents in goldfish (*Carassius auratus*). *FEMS Microbiol. Ecol.* **2011**, *78*, 285–296. [CrossRef]
98. Estruch, G.; Collado, M.C.; Peñaranda, D.S.; Tomás Vidal, A.; Jover Cerdá, M.; Pérez Martínez, G.; Martinez-Llorens, S. Impact of fishmeal replacement in diets for gilthead sea bream (*Sparus aurata*) on the gastrointestinal microbiota determined by pyrosequencing the 16S rRNA Gene. *PLoS ONE* **2015**, *10*, e0136389. [CrossRef]
99. Zhou, Z.; Liu, Y.; Shi, P.; He, S.; Yao, B.; Ringø, E. Molecular characterization of the autochthonous microbiota in the gastrointestinal tract of adult yellow grouper (*Epinephelus awoara*) cultured in cages. *Aquaculture* **2009**, *286*, 184–189. [CrossRef]
100. Xing, M.; Hou, Z.; Yuan, J.; Liu, Y.; Qu, Y.; Liu, B. Taxonomic and functional metagenomic profiling of gastrointestinal tract microbiome of the farmed adult turbot (*Scophthalmus maximus*). *FEMS Microbiol. Ecol.* **2013**, *86*, 432–443. [CrossRef]
101. Parshukov, A.N.; Kashinskaya, E.N.; Simonov, E.P.; Hlunov, O.V.; Izvekova, G.I.; Andree, K.B.; Solovyev, M.M. The isolation of the intestinal gut microbiota of farmed rainbow trout, *Oncorhynchus mykiss* (Walbaum), depending on the infection status of the fish. *Appl. Microbiol.* **2019**, *127*, 379–395. [CrossRef] [PubMed]
102. Wu, S.; Tian, J.; Wang, G.; Li, W.; Zou, H. Characterization of bacterial community in the stomach of yellow catfish (*Pelteobagrus fulvidraco*). *World J. Microbiol. Biotechnol.* **2012**, *28*, 2165–2174. [CrossRef] [PubMed]
103. Zhao, N.; Guo, J.; Zhang, B.; Liu, K.; Liu, Y.; Shen, Y.; Li, J. Heterogeneity of the tissue-specific mucosal microbiome of normal grass carp (*Ctenopharyngodon idella*). *Mar. Biotechnol.* **2022**, *24*, 366–379. [CrossRef] [PubMed]
104. Semova, I.; Carten, J.D.; Stombaugh, J.; Mackey, L.C.; Knight, R.; Farber, S.A.; Rawls, J.F. Microbiota regulate intestinal absorption and metabolism of fatty acids in the zebrafish. *Cell Host Microbe* **2012**, *12*, 277–288. [CrossRef]
105. Zhou, Z.; Shi, P.; He, S.; Liu, Y.; Huang, G.; Yao, B.; Ringø, E. Identification of adherent microbiota in the stomach and intestine of emperor red snapper (*Lutjanus sebae* Cuvier) using 16S rDNA-DGGE. *Aquac. Res.* **2009**, *40*, 1213–1218. [CrossRef]
106. Amenyogbe, E.; Chen, G.; Wang, Z.; Huang, J.; Huang, B.; Li, H. The exploitation of probiotics, prebiotics and synbiotics in aquaculture: Present study, limitations and future directions. A review. *Aquac. Int.* **2020**, *28*, 1017–1041. [CrossRef]
107. Chen, J.; Xie, S. Overview of sulfonamide biodegradation and the relevant pathways and microorganisms. *Sci. Total Environ.* **2018**, *640–641*, 1465–1477. [CrossRef]
108. Zhang, Q.-Q.; Ying, G.-G.; Pan, C.-G.; Liu, Y.-S.; Zhao, J.-L. Comprehensive evaluation of antibiotics emission and fate in the river basins of China: Source analysis, multimedia modeling, and linkage to bacterial resistance. *Environ. Sci. Technol.* **2015**, *49*, 6772–6782. [CrossRef] [PubMed]
109. Thiele-Bruhn, S. Pharmaceutical antibiotic compounds in soils—A review. *J. Plant Nutr. Soil Sci.* **2003**, *166*, 145–167. [CrossRef]
110. Bates, J.M.; Mittge, E.; Kuhlman, J.; Baden, K.N.; Cheesman, S.E.; Guillemin, K. Distinct signals from the microbiota promote different aspects of zebrafish gut differentiation. *Dev. Biol.* **2006**, *297*, 374–386. [CrossRef]
111. Yukgehnaish, K.M.; Kumar, P.; Sivachandran, P.; Marimuthu, K.; Arshad, A.; Paray, B.A.; Arockiaraj, J. Gut microbiota metagenomics in aquaculture: Factors influencing gut microbiome and its physiological role in fish. *Rev. Aquac.* **2020**, *12*, 1903–1927. [CrossRef]
112. Van Doan, H.; Hoseinifar, S.H.; Ringø, E.; Ángeles Esteban, M.; Dadar, M.; Dawood, M.A.O.; Faggio, C. Host-associated probiotics: A key factor in sustainable aquaculture. *Rev. Fish. Sci. Aquac.* **2020**, *28*, 16–42. [CrossRef]
113. Dawood, M.A.O.; Koshio, S.; Abdel-Daim, M.M.; Van Doan, H. Probiotic application for sustainable aquaculture. *Rev. Aquac.* **2019**, *11*, 907–924. [CrossRef]
114. Chauhan, A.; Singh, R. Probiotics in aquaculture: A promising emerging alternative approach. *Symbiosis* **2019**, *77*, 99–113. [CrossRef]
115. Elsabagh, M.; Mohamed, R.; Moustafa, E.M.; Hamza, A.; Farrag, F.; Decamp, O.; Dawood, M.A.O.; Eltholth, M. Assessing the impact of Bacillus strains mixture probiotic on water quality, growth performance, blood profile and intestinal morphology of Nile tilapia, *Oreochromis niloticus*. *Aquac. Nutr.* **2018**, *24*, 1613–1622. [CrossRef]
116. Wang, Y.-B.; Tian, Z.-Q.; Yao, J.-T.; Li, W. Effect of probiotics, *Enteroccus faecium*, on tilapia (*Oreochromis niloticus*) growth performance and immune response. *Aquaculture* **2008**, *277*, 203–207. [CrossRef]
117. Makled, S.O.; Hamdan, A.M.; El-Sayed, A.-F.M.; Hafez, E.E. Evaluation of marine psychrophile, *Psychrobacter namhaensis SO89*, as a probiotic in Nile tilapia (*Oreochromis niloticus*) diets. *Fish Shellfish Immunol.* **2017**, *61*, 194–200. [CrossRef]
118. Soleimani, N.; Hoseinifar, S.H.; Merrifield, D.L.; Barati, M.; Abadi, Z.H. Dietary supplementation of fructooligosaccharide (FOS) improves the innate immune response, stress resistance, digestive enzyme activities and growth performance of Caspian roach (*Rutilus rutilus*) fry. *Fish Shellfish Immunol.* **2012**, *32*, 316–321. [CrossRef]
119. Sumon, M.S.; Ahmmed, F.; Khushi, S.S.; Ahmmed, M.K.; Rouf, M.A.; Chisty, M.A.H.; Sarower, M.G. Growth performance, digestive enzyme activity and immune response of *Macrobrachium rosenbergii* fed with probiotic *Clostridium butyricum* incorporated diets. *J. King Saud Univ. Sci.* **2018**, *30*, 21–28. [CrossRef]

120. Ibrahem, M.D.; Fathi, M.; Mesalhy, S.; Abd El-Aty, A.M. Effect of dietary supplementation of inulin and vitamin C on the growth, hematology, innate immunity, and resistance of Nile tilapia (*Oreochromis niloticus*). *Fish Shellfish Immunol.* **2010**, *29*, 241–246. [CrossRef]
121. Newaj-Fyzul, A.; Adesiyun, A.A.; Mutani, A.; Ramsubhag, A.; Brunt, J.; Austin, B. *Bacillus subtilis AB1* controls Aeromonas infection in rainbow trout (*Oncorhynchus mykiss*, Walbaum). *J. Appl. Microbiol.* **2007**, *103*, 1699–1706. [CrossRef]
122. Ramesh, D.; Vinothkanna, A.; Rai, A.K.; Vignesh, V.S. Isolation of potential probiotic *Bacillus spp.* and assessment of their subcellular components to induce immune responses in *Labeo rohita* against *Aeromonas hydrophila*. *Fish Shellfish Immunol.* **2015**, *45*, 268–276. [CrossRef] [PubMed]
123. Torrecillas, S.; Makol, A.; Benítez-Santana, T.; Caballero, M.J.; Montero, D.; Sweetman, J.; Izquierdo, M. Reduced gut bacterial translocation in European sea bass (*Dicentrarchus labrax*) fed mannan oligosaccharides (MOS). *Fish Shellfish Immunol.* **2011**, *30*, 674–681. [CrossRef] [PubMed]
124. Geraylou, Z.; Souffreau, C.; Rurangwa, E.; D'Hondt, S.; Callewaert, L.; Courtin, C.M.; Delcour, J.A.; Buyse, J.; Ollevier, F. Effects of arabinoxylan-oligosaccharides (AXOS) on juvenile Siberian sturgeon (*Acipenser baerii*) performance, immune responses and gastrointestinal microbial community. *Fish Shellfish Immunol.* **2012**, *33*, 718–724. [CrossRef] [PubMed]
125. Chang, C.S.; Huang, S.L.; Chen, S.; Chen, S.N. Innate immune responses and efficacy of using mushroom beta-glucan mixture (MBG) on orange-spotted grouper, *Epinephelus coioides*, aquaculture. *Fish Shellfish Immunol.* **2013**, *35*, 115–125. [CrossRef] [PubMed]
126. Ebrahimi, G.; Ouraji, H.; Khalesi, M.K.; Sudagar, M.; Barari, A.; Zarei Dangesaraki, M.; Jani Khalili, K.H. Effects of a prebiotic, Immunogen, on feed utilization, body composition, immunity, and resistance to *Aeromonas hydrophila* infection in the common carp *Cyprinus carpio* (Linnaeus) fingerlings. *J. Anim. Physiol. Anim. Nutr.* **2012**, *96*, 591–599. [CrossRef]
127. Lin, S.; Mao, S.; Guan, Y.; Luo, L.; Pan, Y. Effects of dietary chitosan oligosaccharides and *Bacillus coagulans* on the growth, innate immunity, and resistance of koi (*Cyprinus carpio koi*). *Aquaculture* **2012**, *342*, 36–41. [CrossRef]
128. Cui, L.; Xu, W.; Ai, Q.; Wang, D.; Mai, K. Effects of dietary chitosan oligosaccharide complex with rare earth on growth performance and innate immune response of turbot, *Scophthalmus maximus* L. *Aquac. Res.* **2013**, *44*, 683–690. [CrossRef]
129. Gu, M.; Pan, S.; Li, Q.; Qi, Z.; Deng, W.; Bai, N. Chitosan and chitooligosaccharides attenuate soyabean meal-induced intestinal inflammation of turbot (*Scophthalmus maximus*): Possible involvement of NF-κB, activator protein-1 and mitogen-activated protein kinases pathways. *Brit. J. Nutr.* **2021**, *126*, 1651–1662. [CrossRef]
130. Pérez-Jiménez, G.M.; Peña-Marín, E.S.; Maytorena-Verdugo, C.I.; Sepúlveda-Quiroz, C.A.; Jiménez-Martínez, L.D.; De la Rosa-García, S.; Asencio-Alcudia, G.G.; Martinez, R.; Tovar-Ramirez, D.; Galaviz, M.; et al. Incorporation of fructooligosaccharides in diets influence growth performance, digestive enzyme activity, and expression of intestinal barrier function genes in tropical gar (*Atractosteus tropicus*) larvae. *Fishes* **2022**, *7*, 137. [CrossRef]
131. Zhang, C.N.; Li, X.F.; Xu, W.N.; Jiang, G.Z.; Lu, K.L.; Wang, L.N.; Liu, W.B. Combined effects of dietary fructooligosaccharide and *Bacillus licheniformis* on innate immunity, antioxidant capability and disease resistance of triangular bream (*Megalobrama terminalis*). *Fish Shellfish Immunol.* **2013**, *35*, 13801386. [CrossRef] [PubMed]
132. Poolsawat, L.; Li, X.; Yang, H.; Yang, P.; Kabir Chowdhury, M.A.; Yusuf, A.; Leng, X. The potentials of fructooligosaccharide on growth, feed utilization, immune and antioxidant parameters, microbial community and disease resistance of tilapia (*Oreochromis niloticus* × *O. aureus*). *Aquac. Res.* **2020**, *51*, 4430–4442. [CrossRef]
133. Grisdale-Helland, B.; Helland, S.J.; Gatlin, D.M. III. The effects of dietary supplementation with mannanoligosaccharide, fructooligosaccharide, or galactooligosaccharide on the growth and feed utilization of Atlantic salmon (*Salmo salar*). *Aquaculture* **2008**, *283*, 163–167. [CrossRef]
134. Ye, J.D.; Wang, K.; Li, F.D.; Sun, Y.Z. Single or combined effects of fructo- and mannan oligosaccharide supplements and *Bacillus clausii* on the growth, feed utilization, body composition, digestive enzyme activity, innate immune response and lipid metabolism of Japanese flounder *Paralichthys olivaceus*. *Aquac. Nutr.* **2011**, *17*, e902e11.
135. Akrami, R.; Iri, Y.; Khoshbavar Rostami, H.; Razeghi Mansour, M. Effect of dietary supplementation of fructooligosaccharide (FOS) on growth performance, survival, lactobacillus bacterial population and hemato-immunological parameters of stellate sturgeon (*Acipenser stellatus*) juvenile. *Fish Shellfish Immunol.* **2013**, *35*, 1235–1239. [CrossRef]
136. Zhang, Q.; Yu, H.; Tong, T.; Tong, W.; Dong, L.; Xu, M.; Wang, Z. Dietary supplementation of *Bacillus subtilis* and fructooligosaccharide enhance the growth, non-specific immunity of juvenile ovate pompano, *Trachinotus ovatus* and its disease resistance against *Vibrio vulnificus*. *Fish Shellfish Immunol.* **2014**, *38*, 7–14. [CrossRef]
137. Yousefi, S.; Hoseinifar, S.H.; Paknejad, H.; Hajimoradloo, A. The effects of dietary supplement of galactooligosaccharide on innate immunity, immune related genes expression and growth performance in zebrafish (*Danio rerio*). *Fish Shellfish Immunol.* **2018**, *73*, 192–196. [CrossRef]
138. Ajdari, A.; Ghafarifarsani, H.; Hoseinifar, S.H.; Javahery, S.; Narimanizad, F.; Gatphayak, K.; Van Doan, H. Effects of dietary supplementation of PrimaLac, inulin, and Biomin Imbo on growth performance, antioxidant, and innate immune responses of common carp (*Cyprinus carpio*). *Aquac. Nutr.* **2022**, *2022*, 8297479. [CrossRef]
139. Ahmdifar, E.; Akrami, R.; Ghelichi, A.; Mohammadi Zarejabad, A. Effects of different dietary prebiotic inulin levels on blood serum enzymes, hematologic, and biochemical parameters of great sturgeon (*Huso huso*) juveniles. *Comp. Clin. Pathol.* **2011**, *20*, 447–451. [CrossRef]

140. Cerezuela, R.; Cuesta, A.; Meseguer, J.; Esteban, M. Effects of dietary inulin and heat-inactivated *Bacillus subtilis* on gilthead seabream (*Sparus aurata* L.) innate immune parameters. *Benef. Microb.* **2012**, *3*, 77–81. [CrossRef]
141. Syed Raffic Ali, S.; Ambasankar, K.; Praveena, P.E.; Nandakumar, S.; Saiyad Musthafa, M. Effect of dietary prebiotic inulin on histology, immuno-haematological and biochemical parameters of Asian seabass (*Lates calcarifer*). *Aquac. Res.* **2018**, *49*, 2732–2740. [CrossRef]
142. Rairakhwada, D.; Pal, A.K.; Bhathena, Z.P.; Sahu, N.P.; Jha, A.; Mukherjee, S.C. Dietary microbial levan enhances cellular non-specific immunity and survival of common carp (*Cyprinus carpio*) juveniles. *Fish Shellfish Immunol.* **2007**, *22*, 477–486. [CrossRef] [PubMed]
143. Gupta, S.K. Ameliorative and protective effects of prebiotic, microbial levan in common carp (*Cyprinus carpio*) fry under experimental exposure to fipronil. *Int. J. Aquat. Biol.* **2021**, *9*, 134–147.
144. Gupta, S.K.; Pal, A.K.; Sahu, N.P.; Dalvi, R.; Kumar, V.; Mukherjee, S.C. Microbial levan in the diet of *Labeo rohita* Hamilton juveniles: Effect on non-specific immunity and histopathological changes after challenge with *Aeromonas hydrophila*. *J. Fish Dis.* **2008**, *31*, 649–657. [CrossRef]
145. Huang, M.Y.; Chang, C.I.; Chang, C.C.; Tseng, L.W.; Pan, C.L. Effects of dietary levan on growth performance, nonspecific immunity, pathogen resistance, and body composition of orange-spotted grouper (*Epinephelus coioides* H.). *Aquac. Res.* **2014**, *46*, 2752–2762. [CrossRef]
146. Staykov, Y.; Spring, P.; Denev, S.; Sweetman, J. Effect of a mannan oligosaccharide on the growth performance and immune status of rainbow trout (*Oncorhynchus mykiss*). *Aquac. Int.* **2007**, *15*, 153–161. [CrossRef]
147. Rodriguez-Estrada, U.; Satoh, S.; Haga, Y.; Fushimi, H.; Sweetman, J. Effects of single and combined supplementation of *Enterococcus faecalis*, mannan oligosaccharide and polyhydrobutyrate acid on growth performance and immune response of rainbow trout (*Oncorhynchus mykiss*). *Aquac. Sci.* **2009**, *57*, 609–617.
148. Rodríguez-Estrada, U.; Satoh, S.; Haga, Y.; Fushimi, H.; Sweetman, J. Effects of inactivated *Enterococcus faecalis* and mannan oligosaccharide and their combination on growth, immunity, and disease protection in rainbow trout. *N. Am. J. Aquacult.* **2013**, *75*, 416–428. [CrossRef]
149. Yoshida, T.; Kruger, R.; Inglis, V. Augmentation of nonspecific protection in African catfish, *Clarias gariepinus* (Burchell), by the long-term oral administration of immunostimulants. *J. Fish Dis.* **1995**, *8*, 195–198. [CrossRef]
150. Talpur, A.D.; Munir, M.B.; Mary, A.; Hashim, R. Dietary probiotics and prebiotics improved food acceptability, growth performance, haematology and immunological parameters, and disease resistance against *Aeromonas hydrophila* in snakehead (*Channa striata*) fingerlings. *Aquaculture* **2014**, *426*, 14–20. [CrossRef]
151. Akter, M.; Hashim, R.; Sutriana, A.; Nor, S.A.M. Influence of mannan oligosaccharide supplementation on haematological and immunological responses and disease resistance of striped catfish (*Pangasianodon hypophthalmus* Sauvage, 1878) juveniles. *Aquac. Int.* **2019**, *27*, 1535–1551. [CrossRef]
152. Zhou, Q.-C.; Buentello, J.A.; Gatlin, D.M. Effects of dietary prebiotics on growth performance, immune response and intestinal morphology of red drum (*Sciaenops ocellatus*). *Aquaculture* **2010**, *309*, 253–257. [CrossRef]
153. Lokesh, J.; Fernandes, J.M.O.; Korsnes, K.; Bergh, Ø.; Brinchmann, M.F.; Kiron, V. Transcriptional regulation of cytokines in the intestine of Atlantic cod fed yeast derived mannan oligosaccharide or b-Glucan and challenged with *Vibrio anguillarum*. *Fish Shellfish Immunol.* **2012**, *33*, 626–631. [CrossRef]
154. El-Nokrashy, A.M.; El-Banna, R.A.; Edrise, B.M.; Abdel-Rahim, M.M.; Jover-Cerdá, M.; Tomás-Vidal, A.; Prince, A.; Davies, S.; El-Haroun, E.; Goda, A.S. Impact of nucleotide enriched diets on the production of gilthead seabream, *Sparus aurata* fingerlings by modulation of liver mitochondrial enzyme activitity, antioxidant status, immune gene expression, and gut microbial ecology. *Aquaculture* **2021**, *535*, 736398. [CrossRef]
155. Lee, P.P.; Lin, Y.H.; Chen, M.C.; Cheng, W. Dietary administration of sodium alginate ameliorated stress and promoted immune resistance of grouper Epinephelus coioides under cold stress. *Fish Shellfish Immunol.* **2017**, *65*, 127–135. [CrossRef]
156. Yang, P.; Hu, H.; Liu, Y.; Li, Y.; Ai, Q.; Xu, W.; Zhang, W.; Zhang, Y.; Zhang, Y.; Mai, K. Dietary stachyose altered the intestinal microbiota profile and improved the intestinal mucosal barrier function of juvenile turbot, *Scophthalmus maximus* L. *Aquaculture* **2018**, *486*, 98–106. [CrossRef]
157. Pérez-Sánchez, J.; Benedito-Palos, L.; Estensoro, I.; Petropoulos, Y.; Calduch-Giner, J.A.; Browdy, C.L.; Sitjà-Bobadilla, A. Effects of dietary NEXT ENHANCE®150 on growth performance and expression of immune and intestinal integrity related genes in gilthead sea bream (*Sparus aurata* L.). *Fish Shellfish Immunol.* **2015**, *44*, 117–128. [CrossRef]
158. Firmino, J.P.; Fernández-Alacid, L.; Vallejos-Vidal, E.; Salomón, R.; Sanahuja, I.; Tort, L.; Ibarz, A.; Reyes-Lopez, F.E.; Gisbert, E. Carvacrol, thymol, and garlic essential oil promote skin innate immunity in gilthead seabream (*Sparus aurata*) through the multifactorial modulation of the secretory pathway and enhancement of mucus protective capacity. *Front. Immunol.* **2021**, *12*, 633621. [CrossRef] [PubMed]
159. Shukry, M.; El-Kader, M.F.A.; Hendam, B.M.; Dawood, M.A.O.; Farrag, F.A.; Aboelenin, S.M.; Soliman, M.M.; Abdel-Latif, H.M. Dietary *Aspergillus oryzae* modulates serum biochemical indices, immune responses, oxidative stress, and transcription of HSP70 and cytokine genes in *Nile tilapia* exposed to salinity stress. *Animals* **2021**, *11*, 1621. [CrossRef] [PubMed]
160. Ramesh, D.; Souissi, S.; Ahamed, T.S. Effects of the potential probiotics *Bacillus aerophilus KADR3* in inducing immunity and disease resistance in *Labeo rohita*. *Fish Shellfish Immunol.* **2017**, *70*, 408–415. [CrossRef] [PubMed]

161. Nandi, A.; Banerjee, G.; Dan, S.K.; Ghosh, K.; Ray, A.K. Evaluation of in vivo probiotic efficiency of *Bacillus amyloliquefaciens* in *Labeo rohita* challenged by pathogenic strain of *Aeromonas hydrophila MTCC 1739*. *Probiotics Antimicrob. Proteins* **2018**, *10*, 391–398. [CrossRef] [PubMed]
162. Truong Thy, H.T.; Tri, N.N.; Quy, O.M.; Fotedar, R.; Kannika, K.; Unajak, S.; Areechon, N. Effects of the dietary supplementation of mixed probiotic spores of *Bacillus amyloliquefaciens 54A*, and *Bacillus pumilus 47B* on growth, innate immunity and stress responses of striped catfish (*Pangasianodon hypophthalmus*). *Fish Shellfish Immunol.* **2017**, *60*, 391–399. [CrossRef] [PubMed]
163. Zhang, J.; Huang, M.; Feng, J.; Chen, Y.; Li, M.; Chang, X. Effects of dietary *Bacillus licheniformis* on growth performance, intestinal morphology, intestinal microbiome, and disease resistance in common carp (*Cyprinus carpio* L.). *Aquac. Int.* **2021**, *29*, 1343–1358. [CrossRef]
164. Gobi, N.; Vaseeharan, B.; Chen, J.-C.; Rekha, R.; Vijayakumar, S.; Anjugam, M.; Iswarya, A. Dietary supplementation of probiotic *Bacillus licheniformis Dahb1* improves growth performance, mucus and serum immune parameters, antioxidant enzyme activity as well as resistance against *Aeromonas hydrophila* in tilapia *Oreochromis mossambicus*. *Fish Shellfish Immunol.* **2018**, *74*, 501–508. [CrossRef]
165. Tarnecki, A.M.; Wafapoor, M.; Phillips, R.N.; Rhody, N.R. Benefits of a Bacillus probiotic to larval fish survival and transport stress resistance. *Sci. Rep.* **2019**, *9*, 4892. [CrossRef]
166. Weifen, L.; Xiaoping, Z.; Wenhui, S.; Bin, D.; Quan, L.; Luoqin, F.; Jiajia, Z.; Yue, W.; Dongyou, Y. Effects of Bacillus preparations on immunity and antioxidant activities in grass carp (*Ctenopharyngodon idellus*). *Fish Physiol. Biochem.* **2012**, *38*, 1585–1592. [CrossRef]
167. Gobi, N.; Malaikozhundan, B.; Sekar, V.; Shanthi, S.; Vaseeharan, B.; Jayakumar, R.; Nazar, A.K. GFP tagged *Vibrio parahaemolyticus Dahv2* infection and the protective effects of probiotic *Bacillus licheniformis Dahb1* on the growth, immune and antioxidant responses in Pangasius hypophthalmus. *Fish Shellfish Immunol.* **2016**, *52*, 230–238. [CrossRef]
168. Mukherjee, A.; Chandra, G.; Ghosh, K. Single or conjoint application of autochthonous Bacillus strains as potential probiotics: Effects on growth, feed utilization, immunity and disease resistance in rohu, *Labeo rohita* (Hamilton). *Aquaculture* **2019**, *512*, 734302. [CrossRef]
169. Srisapoome, P.; Areechon, N. Efficacy of viable Bacillus pumilus isolated from farmed fish on immune responses and increased disease resistance in Nile tilapia (*Oreochromis niloticus*): Laboratory and on-farm trials. *Fish Shellfish Immunol.* **2017**, *67*, 199–218. [CrossRef] [PubMed]
170. Liu, S.; Wang, S.; Cai, Y.; Li, E.; Ren, Z.; Wu, Y.; Guo, W.; Sun, Y.; Zhou, Y. Beneficial effects of a host gut-derived probiotic, *Bacillus pumilus*, on the growth, non-specific immune response and disease resistance of juvenile golden pompano, *Trachinotus ovatus*. *Aquaculture* **2020**, *514*, 734446. [CrossRef]
171. Sun, Y.-Z.; Yang, H.-L.; Huang, K.-P.; Ye, J.-D.; Zhang, C.-X. Application of autochthonous *Bacillus bioencapsulated* in copepod to grouper *Epinephelus coioides* larvae. *Aquaculture* **2013**, *392*, 44–50. [CrossRef]
172. Standen, B.T.; Peggs, D.L.; Rawling, M.D.; Foey, A.; Davies, S.J.; Santos, G.A.; Merrifield, D.L. Dietary administration of a commercial mixed-species probiotic improves growth performance and modulates the intestinal immunity of tilapia, *Oreochromis niloticus*. *Fish Shellfish Immunol.* **2016**, *49*, 427–435. [CrossRef] [PubMed]
173. Lee, S.; Katya, K.; Park, Y.; Won, S.; Seong, M.; Bai, S.C. Comparative evaluation of dietary probiotics *Bacillus subtilis WB60* and *Lactobacillus plantarum KCTC3928* on the growth performance, immunological parameters, gut morphology and disease resistance in Japanese eel, *Anguilla japonica*. *Fish Shellfish Immunol.* **2017**, *61*, 201–210. [CrossRef] [PubMed]
174. Ghanei-Motlagh, R.; Mohammadian, T.; Gharibi, D.; Khosravi, M.; Mahmoudi, E.; Zarea, M.; El-Matbouli, M.; Menanteau-Ledouble, S. Quorum quenching probiotics modulated digestive enzymes activity, growth performance, gut microflora, haemato-biochemical parameters and resistance against *Vibrio harveyi* in Asian seabass (*Lates calcarifer*). *Aquaculture* **2021**, *531*, 735874. [CrossRef]
175. Kuebutornye, F.K.A.; Wang, Z.; Lu, Y.; Abarike, E.D.; Sakyi, M.E.; Li, Y.; Xie, C.X.; Hlordzi, V. Effects of three host-associated Bacillus species on mucosal immunity and gut health of Nile tilapia, *Oreochromis niloticus* and its resistance against *Aeromonas hydrophila* infection. *Fish Shellfish Immunol.* **2020**, *97*, 83–95. [CrossRef]
176. Saputra, F.; Shiu, Y.L.; Chen, Y.C.; Puspitasari, A.W.; Danata, R.H.; Liu, C.H.; Hu, S.Y. Dietary supplementation with xylanase-expressing *B. áamyloliquefaciens R8* improves growth performance and enhances immunity against *Aeromonas hydrophila* in Nile tilapia (*Oreochromis niloticus*). *Fish Shellfish Immunol.* **2016**, *58*, 397–405. [CrossRef]
177. Xia, Y.; Wang, M.; Gao, F.; Lu, M.; Chen, G. Effects of dietary probiotic supplementation on the growth, gut health and disease resistance of juvenile Nile tilapia (*Oreochromis niloticus*). *Anim. Nutr.* **2020**, *6*, 69–79. [CrossRef]
178. Bandyopadhyay, P.; Das Mohapatra, P.K. Effect of a probiotic bacterium *Bacillus circulans PB7* in the formulated diets: On growth, nutritional quality and immunity of *Catla catla* (Ham.). *Fish Physiol. Biochem.* **2009**, *35*, 467–478. [CrossRef] [PubMed]
179. Niu, K.-M.; Khosravi, S.; Kothari, D.; Lee, W.-D.; Lim, J.-M.; Lee, B.-J.; Kim, K.-W.; Lim, S.-G.; Lee, S.-M.; Kim, S.-K. Effects of dietary multi-strain probiotics supplementation in a low fishmeal diet on growth performance, nutrient utilization, proximate composition, immune parameters, and gut microbiota of juvenile olive flounder (*Paralichthys olivaceus*). *Fish Shellfish Immunol.* **2019**, *93*, 258–268. [CrossRef] [PubMed]
180. Mohammadi, G.; Adorian, T.J.; Rafiee, G. Beneficial effects of *Bacillus subtilis* on water quality, growth, immune responses, endotoxemia and protection against lipopolysaccharide-induced damages in *Oreochromis niloticus* under biofloc technology system. *Aquac. Nutr.* **2020**, *26*, 1476–1492. [CrossRef]

181. Hassaan, M.S.; Soltan, M.A.; Jarmołowicz, S.; Abdo, H.S. Combined effects of dietary malic acid and *Bacillus subtilis* on growth, gut microbiota and blood parameters of Nile tilapia (*Oreochromis niloticus*). *Aquac. Nutr.* **2018**, *24*, 83–93. [CrossRef]
182. Meidong, R.; Khotchanalekha, K.; Doolgindachbaporn, S.; Nagasawa, T.; Nakao, M.; Sakai, K.; Tongpim, S. Evaluation of probiotic *Bacillus aerius B81e* isolated from healthy hybrid catfish on growth, disease resistance and innate immunity of Plamong *Pangasius bocourti*. *Fish Shellfish Immunol.* **2018**, *73*, 1–10. [CrossRef]
183. Reyes-Becerril, M.; Angulo, C.; Angulo, M.; Esteban, M.Á. Probiotic properties of *Debaryomyces hansenii BCS004* and their immunostimulatory effect in supplemented diets for gilthead seabream (*Sparus aurata*). *Aquac. Res.* **2021**, *52*, 2715–2726. [CrossRef]
184. Akbari, H.; Shekrabi, S.P.H.; Soltani, M.; Mehrgan, M.S. Effects of potential probiotic *Enterococcus casseliflavus* (EC-001) on growth performance, immunity, and resistance to *Aeromonas hydrophila* infection in Common carp (*Cyprinus carpio*). *Probiotics Antimicrob. Proteins* **2021**, *13*, 1316–1325. [CrossRef]
185. Baños, A.; Ariza, J.J.; Nuñez, C.; Gil-Martínez, L.; García-López, J.D.; Martínez-Bueno, M.; Valdivia, E. Effects of *Enterococcus faecalis UGRA10* and the enterocin AS-48 against the fish pathogen *Lactococcus garvieae*. Studies in vitro and in vivo. *Food Microbiol.* **2019**, *77*, 69–77. [CrossRef]
186. Tarkhani, R.; Imani, A.; Hoseinifar, S.H.; Moghanlou, K.S.; Manaffar, R. The effects of host-associated *Enterococcus faecium CGMCC1.2136* on serum immnune parameters, digestive enzymes activity and growth performance of the Caspian roach (*Rutilus caspicus*) fingerlings. *Aquaculture* **2020**, *519*, 734741. [CrossRef]
187. Picchietti, S.; Fausto, A.M.; Randelli, E.; Carnevali, O.; Taddei, A.R.; Buonocore, F.; Scaigliati, G.; Abelli, L. Early treatment with *Lactobacillus delbrueckii* strain induces an increase in intestinal T-cells and granulocytes and modulates immune-related genes of larval *Dicentrarchus labrax* (L.). *Fish Shellfish Immunol.* **2009**, *26*, 368–376. [CrossRef]
188. Mohammadi, G.; Rafiee, G.; Tavabe, K.R.; Abdel-Latif, H.M.; Dawood, M.A. The enrichment of diet with beneficial bacteria (single-or multi-strain) in biofloc system enhanced the water quality, growth performance, immune responses, and disease resistance of Nile tilapia (*Oreochromis niloticus*). *Aquaculture* **2021**, *539*, 736640. [CrossRef]
189. Hooshyar, Y.; Abedian Kenari, A.; Paknejad, H.; Gandomi, H. Effects of *Lactobacillus rhamnosus ATCC* 7469 on different parameters related to health status of rainbow trout (*Oncorhynchus mykiss*) and the protection against *Yersinia ruckeri*. *Probiotics Antimicrob. Proteins* **2020**, *12*, 1370–1384. [CrossRef] [PubMed]
190. Hoseinifar, S.H.; Roosta, Z.; Hajimoradloo, A.; Vakili, F. The effects of *Lactobacillus acidophilus* as feed supplement on skin mucosal immune parameters, intestinal microbiota, stress resistance and growth performance of black swordtail (*Xiphophorus helleri*). *Fish Shellfish Immunol.* **2015**, *42*, 533–538. [CrossRef] [PubMed]
191. Al-Dohail, M.A.; Hashim, R.; Aliyu-Paiko, M. Effects of the probiotic, *Lactobacillus acidophilus*, on the growth performance, haematology parameters and immunoglobulin concentration in African Catfish (*Clarias gariepinus*, Burchell 1822) fingerling. *Aquac. Res.* **2009**, *40*, 1642–1652. [CrossRef]
192. Adeshina, I.; Abubakar, M.I.O.; Ajala, B.E. Dietary supplementation with *Lactobacillus acidophilus* enhanced the growth, gut morphometry, antioxidant capacity, and the immune response in juveniles of the common carp, *Cyprinus carpio*. *Fish Physiol. Biochem.* **2020**, *46*, 1375–1385. [CrossRef]
193. Zhang, C.-N.; Zhang, J.-L.; Guan, W.-C.; Zhang, X.-F.; Guan, S.-H.; Zeng, Q.-H.; Cheng, G.-F.; Cui, W. Effects of *Lactobacillus delbrueckii* on immune response, disease resistance against *Aeromonas hydrophila*, antioxidant capability and growth performance of *Cyprinus carpio* Huanghe var. *Fish Shellfish Immunol.* **2017**, *68*, 84–91. [CrossRef]
194. Ahmadifar, E.; Moghadam, M.S.; Dawood, M.A.O.; Hoseinifar, S.H. *Lactobacillus fermentum* and/or ferulic acid improved the immune responses, antioxidative defence and resistance against Aeromonas hydrophila in common carp (*Cyprinus carpio*) fingerlings. *Fish Shellfish Immunol.* **2019**, *94*, 916–923. [CrossRef]
195. Hamdan, A.M.; El-Sayed, A.F.M.; Mahmoud, M.M. Effects of a novel marine probiotic, *Lactobacillus plantarum AH 78*, on growth performance and immune response of Nile tilapia (*Oreochromis niloticus*). *J. Appl. Microbiol.* **2016**, *120*, 1061–1073. [CrossRef]
196. Van Doan, H.; Kurian, A.; Hoseinifar, S.H.; Sel-audom, M.; Jaturasitha, S.; Tongsiri, S.; Ringø, E. Dietary inclusion of orange peels derived pectin and *Lactobacillus plantarum* for Nile tilapia (*Oreochromis niloticus*) cultured under indoor biofloc systems. *Aquaculture* **2019**, *508*, 98–105. [CrossRef]
197. Van Doan, H.; Hoseinifar, S.H.; Khanongnuch, C.; Kanpiengjai, A.; Unban, K.; Van Kim, V.; Srichaiyo, S. Host-associated probiotics boosted mucosal and serum immunity, disease resistance and growth performance of Nile tilapia (*Oreochromis niloticus*). *Aquaculture* **2018**, *491*, 94–100. [CrossRef]
198. Xia, Y.; Lu, M.; Chen, G.; Cao, J.; Gao, F.; Wang, M.; Liu, Z.; Zhang, D.; Zhu, H.; Yi, M. Effects of dietary *Lactobacillus rhamnosus JCM1136* and *Lactococcus lactis subsp. Lactis JCM5805* on the growth, intestinal microbiota, morphology, immune response and disease resistance of juvenile Nile tilapia, *Oreochromis niloticus*. *Fish Shellfish Immunol.* **2018**, *76*, 368–379. [CrossRef] [PubMed]
199. Sayed Hassani, M.H.; Jourdehi, A.Y.; Zelti, A.H.; Masouleh, A.S.; Lakani, F.B. Effects of commercial superzist probiotic on growth performance and hematological and immune indices in fingerlings *Acipenser baerii*. *Aquac. Int.* **2020**, *28*, 377–387. [CrossRef]
200. Abdelfatah, E.N.; Mahboub, H.H.H. Studies on the effect of *Lactococcus garvieae* of dairy origin on both cheese and Nile tilapia (*O. niloticus*). *Int. J. Vet. Sci. Med.* **2018**, *6*, 201–207. [CrossRef] [PubMed]
201. Nguyen, T.L.; Park, C.-I.; Kim, D.-H. Improved growth rate and disease resistance in olive flounder, *Paralichthys olivaceus*, by probiotic *Lactococcus lactis WFLU12* isolated from wild marine fish. *Aquaculture* **2017**, *471*, 113–120. [CrossRef]

202. Feng, J.; Chang, X.; Zhang, Y.; Zhang, X.Y.; Nie, G. Effects of *Lactococcus lactis* from *Cyprinus carpio* L. as probiotics on growth performance, innate immune response and disease resistance against *Aeromonas hydrophila*. *Fish Shellfish Immunol.* **2019**, *93*, 73–81. [CrossRef]
203. Sewaka, M.; Trullas, C.; Chotiko, A.; Rodkhum, C.; Chansue, N.; Boonanuntanasam, S.; Pirarat, N. Efficacy of synbiotic *Jerusalem artichoke* and *Lactobacillus rhamnosus GG*-supplemented diets on growth performance, serum biochemical parameters, intestinal morphology, immune parameters and protection against *Aeromonas veronii* in juvenile red tilapia (*Oreochromis spp.*). *Fish Shellfish Immunol.* **2019**, *86*, 260–268.
204. Feng, J.; Li, D.; Liu, L.; Tang, Y.; Du, R. Characterization and comparison of the adherence and immune modulation of two gut Lactobacillus strains isolated from *Paralichthys olivaceus*. *Aquaculture* **2018**, *499*, 381–388. [CrossRef]
205. Dowidar, M.F.; Abd ElAzeem, S.; Khater, A.M.; Awad Somayah, M.; Metwally, S.A. Improvement of growth performance, immunity and disease resistance in Nile tilapia, *Oreochromis niloticus*, by using dietary probiotics supplementation. *J. Anim. Sci. Vet. Med.* **2018**, *3*, 35–46. [CrossRef]
206. Chen, S.W.; Liu, C.H.; Hu, S.Y. Dietary administration of probiotic *Paenibacillus ehimensis NPUST1* with bacteriocin-like activity improves growth performance and immunity against Aeromonas hydrophila and *Streptococcus iniae* in Nile tilapia (*Oreochromis niloticus*). *Fish Shellfish Immunol.* **2019**, *84*, 695–703. [CrossRef]
207. Gupta, A.; Gupta, P.; Dhawan, A. *Paenibacillus polymyxa* as a water additive improved immune response of *Cyprinus carpio* and disease resistance against *Aeromonas hydrophila*. *Aquac. Rep.* **2016**, *4*, 86–92. [CrossRef]
208. Ferguson, R.M.W.; Merrifield, D.L.; Harper, G.M.; Rawling, M.D.; Mustafa, S.; Picchietti, S.; Balcazar, J.L.; Davies, S.J. The effect of *Pediococcus acidilactici* on the gut microbiota and immune status of on-growing red tilapia (*Oreochromis niloticus*). *J. Appl. Microbiol.* **2010**, *109*, 851–862. [CrossRef]
209. Standen, B.T.; Rawling, M.D.; Davies, S.J.; Castex, M.; Foey, A.; Gioacchini, G.; Carnevali, O.; Merrifield, D.L. Probiotic *Pediococcus acidilactici* modulates both localised intestinal- and peripheral-immunity in tilapia (*Oreochromis niloticus*). *Fish Shellfish Immunol.* **2013**, *35*, 1097–1104. [CrossRef] [PubMed]
210. Hoseinifar, S.H.; Hossein, M.; Paknejad, H.; Safari, R.; Jafar, A.; Yousefi, M.; Doan, H.V.; Mozanzadeh, M.T. Enhanced mucosal immune responses, immune related genes and growth performance in common carp (*Cyprinus carpio*) juveniles fed dietary *Pediococcus acidilactici MA18/5M* and raffinose. *Dev. Comp. Immunol.* **2019**, *94*, 59–65. [CrossRef]
211. Gong, L.; He, H.; Li, D.; Cao, L.; Khan, T.A.; Li, Y.; Pan, L.; Yan, L.; Ding, X.; Sun, Y.; et al. A new isolate of *Pediococcus pentosaceus* (SL001) with antibacterial activity against fish pathogens and potency in facilitating the immunity and growth performance of grass carps. *Front. Microbiol.* **2019**, *10*, 1384. [CrossRef] [PubMed]
212. Eissa, N.; Abou-ElGheit, E. Dietary supplementation impacts of potential non-pathogenic isolates on growth performance, hematological parameters and disease resistance in Nile tilapia (*Oreochromis niloticus*). *J. Vet. Adv.* **2014**, *4*, 712–719. [CrossRef]
213. Boonanuntanasarn, S.; Ditthab, K.; Jangprai, A.; Nakharuthai, C. Effects of microencapsulated *Saccharomyces cerevisiae* on growth, hematological indices, blood chemical, and immune parameters and intestinal morphology in striped catfish, *Pangasianodon hypophthalmus*. *Probiotics Antimicrob. Proteins* **2019**, *11*, 427–437. [CrossRef] [PubMed]
214. Abdel-Tawwab, M.; Abdel-Rahman, A.M.; Ismael, N.E. Evaluation of commercial live bakers' yeast, *Saccharomyces cerevisiae* as a growth and immunity promoter for Fry Nile tilapia, *Oreochromis niloticus* (L.) challenged in situ with *Aeromonas hydrophila*. *Aquaculture* **2008**, *280*, 185–189. [CrossRef]
215. Cordero, H.; Morcillo, P.; Cuesta, A.; Brinchmann, M.F.; Esteban, M.A. Differential proteome profile of skin mucus of gilthead seabream (*Sparus aurata*) after probiotic intake and/or overcrowding stress. *J. Proteom.* **2016**, *132*, 41–50. [CrossRef]
216. Wu, Z.-Q.; Jiang, C.; Ling, F.; Wang, G.-X. Effects of dietary supplementation of intestinal autochthonous bacteria on the innate immunity and disease resistance of grass carp (*Ctenopharyngodon idellus*). *Aquaculture* **2015**, *438*, 105–114. [CrossRef]
217. Schaeck, M.; Reyes-López, F.E.; Vallejos-Vidal, E.; Van Cleemput, J.; Duchateau, L.; Van den Broeck, W.; Tort, L.; Decostere, A. Cellular and transcriptomic response to treatment with the probiotic candidate *Vibrio lentus* in gnotobiotic sea bass (*Dicentrarchus labrax*) larvae. *Fish Shellfish Immunol.* **2017**, *63*, 147–156. [CrossRef]
218. Kahyani, F.; Pirali-Kheirabadi, E.; Shafiei, S.; Shenavar Masouleh, A. Effect of dietary supplementation of potential probiotic *Weissella confusa* on innate immunity, immune-related genes expression, intestinal microbiota and growth performance of rainbow trout (*Oncorhynchus mykiss*). *Aquac. Nutr.* **2021**, *27*, 1411–1420. [CrossRef]
219. De Souza, E.M.; de Souza, R.C.; Melo, J.F.B.; da Costa, M.M.; de Souza, A.M.; Copatti, C.E. Evaluation of the effects of *Ocimum basilicum* essential oil in Nile tilapia diet: Growth, biochemical, intestinal enzymes, haematology, lysozyme and antimicrobial challenges. *Aquaculture* **2019**, *504*, 7–12. [CrossRef]
220. Sun, Y.-Z.; Yang, H.-L.; Ma, R.-L.; Song, K.; Li, J.-S. Effect of *Lactococcus lactis* and *Enterococcus faecium* on growth performance, digestive enzymes and immune response of grouper *Epinephelus coioides*. *Aquac. Nutr.* **2012**, *18*, 281–289. [CrossRef]
221. Cerezuela, R.; Fumanal, M.; Tapia-Paniagua, S.T.; Meseguer, J.; Moriñigo, M.Á.; Esteban, M.Á. Changes in intestinal morphology and microbiota caused by dietary administration of inulin and *Bacillus subtilis* in gilthead sea bream (*Sparus aurata* L.) specimens. *Fish Shellfish Immunol.* **2013**, *34*, 1063–1070. [CrossRef] [PubMed]
222. Abdel-Aziz, M.; Bessat, M.; Fadel, A.; Elblehi, S. Responses of dietary supplementation of probiotic effective microorganisms (EMs) in *Oreochromis niloticus* on growth, hematological, intestinal histopathological, and antiparasitic activities. *Aquac. Int.* **2020**, *28*, 947–963. [CrossRef]

223. Amoah, K.; Dong, X.; Tan, B.; Zhang, S.; Chi, S.; Yang, Q.; Liu, H.; Yang, Y.; Zhang, H. Effects of three probiotic strains (*Bacillus coagulans, B. licheniformis* and *Paenibacillus polymyxa*) on growth, immune response, gut morphology and microbiota, and resistance against *Vibrio harveyi* of northern whitings, *Sillago sihama* Forsskål (1775). *Anim. Feed Sci. Technol.* **2021**, *277*, 114958. [CrossRef]
224. Selim, K.M.; Reda, R.M. Beta-glucans and mannan oligosaccharides enhance growth and immunity in Nile tilapia. *N. Am. J. Aquac.* **2015**, *77*, 22–30. [CrossRef]
225. Munir, M.B.; Hashim, R.; Chai, Y.H.; Marsh, T.L.; Nor, S.A.M. Dietary prebiotics and probiotics influence growth performance, nutrient digestibility and the expression of immune regulatory genes in snakehead (*Channa striata*) fingerlings. *Aquaculture* **2016**, *460*, 59–68. [CrossRef]
226. Mohammadian, T.; Ghanei-Motlagh, R.; Molayemraftar, T.; Mesbah, M.; Zarea, M.; Mohtashamipour, H.; Nejad, A.J. Modulation of growth performance, gut microflora, non-specific immunity and gene expression of proinflammatory cytokines in shabout (*Tor grypus*) upon dietary prebiotic supplementation. *Fish Shellfish Immunol.* **2021**, *112*, 38–45. [CrossRef]
227. Sutriana, A.; Hashim, R.; Akter, M.N.; Mohd Nor, S.A. Galactooligosaccharide and a combination of yeast and β-glucan supplements enhance growth and improve intestinal condition in striped catfish *Pangasianodon hypophthalmus* fed soybean meal diets. *Fish Sci.* **2018**, *84*, 523–533. [CrossRef]
228. Hossain, M.S.; Koshio, S.; Ishikawa, M.; Yokoyama, S.; Sony, N.M.; Ono, S.; Fujieda, T. Comparison of the effects of inosine and inosine monophosphate on growth, immune response, stress resistance and gut morphology of juvenile red sea bream, *Pagrus major*. *Aquaculture* **2016**, *458*, 64–74. [CrossRef]
229. Lu, Z.-Y.; Feng, L.; Jiang, W.-D.; Wu, P.; Liu, Y.; Kuang, S.-Y.; Tang, L.; Zhou, X.-Q. Mannan oligosaccharides improved growth performance and antioxidant capacity in the intestine of on-growing grass carp (*Ctenopharyngodon idella*). *Aquac. Rep.* **2020**, *17*, 100313. [CrossRef]
230. Cheng, Z.; Buentello, A.; Gatlin, D.M. Dietary nucleotides influence immune responses and intestinal morphology of red drum *Sciaenops ocellatus*. *Fish Shellfish Immunol.* **2011**, *30*, 143–147. [CrossRef] [PubMed]
231. Bunnoy, A.; Na-Nakorn, U.; Srisapoome, P. Probiotic effects of a novel strain, *Acinetobacter KU011TH*, on the growth performance, immune responses, and resistance against *Aeromonas hydrophila* of bighead catfish (*Clarias macrocephalus* Günther, 1864). *Microorganisms* **2019**, *7*, 613. [CrossRef] [PubMed]
232. Giannenas, I.; Karamaligas, I.; Margaroni, M.; Pappas, I.; Mayer, E.; Encarnacao, P.; Karagouni, E. Effect of dietary incorporation of a multi-strain probiotic on growth performance and health status in rainbow trout (*Oncorhynchus mykiss*). *Fish Physiol. Biochem.* **2015**, *41*, 119–128. [CrossRef] [PubMed]
233. Maas, R.M.; Verdegem, M.C.; Debnath, S.; Marchal, L.; Schrama, J.W. Effect of enzymes (phytase and xylanase), probiotics (*B. amyloliquefaciens*) and their combination on growth performance and nutrient utilisation in Nile tilapia. *Aquaculture* **2021**, *533*, 736226. [CrossRef]
234. Wang, Y. Use of probiotics *Bacillus coagulans, Rhodopseudomonas palustris* and *Lactobacillus acidophilus* as growth promoters in grass carp (*Ctenopharyngodon idella*) fingerlings. *Aquac. Nutr.* **2011**, *17*, e372–e378. [CrossRef]
235. Yanbo, W.; Zirong, X. Effect of probiotics for common carp (*Cyprinus carpio*) based on growth performance and digestive enzyme activities. *Anim. Feed Sci. Technol.* **2006**, *127*, 283–292. [CrossRef]
236. Asaduzzaman, M.; Sofia, E.; Shakil, A.; Haque, N.F.; Khan, M.N.A.; Ikeda, D.; Kinoshita, S.; Abol-Munafi, A.B. Host gut derived probiotic bacteria promote hypertrophic muscle progression and upregulate growth-related gene expression of slow-growing Malaysian mahseer *Tor tambroides*. *Aquac. Rep.* **2018**, *9*, 37–45. [CrossRef]
237. Jang, W.J.; Lee, J.M.; Hasan, M.T.; Lee, B.-J.; Lim, S.G.; Kong, I.-S. Effects of probiotic supplementation of a plant-based protein diet on intestinal microbial diversitydigestive enzyme activity, intestinal structure, and immunity in olive flounder (*Paralichthys olivaceus*). *Fish Shellfish Immunol.* **2019**, *92*, 719–727. [CrossRef]
238. Munirasu, S.; Ramasubramanian, V.; Arunkumar, P. Effect of probiotics diet on growth and biochemical performance of freshwater fish *Labeo rohita* fingerlings. *J. Entomol. Zool. Stud.* **2017**, *5*, 1374–1379.
239. Adeoye, A.A.; Yomla, R.; Jaramillo-Torres, A.; Rodiles, A.; Merrifield, D.L.; Davies, S.J. Combined effects of exogenous enzymes and probiotic on Nile tilapia (*Oreochromis niloticus*) growth, intestinal morphology and microbiome. *Aquaculture* **2016**, *463*, 61–70. [CrossRef]
240. Haque, M.M.; Hasan, N.A.; Eltholth, M.M.; Saha, P.; Mely, S.S.; Rahman, T.; Murray, F.J. Assessing the impacts of in-feed probiotic on the growth performance and health condition of pangasius (*Pangasianodon hypophthalmus*) in a farm trial. *Aquac. Rep.* **2021**, *20*, 100699. [CrossRef] [PubMed]
241. Silva, V.V.; Salomao, R.A.S.; Mareco, E.A.; Dal Pai, M.; Santos, V.B. Probiotic additive affects muscle growth of Nile tilapia (*Oreochromis niloticus*). *Aquac. Res.* **2021**, *52*, 2061–2069. [CrossRef]
242. Mohammadian, T.; Nasirpour, M.; Tabandeh, M.R.; Mesbah, M. Synbiotic effects of b-glucan, mannan oligosaccharide and *Lactobacillus casei* on growth performance, intestine enzymes activities, immunehematological parameters and immune-related gene expression in common carp, *Cyprinus carpio*: An experimental infection with *Aeromonas hydrophila*. *Aquaculture* **2019**, *511*, 634197.
243. Muchlisin, Z.A.; Murda, T.; Yulvizar, C.; Dewiyanti, I.; Fadli, N.; Afrido, F.; Muhammadar, A.A. Growth performance and feed utilization of keureling fish *Tor tambra* (Cyprinidae) fed formulated diet supplemented with enhanced probiotic. *F1000Research* **2017**, *6*, 137. [CrossRef]

244. Selvaraj, R.; Bogar, T. Effect of probiotic supplemented feed on growth performance of molly fish (*Poecilia sphenops*) in Recirculating aquaculture system. *Pharma Innov. J.* **2019**, *8*, 533–537.
245. Abd El-Rhman, A.M.; Khattab, Y.A.; Shalaby, A.M. *Micrococcus luteus* and Pseudomonas species as probiotics for promoting the growth performance and health of Nile tilapia, *Oreochromis niloticus*. *Fish Shellfish Immunol.* **2009**, *27*, 175–180. [CrossRef]
246. Sankar, H.; Philip, B.; Philip, R.; Singh, I.S.B. Effect of probiotics on digestive enzyme activities and growth of cichlids, *Etroplus suratensis* (Pearl spot) and *Oreochromis mossambicus* (Tilapia). *Aquac. Nutr.* **2017**, *23*, 852–864. [CrossRef]
247. Islam, S.M.; Rohani, M.F.; Shahjahan, M. Probiotic yeast enhances growth performance of Nile tilapia (*Oreochromis niloticus*) through morphological modifications of intestine. *Aquac. Rep.* **2021**, *21*, 100800. [CrossRef]
248. Jahan, N.; Islam, S.M.; Rohani, M.F.; Hossain, M.T.; Shahjahan, M. Probiotic yeast enhances growth performance of rohu (*Labeo rohita*) through upgrading hematology, and intestinal microbiota and morphology. *Aquaculture* **2021**, *545*, 737243. [CrossRef]
249. Banu, M.R.; Akter, S.; Islam, M.R.; Mondol, M.N.; Hossain, M.A. Probiotic yeast enhanced growth performance and disease resistance in freshwater catfish gulsa tengra, *Mystus cavasius*. *Aquac. Rep.* **2020**, *16*, 100237. [CrossRef]
250. Dharmaraj, S.; Dhevendaran, K. Evaluation of Streptomyces as a probiotic feed for the growth of ornamental fish *Xiphophorus helleri*. *Food Technol. Biotechnol.* **2010**, *48*, 497–504.
251. Rohani, M.F.; Islam, S.M.M.; Hossain, M.K.; Ferdous, Z.; Siddik, M.A.B.; Nuruzzaman, M.; Padeniya, U.; Brown, C.; Shahjahan, M. Probiotics, prebiotics and synbiotics improved the functionality of aquafeed: Upgrading growth, reproduction, immunity and disease resistance in fish. *Fish Shellfish Immunol.* **2022**, *120*, 569–589. [CrossRef] [PubMed]
252. Mohapatra, S.; Chakraborty, T.; Prusty, A.K.; PaniPrasad, K.; Mohanta, K.N. Beneficial effects of dietary probiotics mixture on hemato-immunology and cell apoptosis of *Labeo rohita* fingerlings reared at higher water temperatures. *PLoS ONE* **2014**, *9*, e100929. [CrossRef]
253. Afsa, S.; De Marco, G.; Giannetto, A.; Parrino, V.; Cappello, T.; ben Mansour, H.; Maisano, M. Histological endpoints and oxidative stress transcriptional responses in the Mediterranean mussel *Mytilus galloprovincialis* exposed to realistic doses of salicylic acid. *Environ. Toxicol. Pharmacol.* **2022**, *92*, 103855. [CrossRef] [PubMed]
254. Afsa, S.; De Marco, G.; Cristaldi, A.; Giannetto, A.; Galati, M.; Billè, B.; Oliveri Conti, G.; ben Mansour, H.; Ferrante, M.; Cappello, T. Single and combined effects of caffeine and salicylic acid on mussel *Mytilus galloprovincialis*: Changes at histomorphological, molecular and biochemical levels. *Environ. Toxicol. Pharmacol.* **2023**, *101*, 104167. [CrossRef]
255. De Marco, G.; Conti, G.O.; Giannetto, A.; Cappello, T.; Galati, M.; Iaria, C.; Pulvirenti, E.; Capparucci, F.; Mauceri, A.; Ferrante, M.; et al. Embryotoxicity of polystyrene microplastics in zebrafish *Danio rerio*. *Environ. Res.* **2022**, *208*, 112552. [CrossRef]
256. Noureen, A.; De Marco, G.; Rehman, N.; Jabeen, F.; Cappello, T. Ameliorative hematological and histomorphological effects of dietary *Trigonella foenum-graecum* seeds in common carp (*Cyprinus carpio*) exposed to copper oxide nanoparticles. *Int. J. Environ. Res. Public Health* **2022**, *19*, 13462. [CrossRef]
257. Noureen, A.; Jabeen, F.; Wajid, A.; Kazim, M.Z.; Safdar, N.; Cappello, T. Natural bioactive phytocompounds to reduce toxicity in common carp *Cyprinus carpio*: A challenge to environmental risk assessment of nanomaterials. *Water* **2023**, *15*, 1152. [CrossRef]
258. Abid, A.; Davies, S.J.; Waines, P.; Emery, M.; Castex, M.; Gioacchini, G.; Carnevali, O.; Bickerdike, R.; Romero, J.; Merrifield, D.L. Dietary synbiotic application modulates Atlantic salmon (*Salmo salar*) intestinal microbial communities and intestinal immunity. *Fish Shellfish Immunol.* **2013**, *35*, 1948–1956. [CrossRef] [PubMed]
259. Milián-Sorribes, M.C.; Martínez-Llorens, S.; Cruz-Castellón, C.; Jover-Cerdá, M.; Tomás-Vidal, A. Effect of fish oil replacement and probiotic addition on growth, body composition and histological parameters of yellowtail (*Seriola dumerili*). *Aquac. Nutr.* **2021**, *27*, 3–16. [CrossRef]
260. Darafsh, F.; Soltani, M.; Abdolhay, H.A.; Shamsaei Mehrejan, M. Improvement of growth performance, digestive enzymes and body composition of Persian sturgeon (*Acipenser persicus*) following feeding on probiotics: *Bacillus licheniformis*, *Bacillus subtilis* and *Saccharomyces cerevisiae*. *Aquac. Res.* **2020**, *51*, 957–964. [CrossRef]
261. González-Félix, M.L.; Gatlin, D.M.; Urquidez-Bejarano, P.; de la Reé-Rodríguez, C.; Duarte-Rodríguez, L.; Sánchez, F.; Casas-Reyes, A.; Yamamoto, F.Y.; Ochoa-Leyva, A.; Perez-Velazquez, M. Effects of commercial dietary prebiotic and probiotic supplements on growth, innate immune responses, and intestinal microbiota and histology of *Totoaba macdonaldi*. *Aquaculture* **2018**, *491*, 239–251. [CrossRef]
262. Guerreiro, I.; Couto, A.; Pérez-Jiménez, A.; Oliva-Teles, A.; Enes, P. Gut morphology and hepatic oxidative status of European sea bass (*Dicentrarchus labrax*) juveniles fed plant feedstuffs or fishmeal-based diets supplemented with short-chain fructo-oligosaccharides and xylo-oligosaccharides. *Br. J. Nutr.* **2015**, *114*, 1975–1984. [CrossRef] [PubMed]
263. Standen, B.T.; Rodiles, A.; Peggs, D.L.; Davies, S.J.; Santos, G.A.; Merrifield, D.L. Modulation of the intestinal microbiota and morphology of tilapia, *Oreochromis niloticus*, following the application of a multi-species probiotic. *Appl. Microbiol. Biotechnol.* **2015**, *99*, 8403–8417. [CrossRef]
264. Pirarat, N.; Pinpimai, K.; Endo, M.; Katagiri, T.; Ponpornpisit, A.; Chansue, N.; Maita, M. Modulation of intestinal morphology and immunity in nile tilapia (*Oreochromis niloticus*) by *Lactobacillus rhamnosus GG*. *Res. Vet. Sci.* **2011**, *91*, e92–e97. [CrossRef]
265. Yilmaz, S.; Yilmaz, E.; Dawood, M.A.O.; Ringø, E.; Ahmadifar, E.; Abdel-Latif, H.M.R. Probiotics, prebiotics, and synbiotics used to control vibriosis in fish: A review. *Aquaculture* **2022**, *547*, 737514. [CrossRef]
266. Mugwanya, M.; Dawood, M.A.O.; Kimera, F.; Sewilam, H. Updating the role of probiotics, prebiotics, andsSynbiotics for Tilapia aquaculture as leading candidates for food sustainability: A Review. *Probiotics Antimicrob. Proteins* **2022**, *14*, 130–157. [CrossRef]

267. Jiang, H.; Bian, Q.; Zeng, W.; Ren, P.; Sun, H.; Lin, Z.; Tang, Z.; Zhou, X.; Wang, Q.; Wang, Y.; et al. Oral delivery of *Bacillus subtilis* spores expressing grass carp reovirus VP4 protein produces protection against grass carp reovirus infection. *Fish Shellfish Immunol.* **2019**, *84*, 768–780. [CrossRef]
268. Van Doan, H.; Hoseinifar, S.H.; Tapingkae, W.; Seel-audom, M.; Jaturasitha, S.; Dawood, M.A.O.; Wongmaneeprateep, S.; Thu, T.T.N.; Esteban, M.Á. Boosted growth performance, mucosal and serum immunity, and disease resistance Nile Tilapia (*Oreochromis niloticus*) fingerlings using corncob-derived xylooligosaccharide and *Lactobacillus plantarum CR1T5*. *Probiotics Antimicrob. Proteins* **2020**, *12*, 400–411. [CrossRef] [PubMed]
269. Pérez-Sánchez, T.; Ruiz-Zarzuela, I.; de Blas, I.; Balcázar, J.L. Probiotics in aquaculture: A current assessment. *Rev. Aquac.* **2014**, *6*, 133–146. [CrossRef]

animals

MDPI

Review

How Do Alternative Protein Resources Affect the Intestine Morphology and Microbiota of Atlantic Salmon?

Lucia Aidos [1], Giorgio Mirra [2], Margherita Pallaoro [1], Valentina Rafaela Herrera Millar [3], Giuseppe Radaelli [2], Chiara Bazzocchi [1], Silvia Clotilde Modina [1] and Alessia Di Giancamillo [3,*]

[1] Department of Veterinary Medicine and Animal Science, University of Milan, 26900 Lodi, Italy; lucia.aidos@unimi.it (L.A.); margherita.pallaoro@unimi.it (M.P.); chiara.bazzocchi@unimi.it (C.B.); silvia.modina@unimi.it (S.C.M.)

[2] Department of Comparative Biomedicine and Food Science, University of Padua, 35122 Padova, Italy; giorgio.mirra@studenti.unimi.it (G.M.); giuseppe.radaelli@unipd.it (G.R.)

[3] Department of Biomedical Sciences for Health, University of Milan, 20133 Milan, Italy; valentina.herrera@unimi.it

* Correspondence: alessia.digiancamillo@unimi.it

Simple Summary: The Atlantic salmon industry is expected to grow globally by 2–3% by the year 2030. The main protein source used in salmon feeds is fishmeal, which is produced from wild-caught marine fish, whose natural stocks are at risk of depletion. It is of major importance, therefore, to select alternative feed ingredients that present an adequate protein profile, guarantee a good health status and growth performance of Atlantic salmon, and at the same time are environmentally sustainable.

Abstract: The availability and cost of fishmeal constitute a bottleneck in Atlantic salmon production expansion. Fishmeal is produced from wild fish species and constitutes the major feed ingredient in carnivorous species such as the Atlantic salmon. These natural stocks are at risk of depletion and it is therefore of major importance to find alternative protein sources that meet the nutritional requirements of the Atlantic salmon, without compromising the animals' health. Terrestrial animal by-products have been used in aquaculture feed, but their use is limited by the lack of several essential amino acids and consumer acceptance. In the case of plant ingredients, it is necessary to take into account both their concentration and the extraction methodologies, since, if not dosed correctly, they can cause macro- and microscopic alterations of the structure of the gastrointestinal tract and can also negatively modulate the microbiota composition. These alterations may compromise the digestive functions, growth of the animal, and, ultimately, its well-being. An updated revision of alternative protein sources is provided, with the respective impact on the intestine health in terms of both morphology and microbiota composition. Such information may constitute the premise for the choice and development of Atlantic salmon feeds that guarantee fish health and growth performance without having a significant impact on the surrounding environment, both in terms of depletion of the fish's natural stocks and in terms of pressure on the terrestrial agriculture. The sustainability of aquaculture should be a priority when choosing next-generation ingredients.

Keywords: Atlantic salmon; fishmeal replacement; intestine health; intestine morphology microbiota

1. The Salmonid's Gastrointestinal Tract (GIT)

Due to a large number of species, it is difficult to settle on a general nomenclature for the digestive system anatomy in teleost fishes. It has been suggested that there is a broad division of the anatomy into the head-intestine, fore-intestine, also named anterior intestine, mid-intestine (MI), and distal-intestine (DI), even if the limits between these regions are difficult to assess. This anatomical nomenclature has been used by Løkka et al. [1] in a broad study on the intestinal morphology of the wild Atlantic salmon, and the same nomenclature will be used in this review. The following GIT description corresponds to a healthy animal.

The teleost's gastrointestinal tract (GIT) varies in terms of intestinal length, level of looping, number of pyloric caeca, and the presence or absence of a proper stomach [2]. The head-intestine is composed of two elements, the oral or buccal cavity and the gill cavity, also named branchial or pharyngeal, which play a role in food transport and respiration. The anterior intestine includes the esophagus and the stomach, the latter is well-developed in carnivore fish, such as the Atlantic salmon. The MI is further divided into a first (with openings to the pyloric caeca, PC) and a second segment, in which the latter is often addressed as the correspondent mammalian ileum. The DI is very short and it is proposed as the equivalent mammalian colon and anus [3]. In this review, the head- and the anterior intestine are not considered, as the available literature does not refer to effects on these tracts that can be related to feed. For this reason, the description starts with the MI and then with DI.

1.1. Macroscopic Aspects of the Intestine

1.1.1. Mid Intestine (MI)

The MI is the longest portion of the intestine and the luminal surface is remarkably folded, with prominent longitudinal major or primary mucosal folds, which are further subdivided into a series of irregular and well-circumscribed folds called minor or secondary folds. In Atlantic salmon, the first segment of the MI likely is the main site of nutrient absorption of the GIT and the folds increase the absorptive area, similar to what happens with the villi in the mammalian small intestine [4]. However, if in teleosts the folds of the first segment of the MI are longitudinal or with a branching, net-like pattern, in Atlantic salmon these appear longitudinally arranged, but the explanation for this has not been stated [1]. Pyloric caeca are present in the first segment of the MI and consist of blind extensions of the intestine that present small folds with a longitudinal orientation to the length of the caeca. The functions of the pyloric caeca include the secretion of digestive enzymes and the absorption of nutrients [5]. In Atlantic salmon, pyloric caeca account for 87% of the total intestinal length and 64% of the total post-gastric intestinal mass [6]. The second segment of the MI is characterized by macroscopically visible circular folds and it shows a higher diameter when compared to the first segment of the MI. Although the function of the circular folds is not fully known, a defecatory function has been hypothesized by Burnstock [7]. In Atlantic salmon, the second segment of the MI seems to have an important role in the antigenic uptake and it is immunologically more active than the other segments of the GIT, as reviewed by Bjørgen et al. [3]. Many authors showed that the wall of the second segment of the MI is thinner than in the first segment [1]. The thin wall and the broad folding of this region indicate that few mechanical processes take place.

1.1.2. Distal Intestine (DI)

The DI is the shortest region of the intestine and its beginning is usually marked by an increase in diameter. The luminal surface presents small folds without an evident orientation, similar to the first segment of the MI [2,4]. Moreover, the wall of the DI in teleosts presents a thicker muscular layer than the other intestinal regions. This fact was confirmed in Atlantic salmon post-smolts, but not in spawning individuals, but the authors did not present an explanation for this fact [1]. The distal intestine has an important role in osmoregulation [8] and it is also a major site of absorption of intact proteins in all teleost fishes [9].

1.2. Microscopic Aspects of the Intestine: Mid and Distal Intestine (MI and DI)

A description of the histology of all segments of the intestine of Atlantic salmon has been explained extensively elsewhere [1]. Therefore, only a summary of the intestine micro-anatomy is provided here. In vertebrates, the intestinal wall is composed of four layers: the innermost layer is the tunica mucosa, underneath there is the submucosa, followed by the muscularis, and the outer layer is the serosa [10]. Even if in teleosts the same nomenclature is used, there is some discussion on which tissues belong to the mucosal layer and which ones belong to the submucosa. Figure 1 shows the heterogeneous cell populations that can be found in the mucosa.

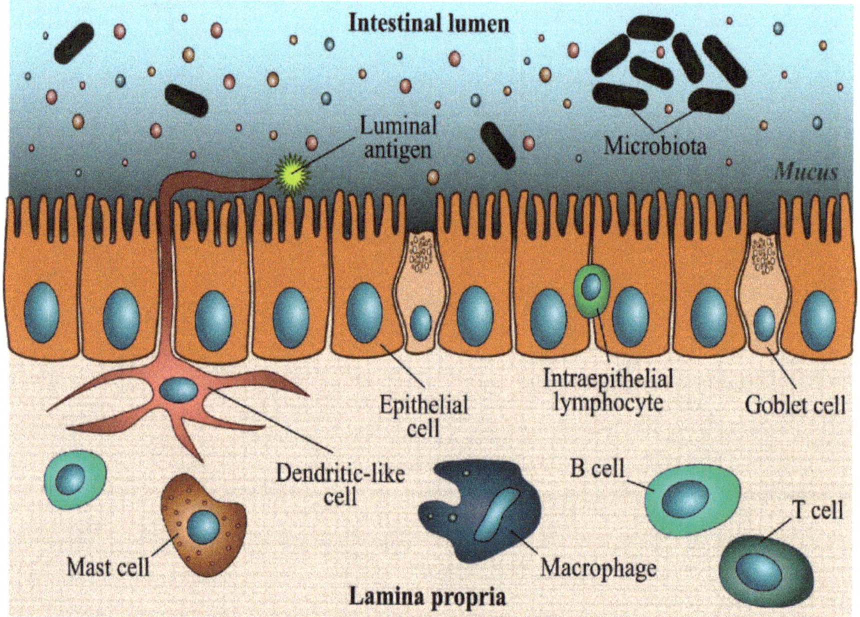

Figure 1. Intestinal mucosa in Atlantic salmon: it is possible to observe goblet cells and intra-epithelial lymphocytes in-between the epithelial cells. The lamina propria presents multiple groups of immune cells, such as dendritic-like cells, mast cells, macrophages, and B and T cells.

The inner mucosal layer is made of one layer of cells, and an underlying lamina propria of connective tissue. The epithelium is composed of a monolayer of four cell types: enterocytes, goblet, endocrine, and immune cells [11–13]. The epithelium has an important role in nutrient transport, osmoregulation, and protection from environmental stressors [14,15]. Enterocytes are columnar cells that present a brush border, which is a microvilli-covered surface, and are responsible for the absorption of nutrients. Tight junctions are structures that form contact points between enterocytes and are responsible for regulating the permeability of ions, (macro)molecules, and cells via the paracellular pathway. Goblet cells are mucus-producing cells that constitute the inner mucus layer. Endocrine cells respond to the presence or absence of food in the intestine by secreting peptide hormones. The intraepithelial lymphocytes are the first immune cell line in the intestine that is involved in defense actions (Figure 1). These lymphocytes, which are located between epithelial cells, present mobility and control the space between epithelial cells above the basement membrane [16]. Noteworthy, mammals present villi which are small, finger-like projections made up of cells that line the small intestine. In fish, these structures are completely absent throughout all the tracts. However, it is important to stress that some authors define the folds as villi [17] regarding the presence of crypt-like structures at the base, similar to those described in mammals.

As described by Bjørgen et al. [3] for all the intestinal segments, the lamina propria is located underneath the basal membrane below the epithelium and it is composed of connective tissue containing leukocytes. Precisely, in Atlantic salmon in the lamina propria, it is possible to find mast cells, macrophages, dendritic cells (antigen-presenting cells),

B cells (IgM-positive cells), and T cells [18] (Figure 1). Mast cells play a crucial role in the innate and adaptive immune responses, acting against allergens, bacteria, toxins, parasites, neuropeptides, and stress by increasing epithelial secretion and peristalsis, and by releasing pro-inflammatory mediators. Intestine macrophages regulate the inflammatory reaction to bacteria and antigens that can pass the epithelium. Macrophages also have a role in the protection of the mucosa against harmful pathogens and in dead cell scavenging. Dendritic cells are responsible for the regulation of tolerance to food antigens and are also necessary for the optimal response to intestinal pathogens. B and T lymphocytes intervene in the specific responses of adaptive immunity, although in different modes: B cells express their antigen receptor on the cell surface and secrete them as an immunoglobulin (Ig) or antibody, while the T cell receptor is always bound to the cell surface [19].

The lamina propria lies on a thick layer of connective tissue, rich in collagen, called the stratum compactum: it is a typical characteristic of carnivorous fishes, and acts as a prevention of stomach distention during food ingestion. This layer is surrounded by the stratum granulosum, which presents a large number of mast cells. On the other hand, the submucosa presents a deeper, looser connective tissue than the lamina propria. The muscular layer, internally, presents a circular orientation of the muscle fibers, whereas externally the muscle fibers are disposed longitudinally.

As described by several authors [20,21], Paneth cells, which are highly specialized secretory epithelial cells that contain antimicrobial peptides that regulate the composition of the intestinal flora, seem to be missing in Atlantic salmon. In addition, as already mentioned, the crypts of Lieberkühn, which are intestinal glands, are absent [22]. Løkka et al. [1] found high proliferation rates of the columnar cells at the base of the primary and secondary folds. This suggests that the folds in teleosts have corresponding functions to the crypts in mammals, which are involved in cell proliferation, differentiation, and migration.

1.3. Microbiota

The intestine microbiota has a crucial role in all vertebrates [23], as commensal bacteria can influence important factors such as growth, development, homeostasis, and well-being [24]. In fish, as in the case of mammals, the intestine microbiota is composed of autochthonous or allochthonous bacteria [25]. The autochthonous bacteria are derived from the colonization of the host's intestinal epithelial surface, while the allochthonous bacteria are transient [26].

Although the microbiota in fish has been studied since the beginning of the last century, information is still limited. Interactions between the host, diet, and environmental conditions contribute to altering the intestine microbiota [27], causing an imbalance in nutrient digestion and utilization [28,29], enzyme production [30], and also in immune status, which can decrease disease resistance [31,32]. It is therefore of great importance to evaluate the impact of food on the intestine microbiota.

2. Nutritional Needs of Salmon—"The Protein Problem"

Atlantic salmon requires the same nutrients (protein, amino acids, essential fatty acids—EFA, vitamins, and minerals) as other carnivorous fish species for normal growth, reproduction, and immune and metabolic functions, but the quantities and proportions are species-specific [33]. Precisely, dietary protein and amino acids supply are major factors influencing the productivity of farmed salmon [34]. Salmon fry shows better growth performances with high protein diets (~50%), while grow-out diets contain normally 42–48% protein. More than a general protein requirement, fish need essential amino acids, such as arginine, histidine, isoleucine, leucine, lysine, methionine, phenylalanine, threonine, tryptophan, and valine for regular growth. An unbalanced level of specific amino acids in the diet may result in mild or severe amino acid deficiency, mainly when fish are subjected to some environmental and physiological stress, even if there are certain essential amino acids such as leucine that, when present in high amounts in the diet, may be toxic [33].

When the fishmeal (FM) inclusion level in the diet of salmon is reduced and replaced with other protein sources, there may be an imbalance of essential amino acids. This is why we have "the protein problem".

3. Fishmeal (FM) Dietary Replacement—Looking for Potential Candidates

In Atlantic salmon, the main used protein source ingredients are FM (anchovies, pilchards, mackerel, herring, and blue whiting), plant protein products (soybean meal, corn gluten meal, canola meal, pea meal), animal by-products meal (poultry by-product meal, meat meal, blood meal, hydrolyzed feather meal), and crustacean meal (krill, shrimp, crab). Among these, FM has been the preferred one because of its high protein content, excellent amino acid profile, high nutrient digestibility, and absence of antinutrients [35,36]. Atlantic salmon feed is usually composed of about 40–60% of FM. Indeed, in 2013, the aquaculture industry consumed about 56% of the world's supplies of FM [37], and the demands are expected to increase as the industry grows. Taking into account that 80% of the world's fish stocks are stated as fully or over-exploited [38], further growth in aquaculture production will be limited by FM availability and by its increasing price. Alternative feed ingredients will, therefore, be extremely necessary [39]. The possible options are vegetable proteins, by-products from fish and terrestrial animal processing industries, organisms from lower trophic levels, and industrially produced bio-protein, such as bacteria and algae [40].

However, it is important to bear in mind that the replacement of FM with terrestrial plant ingredients may induce changes and cause negative effects on the GIT. The use of plant ingredients is limited in carnivorous fish species due to the presence of starch and structural carbohydrates, and a wide range of anti-nutritional factors (ANF). [41–44]. Despite this, plant ingredients remain the most attractive protein sources due to their high protein content and availability. For instance, among plant protein sources, field peas are a good candidate in species such as Atlantic salmon [45], rainbow trout [46], and European sea bass [47], even if the protein content is lower than in other plant protein sources, such as oil seeds. Like soya, peas also contain several ANF but, as observed by Francis et al. [35], their content is lower when compared with other legumes.

The intestine's health largely depends on the impact of the diet and can be assessed by evaluating histological parameters based on structural changes in the intestine itself. The most frequent alterations in the intestine include the reduction in the height of the folds, the simultaneous widening of the lamina propria, an increase in the presence of intra-epithelial lymphocytes, and a large amount of apical vacuolization of the columnar cells; all these histological signs reflect the reaction to the exposure to pathogens or stressors [48]. The impact of several protein sources on the structure of the DI and MI is summarized in Table 1 and will be further described in detail.

Diet can also influence the microbiota status, and many alternative plant-based proteins were investigated for their negative impact on the intestine [49]. The state of intestine health and the microbiota reflect the correct diet and the potential use that Atlantic salmon can make of it. Table 1 summarizes the studies that have focused on how different fishmeal substitutes have affected intestinal health, and Table 2 the effects on intestinal microbiota.

Table 1. Impact of alternative protein sources on the morpho-functionality of the mid and distal intestine in Atlantic salmon. Negative effect, −; positive effect, +; no effect, =; increase, ↑; decrease, ↓; * addition of probiotics. A higher number of these symbols, give an indication of severity degree of the respective impact.

Protein Source	Inclusion Level	Mucosa Folds	Mucosa Thickness	Mucosa Cells	Digestive/Absorption Capacity	Inflammatory Process	Overall Histology	References
Soybean Meal	10%				↓			[50]
	10%	−	−	−		↑		[51]
	15%	−−		−−	↓↓			[50]
	20%	−−−		−−−	↓↓↓			[50]
	20%	−−	−−	−−	↓	↑↑		[51]
	20%		−			↑		[52]
	20%	−	−					[53]
	20% *		+	+				[54]
	25%	−−−−		−−−−	↓↓↓↓			[50]
	30%		−		↓	↑	−	[55]
	30%	−		−				[49]
	32%	−	−		↓	↑↑	−	[56,57]
	35%	−	−	−		↑		[58]
	35%	−−−−		−−−−	↓↓↓↓↓			[50]
	40%	−−	−−	−		−		[57]
	50%	−	−	−		↑		[59]
HP48	17.8%	−	−	−				[60]
Whole Field Pea	20%	=	=	=		=	=	[45]
Pea Protein Concentrate	20%	=			↓		=	[61]
	35%	−	−	−	↓	↑	−	[43]
	44%	=	=	=			=	[62]
Dehulled Faba	20%	=	=	=		=	=	[45]
Faba Bean Concentrate	17%	−	−	−	↓	↑		[63]
	5–20%	=	=	=		↓	=	[64]
	35%	−−			↓↓	↑↑		[63]
	45%	=	=	−		↑	−	[64]
Plant Proteins	25%	=	=	=		↑		[65]
Poultry Meal	20%		−			=		[66]
	58%	=	=	=			=	[67]
Feather Meal	20%			−		=		[62]
Insect Meal	5–15%	=	=				=	[68]
	5–15%	=					=	[69]
	6.25–25%	=		=		=	+	[70]
	15%	−		−		↑	+	[71]
	25%	=	=	=			=	[17]
	85%	=	=	=			=	[71]

Table 1. *Cont.*

Protein Source	Inclusion Level	Mucosa Folds	Mucosa Thickness	Mucosa Cells	Digestive/Absorption Capacity	Inflammatory Process	Overall Histology	References
Microalgae								
Nannochloropsis oceanica	20%	=	=	=	↓	=	=	[72]
Desmodesmus sp.	20%	=	=	=			=	[73]
Yeasts								
Candida utilis	20%	+						[74]
	25%	=	=	=			=	[23]

Table 2. Effects of different protein sources on gut microbiota in Atlantic Salmon. +, presence.

Protein Source	Concentration	Intestinal Tract	Effects on Microbiota (MB)	Microbial Population Mostly Influenced	Enteritis	References
Soybean meal	5–10%	PI and DI mucosa	Yes	Presence of bacteria not normally associated with marine fish (*Escherichia* and *Propionibacterium*)		[75]
	20%	MI and DI mucosa	No	*Tenericutes* dominant Phylum *Proteobacteria*, *Firmicutes* and *Bacteroidetes*. Lactic acid bacteria increase	+	[52]
	20%	MI and DI digesta	Yes	Abundance of *Firmicutes* and *Bacteroides*; increase in *Proteobacteria*	+	[76]
	25%	MI and DI mucosa and digesta	Yes	Increase in *Enterococcus* spp., mostly *Enterococcus faecalis*		[41]
	30%	DI mucosa and digesta	Yes	*Proteobacteria*, *Firmicutes*, and *Bacteroidetes* Abundance of lactic acid bacteria	+	[49]
	30%	DI digesta	No	Dominated by lactic acid bacteria		[77]
	50%	DI digesta	Yes	*Shewanella* dominant bacteria. *Lactococcus*, *Aeromonas*, *Pseudomonas*. Lactic acid bacteria may modulate intestinal inflammation		[59]
Pea protein concentrate	20%	Allochthonous bacteria: Fecal samples Autochthonous Bacteria: PI and DI mucosa	Yes	*Corynebacteriaceaein* dominant in the allochthonous bacterial population and *Lactobacillaceaein* in the autochthonous		[62]
Poultry by-product meal	20%	Allochthonous bacteria: Fecal samples Autochthonous Bacteria: PI and DI mucosa	Yes	*Corynebacteriaceaein* dominant in the allochthonous bacterial population and *Lactobacillaceaein* in the autochthonous		[62]
Insect meal	60%	DI digesta and mucosa	yes	Increasing of *Actinomyces*, *Bacillus*, *Brevibacterium*, *Corynebacterium 1* and *Enterococcus*		[78]
Yeast *Cyberlindnera jadinii*	10% 20%	MI and DI mucosa	yes	*Weissella* and *Leuconostoc* slightly increased. *Lactobacillus* decrease		[79]

3.1. Soybean Meal (SBM) and Soy Protein Concentrate

The inclusion of soybean in feeds for animals, both land and aquatic, has been extensively studied [52,55–60,75,80–84]. Its low price and wide availability have made soybean a perfect candidate for the replacement of FM, and, as a whole, the literature has shown that the MI does not show significant alterations, regardless of the diet. On the other hand, the DI morphology was affected in fish fed the full-fat SBM, with an increased number of goblet cells in the epithelium, enterocytes with low or no absorptive vacuoles and shorter microvilli, and a thicker lamina propria. The condition of the inflammatory response of the mucosa of the DI after the inclusion of SBM is known as SBM-induced enteritis. The actual causative components and SBM mechanisms of action need further research. It is known that the etiology and development of SBM-induced enteritis are related to SBM inclusion levels and the commercial source used for the diet formulation. This can have a great impact on the severity of the disorder, mainly affecting the endocytosis process. Uran et al. [80] evidenced that the cytological endocytosis block is directly related to the disappearance of the supranuclear vacuoles, which can be considered as the most striking feature in the onset of enteritis.

The negative effects of SBM-induced enteritis do not appear to affect all fish species investigated [84]. In particular, carnivorous fish species such as Atlantic salmon suffer from intestinal inflammation after being fed high levels of plant components that contain anti-nutritional factors such as protease inhibitors, oligosaccharides, lectins, phytic acids, and saponins, contained in SMB [41–44,85].

Below is an overview of the literature from the last few years on intestine health and microbiota.

Intestine health—no effects: The effect of feeding conventional and genetically modified (GM) soya plant products could be an economic alternative and it was evaluated by Sanden et al. [86] in farmed Atlantic salmon parr by investigating intestinal indices (relative weight of each intestine portion), histology, and cell proliferation level. No changes were observed in the intestine of Atlantic salmon parr that could be associated with the inclusion of genetically-modified plant material in the feed.

Intestine health—negative effects: Baeverfjord and Krogdahl [57] performed one of the first studies on the changes in the intestine morphology in Atlantic salmon fed diets with FM or SBM. Fish were fed with an SBM diet (40% protein replacement) for 3 weeks and the DI presented the following negative features, typical of SBM-induced enteritis: (1) shorter primary and secondary mucosal folds; (2) loss of the supranuclear vacuolization of the absorptive cells of the intestinal epithelium; (3) increased amount of the connective tissue, resulting in an increase in the width of the central stroma in the mucosal foldings; (4) widening of the lamina propria; (5) great infiltration of inflammatory cells in the lamina propria. The authors observed, as well, that the inflammatory cells were composed of different cell populations: lymphocytes, macrophages and polymorphonuclear leucocytes, and diffuse immunoglobulin M (IgM). Moreover, the vasculature of the lamina propria appeared frequently dilated. This study constituted the first of a series of other studies where various inclusion levels and forms of SBM were investigated. In addition, Bakke-McKellep et al. [58] studied the impact of replacing FM with 35% SBM in the diet (both genetically modified and not genetically modified) and found that the DI was the only tissue that showed a significant negative variation in terms of lamina propria thickness, fold length, and in the number of the supranuclear vacuoles, regardless of the SBM type. With the same inclusion level of SBM, another study with Atlantic salmon showed an average reduction of 23% of the mucosal wet weight of the DI, as well as SBM-induced changes [55]. In the same study, the authors observed that SBM leads to a decrease in carrier-mediated transport and an increase in the permeability of the DI epithelium, with a diminished capacity of the DI to absorb nutrients. A recent study studied the effects of a soybean compound named HP48 which is made from solvent-extracted peeled soybeans [60]. The authors found that the HP48 diet causes negative morphological changes in the DI, such as the widening of the lamina propria and stratum granulosum, disrupted mucosal folds, and the loss of absorptive vacuoles.

Krogdahl [50] observed that in fish fed a diet containing FM as the only protein source, no relevant pathological changes were observed in the intestine; on the contrary, the authors observed that the inclusion of SBM from 10 to 35% of total protein resulted in a graded morphological response: an increase in the vacuolization of absorptive cells, height and width of mucosal folds, and cell infiltration of the submucosa and lamina propria.

The effect of the various inclusion levels of SBM in Atlantic salmon was assessed by Urán et al. [51], where fish were fed diets with 0%, 10%, and 20% SBM for 57 days. Already from day 5 in fish fed the 20% diet and from day 7 in fish fed the 10% diet, there were morphological changes. These consisted of a decrease in the supranuclear vacuoles, damaged and smaller mucosa folds, an increase in cell infiltration in the submucosa, an increase in the goblet cells, and a widening of the lamina propria and submucosa. These changes were greater in fish fed the 20% diet, and, for both diets, changes were more evident over time. Urán et al. [51] also found an evident reduction in microvilli after only 7 days of 20% SBM feeding, indicating a negative impact on the uptake capacity of the enterocytes of the DI. The already mentioned changes are in agreement with a more recent study by Booman et al. [52] where Atlantic salmon fed 20% SBM showed an increase in submucosa thickness and infiltrating cells.

Serious enteritis was described in the DI of the Atlantic salmon fed a diet with an SBM inclusion level of 32% by evaluating the following criteria: presence and size of supranuclear vacuoles, degree of widening of the lamina propria of simple folds, amount of connective tissue between the base folds and stratum compactum, and degree of thickening of the mucosal folds [56].

To investigate the role of the intestinal microbiota in SBM enteropathy, the effect of supplementing a high inclusion level of SBM (50%) with lactic acid bacteria (LAB) was assessed in Atlantic salmon [59]. As expected, in fish fed with non-supplemented SBM, the height of the mucosal folds, the presence of supranuclear vacuoles, the number of goblet cells, and the thickness of the lamina propria were all negatively affected. The supplementation of LAB to the SBM diet decreased the inflammation status, and all histological parameters of inflammation were reduced as well. The results of this study suggest that LAB may modulate intestinal inflammation in fish. Moreover, there is evidence that enteritis caused by the inclusion of SBM in the diet may increase fish susceptibility to bacterial disease [87]. A recent study on Atlantic salmon, aiming to compare several protein sources, found that fish that developed SBM-induced enteritis presented a lower expression of the muc2 mucin, the major protein component of mucus that lines the epithelia of the intestine, compromising the intestinal barrier status [88].

Thereby, it was clear that SBM at inclusion levels from 10 up to 50% led to an inflammation process in the DI of Atlantic salmon. The attention started to be focused on the causes of this SBM-induced enteritis and how to overcome these.

Some authors hypothesize that the causative agent of enteritis may presumably be one or more of the alcohol-soluble components of SBM [50]. In a study by Knudsen et al. [89], soybean molasses were separated into three subfractions, and all possible combinations of these subfractions were fed to Atlantic salmon. Fish fed the water phase subfraction displayed a normal morphology, but fish fed a combination of the butanol phase and the precipitate displayed the same morphological changes as fish that were fed soybean molasses. The authors concluded that the presence of soya saponins, or other components that show the same solubility pattern, is associated with the inflammatory reaction. The extrusion process, a technology that uses a high temperature and short time regime, together with intense mechanical shear [90] seems to inactivate the anti-nutritional factors present in SBM, but, still, research is ongoing to find other soya forms that do not produce negative effects such as enteritis.

Intestine health—positive effects: Catalán et al. [91] studied the effect of an alternative form of soybean on Atlantic salmon intestine health: the fermented SBM-based diet. The authors observed a boost in health and growth physiology in fish, promoted by intestinal lactic acid bacteria and an increase in the PI trans-cellular uptake of water. A similar diet could be tested on the DI, which seems to be the tract of the intestine most affected by SBM [41,50,55,56].

Krogdahl et al. [92] tested a new form of a non-genetically modified soybean, named Triple Null (TN), which is devoid of the Kunitz trypsin inhibitor, lectin, and allergens. The authors found that animals fed TN showed DI enteritis and a higher expression of pro-inflammatory genes, as well as two stress-related genes, when compared with the soy protein concentrate diet. The authors concluded that, most likely, the extrusion process used for producing feed seems sufficient to inactivate proteinaceous antinutrients, the removal of which does not affect the nutritional value of the feed for Atlantic salmon. The addition of probiotics to a 20% SBM diet has been evaluated on post-smolt Atlantic salmon [54]. The authors observed an increase in intestine supranuclear vacuoles, and reduction in the lamina propria width in animals fed the probiotics. It was concluded that, most probably, this addition may prevent enteritis.

Microbiota: Research on the effects of SBM on microbiota is still recent compared to that on the intestine morphology, and, for this reason, it is difficult to conclude the effective relationship between soybean and microbiota; however, Bakke-McKellep et al. [93] found that 25% SBM can alter the microbiota with an increase in the number and diversity of bacteria, especially Enterococcus spp., and mostly Enterococcus faecalis. The same authors found no relationship between microbiota alteration and SBM-induced enteritis. On the contrary, Reveco et al., [76] suggest that the modulation of the bacterial community with 20% of SBM could lead to SBM-induced enteritis; in this study, the authors report an abundance of Bacteridoites and Firmicutes, and also an increase in Proteobacteria. Similarly, Gajardo et al. [49] studied the inclusion of 15% soy protein concentrate, mixed with wheat gluten, and found an abundance of lactic acid bacteria in the mucosa-associated microbiota and alterations in the histomorphology, typical of SBM-induced enteritis; the phyla mainly represented in microbiota were Proteobacteria, Firmicutes, and Bacteroidetes. Although the phyla found were similar, Booman et al., 2018 [52] did not observe any relationship between 20% SBM and microbiota alteration in Atlantic salmon. In this study, the analysis showed that Tenericutes was the dominant phylum, followed by Proteobacteria, Firmicutes, and Bacteroidetes. Navarrete et al., 2013 [59] found that some bacterial groups, such as Shewanella, Lactococcus, Aeromonas, and Pseudomonas were specific to certain diets; however, there was no correlation between SBM-induced inflammation (50% inclusion) and different bacterial groups. The authors also found that the inclusion of lactic acid bacteria may modulate intestinal inflammation. Finally, the inclusion of 5–10% of soybean protein concentrate in the diet did not show better results; Green et al. [75] confirmed the hypothesis that soybean protein concentrate can cause intestinal disorders as a consequence of the intestinal microbiota alteration, even if the authors did not observe any sign of inflammation. The authors also described that soybean protein concentrate could have an effect on raising the bacterial diversity of the intestinal tract, leading to the presence of unusual bacteria, like Escherichia and Propionibacterium, that are not usually associated with marine fish. The impact of yeast inclusion on the modulation of the intestinal microbiota of fish fed SBM was studied by Agboola et al. [77], and it was concluded that the addition of inactivated yeasts did not lead to an alteration in the modulation caused by the SBM diet.

3.2. Peas and Pea Protein Concentrate (PPC)

Despite a lower content of protein when compared to other plant protein sources, field peas seem to be a good candidate as a protein source for fish [45–47]. Peas also contain several anti-nutritional factors, but their level is low compared with other legumes [35]. Supplementation with peas has been tested in Atlantic salmon and its effect was assessed in terms of intestine integrity.

Intestine health—no effects: No histological changes were observed in the DI of Atlantic salmon fed extruded whole field peas at an inclusion level of 20% [45]. A study showed that PPC could replace 20% of the FM protein in the feed for Atlantic salmon without adverse effects on intestine health; no changes in the size of the intestine and no significant differences in the histology of the DI among dietary groups were detected [61]. These results were recently confirmed by [62], where in fish fed a PPC diet at an inclusion level of 4%, there were no signs of DI enteritis.

Intestine health—negative effects: Penn et al. [43] tested several diets with different inclusion levels of plant proteins. In fish-fed diets containing 35% of PPC, there were evident histological changes in the DI: shorter simple mucosal folds, wider lamina propria with concomitant leukocyte infiltration, reduced to absent vacuolization, and apical displacement of nuclei. It was also evident that there was a reduction in the brush border enzyme activities.

Microbiota: Only one work analyzing the relationship between PPC and microbiota was found in the literature. Hartviksen [66] observed an increase in the population level of autochthonous bacteria associated with the ID mucosa, although the reason for this was not clear. In all diets, the dominant phyla were Corynebacteriaceae in the allochthonous population and Lactobacillaceae in the autochthonous bacterial population. The same authors found no serious signs of enteritis.

3.3. *Faba Bean and Faba Bean Protein Concentrate (BPC)*

Whole crushed faba beans have a lower protein content (~25%) when compared with soybean (~35%) [94]. As for peas, further faba bean processing may lead to ingredients with a higher protein concentration, more interesting for FM replacement: crushed faba beans (28% protein) and air-classified, de-hulled, crushed faba bean protein concentrate (BPC, 61% protein).

Intestine health—no effects: No histological changes were observed in the DI of Atlantic salmon fed either whole or dehulled faba beans at an inclusion level of 20% [45].

Intestine health—negative effects: The same author tested the effect of two levels (17 and 35%) of a wet-processed faba bean isolate with a high protein content (~80% crude protein), and low levels of anti-nutritional factors [63]. The DI of fish fed the diet with the highest content of faba bean showed a clear deterioration of the absorptive vacuoles with an increasing level of supplementation of the wet processed faba. The lamina propria appeared significantly wider, which was probably due to the increased infiltration of eosinophilic granulocytes. These signs of progressing enteritis were similar to those observed in SBM-induced enteritis. The authors concluded that protein derived from faba beans may replace FM at a certain level in feeds for Atlantic salmon.

Intestine health—positive effects: Inclusion levels ranging from 50 to 200 g/kg of feed, which partially replaced either SPC or FM, showed the greatest potential in terms of fish growth and no negative effects in terms of intestine inflammation. This was evaluated through a histological scoring of the following parameters: sub-epithelial mucosa, mucosal folds, lamina propria, eosinophilic granulocytes, goblet cells, and supranuclear vacuoles. High inclusion levels of BPC (450 g/kg) caused mild intestinal inflammation, although this was not as severe as the inflammation seen in fish fed higher SBM diets. The authors concluded that BPC at these inclusion levels represents a valuable alternative for FM in Atlantic salmon.

Microbiota: As for the effects of the use of faba beans in diets on intestinal microbiota, no information was found in the literature.

3.4. Other Plant Protein Sources (PPS)

Other plant ingredients were evaluated as protein sources: wheat, maize meal, sunflower meal, rapeseed meal, and wheat gluten.

Intestine health—no effects: The addition of a seaweed meal derived from brown seaweeds (*Laminaria* sp., kelp) was tested in Atlantic salmon at inclusion levels from 3 to 100% [95]. Fish fed the 3 or 10% inclusion levels diets showed higher whole intestine and intestinal weights and lengths, suggesting an increased holding time and a larger surface area for digestion and nutrient absorption. Two commercially available algae-derived products (dry algae meal), Verdemin (derived from *Ulva ohnoi*) and Rosamin (derived from diatom *Entomoneis* spp.), were tested in the diet of Atlantic Salmon [96]. No positive or negative effects were found on fish growth, but the authors did not investigate the effects of this algae inclusion on intestine morphology and health.

Intestine health—negative effects: In Atlantic salmon fed 25% of a high plant protein replacement diet, there was inflammatory infiltration in the submucosa and lamina propria of the DI, but at a low level [65]. Wheat gluten was investigated in two inclusion levels in the diet of Atlantic salmon, 15 or 30% [97]. The authors found shifts in the expression of a large number of genes responsible for immunity and tissue structure and the integrity of the intestine transcriptome in fish fed wheat gluten, indicating the nutritional stress of fish. A high wheat gluten inclusion in the diet produced a negative impact on intestinal health, and symptoms were found to be similar to those of gluten sensitivity in humans.

Microbiota: No information was found in the available literature on the impact of these plant protein sources on the intestinal microbiota.

3.5. Animal Protein: Poultry by-Product Meal (PBM)

In this review, until now, only plant protein sources have been addressed as replacement ingredients for FM, but there are animal protein sources that may constitute valid alternatives as well. A good example is the poultry by-product meal (PBM), which is a result of the poultry processing industry. It presents a high protein content with a favorable profile of essential amino acids for fish production [98]. These characteristics, together with its low price and wide availability, make PBM an ideal candidate for replacing FM in aquaculture feeds. However, there are not many studies regarding the impact of the replacement of FM with PBM in the intestines of Atlantic salmon.

Intestine health—negative effects: PBM and feather meal were given to Atlantic salmon at an inclusion level of 20% [62]. None of the fish fed these diets showed severe signs of inflammation. It has to be pointed out that PBM resulted in a significantly decreased submucosa width, and feather meal resulted in a decrease in goblet cells.

Intestine health—positive effects: A diet containing 58% of PBM was given to fish in a 40-day trial [67]. There were no changes in the DI histology, but there was a decrease in the chyme water content and an increase in the expression of the tight junction protein occluding. This is a contradictory result and the authors hypothesize that the decrease in the chyme water content was probably related to the low lipid digestibility.

Microbiota: Hartviksen et al. [62] observed that Atlantic salmon fed with PBM and feather meal influenced the microbiota composition of the intestine; it seems that PBM modulates the allochthonous bacteria composition, by increasing its population in the DI. In addition, in this diet, the main represented phyla were *Corynebacteriaceaein* for the allochthonous and *Lactobacillaceaein* for the autochthonous bacteria population.

3.6. Animal Protein: Insect Meal (IM)

3.6.1. Black Soldier Fly (BSF-Hermetia Illucens)

There is an animal protein source for which interest as a feed ingredient for terrestrial and aquatic animals continues to grow every year: insect meal (IM). Among these, the black soldier fly (BSF-*Hermetia illucens*) larvae are considered an important potential species to be used for animal feed [99,100]. This species has been used as a protein source in animal feed since the 1970s, mainly because of its capacity to convert food waste, such as vegetables, fruit, factory waste, and animal tissues, into high-quality protein. BSF larvae contain about 40% protein and present a balanced profile of essential amino acids [101].

Intestine health—no effects: Intestine health seemed to be maintained in fish fed a test diet where all the FM was replaced with BSF larvae meal [71]. Fish fed insect meal presented a higher weight of the DI and less severe enterocytes steatosis (abnormal retention of fat in the cells). There were, however, inflammatory morphological changes in all the examined intestinal segments, which are similar to those to those caused by SBM, but, overall, the authors concluded that intestinal health was not compromised by the IM.

Recently, a research group assessed the partial replacement of FM with BSF larvae in Atlantic salmon fry [69] and in pre-smolts [68]. According to the authors, it is safe to replace FM with BSF up to 10% in fry, as no negative effects on growth, feed utilization, and intestine structure were observed; the mucosa folds and thickness, as well as the overall histology, were not influenced by the BSF larvae meal inclusion. According to the authors, for pre-smolts, it is possible to increase the replacement level up to 15% without compromising growth performance, feed utilization, somatic indices, and the histomorphology of the intestine. Indeed, villus height and area, and the muscular thickness of the DI appeared unaltered when using a BSF larvae meal up to 15% inclusion level. The villus width was higher in fish fed the insect meal diet, but the authors claim that it is not necessarily a negative result, as it may imply an increase in the number and measure of the enterocytes, which provide a greater absorption surface area.

Intestine health—positive effects: Belghit et al. [102] studied the effect of the increasing supplementation of IM, produced from BSF larvae, in the diet of Atlantic salmon. The partial or complete replacement of the FM protein with insect meal in the diet did not show negative effects on the feed intake or on the growth performance of fish. Unfortunately, in this study, the histomorphology of the intestine was not evaluated, but the fact that the inclusion of insect meal did not affect the digestibility of proteins, lipids, amino acids, and fatty acids indicates that most probably the intestine architecture was not significantly altered. This is confirmed by other studies where several levels of insect meal produced from BSF larvae were tested on Atlantic salmon. For instance, Weththasinghe et al. [70] showed that replacing conventional protein sources with low to moderate levels of insect meal (6.25% and 12.5%) leads to an evident reduction in enterocyte steatosis in the pyloric caeca. Even with high replacement levels, such as 25%, an improvement in the DI structure was observed; in all of the experimental groups, an increase in the width and infiltration of the submucosa and lamina propria by inflammatory cells was observed. There were no signs of enteritis in the DI in fish fed the IM diet. Likewise, Li et al. [103] confirmed that the pyloric caeca mucosa clearly showed less hyper-vacuolization (indicative of lipid accumulation in enterocytes) in the IM-fed fish. As for the DI, the authors assessed the shortening and fusion of mucosal folds, cellular infiltration within the lamina propria and submucosa, enterocyte vacuolization, and nucleus position disparity. None of these parameters were affected by the inclusion of IM in the diet. Lock et al. [17] noticed that there were no differences in the DI structure between the several inclusion levels of BSF larvae (from 25 to 100%); the DI epithelium showed a normal morphology with no loss of villi architecture, vacuole formation, or necrosis in the tips of the villi.

Microbiota: Another indication of positive use of the BSF was found in a study by Li et al. [78], where Atlantic salmon were fed a commercial reference diet and an IM-based test diet to compare the intestinal microbiota between the two diets. The authors concluded that fish fed the IM diet showed deep changes in the intestinal microbiota, confirming the results of a previous study of the same authors, where an IM diet altered the DI microbiota in terms of higher microbial richness and diversity. It was found that bacterial communities such as *Actinomyces, Bacillus, Brevibacterium, Corynebacterium 1*, and *Enterococcus* were increased in the IM diet [104] The authors state, though, that further studies would be necessary to evaluate if and how much these alterations can be accountable to differences in microbiota feed and dietary nutrients. This could be useful also to understand the impact of these changes on fish physiology and health.

3.6.2. Yellow Mealworm (Tenebrio Molitor)

Another candidate as IM for aquaculture feeds is the yellow mealworm (*Tenebrio molitor*). Like the BSF larvae, the *T. molitor* also has a substantial content of proteins and lipids, and a good amino acid and essential fatty acid profile. Shafique et al. [105] made an interesting review on the use of this insect species in aquafeed; unfortunately, there is no information on Atlantic salmon, but the results obtained up to now in other salmonids seem promising in terms of growth, biometric indices, and body composition. As the authors mention, further studies are necessary in order to conduct a thorough evaluation of the impact of the use of this ingredient in terms of intestine health in fish. The impact on the intestine microbiota from the addition of the yellow mealworm in the diets of Atlantic salmon has not been studied to date.

3.7. Single-Cell Protein (SCP)

Single-cell protein (SCP) also appears to be an excellent alternative to FM and to plant ingredients due to its adequate nutritional profile and the fact that it can be cultivated in any location and climate. Moreover, it presents the advantage of not stressing the existing agricultural systems. SCP products can be produced from several microbial sources, such as microalgae, yeast and other fungi, and bacteria [106].

3.7.1. Microalgae

Microalgae are rich in high-quality proteins, essential amino acids, polyunsaturated fatty acids, polysaccharides, vitamins, minerals, and pigments [107].

Intestine health—no effects: The effect of the replacement of FM with the microalga *Nannochloropsis oceania* was assessed in terms of growth performance, feed utilization, and intestine health [108]. The histomorphology of the DI was not altered by the supplementation with this microalga in the feed, and there were no signs of inflammation of the intestine, even if the expression of the pro-inflammatory gene il17d was higher in fish fed 10% alga. The results of this study, altogether, indicate that the inclusion of this microalgae at low levels (c.a. 10%) does not produce negative effects on weight gain, specific growth rate, and health parameters, making it a good candidate for FM partial replacement. The same author examined the effect of pre-extruding two microalgae, *Nannochloropsis oceanica*, and *Tetraselmis* sp., on the growth, the fatty acid content in the flesh, and the health of Atlantic salmon [72]. First of all, neither *Nannochloropsis* nor *Tetraselmis*, at high inclusion levels (30%) show signs of intestine inflammation other than enterocyte vacuolization. For both algae, the expression of *muc2*, a gene that encodes the mucine protein which covers the whole intestine, in the DI was not significantly altered. Microalgae present rigid cell walls that may lower the digestibility and nutrient bioavailability in carnivorous fish, but the extrusion process may contribute to its breakdown and therefore to better nutrient utilization. It has been hypothesized that certain food additives may contribute to a better nutrient utilization of microalgae by fish. Gong et al. [109] tested the influence of two commercial food additives in food containing 10% of *N. oceanica*. The addition of the feed additives produced no significant effects on growth, feed utilization, or the histomorphology of the DI. Even if there was an increase in the cell proliferation of the mucosa folds in fish fed the microalga alone, as well as the microalga-commercial additive combination, the authors concluded that the tested feed additives did not show a distinct advantage for Atlantic salmon.

Intestine health—positive effects: Kiron et al. [73] investigated the inclusion of a defatted marine alga, the *Desmodesmus* sp., in Atlantic salmon diets. The expression of immune and inflammatory marker genes and the morphological observations did not reveal any abnormalities in the intestinal health of the fish fed microalga. Histological analyses revealed a normal lamina propria and goblet cells. The authors concluded that it is possible to include defatted *Desmodesmus* sp. at 20% in the feeds of Atlantic salmon. Grammes et al. [110] investigated the possibility that the addition of certain microbial ingredients such as yeasts or microalgae may relieve SBM-induced enteritis in Atlantic salmon. The results showed that the supplementation of a 20% SBM feed with the yeast *Candida utilis* or the microalgae *Chlorella vulgaris* resulted in a healthy fish intestine, indicating it was highly effective to counteract enteritis.

3.7.2. Yeasts

Yeasts are becoming popular as ingredients in fish feeds because of their protein content but also because they potentially support better growth and good fish health, as reviewed by Agboola et al. [111].

Intestine health—no effects: The inclusion of 25% *Candida utilis* yeast in the feed during the delicate stage of smoltification did not produce any alterations in the morphology of the DI, in terms of the length of the simple folds, mucosa thickness, or intestinal epithelial cells [23]. The replacement of 40% of the crude protein from FM with the yeasts *Candida utilis*, *Kluyveromyces marxianus*, or *Saccharomyces cerevisiae* was evaluated [112]. Fish fed the *S. cerevisiae* and *K. marxianus* diets presented a heavier DI than fish fed the FM diet, but no other assessments were performed on the intestine. Overall, the authors concluded that *C. utilis* and *K. marxianus* can be used as suitable protein sources for Atlantic salmon, while *S. cerevisiae* seems to be less promising.

Reveco-Urzua et al. [53] aimed to evaluate the intestinal tissue responses of fish fed inactive dry *Candida utilis* yeast biomass, SBM, or a combination of an SBM-based diet with several inclusion levels of *Candida utilis*. The authors found that the inclusion of *C. utilis* in an FM-based diet did not produce any changes in terms of DI morphology, immune cell population, or gene expression. When combined with SBM, low inclusion levels of *C. utilis* modulated immune cell populations in the DI and reduced the severity of SBM-induced enteritis. However, this effect on enteritis was not confirmed by a study on parr, where the severity of the reduced height of the intestine folds and the decreased presence of supranuclear vacuoles was not counteracted by the addition of *C. utilis* in the feed [74].

Microbiota: It has been observed that an inclusion level of up to 20% of torula yeast in Atlantic salmon feed produced no alterations in growth performance, and showed benefits in the intestine microbial community [79].

However, at inclusion levels higher than 20%, not only were there no growth benefits, but there were adverse changes in the intestine microbiome, for example, an increase in the bacterial community, such as *Weissella* and *Leuconostoc*, and also a decrease in lactic acid bacteria (*Lactococcus*).

4. Conclusions

The development of more sustainable aquaculture from the economic and environmental point of view should be focused on the development of diets with new, sustainable sources of proteins to fully or partially replace fishmeal. The future elected protein sources must meet the nutritional requirements of carnivorous fish species such as the Atlantic salmon without compromising the growth and intestine health of fish, which are usually endorsed by the anti-nutritional factors present in many plant ingredients. It appears quite evident from this review that even at low inclusion levels, soybean meal causes negative effects on the distal intestine of Atlantic salmon. Nevertheless, soybean meal remains an attractive alternative to fishmeal, taking into account its protein content, market availability, and cost, although other protein sources should be seriously taken into account when formulating feeds. Ingredients with an adequate content of protein, which cause mild or no effects on the distal intestine even at high inclusion levels, such as microalgae, insects, and single-cell proteins, could be also interesting choices for feed formulations.

The negative effects on morphology caused by alternative protein sources seem to be greater than the positive ones. However, concerning carnivorous fish such as the Atlantic salmon, if an alternative protein source produces no negative effects on fish health, then this is an overall positive effect from a sustainability point of view. This review shows that this is a "hot topic" nowadays. There is a lot of research on new sustainable protein sources, but this research is not always easy to confront as there are a great variety of inclusion levels, trial lengths, and stages of fish development. In the current scenario, more research still needs to be performed in order to reach a balanced alternative to fishmeal, with the right inclusion level used in the right fish stage of development. Indeed, regarding microbiota, it is now starting to be studied, and from this review it can be seen that there are not many works yet and therefore it is not yet possible to draw clear conclusions.

Finally, using alternative protein sources could have a positive impact, as well, on the preservation of the stocks of marine species used for the production of fishmeal. It may not be ethically correct to harvest fish that is used to produce aquaculture feed, to produce fish that will be used as food for humans [39,113]. Thus, it is of major importance to increase efforts to replace fishmeal in feeds for fish in general, and for Atlantic salmon in particular, and to pursue strategies that minimize the impacts of fishmeal reduction in the diet and on fish fillet quality. The alternative protein sources in aquafeeds must guarantee fish health but also a final product that is suitable from a nutritional point of view, and at the same time well accepted by the consumers.

Author Contributions: Conceptualization, A.D.G., S.C.M. and L.A.; investigation, A.D.G., S.C.M., C.B., G.R., V.R.H.M., M.P., G.M. and L.A.; writing—original draft preparation, L.A. and G.M.; writing—review and editing, A.D.G., S.C.M., G.R., C.B., V.R.H.M., M.P., G.M. and L.A. All authors have read and agreed to the published version of the manuscript.

Funding: This research received no external funding.

Institutional Review Board Statement: Not applicable.

Informed Consent Statement: Not applicable.

Acknowledgments: The authors wish to thank Fabrizio De Luca for the drawing production.

Conflicts of Interest: The authors declare no conflict of interest.

References

1. Løkka, G.; Austbø, L.; Falk, K.; Bjerkås, I.; Koppang, E.O. Intestinal Morphology of the Wild Atlantic Salmon (*Salmo salar*). *J. Morphol.* **2013**, *274*, 859–876. [CrossRef] [PubMed]
2. Kapoor, B.G.; Smit, H.; Verighina, I.A. The Alimentary Canal and Digestion in Teleosts. In *Advances in Marine Biology*; Russell, F.S., Yonge, M., Eds.; Academic Press: Cambridge, MA, USA, 1976; Volume 13, pp. 109–239.
3. Bjørgen, H.; Li, Y.; Kortner, T.M.; Krogdahl, Å.; Koppang, E.O. Anatomy, Immunology, Digestive Physiology and Microbiota of the Salmonid Intestine: Knowns and Unknowns under the Impact of an Expanding Industrialized Production. *Fish Shellfish Immunol.* **2020**, *107*, 172–186. [CrossRef] [PubMed]
4. Harder, W. The Digestive Tract. In *Anatomy of Fishes. Part 1: Text*; E. Schweizerbart'sche Verlagsbuchhandlung: Stuttgart, Germany, 1975; Volume 62, pp. 128–186.
5. Rust, M.B. 7—Nutritional Physiology. In *Fish Nutrition*, 3rd ed.; Halver, J.E., Hardy, R.W., Eds.; Academic Press: San Diego, CA, USA, 2003; pp. 367–452. ISBN 978-0-12-319652-1.
6. Denstadli, V.; Vegusdal, A.; Krogdahl, Å.; Bakke-McKellep, A.M.; Berge, G.M.; Holm, H.; Hillestad, M.; Ruyter, B. Lipid Absorption in Different Segments of the Gastrointestinal Tract of Atlantic Salmon (*Salmo salar* L.). *Aquaculture* **2004**, *240*, 385–398. [CrossRef]
7. Burnstock, G. The Morphology of the Gut of the Brown Trout (*Salmo trutta*). *J. Cell Sci.* **1959**, *s3-100*, 183–198. [CrossRef]
8. Sundell, K.; Jutfelt, F.; Ágústsson, T.; Olsen, R.-E.; Sandblom, E.; Hansen, T.; Björnsson, B.T. Intestinal Transport Mechanisms and Plasma Cortisol Levels during Normal and Out-of-Season Parr–Smolt Transformation of Atlantic Salmon. *Salmo salar. Aquac.* **2003**, *222*, 265–285. [CrossRef]
9. Sire, M.F.; Vernier, J.-M. Intestinal Absorption of Protein in Teleost Fish. *Comp. Biochem. Physiol. A Physiol.* **1992**, *103*, 771–781. [CrossRef]
10. Kent, G.; Carr, R. Digestive System. In *Comparative Anatomy of the Vertebrates*; Kent, G.C., Carr, R.K., Eds.; McGraw-Hill Publishing: New York, NY, USA, 2001; pp. 265–291.
11. Jutfelt, F. Barrier Function of the Gut. In *Encyclopedia of Fish Physiology from Genome to Environment*; Academic Press: Cambridge, MA, USA, 2011.
12. Takei, Y.; Loretz, C.A. 7—The Gastrointestinal Tract as an Endocrine/Neuroendocrine/Paracrine Organ: Organization, Chemical Messengers and Physiological Targets. In *Fish Physiology*; Grosell, M., Farrell, A.P., Brauner, C.J., Eds.; The Multifunctional Gut of Fish; Academic Press: Cambridge, MA, USA, 2010; Volume 30, pp. 261–317.

13. Wilson, J.M.; Castro, L.F.C. 1—Morphological Diversity of the Gastrointestinal Tract in Fishes. In *Fish Physiology*; Grosell, M., Farrell, A.P., Brauner, C.J., Eds.; The Multifunctional Gut of Fish; Academic Press: Cambridge, MA, USA, 2010; Volume 30, pp. 1–55.
14. Grosell, M. Intestinal Anion Exchange in Marine Teleosts Is Involved in Osmoregulation and Contributes to the Oceanic Inorganic Carbon Cycle. *Acta Physiol.* **2011**, *202*, 421–434. [CrossRef]
15. Pascoli, F.; Negrato, E.; Di Giancamillo, A.; Bertotto, D.; Domeneghini, C.; Simontacchi, C.; Mutinelli, F.; Radaelli, G. Evaluation of Oxidative Stress Biomarkers in Zosterisessor Ophiocephalus from the Venice Lagoon, Italy. *Aquat. Toxicol.* **2011**, *101*, 512–520. [CrossRef]
16. Poussier, P.; Edouard, P.; Lee, C.; Binnie, M.; Julius, M. Thymus-Independent Development and Negative Selection of T Cells Expressing T Cell Receptor Alpha/Beta in the Intestinal Epithelium: Evidence for Distinct Circulation Patterns of Gut- and Thymus-Derived T Lymphocytes. *J. Exp. Med.* **1992**, *176*, 187–199. [CrossRef]
17. Lock, E.R.; Arsiwalla, T.; Waagbø, R. Insect Larvae Meal as an Alternative Source of Nutrients in the Diet of Atlantic Salmon (*Salmo salar*) Postsmolt. *Aquac. Nutr.* **2016**, *22*, 1202–1213. [CrossRef]
18. Bjørgen, H.; Hellberg, H.; Løken, O.M.; Gunnes, G.; Koppang, E.O.; Dale, O.B. Tumor Microenvironment and Stroma in Intestinal Adenocarcinomas and Associated Metastases in Atlantic Salmon Broodfish (*Salmo salar*). *Vet. Immunol. Immunopathol.* **2019**, *214*, 109891. [CrossRef]
19. Tafalla, C.; Leal, E.; Yamaguchi, T.; Fischer, U. T Cell Immunity in the Teleost Digestive Tract. *Dev. Comp. Immunol.* **2016**, *64*, 167–177. [CrossRef]
20. Ng, A.N.Y.; de Jong-Curtain, T.A.; Mawdsley, D.J.; White, S.J.; Shin, J.; Appel, B.; Dong, P.D.S.; Stainier, D.Y.R.; Heath, J.K. Formation of the Digestive System in Zebrafish: III. Intestinal Epithelium Morphogenesis. *Dev. Biol.* **2005**, *286*, 114–135. [CrossRef] [PubMed]
21. Wallace, K.N.; Akhter, S.; Smith, E.M.; Lorent, K.; Pack, M. Intestinal Growth and Differentiation in Zebrafish. *Mech. Dev.* **2005**, *122*, 157–173. [CrossRef] [PubMed]
22. Whittamore, J.M. Osmoregulation and Epithelial Water Transport: Lessons from the Intestine of Marine Teleost Fish. *J. Comp. Physiol. B* **2012**, *182*, 1–39. [CrossRef] [PubMed]
23. Sahlmann, C.; Djordjevic, B.; Lagos, L.; Mydland, L.T.; Morales-Lange, B.; Øvrum Hansen, J.; Ånestad, R.; Mercado, L.; Bjelanovic, M.; Press, C.M.; et al. Yeast as a Protein Source during Smoltification of Atlantic Salmon (*Salmo salar* L.), Enhances Performance and Modulates Health. *Aquaculture* **2019**, *513*, 734396. [CrossRef]
24. Dhanasiri, A.K.S.; Brunvold, L.; Brinchmann, M.F.; Korsnes, K.; Bergh, Ø.; Kiron, V. Changes in the Intestinal Microbiota of Wild Atlantic Cod *Gadus Morhua* L. Upon Captive Rearing. *Microb. Ecol.* **2011**, *61*, 20–30. [CrossRef] [PubMed]
25. Ringø, E.; Zhou, Z.; Vecino, J.L.G.; Wadsworth, S.; Romero, J.; Krogdahl, Å.; Olsen, R.e.; Dimitroglou, A.; Foey, A.; Davies, S.; et al. Effect of Dietary Components on the Gut Microbiota of Aquatic Animals. A Never-Ending Story? *Aquac. Nutr.* **2016**, *22*, 219–282. [CrossRef]
26. Ringø, E.; Olsen, R.E.; Mayhew, T.M.; Myklebust, R. Electron Microscopy of the Intestinal Microflora of Fish. *Aquaculture* **2003**, *227*, 395–415. [CrossRef]
27. Wang, J.; Jaramillo-Torres, A.; Li, Y.; Kortner, T.M.; Gajardo, K.; Brevik, Ø.J.; Jakobsen, J.V.; Krogdahl, Å. Microbiota in Intestinal Digesta of Atlantic Salmon (*Salmo salar*), Observed from Late Freshwater Stage until One Year in Seawater, and Effects of Functional Ingredients: A Case Study from a Commercial Sized Research Site in the Arctic Region. *Anim. Microbiome* **2021**, *3*, 14. [CrossRef]
28. Falcinelli, S.; Picchietti, S.; Rodiles, A.; Cossignani, L.; Merrifield, D.L.; Taddei, A.R.; Maradonna, F.; Olivotto, I.; Gioacchini, G.; Carnevali, O. *Lactobacillus Rhamnosus* Lowers Zebrafish Lipid Content by Changing Gut Microbiota and Host Transcription of Genes Involved in Lipid Metabolism. *Sci. Rep.* **2015**, *5*, 9336. [CrossRef] [PubMed]
29. Semova, I.; Carten, J.D.; Stombaugh, J.; Mackey, L.C.; Knight, R.; Farber, S.A.; Rawls, J.F. Microbiota Regulate Intestinal Absorption and Metabolism of Fatty Acids in the Zebrafish. *Cell Host Microbe* **2012**, *12*, 277–288. [CrossRef] [PubMed]
30. Ray, A.K.; Ghosh, K.; Ringø, E. Enzyme-Producing Bacteria Isolated from Fish Gut: A Review. *Aquac. Nutr.* **2012**, *18*, 465–492. [CrossRef]
31. Austin, B. The Bacterial Microflora of Fish, Revised. *Sci. World J.* **2006**, *6*, 931–945. [CrossRef]
32. López Nadal, A.; Ikeda-Ohtsubo, W.; Sipkema, D.; Peggs, D.; McGurk, C.; Forlenza, M.; Wiegertjes, G.F.; Brugman, S. Feed, Microbiota, and Gut Immunity: Using the Zebrafish Model to Understand Fish Health. *Front. Immunol.* **2020**, *11*, 114.
33. FAO: Nutritional Requirements. Available online: https://www.fao.org/fishery/affris/species-profiles/atlantic-salmon/nutritional-requirements/en/ (accessed on 19 May 2023).
34. Lall, S.P.; Anderson, S. Amino Acid Nutrition of Salmonids: Dietary Requirements and Bioavailablity. *Cah. Options Mediterr.* **2005**, *63*, 73–90.
35. Francis, G.; Makkar, H.P.S.; Becker, K. Antinutritional Factors Present in Plant-Derived Alternate Fish Feed Ingredients and Their Effects in Fish. *Aquaculture* **2001**, *199*, 197–227. [CrossRef]
36. Gatlin, D.M.; Barrows, F.T.; Brown, P.; Dabrowski, K.; Gaylord, T.G.; Hardy, R.W.; Herman, E.; Hu, G.; Krogdahl, Å.; Nelson, R.; et al. Expanding the Utilization of Sustainable Plant Products in Aquafeeds: A Review. *Aquac. Res.* **2007**, *38*, 551–579. [CrossRef]
37. FAO (Food and Agriculture Organization of the United Nations). Fisheries and Aquaculture Department—Species Fact Sheets. Available online: www.fao.org (accessed on 19 May 2023).

38. Crespi, V. *The State of World Fisheries and Aquaculture*; FAO: Rome, Italy, 2020; ISBN 978-92-5-132692-3.
39. Tacon, A.G.J.; Metian, M. Global Overview on the Use of Fish Meal and Fish Oil in Industrially Compounded Aquafeeds: Trends and Future Prospects. *Aquaculture* **2008**, *285*, 146–158. [CrossRef]
40. Gillund, F.; Myhr, A.I. Perspectives on Salmon Feed: A Deliberative Assessment of Several Alternative Feed Resources. *J. Agric. Environ. Ethics* **2010**, *23*, 527–550. [CrossRef]
41. Bakke-McKellep, A.M.; Koppang, E.O.; Gunnes, G.; Sanden, M.; Hemre, G.-I.; Landsverk, T.; Krogdahl, A. Histological, Digestive, Metabolic, Hormonal and Some Immune Factor Responses in Atlantic Salmon, *Salmo salar* L., Fed Genetically Modified Soybeans. *J. Fish Dis.* **2007**, *30*, 65–79. [CrossRef] [PubMed]
42. Dale, O.B.; Tørud, B.; Kvellestad, A.; Koppang, H.S.; Koppang, E.O. From Chronic Feed-Induced Intestinal Inflammation to Adenocarcinoma with Metastases in Salmonid Fish. *Cancer Res.* **2009**, *69*, 4355–4362. [CrossRef] [PubMed]
43. Penn, M.H.; Bendiksen, E.Å.; Campbell, P.; Krogdahl, Å. High Level of Dietary Pea Protein Concentrate Induces Enteropathy in Atlantic Salmon (*Salmo salar* L.). *Aquaculture* **2011**, *310*, 267–273. [CrossRef]
44. Urán, P.A.; Schrama, J.W.; Rombout, J.H.W.M.; Obach, A.; Jensen, L.; Koppe, W.; Verreth, J.A.J. Soybean Meal-Induced Enteritis in Atlantic Salmon (*Salmo salar* L.) at Different Temperatures. *Aquac. Nutr.* **2008**, *14*, 324–330. [CrossRef]
45. Aslaksen, M.A.; Kraugerud, O.F.; Penn, M.; Svihus, B.; Denstadli, V.; Jørgensen, H.Y.; Hillestad, M.; Krogdahl, Å.; Storebakken, T. Screening of Nutrient Digestibilities and Intestinal Pathologies in Atlantic Salmon, *Salmo salar*, Fed Diets with Legumes, Oilseeds, or Cereals. *Aquaculture* **2007**, *272*, 541–555. [CrossRef]
46. Thiessen, D.L.; Campbell, G.L.; Adelizi, P.D. Digestibility and Growth Performance of Juvenile Rainbow Trout (*Oncorhynchus Mykiss*) Fed with Pea and Canola Products. *Aquac. Nutr.* **2003**, *9*, 67–75. [CrossRef]
47. Gouveia, A.; Davies, S.J. Inclusion of an Extruded Dehulled Pea Seed Meal in Diets for Juvenile European Sea Bass (*Dicentrarchus Labrax*). *Aquaculture* **2000**, *182*, 183–193. [CrossRef]
48. Shubin, A.V.; Demidyuk, I.V.; Komissarov, A.A.; Rafieva, L.M.; Kostrov, S.V. Cytoplasmic Vacuolization in Cell Death and Survival. *Oncotarget* **2016**, *7*, 55863–55889. [CrossRef]
49. Gajardo, K.; Jaramillo-Torres, A.; Kortner, T.M.; Merrifield, D.L.; Tinsley, J.; Bakke, A.M.; Krogdahl, Å. Alternative Protein Sources in the Diet Modulate Microbiota and Functionality in the Distal Intestine of Atlantic Salmon (*Salmo salar*). *Appl. Environ. Microbiol.* **2016**, *83*, e02615-16. [CrossRef]
50. Krogdahl, Å.; Bakke-McKellep, A.M.; Baeverfjord, G. Effects of Graded Levels of Standard Soybean Meal on Intestinal Structure, Mucosal Enzyme Activities, and Pancreatic Response in Atlantic Salmon (*Salmo salar* L.). *Aquac. Nutr.* **2003**, *9*, 361–371. [CrossRef]
51. Urán, P.A.; Schrama, J.W.; Rombout, J.H.W.M.; Taverne-Thiele, J.J.; Obach, A.; Koppe, W.; Verreth, J.A.J. Time-Related Changes of the Intestinal Morphology of Atlantic Salmon, *Salmo salar* L., at Two Different Soybean Meal Inclusion Levels. *J. Fish Dis.* **2009**, *32*, 733–744. [CrossRef] [PubMed]
52. Booman, M.; Forster, I.; Vederas, J.C.; Groman, D.B.; Jones, S.R.M. Soybean Meal-Induced Enteritis in Atlantic Salmon (*Salmo salar*) and Chinook Salmon (*Oncorhynchus tshawytscha*) but Not in Pink Salmon (*O. Gorbuscha*). *Aquaculture* **2018**, *483*, 238–243. [CrossRef]
53. Reveco-Urzua, F.E.; Hofossæter, M.; Kovi, M.R.; Mydland, L.T.; Ånestad, R.; Sørby, R.; Press, C.M.; Lagos, L.; Øverland, M. Candida Utilis Yeast as a Functional Protein Source for Atlantic Salmon (*Salmo salar* L.): Local Intestinal Tissue and Plasma Proteome Responses. *PLoS ONE* **2019**, *14*, e0218360. [CrossRef] [PubMed]
54. Nimalan, N.; Sørensen, S.L.; Fečkaninová, A.; Koščová, J.; Mudroňová, D.; Gancarčíková, S.; Vatsos, I.N.; Bisa, S.; Kiron, V.; Sørensen, M. Supplementation of Lactic Acid Bacteria Has Positive Effects on the Mucosal Health of Atlantic Salmon (*Salmo salar*) Fed Soybean Meal. *Aquac. Rep.* **2023**, *28*, 101461. [CrossRef]
55. Nordrum, S.; Bakke-McKellep, A.M.; Krogdahl, A.; Buddington, R.K. Effects of Soybean Meal and Salinity on Intestinal Transport of Nutrients in Atlantic Salmon (*Salmo salar* L.) and Rainbow Trout (*Oncorhynchus mykiss*). *Comp. Biochem. Physiol. B Biochem. Mol. Biol.* **2000**, *125*, 317–335. [CrossRef]
56. Refstie, S.; Baeverfjord, G.; Seim, R.R.; Elvebø, O. Effects of Dietary Yeast Cell Wall β-Glucans and MOS on Performance, Gut Health, and Salmon Lice Resistance in Atlantic Salmon (*Salmo salar*) Fed Sunflower and Soybean Meal. *Aquaculture* **2010**, *305*, 109–116. [CrossRef]
57. Baeverfjord, G.; Krogdahl, A. Development and Regression of Soybean Meal Induced Enteritis in Atlantic Salmon, *Salmo salar* L., Distal Intestine: A Comparison with the Intestines of Fasted Fish. *J. Fish Dis.* **1996**, *19*, 375–387. [CrossRef]
58. Bakke-McKellep, A.M.; Press, C.M.; Baeverfjord, G.; Krogdahl, Å.; Landsverk, T. Changes in Immune and Enzyme Histochemical Phenotypes of Cells in the Intestinal Mucosa of Atlantic Salmon, *Salmo salar* L., with Soybean Meal-Induced Enteritis. *J. Fish Dis.* **2000**, *23*, 115–127. [CrossRef]
59. Navarrete, P.; Fuentes, P.; la Fuente, L.D.; Barros, L.; Magne, F.; Opazo, R.; Ibacache, C.; Espejo, R.; Romero, J. Short-Term Effects of Dietary Soybean Meal and Lactic Acid Bacteria on the Intestinal Morphology and Microbiota of Atlantic Salmon (*Salmo salar*). *Aquac. Nutr.* **2013**, *19*, 827–836. [CrossRef]
60. Nordvi, M.F.; Løvmo, S.D.; Whatmore, P.; Sundh, H.; Sigholt, T.; Olsen, R.E. Low Intestinal Inflammation Model (HP48) in Atlantic Salmon (*Salmo salar*) and Inflammatory Mitigation by Bactocell. *Aquaculture* **2023**, *563*, 738920. [CrossRef]
61. Øverland, M.; Sørensen, M.; Storebakken, T.; Penn, M.; Krogdahl, Å.; Skrede, A. Pea Protein Concentrate Substituting Fish Meal or Soybean Meal in Diets for Atlantic Salmon (*Salmo salar*)—Effect on Growth Performance, Nutrient Digestibility, Carcass Composition, Gut Health, and Physical Feed Quality. *Aquaculture* **2009**, *288*, 305–311. [CrossRef]

62. Hartviksen, M.; Vecino, J.L.G.; Ringø, E.; Bakke, A.-M.; Wadsworth, S.; Krogdahl, Å.; Ruohonen, K.; Kettunen, A. Alternative Dietary Protein Sources for Atlantic Salmon (*Salmo salar* L.) Effect on Intestinal Microbiota, Intestinal and Liver Histology and Growth. *Aquac. Nutr.* **2014**, *20*, 381–398. [CrossRef]
63. De Santis, C.; Martin, S.A.M.; Dehler, C.E.; Iannetta, P.P.M.; Leeming, D.; Tocher, D.R. Influence of Dietary Inclusion of a Wet Processed Faba Bean Protein Isolate on Post-Smolt Atlantic Salmon (*Salmo salar*). *Aquaculture* **2016**, *465*, 124–133. [CrossRef]
64. De Santis, C.; Ruohonen, K.; Tocher, D.R.; Martin, S.A.M.; Król, E.; Secombes, C.J.; Bell, J.G.; El-Mowafi, A.; Crampton, V.O. Atlantic Salmon (*Salmo salar*) Parr as a Model to Predict the Optimum Inclusion of Air Classified Faba Bean Protein Concentrate in Feeds for Seawater Salmon. *Aquaculture* **2015**, *444*, 70–78. [CrossRef]
65. Tacchi, L.; Secombes, C.J.; Bickerdike, R.; Adler, M.A.; Venegas, C.; Takle, H.; Martin, S.A. Transcriptomic and Physiological Responses to Fishmeal Substitution with Plant Proteins in Formulated Feed in Farmed Atlantic Salmon (*Salmo salar*). *BMC Genom.* **2012**, *13*, 363. [CrossRef]
66. Hartviksen, M.; Bakke, A.M.; Vecino, J.G.; Ringø, E.; Krogdahl, Å. Evaluation of the Effect of Commercially Available Plant and Animal Protein Sources in Diets for Atlantic Salmon (*Salmo salar* L.): Digestive and Metabolic Investigations. *Fish Physiol. Biochem.* **2014**, *40*, 1621–1637. [CrossRef]
67. Hu, H.; Kortner, T.M.; Gajardo, K.; Chikwati, E.; Tinsley, J.; Krogdahl, Å. Intestinal Fluid Permeability in Atlantic Salmon (*Salmo salar* L.) Is Affected by Dietary Protein Source. *PLoS ONE* **2016**, *11*, e0167515. [CrossRef]
68. Mikołajczak, Z.; Mazurkiewicz, J.; Rawski, M.; Kierończyk, B.; Józefiak, A.; Świątkiewicz, S.; Józefiak, D. Black Soldier Fly Full-Fat Meal in Atlantic Salmon Nutrition—Part B: Effects on Growth Performance, Feed Utilization, Selected Nutriphysiological Traits and Production Sustainability in Pre-Smolts. *Ann. Anim. Sci.* **2022**. *ahead-of-print*. [CrossRef]
69. Mikołajczak, Z.; Mazurkiewicz, J.; Rawski, M.; Kierończyk, B.; Józefiak, A.; Świątkiewicz, S.; Józefiak, D. Black Soldier Fly Full-Fat Meal in Atlantic Salmon Nutrition—Part A: Effects on Growth Performance, Feed Utilization, Selected Nutriphysiological Traits and Production Sustainability in Fries. *Ann. Anim. Sci.* **2022**. *ahead-of-print*. [CrossRef]
70. Weththasinghe, P.; Lagos, L.; Cortés, M.; Hansen, J.Ø.; Øverland, M. Dietary Inclusion of Black Soldier Fly (*Hermetia illucens*) Larvae Meal and Paste Improved Gut Health but Had Minor Effects on Skin Mucus Proteome and Immune Response in Atlantic Salmon (*Salmo salar*). *Front. Immunol.* **2021**, *12*, 599530. [CrossRef]
71. Li, Y.; Kortner, T.M.; Chikwati, E.M.; Belghit, I.; Lock, E.-J.; Krogdahl, Å. Total Replacement of Fish Meal with Black Soldier Fly (*Hermetia Illucens*) Larvae Meal Does Not Compromise the Gut Health of Atlantic Salmon (*Salmo salar*). *Aquaculture* **2020**, *520*, 734967. [CrossRef]
72. Sørensen, S.L.; Ghirmay, A.; Gong, Y.; Dahle, D.; Vasanth, G.; Sørensen, M.; Kiron, V. Growth, Chemical Composition, Histology and Antioxidant Genes of Atlantic Salmon (*Salmo salar*) Fed Whole or Pre-Processed *Nannochloropsis Oceanica* and *Tetraselmis* Sp. *Fishes* **2021**, *6*, 23. [CrossRef]
73. Kiron, V.; Sørensen, M.; Huntley, M.; Vasanth, G.K.; Gong, Y.; Dahle, D.; Palihawadana, A.M. Defatted Biomass of the Microalga, Desmodesmus Sp., Can Replace Fishmeal in the Feeds for Atlantic Salmon. *Front. Mar. Sci.* **2016**, *3*, 67. [CrossRef]
74. Hansen, J.; Hofossæter, M.; Sahlmann, C.; Ånestad, R.; Reveco-Urzua, F.E.; Press, C.M.; Mydland, L.T.; Øverland, M. Effect of Candida Utilis on Growth and Intestinal Health of Atlantic Salmon (*Salmo salar*) Parr. *Aquaculture* **2019**, *511*, 734239. [CrossRef]
75. Green, T.J.; Smullen, R.; Barnes, A.C. Dietary Soybean Protein Concentrate-Induced Intestinal Disorder in Marine Farmed Atlantic Salmon, *Salmo salar* Is Associated with Alterations in Gut Microbiota. *Vet. Microbiol.* **2013**, *166*, 286–292. [CrossRef] [PubMed]
76. Reveco, F.E.; Øverland, M.; Romarheim, O.H.; Mydland, L.T. Intestinal Bacterial Community Structure Differs between Healthy and Inflamed Intestines in Atlantic Salmon (*Salmo salar* L.). *Aquaculture* **2014**, *420–421*, 262–269. [CrossRef]
77. Agboola, J.O.; Rocha, S.D.C.; Mensah, D.D.; Hansen, J.Ø.; Øyås, O.; Lapeña, D.; Mydland, L.T.; Arntzen, M.Ø.; Horn, S.J.; Øverland, M. Effect of Yeast Species and Processing on Intestinal Microbiota of Atlantic Salmon (*Salmo salar*) Fed Soybean Meal-Based Diets in Seawater. *Anim. Microbiome* **2023**, *5*, 21. [CrossRef]
78. Li, Y.; Gajardo, K.; Jaramillo-Torres, A.; Kortner, T.M.; Krogdahl, Å. Consistent Changes in the Intestinal Microbiota of Atlantic Salmon Fed Insect Meal Diets. *Anim. Microbiome* **2022**, *4*, 8. [CrossRef]
79. Leeper, A.; Ekmay, R.; Knobloch, S.; Skírnisdóttir, S.; Varunjikar, M.; Dubois, M.; Smárason, B.Ö.; Árnason, J.; Koppe, W.; Benhaïm, D. Torula Yeast in the Diet of Atlantic Salmon *Salmo salar* and the Impact on Growth Performance and Gut Microbiome. *Sci. Rep.* **2022**, *12*, 567. [CrossRef]
80. Urán, P.A.; Aydin, R.; Schrama, J.W.; Verreth, J.A.J.; Rombout, J.H.W.M. Soybean Meal-Induced Uptake Block in Atlantic Salmon *Salmo salar* Distal Enterocytes. *J. Fish Biol.* **2008**, *73*, 2571–2579. [CrossRef]
81. Bruce, T.J.; Neiger, R.D.; Brown, M.L. Gut Histology, Immunology and the Intestinal Microbiota of Rainbow Trout, *Oncorhynchus Mykiss* (Walbaum), Fed Process Variants of Soybean Meal. *Aquac. Res.* **2018**, *49*, 492–504. [CrossRef]
82. Miao, S.; Zhao, C.; Zhu, J.; Hu, J.; Dong, X.; Sun, L. Dietary Soybean Meal Affects Intestinal Homoeostasis by Altering the Microbiota, Morphology and Inflammatory Cytokine Gene Expression in Northern Snakehead. *Sci. Rep.* **2018**, *8*, 113. [CrossRef] [PubMed]
83. Merrifield, D.L.; Dimitroglou, A.; Bradley, G.; Baker, R.T.M.; Davies, S.J. Soybean Meal Alters Autochthonous Microbial Populations, Microvilli Morphology and Compromises Intestinal Enterocyte Integrity of Rainbow Trout, *Oncorhynchus mykiss* (Walbaum). *J. Fish Dis.* **2009**, *32*, 755–766. [CrossRef] [PubMed]

84. Buttle, L.G.; Burrells, A.C.; Good, J.E.; Williams, P.D.; Southgate, P.J.; Burrells, C. The Binding of Soybean Agglutinin (SBA) to the Intestinal Epithelium of Atlantic Salmon, *Salmo salar* and Rainbow Trout, Oncorhynchus Mykiss, Fed High Levels of Soybean Meal. *Vet. Immunol. Immunopathol.* **2001**, *80*, 237–244. [CrossRef] [PubMed]
85. Gajardo, K.; Rodiles, A.; Kortner, T.M.; Krogdahl, Å.; Bakke, A.M.; Merrifield, D.L.; Sørum, H. A High-Resolution Map of the Gut Microbiota in Atlantic Salmon (*Salmo salar*): A Basis for Comparative Gut Microbial Research. *Sci. Rep.* **2016**, *6*, 30893. [CrossRef] [PubMed]
86. Sanden, M.; Berntssen, M.H.G.; Krogdahl, Å.; Hemre, G.-I.; Bakke-McKellep, A.-M. An Examination of the Intestinal Tract of Atlantic Salmon, *Salmo salar* L., Parr Fed Different Varieties of Soy and Maize. *J. Fish Dis.* **2005**, *28*, 317–330. [CrossRef] [PubMed]
87. Krogdahl; Bakke-Mckellep; RØed; Baeverfjord. Feeding Atlantic Salmon *Salmo salar* L. Soybean Products: Effects on Disease Resistance (Furunculosis), and Lysozyme and IgM Levels in the Intestinal Mucosa. *Aquac. Nutr.* **2000**, *6*, 77–84. [CrossRef]
88. Sørensen, S.L.; Park, Y.; Gong, Y.; Vasanth, G.K.; Dahle, D.; Korsnes, K.; Phuong, T.H.; Kiron, V.; Øyen, S.; Pittman, K.; et al. Nutrient Digestibility, Growth, Mucosal Barrier Status, and Activity of Leucocytes From Head Kidney of Atlantic Salmon Fed Marine- or Plant-Derived Protein and Lipid Sources. *Front. Immunol.* **2021**, *11*, 623726. [CrossRef]
89. Knudsen, D.; Urán, P.; Arnous, A.; Koppe, W.; FRØKIÆR†, H. Saponin-Containing Subfractions of Soybean Molasses Induce Enteritis in the Distal Intestine of Atlantic Salmon. *J. Agric. Food Chem.* **2007**, *55*, 2261–2267. [CrossRef]
90. Nikmaram, N.; Leong, S.Y.; Koubaa, M.; Zhu, Z.; Barba, F.J.; Greiner, R.; Oey, I.; Roohinejad, S. Effect of Extrusion on the Anti-Nutritional Factors of Food Products: An Overview. *Food Control.* **2017**, *79*, 62–73. [CrossRef]
91. Catalán, N.; Villasante, A.; Wacyk, J.; Ramírez, C.; Romero, J. Fermented Soybean Meal Increases Lactic Acid Bacteria in Gut Microbiota of Atlantic Salmon (*Salmo salar*). *Probiotics Antimicrob. Proteins* **2018**, *10*, 566–576. [CrossRef]
92. Krogdahl, Å.; Kortner, T.M.; Jaramillo-Torres, A.; Gamil, A.A.A.; Chikwati, E.; Li, Y.; Schmidt, M.; Herman, E.; Hymowitz, T.; Teimouri, S.; et al. Removal of Three Proteinaceous Antinutrients from Soybean Does Not Mitigate Soybean-Induced Enteritis in Atlantic Salmon (*Salmo salar*, L). *Aquaculture* **2020**, *514*, 734495. [CrossRef]
93. Bakke-McKellep, A.M.; Penn, M.H.; Salas, P.M.; Refstie, S.; Sperstad, S.; Landsverk, T.; Ringø, E.; Krogdahl, Å. Effects of Dietary Soyabean Meal, Inulin and Oxytetracycline on Intestinal Microbiota and Epithelial Cell Stress, Apoptosis and Proliferation in the Teleost Atlantic Salmon (*Salmo salar* L.). *Br. J. Nutr.* **2007**, *97*, 699–713. [CrossRef] [PubMed]
94. El-Shemy, H.; Abdel-Rahim, E.; Shaban, O.; Ragab, A.; Carnovale, E.; Fujita, K. Comparison of Nutritional and Antinutritional Factors in Soybean and Fababean Seeds with or without Cortex. *Soil Sci. Plant Nutr.* **2000**, *46*, 515–524.
95. Kamunde, C.; Sappal, R.; Melegy, T.M. Brown Seaweed (AquaArom) Supplementation Increases Food Intake and Improves Growth, Antioxidant Status and Resistance to Temperature Stress in Atlantic Salmon, *Salmo salar*. *PLoS ONE* **2019**, *14*, e0219792. [CrossRef] [PubMed]
96. Norambuena, F.; Hermon, K.; Skrzypczyk, V.; Emery, J.A.; Sharon, Y.; Beard, A.; Turchini, G.M. Algae in Fish Feed: Performances and Fatty Acid Metabolism in Juvenile Atlantic Salmon. *PLoS ONE* **2015**, *10*, e0124042. [CrossRef]
97. Johny, A.; Berge, G.M.; Bogevik, A.S.; Krasnov, A.; Ruyter, B.; Fæste, C.K.; Østbye, T.-K.K. Sensitivity to Dietary Wheat Gluten in Atlantic Salmon Indicated by Gene Expression Changes in Liver and Intestine. *Genes* **2020**, *11*, 1339. [CrossRef] [PubMed]
98. Rawles, S.D.; Thompson, K.R.; Brady, Y.J.; Metts, L.S.; Aksoy, M.Y.; Gannam, A.L.; Twibell, R.G.; Ostrand, S.; Webster, C.D. Effects of Replacing Fish Meal with Poultry By-Product Meal and Soybean Meal and Reduced Protein Level on the Performance and Immune Status of Pond-Grown Sunshine Bass (*Morone chrysops* × *M. Saxatilis*). *Aquac. Nutr.* **2011**, *17*, e708–e721. [CrossRef]
99. Cammack, J.; Tomberlin, J. The Impact of Diet Protein and Carbohydrate on Select Life-History Traits of The Black Soldier Fly *Hermetia Illucens* (L.) (Diptera: *Stratiomyidae*). *Insects* **2017**, *8*, 56. [CrossRef]
100. van Huis, A. Potential of Insects as Food and Feed in Assuring Food Security. *Annu. Rev. Entomol.* **2013**, *58*, 563–583. [CrossRef]
101. Wang, Y.-S.; Shelomi, M. Review of Black Soldier Fly (*Hermetia illucens*) as Animal Feed and Human Food. *Foods* **2017**, *6*, 91. [CrossRef] [PubMed]
102. Belghit, I.; Liland, N.S.; Gjesdal, P.; Biancarosa, I.; Menchetti, E.; Li, Y.; Waagbø, R.; Krogdahl, Å.; Lock, E.-J. Black Soldier Fly Larvae Meal Can Replace Fish Meal in Diets of Sea-Water Phase Atlantic Salmon (*Salmo salar*). *Aquaculture* **2019**, *503*, 609–619. [CrossRef]
103. Li, Y.; Kortner, T.M.; Chikwati, E.M.; Munang'andu, H.M.; Lock, E.-J.; Krogdahl, Å. Gut Health and Vaccination Response in Pre-Smolt Atlantic Salmon (*Salmo salar*) Fed Black Soldier Fly (*Hermetia illucens*) Larvae Meal. *Fish Shellfish Immunol.* **2019**, *86*, 1106–1113. [CrossRef] [PubMed]
104. Li, Y.; Bruni, L.; Jaramillo-Torres, A.; Gajardo, K.; Kortner, T.M.; Krogdahl, Å. Differential Response of Digesta- and Mucosa-Associated Intestinal Microbiota to Dietary Insect Meal during the Seawater Phase of Atlantic Salmon. *Anim. Microbiome* **2021**, *3*, 1–18. [CrossRef]
105. Shafique, L.; Abdel-Latif, H.M.R.; Hassan, F.; Alagawany, M.; Naiel, M.A.E.; Dawood, M.A.O.; Yilmaz, S.; Liu, Q. The Feasibility of Using Yellow Mealworms (*Tenebrio molitor*): Towards a Sustainable Aquafeed Industry. *Animals* **2021**, *11*, 811. [CrossRef] [PubMed]
106. Jones, S.W.; Karpol, A.; Friedman, S.; Maru, B.T.; Tracy, B.P. Recent Advances in Single Cell Protein Use as a Feed Ingredient in Aquaculture. *Curr. Opin. Biotechnol.* **2020**, *61*, 189–197. [CrossRef]
107. Becker, W. Microalgae in Human and Animal Nutrition. In *Handbook of Microalgal Culture*; John Wiley & Sons, Ltd.: Hoboken, NJ, USA, 2003; pp. 312–351. ISBN 978-0-470-99528-0.

108. Sørensen, M.; Gong, Y.; Bjarnason, F.; Vasanth, G.K.; Dahle, D.; Huntley, M.; Kiron, V. Nannochloropsis Oceania-Derived Defatted Meal as an Alternative to Fishmeal in Atlantic Salmon Feeds. *PLoS ONE* **2017**, *12*, e0179907. [CrossRef]
109. Gong, Y.; Sørensen, S.L.; Dahle, D.; Nadanasabesan, N.; Dias, J.; Valente, L.M.P.; Sørensen, M.; Kiron, V. Approaches to Improve Utilization of Nannochloropsis Oceanica in Plant-Based Feeds for Atlantic Salmon. *Aquaculture* **2020**, *522*, 735122. [CrossRef]
110. Grammes, F.; Reveco, F.E.; Romarheim, O.H.; Landsverk, T.; Mydland, L.T.; Øverland, M. *Candida Utilis* and *Chlorella Vulgaris* Counteract Intestinal Inflammation in Atlantic Salmon (*Salmo salar* L.). *PLoS ONE* **2013**, *8*, e83213. [CrossRef]
111. Agboola, J.O.; Øverland, M.; Skrede, A.; Hansen, J.Ø. Yeast as Major Protein-Rich Ingredient in Aquafeeds: A Review of the Implications for Aquaculture Production. *Rev. Aquac.* **2021**, *13*, 949–970. [CrossRef]
112. Øverland, M.; Karlsson, A.; Mydland, L.T.; Romarheim, O.H.; Skrede, A. Evaluation of *Candida utilis, Kluyveromyces marxianus* and *Saccharomyces cerevisiae* Yeasts as Protein Sources in Diets for Atlantic Salmon (*Salmo salar*). *Aquaculture* **2013**, *402–403*, 1–7. [CrossRef]
113. Naylor, R.L. Managing Food Production Systems for Resilience. In *Principles of Ecosystem Stewardship: Resilience-Based Natural Resource Management in a Changing World*; Folke, C., Kofinas, G.P., Chapin, F.S., Eds.; Springer: New York, NY, USA, 2009; pp. 259–280. ISBN 978-0-387-73033-2.

animals

MDPI

Review

A Systematic Review and Meta-Analysis of Basal and Post-Stress Circulating Cortisol Concentration in an Important Marine Aquaculture Fish Species, European Sea Bass, *Dicentrarchus labrax*

Athanasios Samaras

Independent Researcher, 71303 Heraklion, Greece; samaras.thanas@gmail.com

Simple Summary: European sea bass is a species with high economic and societal value in the Mediterranean due to its intensive use in aquaculture. However, it is a species characterized by high cortisol levels that show high variation. The present systematic review and meta-analysis collected and examined all the published data on circulating cortisol in this species. The aim of the study was to analyze all published values in order to provide normal values and ranges of plasma cortisol in this species, both in basal and post-acute stress conditions. Results revealed a very high between-study heterogeneity, while it also calculated the pooled levels of cortisol and their confidence intervals for both basal and post-stress conditions. Moreover, results were analyzed based on various parameters that can potentially affect cortisol levels, including technical, such as assay type and rearing unit, as well as biological, such as body size and anesthesia, influences.

Abstract: Background: European sea bass is a species characterized by high and dispersed cortisol levels. The aim of the present study was to analyze all published data on basal and post-acute stress cortisol levels in this species. Methods: For this systematic review and meta-analysis the Web of Science and Scopus databases were searched for papers reporting plasma or serum cortisol levels in E. sea bass, without language or date restrictions. Data were extracted directly for the reported results and were analyzed separately for basal and post-acute stress levels, as well their standardized mean differences (SMD) using random-effects meta analyses. Results: Of 407 unique records identified, 69 were eligible. Basal cortisol levels had a pooled effect of 88.7 ng mL^{-1} (n = 57), while post-acute stress levels were 385.9 ng mL^{-1} (n = 34). The average SMD between basal and post-stress was calculated to be 3.02 (n = 22). All analyses had a high between-study heterogeneity. Results for basal and post-stress levels were affected by the assay type and anesthesia prior to blood sampling Conclusions: Cortisol levels in E. sea bass are higher than most studied fish species and display large heterogeneity. Application of stress led to elevated cortisol levels in all studies examined. In all cases, sources of between-studies heterogeneity were identified.

Keywords: aquaculture; cortisol; European sea bass; meta-analysis; reference values; stress; welfare

1. Introduction

Cortisol is the major stress hormone in fish [1–3]. It is the final product of the action of the Hypothalamus–Pituitary–Interrenal tissue (HPI) axis, and it has been reported to respond with increased concentrations to various types of stress in order to regulate stress responses [1–3]. Apart from controlling the stress responses, cortisol is a regulatory hormone for both metabolism and osmoregulation in fish [3]. Therefore, it is a crucial hormone in the physiology and biology of fish.

European sea bass, *Dicentrarchus labrax*, is a fish species with high economic value due to the fact that it is one of the main marine aquaculture fish species in the Mediterranean. Although widely cultured, this species shows high cortisol responses to stress compared to other species widely cultivated in the Mediterranean, such as gilthead seabream, *Sparus aurata*, and meage, *Argyrosomus regius* [4,5], as well as high variation in cortisol concentrations in basal (pre-stress) and post-stress conditions [6]. A decade ago, Ellis et al. (2012) [6] reported that a high between-study variation in the basal and post-stress cortisol concentration can be observed in this species, proposing five possible sources of variation. Recent research has shown that most of these sources indeed can add variation, while other factors remain untested still (Table 1).

Table 1. Suggested sources of variation in cortisol levels of E. sea bass presented by Ellis et al. (2012) [6], and the respective factors that have been studied thereinafter.

	Source of Variation	Reference
1.	true differences in basal cortisol concentrations associated with different fish conditions (e.g., body size, age, etc.)	Body size [7]
2.	Genetic/strain differences in cortisol responsiveness coping style/personality/temperament/behavioural syndromes	Genetics [8–11] Coping styles [12,13]
3.	different environmental conditions (e.g., water temperature, salinity, lightning, photoperiod, season, food composition)	Temperature [14,15] Salinity [16,17] Photoperiod, season [18,19] Food composition [20–23]
4.	presence of unrecognized stressors (e.g., poor husbandry conditions, disease) in some studies	-
5.	the analytical method used (RIA, EIA/ELISA, HPLC) and possible errors	ELISA vs. LC-MS/MS [24]

1.1. Rationale for Meta-Analysis

As discussed earlier, cortisol variation in E. sea bass is large, both in the same study (within population [8]) and between different studies. This makes the generalization of conclusions based on a single measurement impossible, since under such conditions it is hard to define accurate reference values. However, cortisol measurement is an important indicator of the physiological status of fish in terms of stress, osmoregulatory, and metabolic regulations [2]. A meta-analysis of data from different independent studies can provide a quantitative statistical way to combine their results. Especially in cases with data that show high heterogeneity, such as cortisol in E. sea bass, a meta-analysis using random effects models allows for important conclusions to be drawn. Moreover, certain meta-analysis statistical tools provide the means to incorporate other effects, such as environmental (husbandry, water quality etc.), biological (body weight), and technical (method used to quantify cortisol), in the analysis. For instance, the assay type used to measure cortisol in each study (ELISA, RIA, HPLC) can be included in a sub-group meta-analysis to provide information on whether the assay type affects the outcome of the study.

1.2. Objectives

The objective of the present study was to provide normal values and ranges of plasma cortisol in E. sea bass, both in basal (i.e., without experimental exposure to any stressors) and post-acute stress conditions, as well as quantify the standardized mean difference between basal and post- stress (both acute and chronic) cortisol levels. Moreover, this study aimed at investigating the effects of (1) cortisol measurement assay type, (2) type of rearing unit, (3) use of anesthesia during the blood sampling, (4) water parameters such as temperature, salinity, dissolved oxygen, pH, (5) fish body weight, (6) stocking density, and, in the cases of post-stress levels, (7) the time after stress that the blood sampling took place.

2. Materials and Methods

The Preferred Reporting Items for Systematic reviews and Meta-Analyses Statement (PRISMA) guidelines to plan, implement, and report this systematic review and meta-analysis have been followed in this study [25]. The PRISMA checklist is available in Supplementary Table S1.

2.1. Identification of Studies

The databases Web of Science and Scopus were assessed using the search terms ("European sea bass" OR "sea bass" OR dicentrarchus OR labrax) AND (cortisol OR glucocorticoid OR corticosteroid) to find peer-reviewed articles reporting cortisol levels in E. sea bass, until the date of search (7 March 2022). Figure 1 shows a flow diagram that summarizes all stages of the systematic review process, including the numbers of studies identified at each stage and any reasons for exclusion. This workflow has not been published and no protocol was prepared.

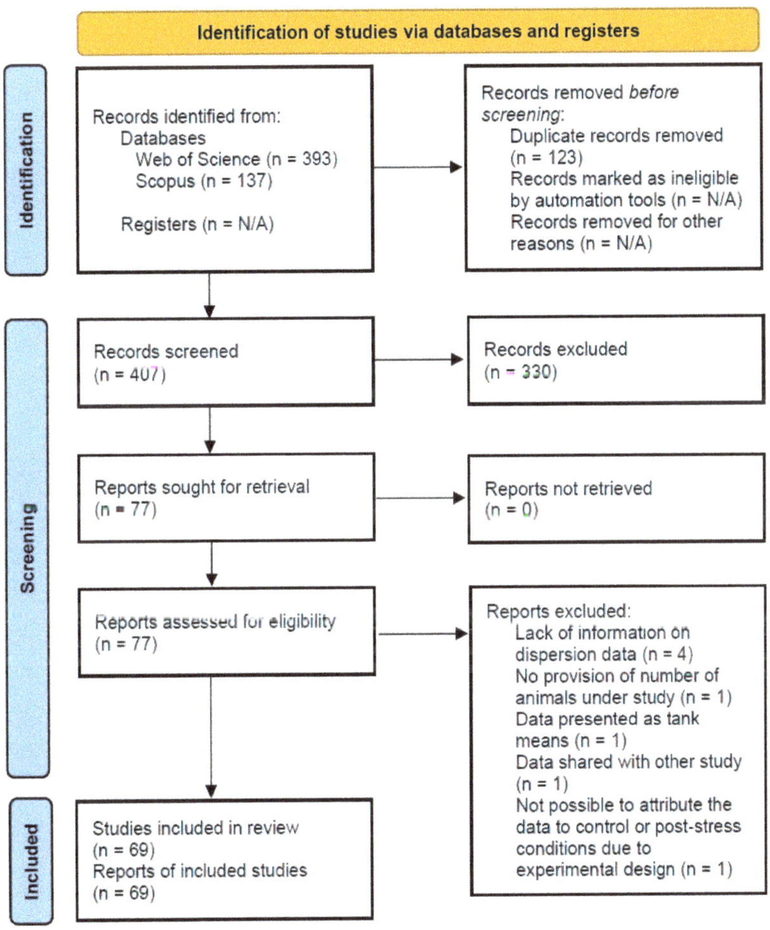

Figure 1. Flow diagram showing phases of systematic review and study selection for meta-analysis.

2.2. Eligibility Criteria

Out of the 407 research items retrieved from the database search after duplicate removal, the criteria used for screening the articles were that: (Scr1) the study concerned E. sea bass, (Scr2) examined fully developed fish, therefore excluding larvae, and (Scr3) reported plasma, or serum cortisol levels with their concentration. At this stage, 330 research items were excluded, and the remaining 77 were subjected to more detailed inspection for eligibility, including only studies that: (Eli1) provided information on the number of animals used in each experimental group, (Eli2) provided information regarding the dispersion of the data, either as Standard Deviation (S.D.) or as Standard Error of the Mean (S.E.M.), (Eli3) used the individuals and not the tanks as experimental units and, therefore, presented the results as an average of individuals and not tanks, and (Eli4) it was possible to attribute the data to control or post-stress conditions. After this stage, 8 research items were excluded ([16,26–32]; Table S2), and the remaining 69 items were used in the meta-analysis.

2.3. Data Extraction

All qualitative information of the studies, such as assay type, rearing system, fish size, and water parameters, were retrieved from the text and tables of the research items. Data from the research items were extracted either from tables or graphs reporting the mean value ± S.D. or S.E.M. In the latter case, the software ImageJ was used for image analysis based on measuring the length of the *y*-axis and the length of the projections of the cortisol mean ± S.D. or S.E.M. In cases where various post-stress time points were presented, the time of the peak response was used. Moreover, when more than one control group was presented and in order not to violate the assumption of independency of data for the meta-analysis by analyzing data from the same research item more than once, the control groups were pooled. All cortisol data were analyzed as ng mL^{-1}. The vast majority of the studies reported this measuring unit (62 out of 69 studies) or its derivatives ug/dL, ug/mL and ng/dL (4 out of 69 studies). There were 3 studies that reported the results as nmol L^{-1} (nM), and their data were converted to ng mL^{-1} by multiplying with the conversion factor 0.36245.

2.4. Coding of Data

Each study was coded for quantitative and qualitative data. Quantitative data included (1) fish body size, (2) water temperature, (3) dissolved oxygen, (4) water salinity, (5) water pH, and (6) stocking density. The qualitative data included (7) assay type used to measure cortisol, (8) rearing system, (9) anesthesia type, and, (10) in cases of post-stress samplings, total time between the application of stress and sampling, defined as classes (e.g., 0–30 min, 30–60 min etc.).

In cases where the quantitative data were reported as a range, the mean value was calculated. For instance, in the study by Tintos et al. 2006 [33] where the body weight was reported to range between 15–20 g, the weight was recorded as the mean between the two values, i.e., 17.5 g.

2.5. Statistical Analysis

All statistical analysis was performed in RStudio [34], using the packages "meta" [35], "dmetar" [36], and "tidyverse" [37]. Since considerable between-study heterogeneity was expected, a random-effects model was used to pool effect sizes. The heterogeneity variance τ^2 was calculated using the restricted maximum likelihood estimator, while the confidence interval around the pooled effect was calculated using the Knapp–Hartung adjustments [38].

Specifically, for the basal and post-stress analysis, the pre-calculated effect sizes were analyzed under the "metagen" function, while analysis of the standardized mean differences (SMD) and their 95% confidence intervals (CIs) between basal and post-stress levels were analyzed under the "metacont" function, using Hedges method to calculate the SMD due to the small number of subjects in most studies [35,39]. Subgroup analysis for the qualitative treatment data was performed using the "byvar" argument and was based on calculating different τ^2 for each subgroup, and subgroups were tested for significant differences using the Q test. Meta-regression on the quantitative influence data was performed using the "metareg" function and risk of bias was assessed by Egger's test using the "metabias" function. The forest plot was created using the "forest" function.

3. Results

3.1. Study Characteristics

The final outcome of the literature search was 69 peer-reviewed studies that concerned circulating cortisol levels in E. sea bass and provided sufficient information on the number of animals used and the population mean values and dispersion (Figure 1). Out of these, 35 studies reported only basal, 12 only post-stress, and 22 both basal and post-stress cortisol concentrations, thus making a total of 57 studies reporting basal levels and 34 studies reporting post-stress levels. Summary characteristics for the research items included in the meta-analysis are presented (Table 2). All 69 studies had reported the weight of the fish.

3.2. Basal Levels

Frequency distribution analysis on the basal cortisol levels resulted in a positively skewed distribution (Figure 2a). The pooled effect size was calculated at 88.7 ng mL^{-1} [95%-CI: 65.5–109.8], with a high between-study heterogeneity (Table 3). Egger's test showed that no significant publication bias was present in the dataset (intercept = 13.5, df = 55; p = 0.525).

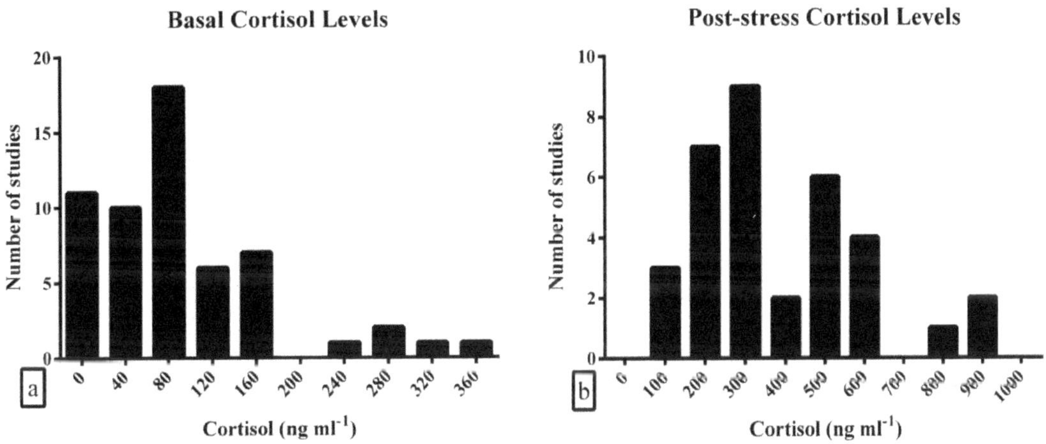

Figure 2. Frequency distribution of meta-analysis data for (**a**) basal and (**b**) post-stress cortisol levels.

Subgroup analysis based on the assay type revealed differences between groups. Due to the low number of studies using HPLC and chemiluminescence/electrochemiluminescence assays, an analysis including only the studies using ELISA and RIA assays was performed and showed that the difference between them was significant (Q_1 = 5.36; p = 0.021), being higher in ELISA than RIA. The use of anesthetic also had a significant effect on cortisol (Q_5 = 12.69; p = 0.026), while no differences were observed between the rearing unit systems (Q_3 = 1.41; p = 0.703).

Meta-regression analysis between effect sizes and quantitative characteristics showed that none of the examined parameters, i.e., fish body weight, water temperature, dissolved oxygen concentration, salinity, pH, and stocking density, affected the results.

Table 2. Studies included in the meta-analysis (n = 69). In Dissolved oxygen, Salinity, pH, and Stocking density a "+" symbol signifies the presence, while a "–" symbol the absence of information in the respective study.

Study	Year	Treatment	Temperature (°C)	Dissolved Oxygen (mg L^{-1})	Salinity	pH	Stocking Density	Assay Type	Rearing System	Anesthetic	Total Time (min)
[5]	2016	Basal/Post-stress	15.0	–	–	–	+	ELISA	Cage	Phenoxyethanol	30–60
[7]	2014	Basal/Post-stress	-	+	+	–	+	ELISA	Open flow	Clove oil	60–90
[10]	2021	Basal/Post-stress	19.0	–	+	–	+	ELISA	Open flow	Phenoxyethanol	60–90
[11]	2020	Post-stress	18.0	–	+	–	+	ELISA	Open flow	Phenoxyethanol	30–60
[12]	2015	Post-stress	20.1	–	+	–	–	ELISA	RAS	Benzocaine	30–60
[13]	2020	Post-stress	-	–	–	–	–	ELISA	Open flow	Benzocaine	30–60
[14]	2018	Basal/Post-stress	-	–	+	+	+	ELISA	RAS	Phenoxyethanol	60–90
[15]	2020	Basal	16.0	–	+	–	–	ELISA	RAS	MS222	
[18]	2011	Basal	-	–	–	–	–	RIA	Pond	Ice slurry/cold water	
[20]	2013	Post-stress	21.2	+	+	–	–	RIA	RAS	Clove oil	
[21]	2020	Basal	18.0	+	–	+	+	HPLC	RAS	Clove oil	
[22]	2020	Post-stress	22.0	+	–	–	–		Open flow	Phenoxyethanol	90–120
[23]	2017	Basal	25.0	–	+	–	–	ELISA	RAS	Phenoxyethanol	
[24]	2021	Post-stress	21.0	–	–	–	–	ELISA	RAS	Benzocaine	60–90
[33]	2006	Basal	20.0	–	+	–	–	ELISA	Open flow	Phenoxyethanol	
[40]	2009	Basal	19.5	–	–	–	–	RIA	Open flow	-	
[41]	2019	Post-stress	20.9	–	+	+	–	ELISA	RAS	Benzocaine	
[42]	2020	Post-stress	21.5	–	+	+	–	ELISA	RAS	Benzocaine	30–60
[43]	2020	Basal	18.0	–	+	+	+	Chemiluminescence	Open flow	Clove oil	
[44]	2010	Basal/Post-stress	-	–	–	–	–	RIA	Pond	MS222	120–240
[45]	2018	Basal	20.5	+	–	–	–	RIA	-	None	
[46]	2021	Basal	17.5	–	–	–	+	ELISA	RAS	MS222	
[47]	2021	Basal	17.0	–	+	–	+	RIA	RAS	MS222	
[48]	2010	Basal	18.0	–	–	–	+	HPLC	Open flow	Clove oil	
[49]	2014	Basal	18.0	–	+	–	+	HPLC	-	Clove oil	
[50]	2019	Basal	18.0	–	+	–	–	ELISA	RAS	Clove oil	
[51]	2011	Basal	20.0	–	+	+	+	ELISA	Open flow	MS222	
[52]	2012	Basal	18.0	–	–	–	–	ELISA	Open flow	MS222	
[53]	1998	Basal	-	–	–	–	–	RIA	Open flow	MS222	
[54]	2020	Basal/Post-stress	21.9	–	+	+	–	ELISA	Open flow	Phenoxyethanol	0–30

Table 2. *Cont.*

Study	Year	Treatment	Temperature (°C)	Dissolved Oxygen (mg L^{-1})	Salinity	pH	Stocking Density	Assay Type	Rearing System	Anesthetic	Total Time (min)
[55]	2010	Basal	21.2	–	+	+	+	RIA	Open flow	Phenoxyethanol	
[56]	2008	Basal/Post-stress	18.5	+	+	–	+	RIA	RAS	Phenoxyethanol	60–90
[57]	2021	Post-stress	21.0	+	+	–	–	RIA	RAS	MS222	0–30
[58]	2008	Basal/Post-stress	20.0	+	+	–	+	ELISA	Open flow	Clove oil	60–90
[59]	2012	Basal	-	–	–	–	–	ELISA	Cage	None	
[60]	2007	Basal	27.3	–	–	–	–	ELISA	Open flow	Phenoxyethanol	
[61]	2011	Basal/Post-stress	-	–	–	–	–	ELISA	Open flow	Phenoxyethanol	90–120
[62]	2019	Basal/Post-stress	25.4	–	+	–	–	ELISA	RAS	Blow to head/decapitation	60–90
[63]	2019	Basal	-	+	+	–	–	ELISA	RAS	Phenoxyethanol	
[64]	2011	Basal	-	–	–	–	–	ELISA	Cage	MS222	
[65]	2001	Basal/Post-stress	14.0	–	+	–	+	RIA	Cage	None	0–30
[66]	2014	Post-stress	20.0	–	+	–	+	ELISA	Open flow	Phenoxyethanol	30–60
[67]	2020	Post-stress	15.7	–	+	–	–	ELISA	Open flow	Phenoxyethanol	120–240
[68]	2019	Basal/Post-stress	12.5	–	–	–	–	ELISA	Cage	Blow to head/decapitation	60–90
[69]	2005	Basal	-	–	–	–	–	RIA	-	Clove oil	
[70]	2011	Basal	20.2	+	+	+	+	RIA	Open flow	MS222	
[71]	2007	Basal/Post-stress	-	–	–	–	+	RIA	Cage	MS222	120–240
[72]	1996	Basal/Post-stress	13.0	–	+	–	–	RIA	Open flow	-	>240
[73]	2010	Basal	27.0	+	+	–	–	Electro-chemiluminescence	Open flow	-	
[74]	2006	Basal/Post-stress	18.0	–	–	–	+	RIA	Open flow	Phenoxyethanol	>240
[75]	2003	Basal/Post-stress	23.0	–	+	–	+	RIA	Open flow	Phenoxyethanol	30–60
[76]	2017	Basal/Post-stress	-	+	+	+	+	ELISA	Cage	Phenoxyethanol	30–60
[77]	2018	Basal/Post-stress	19.0	–	–	–	+	RIA	Open flow	Phenoxyethanol	60–90
[78]	2009	Basal	23.4	–	+	+	+	RIA	RAS	Ice slurry/cold water	
[79]	2013	Basal/Post-stress	22.0	+	+	+	+	ELISA	RAS	MS222	30–60
[80]	2010	Basal/Post-stress	22.0	+	+	+	+	ELISA	RAS	MS222	30–60
[81]	2019	Basal	20.0	–	+	+	–	ELISA	RAS	-	
[82]	2008	Basal/Post-stress	24.0	+	+	–	–	RIA	Pond	Ice slurry/cold water	90–120

Table 2. *Cont.*

Study	Year	Treatment	Temperature (°C)	Dissolved Oxygen (mg L^{-1})	Salinity	pH	Stocking Density	Assay Type	Rearing System	Anesthetic	Total Time (min)
[83]	2006	Basal	20.0	+	+	+	–	ELISA	RAS	None	
[84]	2004	Basal	20.0	+	+	+	–	ELISA	RAS	None	
[85]	2005	Basal	20.0	–	+	+	+	ELISA	RAS	MS222	
[86]	2012	Basal/Post-stress	23.0	+	–	–	+	RIA	Open flow	None	90–120
[87]	2015	Basal	19.0	–	+	–	–	ELISA	Open flow	Phenoxyethanol	
[88]	2006	Basal	17.0	–	+	–	–	RIA	Open flow	Ice slurry/cold water	
[89]	2006	Basal	12.5	–	+	–	+	RIA	-	Blow to head/decapitation	
[90]	2010	Basal	18.0	–	–	–	–	ELISA	Open flow	MS222	
[91]	2009	Basal	14.0	–	+	–	–	RIA	Open flow	Phenoxyethanol	
[92]	2012	Post-stress	-	–	–	–	–	RIA	-	Ice slurry/cold water	120–240
[93]	2010	Basal	18.0	+	–	+	–	ELISA	-	None	

Table 3. Results of the meta-analysis regarding basal cortisol levels, including subgroup analysis. ES: effect size; 95%-CI: 95% confidence intervals; I^2: Higgin's and Thompson's between-study heterogeneity statistic; τ^2: heterogeneity variance.

	n	ES	95%-CI	I^2	τ^2
Whole population	57	88.7	65.5–109.8	100.0%	6235.7
Assay type					
RIA	22	60.8	37.7–83.8	100.0%	2677.7
ELISA	30	108.2	72.9–143.4	100.0%	8787.1
HPLC	3	97.3	−32.5–227.6	98.9%	2696.8
Chemiluminescence Electrochemiluminescence	2	-	-	-	-
Rearing unit					
Open flow	25	81.1	57.6–104.6	100.0%	3168.4
RAS	17	102.9	48.1–157.7	99.9%	11,200.3
Sea cages	7	95.2	0.1–190.3	100.0%	10,575.0
Ponds	3	113.9	−21.6–249.3	99.4%	2955.7
Anesthesia					
Phenoxyethanol	17	71.7	36.9–106.5	99.9%	4509.7
MS222	14	72.2	29.4–115.0	100.0%	5482.7
Clove oil	8	113.9	80.0–147.8	99.3%	1619.8
None	7	142.3	24.5–260.0	100.0%	16,051.0
Ice/Cold water	4	86.2	−19.9–192.3	100.0%	4433.8
Decap/Blow to head	3	25.6	−77.4–128.7	99.9%	1719.6

3.3. Post-Stress Levels

Frequency distribution analysis on the post-stress cortisol levels resulted in a positively skewed distribution (Figure 2b). The pooled effect size was calculated at 385.9 ng mL^{-1} [95%-CI: 310.8–460.9], with a high between-study heterogeneity (Table 4). Egger's test resulted in a marginally significant publication bias in the dataset (intercept = 6.4, df = 32; $p = 0.047$).

Subgroup analysis based on the assay type revealed differences between groups ($Q_1 = 7.76$; $p = 0.005$), being higher in ELISA than RIA. The use of anesthetic also had a significant effect on cortisol ($Q_3 = 17.13$; $p < 0.001$), excluding the "none", "ice/cold water", and "decap/blow to head" groups from the analysis due to the small number of studies in each group. On the other hand, no differences were observed between rearing unit systems ($Q_2 = 5.36$; $p = 0.069$), excluding the ponds due to their small number. Time after stress, excluding the ">240" group due to the low number of studies, showed a significant effect ($Q_4 = 9.99$; $p = 0.041$). Finally, none of the quantitative parameters was related to the effect sizes when the respective meta-regression analysis was performed.

3.4. Standardized Mean Difference between Basal and Post-Stress Cortisol

To assess the difference between basal and post-stress cortisol, 22 studies that included both pre- and post- exposure to acute stress data were used. The pooled SMD was calculated to be 3.02 [95%-CI: 2.46–3.58] (Table 5). The between-study heterogeneity was lower than the ones in the previous analysis, but still significantly large ($I^2 = 81.7\%$, $\tau^2 = 1.12$). Egger's test using the Pustejovsky and Rodgers modification to avoid false positive results that arise with the classical Egger's test on SMDs [94], resulted in a significant publication bias in the dataset (intercept = 4.7, df = 20; $p = 0.010$).

Table 4. Results of the meta-analysis regarding post-stress cortisol levels, including subgroup analysis. ES: effect size; 95%-CI: 95% confidence intervals; I^2: Higgin's and Thompson's between-study heterogeneity statistic; τ^2: heterogeneity variance.

	n	ES	95%-CI	I^2	τ^2
Whole population	34	385.9	310.8–460.9	99.4%	44,598.1
Assay type					
RIA	13	273.8	180.5–367.1	99.6%	21,989.3
ELISA	20	458.6	352.5–564.6	98.9%	49,493.9
Rearing unit					
Open flow	15	382.6	278.8–486.5	99.5%	34,183.3
RAS	11	452.1	261.4–642.8	99.4%	78,819.7
Sea cages	5	269.4	154.6–384.1	90.0%	7570.1
Ponds	2	-	-	-	-
Anesthesia					
Phenoxyethanol	14	308.1	216.9–399.4	99.6%	24,364.3
MS222	5	273.6	50.37–496.8	96.4%	31,117.1
Clove oil	3	521.0	−119.5–1161.4	94.3%	63,010.3
Benzocaine	5	692.1	439.5–944.7	96.7%	39,320.6
None	2	-	-	-	-
Ice/Cold water	2	-	-	-	-
Decap/Blow to head	2	-	-	-	-
Time post-stress (min)					
0–30	3	259.2	125.8–392.5	87.3%	2353.4
30–60	10	400.6	220.7–580.5	99.6%	60,818.3
60–90	9	415.4	247.1–583.7	99.6%	46,929.3
90–120	4	423.1	228.4–617.8	89.7%	12,499.1
120–240	4	267.0	72.9–461.2	97.5%	11,913.2
>240	2	-	-	-	-

Table 5. Results of the meta-analysis regarding the standardized mean difference (SMD) between basal and post-stress cortisol levels, including subgroup analysis. ES: effect size; 95%-CI: 95% confidence intervals; I^2: Higgin's and Thompson's between-study heterogeneity statistic; τ^2: heterogeneity variance.

	n	SMD	95%-CI	I^2	τ^2
Whole population	22	3.02	2.46–3.58	81.7%	1.12
Assay type					
RIA	10	3.00	2.36–3.65	79.8%	0.61
ELISA	12	3.09	2.07–4.11	84.4%	1.86
Rearing unit					
Open flow	10	3.16	2.16–4.16	81.9%	1.21
RAS	5	3.04	1.42–4.67	87.5%	1.45
Sea cages	5	2.90	1.06–4.74	87.1%	1.87
Ponds	2	-	-	-	-
Anesthesia					
Phenoxyethanol	10	3.13	2.24–4.02	82.5%	1.25
MS222	4	3.63	2.11–5.15	36.3%	0.32
Clove oil	2	-	-	-	-
None	2	-	-	-	-
Ice/Cold water	1	-	-	-	-
Decap/Blow to head	2	-	-	-	-

Table 5. *Cont.*

	n	SMD	95%-CI	I^2	τ^2
Time post-stress (min)					
0–30	2	-	-	-	-
30–60	5	3.59	1.59–5.59	91.7%	2.39
60–90	8	3.09	1.86–4.32	81.9%	1.30
90–120	3	2.60	0.78–4.41	19.8%	0.15
120–240	2	-	-	-	-
>240	2	-	-	-	-

It is obvious that stress had an overall effect on cortisol, a result that was observed in every study (Figure 3). The high heterogeneity between the studies can also be observed (Figure 3).

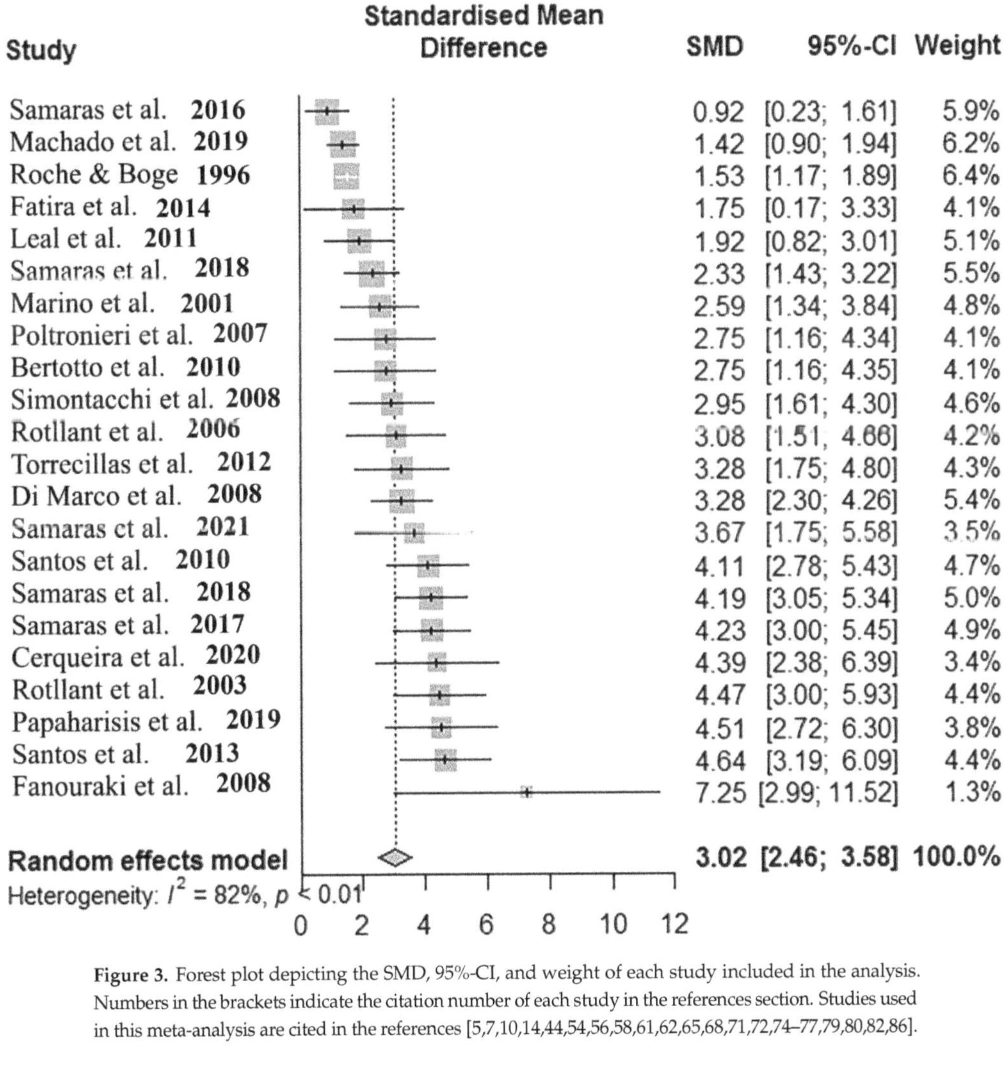

Figure 3. Forest plot depicting the SMD, 95%-CI, and weight of each study included in the analysis. Numbers in the brackets indicate the citation number of each study in the references section. Studies used in this meta-analysis are cited in the references [5,7,10,14,44,54,56,58,61,62,65,68,71,72,74–77,79,80,82,86].

Subgroup analysis based on the assay type revealed no significant differences between groups ($Q_1 = 0.03$; $p = 0.870$), being similar between studies using ELISA rather than RIA assays. No differences were observed between rearing unit systems, excluding ponds due to the low number of studies in this group ($Q_2 = 0.11$; $p = 0.945$). The same was true for the use of anesthetics, analyzing only phenoxyethanol and MS222 ($Q_1 = 0.65$; $p = 0.421$). Finally, time after stress had no significant effect on cortisol response ($Q_2 = 1.56$; $p = 0.458$) excluding the "0–30", "120–240", and "<240" due to low number of studies. Finally, meta-regression analysis between effect sizes and quantitative characteristics showed that none of the examined parameters, i.e., fish body weight, water temperature, dissolved oxygen concentration, salinity, pH, and stocking density, affected the results.

4. Discussion

The study of circulating cortisol concentration in E. sea bass is intriguing due to the fact that this species shows high basal and post-stress levels of cortisol, as well as high variation both within the same population and between different studies [4,6,8]. In fact, out of the studied teleost species, E. sea bass is among the ones with the highest reported cortisol levels, together with the chub, *Leuciscus cephalus*, the latter having been characterized as a cortisol resistant species [95]. This high stress susceptibility has been suggested to be co-responsible for disease outbreaks in this species [96].

In this context, a systemic review of the published cortisol levels of E. sea bass could assist in better understanding whether cortisol levels in this species are indeed high as well as to define possible sources of variation between studies. This meta-analysis led to the conclusion that a high between-studies heterogeneity exists in both basal and post-stress concentrations. The reported basal concentrations were calculated to have a pooled effect size of 88.7 ng mL^{-1} with an 95% confidence interval between 65.5–109.8 ng mL^{-1}, while post-stress concentration had a pooled effect size of 385.9 ng mL^{-1} [95%-CI: 310.8–460.9].

Subgroup meta-analysis revealed some interesting findings. However, these findings should be interpreted with care since there are constraints in the use of subgroups meta-analyses. The most important ones are the small number of studies in a subgroup and the high between-study heterogeneity, since both reduce the statistical power of the analysis. In order to reduce the "small number of studies" effect in the current analyses, subgroups with lower than three studies were excluded from subgroup meta-analysis. On the other hand, although high heterogeneity was observed in the current study, there are no available tools to mitigate its effect on the subgroups analysis.

Having the above constrains in mind, one of the major factors that seemed to affect the heterogeneity was the assay type. In both basal and post-stress levels, the pooled effect size of studies using ELISA assays was significantly higher than studies using RIA assays. It is generally accepted that RIA has a higher efficiency in measuring cortisol compared to ELISA. In many studies, RIA assays are considered as more accurate when it comes to the analysis of fish cortisol [97], human salivary cortisol [98], as well as mice [99] and bird [100] corticosterone. However, there are also studies in mammals that show equal results between ELISA and RIA [101], or even better performance in the ELISA assays [102]. Therefore, it is difficult to definitely conclude which assay type is more accurate in reporting cortisol levels, but the current study supports the notion that the cortisol assay type should be taken into careful consideration when designing a study and when interpreting the results. On the other hand, when the standardized mean differences between basal and post-stress cortisol levels were analyzed, no difference between assay types was observed. This result indicates that although ELISA assays tend to over-estimate cortisol levels, they do so in a similar manner in basal and post-stress concentrations. In other words, both assay types record the magnitude of the response in the same way although ELISA overestimates the absolute values.

The rearing system, on the other hand, seemed not to affect cortisol levels. The most commonly used systems were the open-flow and the RAS, consisting of approximately 3/4 of the total number of studies. To the best of our knowledge, there are no published studies in E. sea bass to directly compare fish welfare between these rearing systems, though the effects of increased stocking density seem to be the same in fish reared in RAS and open flow systems [55,78]. What has been shown to affect welfare in this species is the size of the rearing unit, in either open flow tanks in larval stages [103] or sea cages during on-growing [76].

Regarding means of anesthesia, most studies used chemical anesthetics, mainly phenoxyethanol, followed by MS222. In both basal and post-stress conditions anesthesia treatment significantly affected the results, showing lower values in decapitated or percussively blown-in-the-head fish. It is well known that E. sea bass is a species with a rapid cortisol response [75], and, therefore, immediate killing does not allow cortisol to rise. In that way, minimum cortisol levels are reported when using this method. In basal conditions, the highest levels were reported in fish that were not under anesthesia, i.e., they were conscious, during blood sampling. When anesthetics were used, phenoxyethanol and MS222 resulted in lower levels than the other anesthetics indicating that, when E. sea bass is sampled for cortisol levels under anesthesia, it is preferable to use one of the aforementioned anesthetics. However, it should be noted that studies assessing direct comparisons between conscious and chemically anesthetized fish, using either phenoxyethanol or clove-oil [59] and MS222 [65], have not reported such differences.

In terms of magnitude of the stress response in relation to the time after stress, it is known that, in E. sea bass, cortisol starts to rise at least 6 min after the application of stress [75], reaching maximum levels at 60 to 120 min post-stress when recovery starts to take place [4,7,14]. Grouping of studies based on the post-stress time at which fish were sampled revealed a similar outcome although it should be noted that the differences were not significant. However, this synthesis of results reflected the typical, more-or less, time-course cortisol response of this species [4,5,14,75], even though different stressors, in terms of nature, intensity, and duration were used.

It is acknowledged that there are some limitations in the conclusions of the current meta-analysis. As mentioned before, the first is due to the high between-studies heterogeneity. This is a result of various reasons, including the different aims of the studies, the different assays used, the rearing methods, temperature, and so on. The second lies to the fact that E. sea bass responds very fast to handling [75], and it is, therefore, difficult to ascertain that the reported basal levels have been obtained under similar sampling stress between studies. This is similar to the source of variation #4 proposed by Ellis et al. [6] (Table 1). Third, there was a scarcity of information in environmental data that could affect the cortisol response, such as water temperature, salinity, pH, and dissolved oxygen concentration. Finally, the circadian rhythm [7,19] the seasonality [5,18], and the sex of the fish [104] are additional potential sources of variation in cortisol levels which are hardly taken into consideration—and subsequently not reported—in most published studies, and, therefore, the current study could not include them in the analysis.

5. Conclusions

In conclusion, taking into consideration the limitations discussed above, the current meta-analysis examined 69 studies and calculated a pooled effect for basal and post-stress cortisol levels for E. sea bass. A high between-studies heterogeneity was recorded, with the factors assay type and anesthesia affecting cortisol levels and adding variance to the results. Moreover, a significant effect of acute stress on cortisol levels was observed in all studies examined. On the contrary, no association between cortisol and fish body weight or environmental conditions such as water temperature, dissolved oxygen, salinity pH, stocking density, or the rearing unit was observed. Finally, although it was not possible to directly test for genetic differences between fish, seasonality, circadian rhythms, and sex due to the lack of data in the published studies, the high between-study heterogeneity indicates that these factors may be additional factors causing variation between studies in the examined species.

Supplementary Materials: The following supporting information can be downloaded at: https://www.mdpi.com/article/10.3390/ani13081340/s1, Table S1: PRISMA 202 Checklist; Table S2: Excluded studies due to violation of exclusion criteria.

Funding: This research received no external funding.

Institutional Review Board Statement: Not applicable.

Informed Consent Statement: Not applicable.

Data Availability Statement: Data available upon reasonable request.

Acknowledgments: The author would like to thank Irini Tsikopoulou for her advice in software use.

Conflicts of Interest: The author declares no conflict of interest.

References

1. Wendelaar Bonga, S.E. The stress response in fish. *Physiol. Rev.* **1997**, *77*, 591–625. [CrossRef] [PubMed]
2. Gorissen, M.; Flik, G. The endocrinology of the stress response in fish—An adaptation-physiological view. In *Fish Physiology Biology of Stress in Fish*; Academic Press: Cambridge, MA, USA, 2016; Volume 35, pp. 75–111.
3. Mommsen, T.P.; Vijayan, M.M.; Moon, T.W. Cortisol in teleosts: Dynamics, mechanisms of action, and metabolic regulation. *Rev. Fish Biol. Fish.* **1999**, *9*, 211–268. [CrossRef]
4. Fanouraki, E.; Mylonas, C.C.; Papandroulakis, N.; Pavlidis, M. Species specificity in the magnitude and duration of the acute stress response in Mediterranean marine fish in culture. *Gen. Comp. Endocrinol.* **2011**, *173*, 313–322. [CrossRef] [PubMed]
5. Samaras, A.; Papandroulakis, N.; Costari, M.; Pavlidis, M. Stress and metabolic indicators in a relatively high (European sea bass, *Dicentrarchus labrax*) and a low (meagre, *Argyrosomus regius*) cortisol responsive species, in different water temperatures. *Aquac. Res.* **2016**, *47*, 3501–3515. [CrossRef]
6. Ellis, T.; Yildiz, H.Y.; López-Olmeda, J.; Spedicato, M.T.; Tort, L.; Øverli, Ø.; Martins, C.I.M. Cortisol and finfish welfare. *Fish Physiol. Biochem.* **2012**, *38*, 163–188. [CrossRef] [PubMed]
7. Fatira, E.; Papandroulakis, N.; Pavlidis, M. Diel changes in plasma cortisol and effects of size and stress duration on the cortisol response in European sea bass (*Dicentrarchus labrax*). *Fish Physiol. Biochem.* **2014**, *40*, 911–919. [CrossRef]
8. Samaras, A.; Dimitroglou, A.; Sarropoulou, E.; Papaharisis, L.; Kottaras, L.; Pavlidis, M. Repeatability of cortisol stress response in the European sea bass (*Dicentrarchus labrax*) and transcription differences between individuals with divergent responses. *Sci. Rep.* **2016**, *6*, 34858. [CrossRef]
9. Samaras, A.; Pavlidis, M. Regulation of divergent cortisol responsiveness in European sea bass, *Dicentrarchus labrax* L. *PLoS ONE* **2018**, *13*, e0202195. [CrossRef]
10. Samaras, A.; Dimitroglou, A.; Kollias, S.; Skouradakis, G.; Papadakis, I.E.; Pavlidis, M. Cortisol concentration in scales is a valid indicator for the assessment of chronic stress in European sea bass, *Dicentrarchus labrax* L. *Aquaculture* **2021**, *545*, 737257. [CrossRef]
11. Chatziplis, D.; Oikonomou, S.; Loukovitis, D.; Tsiokos, D.; Samaras, A.; Dimitroglou, A.; Kottaras, L.; Papanna, K.; Papaharisis, L.; Tsigenopoulos, C.; et al. QTL for stress and disease resistance in European sea bass, *Dicentrarchus labrax* L. *Animals* **2020**, *10*, 1668. [CrossRef]
12. Ferrari, S.; Millot, S.; Leguay, D.; Chatain, B.; Bégout, M.L. Consistency in European seabass coping styles: A life-history approach. *Appl. Anim. Behav. Sci.* **2015**, *167*, 74–88. [CrossRef]
13. Ferrari, S.; Rey, S.; Høglund, E.; Øverli, Ø.; Chatain, B.; MacKenzie, S.; Bégout, M.L. Physiological responses during acute stress recovery depend on stress coping style in European sea bass, *Dicentrarchus labrax*. *Physiol. Behav.* **2020**, *216*, 112801. [CrossRef] [PubMed]
14. Samaras, A.; Papandroulakis, N.; Lika, K.; Pavlidis, M. Water temperature modifies the acute stress response of European sea bass, *Dicentrarchus labrax* L. (1758). *J. Therm. Biol.* **2018**, *78*, 84–91. [CrossRef] [PubMed]
15. Islam, M.J.; Kunzmann, A.; Bögner, M.; Meyer, A.; Thiele, R.; James Slater, M. Metabolic and molecular stress responses of European seabass, *Dicentrarchus labrax* at low and high temperature extremes. *Ecol. Indic.* **2020**, *112*, 106118. [CrossRef]
16. Islam, M.J.; Slater, M.J.; Kunzmann, A. What metabolic, osmotic and molecular stress responses tell us about extreme ambient heatwave impacts in fish at low salinities: The case of European seabass, *Dicentrarchus labrax*. *Sci. Total Environ.* **2020**, *749*, 141458. [CrossRef] [PubMed]
17. Islam, M.J.; Kunzmann, A.; Slater, M.J. Extreme winter cold-induced osmoregulatory, metabolic, and physiological responses in European seabass (*Dicentrarchus labrax*) acclimatized at different salinities. *Sci. Total Environ.* **2021**, *771*, 145202. [CrossRef]
18. Pascoli, F.; Lanzano, G.S.; Negrato, E.; Poltronieri, C.; Trocino, A.; Radaelli, G.; Bertotto, D. Seasonal effects on hematological and innate immune parameters in sea bass *Dicentrarchus labrax*. *Fish Shellfish Immunol.* **2011**, *31*, 1081–1087. [CrossRef]
19. Planas, J.; Gutierrez, J.; Fernandez, J.; Carrillo, M.; Canals, P. Annual and daily variations of plasma cortisol in sea bass, *Dicentrarchus labrax* L. *Aquaculture* **1990**, *91*, 171–178. [CrossRef]
20. Benhaïm, D.; Bégout, M.L.; Péan, S.; Manca, M.; Prunet, P.; Chatain, B. Impact of a plant-based diet on behavioural and physiological traits in sea bass (*Dicentrarchus labrax*). *Aquat. Living Resour.* **2013**, *26*, 121–131. [CrossRef]

21. Carbonara, P.; Zupa, W.; Bitetto, I.; Alfonso, S.; Dara, M.; Cammarata, M. Evaluation of the effects of the enriched-organic diets composition on european sea bass welfare through a multi-parametric approach. *J. Mar. Sci. Eng.* **2020**, *8*, 934. [CrossRef]
22. Serradell, A.; Torrecillas, S.; Makol, A.; Valdenegro, V.; Fernández-Montero, A.; Acosta, F.; Izquierdo, M.S.; Montero, D. Prebiotics and phytogenics functional additives in low fish meal and fish oil based diets for European sea bass (*Dicentrarchus labrax*): Effects on stress and immune responses. *Fish Shellfish Immunol.* **2020**, *100*, 219–229. [CrossRef] [PubMed]
23. Azeredo, R.; Machado, M.; Afonso, A.; Fierro-Castro, C.; Reyes-López, F.E.; Tort, L.; Gesto, M.; Conde-Sieira, M.; Míguez, J.M.; Soengas, J.L.; et al. Neuroendocrine and immune responses undertake different fates following tryptophan or methionine dietary treatment: Tales from a teleost model. *Front. Immunol.* **2017**, *8*, 1226. [CrossRef] [PubMed]
24. Sadoul, B.; Alfonso, S.; Cousin, X.; Prunet, P.; Bégout, M.L.; Leguen, I. Global assessment of the response to chronic stress in European sea bass. *Aquaculture* **2021**, *544*, 737072. [CrossRef]
25. Moher, D.; Liberati, A.; Tetzlaff, J.; Altman, D.G.; Altman, D.; Antes, G.; Atkins, D.; Barbour, V.; Barrowman, N.; Berlin, J.A.; et al. Preferred reporting items for systematic reviews and meta-analyses: The PRISMA statement. *PLoS Med.* **2009**, *6*, e1000097. [CrossRef]
26. Karakatsouli, N.; Katsakoulis, P.; Leondaritis, G.; Kalogiannis, D.; Papoutsoglou, S.E.; Chadio, S.; Sakellaridis, N. Acute stress response of European sea bass *Dicentrarchus labrax* under blue and white light. *Aquaculture* **2012**, *364–365*, 48–52. [CrossRef]
27. Lupatsch, I.; Santos, G.A.; Schrama, J.W.; Verreth, J.A.J. Effect of stocking density and feeding level on energy expenditure and stress responsiveness in European sea bass *Dicentrarchus labrax*. *Aquaculture* **2010**, *298*, 245–250. [CrossRef]
28. Santulli, A.; Modica, A.; Messina, C.; Ceffa, L.; Curatolo, A.; Rivas, G.; Fabi, G.; D'Amelio, V. Biochemical responses of European sea bass (*Dicentrarchus labrax* L.) to the stress induced by off shore experimental seismic prospecting. *Mar. Pollut. Bull.* **1999**, *38*, 1105–1114. [CrossRef]
29. Caruso, G.; Genovese, L.; Maricchiolo, G.; Modica, A. Haematological, biochemical and immunological parameters as stress indicators in *Dicentrarchus labrax* and *Sparus aurata* farmed in off-shore cages. *Aquac. Int.* **2005**, *13*, 67–73. [CrossRef]
30. Yildiz, H.Y.; Altunay, S. Physiological stress and innate immune response in gilthead sea bream (*Sparus aurata*) and sea bass (*Dicentrarchus labrax*) exposed to combination of trimethoprim and sulfamethoxazole (TMP-SMX). *Fish Physiol. Biochem.* **2011**, *37*, 401–409. [CrossRef]
31. Goda, A.M.A.S.; Srour, T.M.; Mansour, A.T.; Baromh, M.Z.; Sallam, G.R.; Baromh, A.Z. Assessment of stresful ambient water salinity on growth, feed utilization and hematological indices of european sea bass, *Dicentrarchus labrax*, juveniles. *AACL Bioflux* **2019**, *12*, 553–563.
32. Gornati, R.; Papis, E.; Rimoldi, S.; Terova, G.; Saroglia, M.; Bernardini, G. Rearing density influences the expression of stress-related genes in sea bass (Dicentrarchus labrax, L.). *Gene* **2004**, *341*, 111–118. [CrossRef] [PubMed]
33. Tintos, A.; Míguez, J.M.; Mancera, J.M.; Soengas, J.L. Development of a microtitre plate indirect ELISA for measuring cortisol in teleosts, and evaluation of stress responses in rainbow trout and gilthead sea bream. *J. Fish Biol.* **2006**, *68*, 251–263. [CrossRef]
34. RStudio Team (2020). RStudio: Integrated Development for R. RStudio, PBC, Boston, MA, USA. Available online: http://www.rstudio.com/ (accessed on 2 February 2023).
35. Balduzzi, S.; Rücker, G.; Schwarzer, G. How to perform a meta-analysis with R: A practical tutorial. *Evid. Based Ment. Heal.* **2019**, *22*, 153–160. [CrossRef] [PubMed]
36. Harrer, M.; Cuijpes, P.; Furukawa, T.; Ebert, D.D. dmetar: Companion R Package for the Guide "Doing Meta-Analysis in R" 2019. Available online: https://dmetar.protectlab.org/ (accessed on 2 February 2023).
37. Wickham, H.; Averick, M.; Bryan, J.; Chang, W.; McGowan, L.; François, R.; Grolemund, G.; Hayes, A.; Henry, L.; Hester, J.; et al. Welcome to the Tidyverse. *J. Open Source Softw.* **2019**, *4*, 1686. [CrossRef]
38. Knapp, G.; Hartung, J. Improved tests for a random effects meta-regression with a single covariate. *Stat. Med.* **2003**, *22*, 2693–2710. [CrossRef]
39. Harrer, M.; Cuijpes, P.; Furukawa, T.A.; Ebert, D.D. *Doing Meta-Analysis with R: A Hands-on Guide*; Chapmann & Hall: London, UK; CRC Press: Boca Raton, FL, USA, 2021; ISBN 978-0-367-61007-4.
40. Acerete, L.; Reig, L.; Alvarez, D.; Flos, R.; Tort, L. Comparison of two stunning/slaughtering methods on stress response and quality indicators of European sea bass (*Dicentrarchus labrax*). *Aquaculture* **2009**, *287*, 139–144. [CrossRef]
41. Alfonso, S.; Sadoul, B.; Gesto, M.; Joassard, L.; Chatain, B.; Geffroy, B.; Bégout, M.L. Coping styles in European sea bass: The link between boldness, stress response and neurogenesis. *Physiol. Behav.* **2019**, *207*, 76–85. [CrossRef]
42. Alfonso, S.; Gesto, M.; Sadoul, B. Temperature increase and its effects on fish stress physiology in the context of global warming. *J. Fish Biol.* **2020**, *98*, 1496–1508. [CrossRef]
43. Alfonso, S.; Zupa, W.; Manfrin, A.; Fiocchi, E.; Dioguardi, M.; Dara, M.; Lembo, G.; Carbonara, P.; Cammarata, M.; Cammarata, M. Surgical implantation of electronic tags does not induce medium-term effect: Insights from growth and stress physiological profile in two marine fish species. *Anim. Biotelemetry* **2020**, *8*, 4–9. [CrossRef]
44. Bertotto, D.; Poltronieri, C.; Negrato, E.; Majolini, D.; Radaelli, G.; Simontacchi, C. Alternative matrices for cortisol measurement in fish. *Aquac. Res.* **2010**, *41*, 1261–1267. [CrossRef]
45. Bodur, T.; León-Bernabeu, S.; Navarro, A.; Tort, L.; Afonso, J.M.; Montero, D. Effects of new plant based anesthetics Origanum sp. and Eucalyptus sp. oils on stress and welfare parameters in *Dicentrarchus labrax* and their comparison with clove oil. *Aquaculture* **2018**, *495*, 402–408. [CrossRef]

46. Bögner, D.; Bögner, M.; Schmachtl, F.; Bill, N.; Halfer, J.; Slater, M.J. Hydrogen peroxide oxygenation and disinfection capacity in recirculating aquaculture systems. *Aquac. Eng.* **2021**, *92*, 102140. [CrossRef]
47. Brandts, I.; Balasch, J.C.; Gonçalves, A.P.; Martins, M.A.; Pereira, M.L.; Tvarijonaviciute, A.; Teles, M.; Oliveira, M. Immunomodulatory effects of nanoplastics and humic acids in the European seabass (*Dicentrarchus labrax*). *J. Hazard. Mater.* **2021**, *414*, 125562. [CrossRef] [PubMed]
48. Carbonara, P.; Corsi, I.; Focardi, S.; Lembo, G.; Rochira, S.; Scolamacchia, M.; Spedicato, M.T.; Mckinley, R.S. The effects of stress induced by cortisol administration on the repeatability of swimming performance tests in the European sea bass (*Dicentrarchus labrax* L.). *Mar. Freshw. Behav. Physiol.* **2010**, *43*, 283–296. [CrossRef]
49. Carbonara, P.; Scolamacchia, M.; Spedicato, M.T.; Zupa, W.; Mckinley, R.S.; Lembo, G. Muscle activity as a key indicator of welfare in farmed European sea bass (*Dicentrarchus labrax* L. 1758). *Aquac. Res.* **2015**, *46*, 2133–2146. [CrossRef]
50. Carbonara, P.; Dioguardi, M.; Cammarata, M.; Zupa, W.; Vazzana, M.; Spedicato, M.T.; Lembo, G. Basic knowledge of social hierarchies and physiological profile of reared sea bass *Dicentrarchus labrax* (L.). *PLoS ONE* **2019**, *14*, e0208688. [CrossRef]
51. Caruso, G.; Denaro, M.G.; Caruso, R.; Mancari, F.; Genovese, L.; Maricchiolo, G. Response to short term starvation of growth, haematological, biochemical and non-specific immune parameters in European sea bass (*Dicentrarchus labrax*) and blackspot sea bream (*Pagellus bogaraveo*). *Mar. Environ. Res.* **2011**, *72*, 46–52. [CrossRef]
52. Celi, M.; Vazzana, M.; Sanfratello, M.A.; Parrinello, N. Elevated cortisol modulates Hsp70 and Hsp90 gene expression and protein in sea bass head kidney and isolated leukocytes. *Gen. Comp. Endocrinol.* **2012**, *175*, 424–431. [CrossRef]
53. Rubio, V.C.; Sánchez, E.; Cerdá-Reverter, J.M. Compensatory feeding in the sea bass after fasting and physical stress. *Aquaculture* **2010**, *298*, 332–337. [CrossRef]
54. Cerqueira, M.; Millot, S.; Felix, A.; Silva, T.; Oliveira, G.A.; Oliveira, C.C.V.; Rey, S.; MacKenzie, S.; Oliveira, R. Cognitive appraisal in fish: Stressor predictability modulates the physiological and neurobehavioural stress response in sea bass. *Proc. R. Soc. B Biol. Sci.* **2020**, *287*, 20192922. [CrossRef]
55. D'Orbcastel, E.R.; Lemarié, G.; Breuil, G.; Petochi, T.; Marino, G.; Triplet, S.; Dutto, G.; Fivelstad, S.; Coeurdacier, J.L.; Blancheton, J.P. Effects of rearing density on sea bass (*Dicentrarchus labrax*) biological performance, blood parameters and disease resistance in a flow through system. *Aquat. Living Resour.* **2010**, *23*, 109–117. [CrossRef]
56. Di Marco, P.; Priori, A.; Finoia, M.G.; Massari, A.; Mandich, A.; Marino, G. Physiological responses of European sea bass *Dicentrarchus labrax* to different stocking densities and acute stress challenge. *Aquaculture* **2008**, *275*, 319–328. [CrossRef]
57. Faggion, S.; Bertotto, D.; Babbucci, M.; Dalla Rovere, G.; Franch, R.; Bovolenta, M.; Laureau, S.; Pascoli, F.; Toffan, A.; Bargelloni, L.; et al. Resistance to viral nervous necrosis in European sea bass (*Dicentrarchus labrax* L.): Heritability and relationships with body weight, cortisol concentration, and antibody titer. *Genet. Sel. Evol.* **2021**, *53*, 32. [CrossRef] [PubMed]
58. Fanouraki, E.; Papandroulakis, N.; Ellis, T.; Mylonas, C.C.; Scott, A.P.; Pavlidis, M. Water cortisol is a reliable indicator of stress in European sea bass, *Dicentrarchus labrax*. *Behaviour* **2008**, *145*, 1267–1281.
59. Filiciotto, F.; Buscaino, G.; Buffa, G.; Bellante, A.; Maccarrone, V.; Mazzola, S. Anaesthetic qualities of eugenol and 2-phenoxyethanol and their effect on same haematological parameters in farmed European sea bass (*Dicentrarchus labrax* L.). *J. Anim. Vet. Adv.* **2012**, *11*, 494–502. [CrossRef]
60. Herrero, M.J.; Martínez, F.J.; Míguez, J.M.; Madrid, J.A. Response of plasma and gastrointestinal melatonin, plasma cortisol and activity rhythms of European sea bass (*Dicentrarchus labrax*) to dietary supplementation with tryptophan and melatonin. *J. Comp. Physiol. B Biochem. Syst. Environ. Physiol.* **2007**, *177*, 319–326. [CrossRef]
61. Leal, E.; Fernández-Durán, B.; Guillot, R.; Ríos, D.; Cerdá-Reverter, J.M. Stress-induced effects on feeding behavior and growth performance of the sea bass (*Dicentrarchus labrax*): A self-feeding approach. *J. Comp. Physiol. B Biochem. Syst. Environ. Physiol.* **2011**, *181*, 1035–1044. [CrossRef]
62. Machado, M.; Castro, C.; Oliva-Teles, A.; Costas, B. Interactive effects of dietary vegetable oil and carbohydrate incorporation on the innate immune response of European seabass (*Dicentrarchus labrax*) juveniles subjected to acute stress. *Aquaculture* **2019**, *498*, 171–180. [CrossRef]
63. Machado, M.; Azeredo, R.; Domingues, A.; Fernandez-Boo, S.; Dias, J.; Conceição, L.E.C.; Costas, B. Dietary tryptophan deficiency and its supplementation compromises inflammatory mechanisms and disease resistance in a teleost fish. *Sci. Rep.* **2019**, *9*, 7689. [CrossRef]
64. Maricchiolo, G.; Mirto, S.; Caruso, G.; Caruso, T.; Bonaventura, R.; Celi, M.; Matranga, V.; Genovese, L. Welfare status of cage farmed European sea bass (*Dicentrarchus labrax*): A comparison between submerged and surface cages. *Aquaculture* **2011**, *314*, 173–181. [CrossRef]
65. Marino, G.; Di Marco, P.; Mandich, A.; Finoia, M.G.; Cataudella, S. Changes in serum cortisol, metabolites, osmotic pressure and electrolytes in response to different blood sampling procedures in cultured sea bass (*Dicentrarchus labrax* L.). *J. Appl. Ichthyol.* **2001**, *17*, 115–120. [CrossRef]
66. Millot, S.; Cerqueira, M.; Castanheira, M.-F.; Øverli, Ø.; Oliveira, R.F.; Martins, C.I.M. Behavioural Stress Responses Predict Environmental Perception in European Sea Bass (*Dicentrarchus labrax*). *PLoS ONE* **2014**, *9*, e108800. [CrossRef] [PubMed]
67. Ordóñez-Grande, B.; Guerreiro, P.M.; Sanahuja, I.; Fernández-Alacid, L.; Ibarz, A. Evaluation of an acute osmotic stress in european sea bass via skin mucus biomarkers. *Animals* **2020**, *10*, 1546. [CrossRef]

68. Papaharisis, L.; Tsironi, T.; Dimitroglou, A.; Taoukis, P.; Pavlidis, M. Stress assessment, quality indicators and shelf life of three aquaculture important marine fish, in relation to harvest practices, water temperature and slaughter method. *Aquac. Res.* **2019**, *50*, 2608–2620. [CrossRef]
69. Peruzzi, S.; Varsamos, S.; Chatain, B.; Fauvel, C.; Menu, B.; Falguière, J.C.; Sévère, A.; Flik, G. Haematological and physiological characteristics of diploid and triploid sea bass, *Dicentrarchus labrax* L. *Aquaculture* **2005**, *244*, 359–367. [CrossRef]
70. Petochi, T.; Di Marco, P.; Priori, A.; Finoia, M.G.; Mercatali, I.; Marino, G. Coping strategy and stress response of European sea bass *Dicentrarchus labrax* to acute and chronic environmental hypercapnia under hyperoxic conditions. *Aquaculture* **2011**, *315*, 312–320. [CrossRef]
71. Poltronieri, C.; Maccatrozzo, L.; Simontacchi, C.; Bertotto, D.; Funkenstein, B.; Patruno, M.; Radaelli, G. Quantitative RT-PCR analysis and immunohistochemical localization of HSP70 in sea bass *Dicentrarchus labrax* exposed to transport stress. *Eur. J. Histochem.* **2007**, *51*, 125–136. [CrossRef]
72. Roche, H.; Boge, G. Fish blood parameters as a potential tool for identification of stress caused by environmental factors and chemical intoxication. *Mar. Environ. Res.* **1996**, *41*, 27–43. [CrossRef]
73. Roque, A.; Yildiz, H.Y.; Carazo, I.; Duncan, N. Physiological stress responses of sea bass (*Dicentrarchus labrax*) to hydrogen peroxide (H2O2) exposure. *Aquaculture* **2010**, *304*, 104–107. [CrossRef]
74. Rotllant, J.; Ruane, N.M.; Dinis, M.T.; Canario, A.V.M.; Power, D.M. Intra-adrenal interactions in fish: Catecholamine stimulated cortisol release in sea bass (*Dicentrarchus labrax* L.). *Comp. Biochem. Physiol. Part A Mol. Integr. Physiol.* **2006**, *143*, 375–381. [CrossRef]
75. Rotllant, J.; Ruane, N.M.; Caballero, M.J.; Montero, D.; Tort, L. Response to confinement in sea bass (*Dicentrarchus labrax*) is characterised by an increased biosynthetic capacity of interrenal tissue with no effect on ACTH sensitivity. *Comp. Biochem. Physiol. A Mol. Integr. Physiol.* **2003**, *136*, 613–620. [CrossRef] [PubMed]
76. Samaras, A.; Pavlidis, M.; Lika, K.; Theodoridi, A.; Papandroulakis, N. Scale matters: Performance of European sea bass, *Dicentrarchus labrax*, L. (1758), reared in cages of different volumes. *Aquac. Res.* **2017**, *48*, 990–1005. [CrossRef]
77. Samaras, A.; Santo, C.E.; Papandroulakis, N.; Mitrizakis, N.; Pavlidis, M.; Höglund, E.; Pelgrim, T.N.M.; Zethof, J.; Spanings, F.A.T.; Vindas, M.A.; et al. Allostatic load and stress physiology in European seabass (*Dicentrarchus labrax* L.) and gilthead seabream (*Sparus aurata* L.). *Front. Endocrinol.* **2018**, *9*, 451. [CrossRef] [PubMed]
78. Sammouth, S.; d'Orbcastel, E.R.; Gasset, E.; Lemarié, G.; Breuil, G.; Marino, G.; Coeurdacier, J.L.; Fivelstad, S.; Blancheton, J.P. The effect of density on sea bass (*Dicentrarchus labrax*) performance in a tank-based recirculating system. *Aquac. Eng.* **2009**, *40*, 72–78. [CrossRef]
79. Santos, G.A.; Schrama, J.W.; Capelle, J.; Rombout, J.H.W.M.; Verreth, J.A.J. Effects of dissolved carbon dioxide on energy metabolism and stress responses in European seabass (*Dicentrarchus labrax*). *Aquac. Res.* **2013**, *44*, 1370–1382. [CrossRef]
80. Santos, G.A.; Schrama, J.W.; Mamauag, R.E.P.; Rombout, J.H.W.M.; Verreth, J.A.J. Chronic stress impairs performance, energy metabolism and welfare indicators in European seabass (*Dicentrarchus labrax*): The combined effects of fish crowding and water quality deterioration. *Aquaculture* **2010**, *299*, 73–80. [CrossRef]
81. Silva-Brito, F.; Timóteo, F.; Esteves, Â.; Peixoto, M.J.; Ozorio, R.; Magnoni, L. Impact of the replacement of dietary fish oil by animal fats and environmental salinity on the metabolic response of European Seabass (*Dicentrarchus labrax*). *Comp. Biochem. Physiol. Part-B Biochem. Mol. Biol.* **2019**, *233*, 46–59. [CrossRef]
82. Simontacchi, C.; Poltronieri, C.; Carraro, C.; Bertotto, D.; Xiccato, G.; Trocino, A.; Radaelli, G. Alternative stress indicators in sea bass *Dicentrarchus labrax*, L.J. *Fish Biol.* **2008**, *72*, 747–752. [CrossRef]
83. Teles, M.; Pacheco, M.; Santos, M.A. Biotransformation, stress and genotoxic effects of 17β-estradiol in juvenile sea bass (*Dicentrarchus labrax* L.). *Environ. Int.* **2006**, *32*, 470–477. [CrossRef]
84. Teles, M.; Gravato, C.; Pacheco, M.; Santos, M.A. Juvenile sea bass biotransformation, genotoxic and endocrine responses to β-naphthoflavone, 4-nonylphenol and 17β-estradiol individual and combined exposures. *Chemosphere* **2004**, *57*, 147–158. [CrossRef]
85. Terova, G.; Gornati, R.; Rimoldi, S.; Bernardini, G.; Saroglia, M. Quantification of a glucocorticoid receptor in sea bass (*Dicentrarchus labrax*, L.) reared at high stocking density. *Gene* **2005**, *357*, 144–151. [CrossRef] [PubMed]
86. Torrecillas, S.; Makol, A.; Caballero, M.J.; Montero, D.; Dhanasiri, A.K.S.; Sweetman, J.; Izquierdo, M. Effects on mortality and stress response in European sea bass, *Dicentrarchus labrax* (L.), fed mannan oligosaccharides (MOS) after *Vibrio anguillarum* exposure. *J. Fish Dis.* **2012**, *35*, 591–602. [CrossRef] [PubMed]
87. Tsalafouta, A.; Papandroulakis, N.; Pavlidis, M. Early life stress and effects at subsequent stages of development in European sea bass (*D. labrax*). *Aquaculture* **2015**, *436*, 27–33. [CrossRef]
88. Varsamos, S.; Flik, G.; Pepin, J.F.; Wendelaar Bonga, S.E.; Breuil, G. Husbandry stress during early life stages affects the stress response and health status of juvenile sea bass, *Dicentrarchus labrax*. *Fish Shellfish Immunol.* **2006**, *20*, 83–96. [CrossRef] [PubMed]
89. Varsamos, S.; Xuereb, B.; Commes, T.; Flik, G.; Spanings-Pierrot, C. Pituitary hormone mRNA expression in European sea bass *Dicentrarchus labrax* in seawater and following acclimation to fresh water. *J. Endocrinol.* **2006**, *191*, 473–480. [CrossRef]
90. Vazzana, M.; Vizzini, A.; Sanfratello, M.A.; Celi, M.; Salerno, G.; Parrinello, N. Differential expression of two glucocorticoid receptors in seabass (teleost fish) head kidney after exogeneous cortisol inoculation. *Comp. Biochem. Physiol. A Mol. Integr. Physiol.* **2010**, *157*, 49–54. [CrossRef]

91. Vera, L.M.; Migaud, H. Continuous high light intensity can induce retinal degeneration in Atlantic salmon, Atlantic cod and European sea bass. *Aquaculture* **2009**, *296*, 150–158. [CrossRef]
92. Volckaert, F.a.M.; Hellemans, B.; Batargias, C.; Louro, B.; Massault, C.; Van Houdt, J.K.J.; Haley, C.; de Koning, D.-J.; Canario, A.V.M. Heritability of cortisol response to confinement stress in European sea bass *Dicentrarchus labrax*. *Genet. Sel. Evol.* **2012**, *44*, 15. [CrossRef]
93. Yildiz, H.; Ergonul, M.B. Is prophylactic formalin exposure a stress source for gilthead sea bream (*Sparus aurata*) and sea bass (*Dicentrarchus labrax*)? *Ankara Univ. Vet. Fak. Derg.* **2010**, *57*, 113–118. [CrossRef]
94. Pustejovsky, J.E.; Rodgers, M.A. Testing for funnel plot asymmetry of standardized mean differences. *Res. Synth. Methods* **2019**, *10*, 57–71. [CrossRef]
95. Pottinger, T.G.; Carrick, T.R.; Appleby, A.; Yeomans, W.E. High blood cortisol levels and low cortisol receptor affinity: Is the chub, Leuciscus cephalus, a cortisol-resistant teleost? *Gen. Comp. Endocrinol.* **2000**, *120*, 108–117. [CrossRef] [PubMed]
96. FAO Cultured Aquatic Species Information Programme: Dicentrarchus labrax (Linnaeus, 1758). Available online: http://www.fao.org/fishery/culturedspecies/Dicentrarchus_labrax/en#tcNA0112 (accessed on 24 September 2021).
97. Sink, T.D.; Lochmann, R.T.; Fecteau, K.A. Validation, use, and disadvantages of enzyme-linked immunosorbent assay kits for detection of cortisol in channel catfish, largemouth bass, red pacu, and golden shiners. *Fish Physiol. Biochem.* **2008**, *34*, 95–101. [CrossRef] [PubMed]
98. Raff, H.; Homar, P.J.; Burns, E.A. Comparison of two methods for measuring salivary cortisol. *Clin. Chem.* **2002**, *48*, 207–208. [CrossRef] [PubMed]
99. Rod, A.M.K.; Harkestad, N.; Jellestad, F.K.; Murison, R. Comparison of commercial ELISA assays for quantification of corticosterone in serum. *Sci. Rep.* **2017**, *7*, 6748. [CrossRef]
100. Glucs, Z.E.; Smith, D.R.; Tubbs, C.W.; Scherbisnki, J.J.; Welch, A.; Burnett, J.; Clark, M.; Eng, C.; Finkelstein, M.E. Glucocorticoid measurement in plasma, urates and feathers from California condors (*Gymnogyps califorianus*) in response to a human-induced stressor. *PLoS ONE* **2018**, *13*, e0205565. [CrossRef]
101. Bekhbat, M.; Glasper, E.R.; Rowson, S.A.; Kelly, S.D.; Gretchen, N.N. Measuring corticosterone concentrations over a physiological dynamic range in female rats. *Physiol. Behav.* **2018**, *194*, 73–76. [CrossRef]
102. Nejad, J.G.; Park, K.H.; Forghani, F.; Lee, H.G.; Lee, J.S.; Sung, K.I. Measuring hair and blood cortisol in sheep and dairy cattle using RIA and ELISA assay: A comparison. *Biol. Rhythm Res.* **2020**, *51*, 887–897. [CrossRef]
103. Lika, K.; Pavlidis, M.; Mitrizakis, N.; Samaras, A.; Papandroulakis, N. Do experimental units of different scale affect the biological performance of European sea bass *Dicentrarchus labrax* larvae? *J. Fish Biol.* **2015**, *86*, 1271–1285. [CrossRef]
104. Haddy, J.A.; Pankhurst, N.W. Stress-induced changes in concentrations of plasma sex steroids in black bream. *J. Fish Biol.* **1999**, *55*, 1304–1316. [CrossRef]

MDPI AG
Grosspeteranlage 5
4052 Basel
Switzerland
Tel.: +41 61 683 77 34

Animals Editorial Office
E-mail: animals@mdpi.com
www.mdpi.com/journal/animals

www.ingramcontent.com/pod-product-compliance
Lightning Source LLC
Chambersburg PA
CBHW061835210926
51850CB00006B/895
* 9 7 8 3 7 2 5 8 6 0 0 9 8 *